TOPOLOGY and GROUPOIDS

Ronald Brown MA, DPhil (Oxon)

This revised, updated and extended version published in 2006 by
www.groupoids.org
Deganwy, United Kingdom

Original version first published by McGraw-Hill in 1968 under the title *Elements of Modern Topology*.
Second revised, updated and expanded version published by Ellis Horwood Limited in 1988 under the title *Topology: A Geometric Account of General Topology, Homotopy Types and the Fundamental Groupoid*.

Corrections made to this version: May 17, 2006

Cover design by John Robinson and Ben Dickens. The front cover is based on John's sculpture 'Journey', and the back cover photo includes the sculpture 'Immortality', both based on the Möbius Band. For more information, see the web sites www.popmath.org.uk and www.johnrobinson.com.

ISBN: 1-4196-2722-8
Library of Congress Control Number: 2006901092
To order additional copies, or the e-version, contact the author or
BookSurge LLC
www.booksurge.com
1-886-308-6235
orders@booksurge.com
For licensing of the e-version, contact the author.

Contents

Preface to the first edition

This book is intended as a textbook on point set and algebraic topology at the undergraduate and immediate post-graduate levels. In particular, I have tried to make the point set topology commence in an elementary manner suitable for the student beginning to study the subject.

The choice of topics given here is perhaps unusual, but has the aim of presenting the subject with a geometric flavour and with a coherent outlook. The first consideration has led to the omission of a number of topics important more from the point of view of analysis, such as uniform spaces, convergence, and various alternative kinds of compactness. The second consideration, together with restriction of space, has led to the omission of homology theory, of the theory of manifolds, and of any complete account of simplicial complexes. It has also led to the omission of topics which are important but not exactly germane to this book; for example, I have not included accounts of paracompactness or the theory of continua, nor proofs of the Tychonoff theorem or the Tietze extension theorem.

I felt, on the other hand, that the general direction of this book should be towards homotopy theory, since this subject links naturally with the point set topology and also occupies a central role in modern developments. However, in homotopy theory there are a number of ideas and constructions which are important in many applications (particularly the ideas of adjunction space, cell complex, join, and homotopy extension property) but for which no elementary account has appeared. One of my aims, then, has been to cover this sort of topic, and so to supplement existing accounts, rather than provide a new reference book to contain everything a young geometer ought to know. At the same time, I have tried to show that point set topology has its main value as a language for doing 'continuous geometry'; I believe it is important that the subject be presented to the student in this way, rather than as a self-contained system of axioms, definitions, and theorems which are to be studied for their own sake, or for their interest a generalisations of facts about Euclidean space. For this reason, I have kept the point set topology to the minimum needed for later purposes, and at the same time emphasised general processes of construction which lead to interesting topological spaces, and for which other structures, such as metrics or uniformities, are largely irrelevant.

The language of point set topology is geometric. On the other hand, the notions of category and functor are algebraic: they play in modern mathematics the unifying role which has earlier been given to the notion of a group. Particular kinds of categories are the groupoids, of which in turn groups are special cases. The algebra of groupoids has been developed recently by P. J. Higgins. In 1965, I discovered the utility of this algebra for computing the fundamental group, and this has led me to include in the last four chapters of this book an account of the elements of the algebra of categories and groupoids. The treatment given in these chapters is quite novel, in that this algebra is used in an essential way and not just as a convenient, but not entirely necessary, general language. The writing of these chapters gave me great pleasure as I found the way in which the various topological notions of sum, adjunction space, homotopy, covering space, are modelled by the corresponding notions for groupoids. The most important feature here is probably the way in which the computations of the fundamental group derive from a general property of the fundamental groupoid (that it sends 'nice' pushouts to pushouts). This contrasts with the usual rather *ad hoc* computation via simplicial complexes.

The development of these last four chapters has meant that I have had to cut down on some other topics. For example, I have had to relegate the accounts of k-spaces, infinite cell complexes, and function spaces to the Exercises. A further reason for playing down these topics was that the recent development of the theory of quasi-topological spaces is likely to lead to a radical change in their exposition (cf. [Spa63] and a forthcoming book by Dyer and Eilenberg). I have also relegated to the Exercises the discussion of group-like structures and exact sequences, and have omitted an account of the major properties of the Whitehead product. The reasons behind this were lack of space, the difficulty of presenting computations, and, finally, the feeling that these topics needed reconsideration in the light of the preceding work on groupoids.

The material in this book would more than cover a two-term undergraduate course in point set and algebraic topology. Such a course could include, for the point set topology, all of chapters 1 to 3 and some material from chapters 4 and 5. This could be followed by a course on the fundamental groupoid comprising chapter 6 and parts of chapters 8 or 9; naturally, more could be covered if this were linked with a course on categories and functors.

The last four chapters, with supporting parts of chapters 4 and 5, would probably be suitable for a two-term M.Sc. course. Alternatively, the purely algebraic parts of the last four chapters would cover the elements (but with not enough algebraic applications) of a one-term course on groupoids.

The Appendix contains an account of functions, cardinality, and some 'universal constructions' which are used from chapter 4 onwards. On pages 465 to 469 (of this edition) is a Glossary of those terms from set theory which are used in the book but not defined in the main text.

This book is not self-contained—in particular, it will be assumed at certain points that the reader is familiar with the usual accounts of vector spaces and of groups.

On the other hand, a more 'topological' account of elementary analysis is by no means universal. For this reason, and in order to motivate the axioms for a topological space, I have started the book with an account of the elementary topological notions on the real line $\mathbb{R}$.

In the earlier chapters, I have included among the exercises a fair number of straightforward verifications or simple tests of the reader's understanding. Apart from a few simple definitions and verifications, no results from the exercises are used in the text; this means that occasionally a result from an exercise is actually proved later in the text. However, there are some cross-references among the exercises themselves. Exercises from later chapters are usually more difficult than those from earlier ones—exercises of technical difficulty are marked with an asterisk.

Cross-references are always given in square brackets; thus [6.5.12] refers to 5.12 of §5 of chapter 6. The end of a Proof is denoted by the symbol □.

I would like to thank Dr W. F. Newns for reading through much of the manuscript and for many suggestions which resulted in improved exposition and in the removal of errors. I would also like to thank Professor G. Horrocks and Dr R. M. F. Moss for their comments on some of the chapters. I am indebted to Dr P. J. Higgins for conversations and letters; for sending me notes of a course on groupoids he gave at King's College, London in 1965; and for writing his paper, [Hig63].

I would also like to thank Dr W. Gilbert, P. R. Heath, D. Pitt, T. Poston, Dr E. Rees, and E. E. Thompson for help in proofreading.

RONALD BROWN
Hull, 1968

Preface to the second edition

When the first edition of this book ended its print-run in 1972, I did not take up the publisher's offer to reprint it with corrections, owing to an idea that I could improve on the book in some key aspects of organisation. This idea turned out to be a misconception.

It is true that one particular criticism had some justice, namely the feeling that there was a jump in level of abstraction at chapter 6, which introduces the idea of category. Thus there was the possibility of making a two-volume work, with maybe more complete material in each. The difficulty was the close cross-links, in substance and in spirit, between the two parts. So the present edition makes no attempt to change the original global conception.

A further difficulty in making major changes was that the original manuscript had been written with a vision and energy which by the mid-1970s had turned into a research programme arising out of the work on groupoids which had gone into the text. In particular, the notion of higher homotopy groupoids and higher dimensional Van Kampen theorems began to take shape with the help of colleagues and research students, and I found it impossible to take up the task of reshaping the text.

However, a good deal of the material in the book is still unavailable elsewhere. Also, the categorical spirit in which the book was written, and in particular the use of universal properties, has not penetrated many other books in topology at this level. This is quite surprising in view of the comment in W. S. Massey's excellent 1967 book on algebraic topology ([Mas67]): 'This method of characterising various mathematical structures as solutions to universal mapping problems seems to be one of the truly unifying mathematical principles to have emerged since 1945, and it should be brought into the mathematics curriculum as early as possible.' For examples of the use of universal properties in computer science, see the articles in [PAPR86], which is the proceedings of a conference on category theory and computer programming.

A further surprise to me has been that twenty years after the publication of the first edition the theory of the fundamental groupoid still is not taken by many topologists as a necessary and convenient generalisation of the theory of the fundamental group. Thus no other English language text on topology in print has followed the exposition given here. It therefore seemed essential to make this exposition available for a judgement to be possible on the convenience or elegance of this approach. Also much has happened since the previous edition was published, so that it is now possible to put the methods more in context, and to add extra applications. There is a new section on the Jordan Curve Theorem in which the proof uses the Van Kampen theorem for non-connected spaces in an essential way. There are new sections on covering morphisms of groupoids, and new sections on the fundamental groupoid, and hence on the fundamental group, of an orbit space. The notion of fibration of groupoid is used in an essential way.

The exposition given here lends credence to the view that groupoids form a natural context for discussing a key question in mathematics, the relation between local and global phenomena. This idea is confirmed in the area of differential geometry by the following quotation from the introduction to [Mac87]:

> 'The concept of groupoid is one of the means by which the twentieth century reclaims the original domain of applications of the group concept. The modern rigorous concept of group is far too restrictive for the range of geometrical applications envisaged in the work of Lie. There have thus arisen the concepts of Lie pseudogroup, of differentiable and Lie groupoid, and of principal bundle—as well as various related infinitesimal concepts such as Lie equation, graded Lie algebra and Lie algebroid—by which mathematics seeks to acquire a rigorous language by which to study the geometric phenomena associated with geometrical transformations which are only locally defined.'

It thus seems that the notion of groupoid gives a more flexible and powerful approach to the notion of symmetry.

I will be pleased if the exposition of this book can be improved in radical ways. Young mathematicians should be aware of the temporary nature of mathematical exposition. The attempt to form a 'final view' reminds one of the schoolboy question: what would happen if you laid worms in a straight line from Marble Arch to Picadilly Circus? Answer: one of them would be bound to wiggle and spoil it all.

Some might argue that the groupoid worm has here not only wiggled from its accustomed place in topology, but become altogether too big for its boots, to which a worm, after all, has no rights. But I hope many will find it interesting to trace through this first attempt to answer, in part, the questions: Is it possible to rewrite homotopy theory, substituting the word groupoid for the word group, and making other consequential changes? If this is done, is the result more pleasing?

These questions, both of the form 'What if. . . ?', came to acquire for me a force and an obsession when pursued into the topic of *higher homotopy groupoids*. The scribbling of countless squares and cubes and their compositions lead to a conviction in 1966 that the standard group theory, once it was rephrased as a groupoid theory, had a generalisation to higher dimensions. Gradually, collaborations with Chris Spencer in 1971–1974, with Philip Higgins since 1974, with J.-L. Loday from 1981, and work of my research students, A. Razak Salleh, Keith Dakin, Nick Ashley, David Jones, Graham Ellis and Ghaffar Mosa, at Bangor, and Philip Higgins' research students Jim Howie and John Taylor at London and Durham, made the theory take shape. In this way a worry of the algebraic topologists of the 1930s as to why the higher homotopy groups were abelian, and so less complicated than the fundamental group, came to seem a genuine question. The surprising answer is that the higher homotopy *groupoids* are non-abelian, and are just right for doing many aspects of homotopy theory. In particular, they satisfy a version of the Van Kampen theorem which enables explicit and direct computations to be made. It will be interesting to see if the higher dimensional theory will come to bear a relation to the standard group theory similar to that of many-variable to one-variable calculus.

But the higher dimensional theory is a story in which we cannot embark in this book. We now give the changes that have been made in this new edition.

A section on function spaces in the category of k-spaces has been added to chapter 5. One of the reasons is that the material is quite difficult to find elsewhere. Another reason for its inclusion is that it will suggest to the reader that there are still matters to be decided on the appropriate setting for our intuitive notions of continuity. In any case, the generalisation from spaces to k-spaces makes the proofs if anything simpler.

I am grateful to a number of people for comments, particularly Eldon Dyer and Peter May who suggested that chapter 7 needed clarification, and Daniel Grayson who suggested the notation $[(X, i), (Y, u)]$ now used in chapter 7 to replace the original $X//u$, which was non-standard and too brief. (But this double slash is used in a new context in chapter 9[1].) In

[1] In the new edition, this is chapter 11.

the event, chapter 7 has been completely revised to make use of the term *cofibration* rather than HEP. The idea of *fibration of groupoid*, which came to light towards the end of the writing of the first edition and so appeared there only in an exercise, has now been incorporated into the main text. However, I have not included fibrations of spaces, since to do so would have enlarged the text unduly, or force the omission of material for which no other textbook account is available.

In chapter 8, an error in section 8.2 has been corrected. Also free groupoids have now been used to define the notion of path in a graph. An exercise on the computation of the fundamental group of a union of non-connected spaces has now been incorporated into the text, as an illustration of the methods and because of its intrinsic importance. This result is used in a new section which gives a proof of the Jordan Curve Theorem, and some new results on the Phragmen-Brouwer Property.

In chapter 9[2], section 9.4 on the existence of covering groupoids has been rewritten to give a clearer idea of the notion of action of groupoids on sets. Section 9.5 includes some results on topologising the fundamental groupoid. The relation between covering spaces of X and covering groupoids of πX has been clarified by adding section 9.6, which gives an account of the equivalence of the categories of these objects. This enables an algebraic account of the theory of regular covering spaces. Section 9.7 gives a new account of pullbacks of covering spaces and covering morphisms, using exact sequences. Section 9.8 gives an account of the Nielsen-Schreier and Kurosch subgroup theorems, using groupoid versions, due to Hasse and Higgins, of the more traditional covering space proofs.

Sections 9.9 to 9.10[3] give the first account in any text of the theory of the fundamental groupoid of an orbit space. This uses work of John Taylor and Philip Higgins, and is a good illustration of the utility of the groupoid methods, since a group version is considerably more awkward in the statement of results and in the proofs. I am grateful to Ross Geoghegan who, in a review of a paper by M. A. Armstrong, pointed out the desirability of having Armstrong's results on the computation of the fundamental group of an orbit space available in a text.

The Bibliography has been extended for this edition. It would be impossible to make such a Bibliography complete, and I apologise in advance for any omissions or lack of balance. The main intention has been to show the subject matter of this book as part of a wide mathematical scene; occasionally it is used to give the reader an opportunity to explore some aspects which are not so well represented in texts; other times, the Bibliography is used to acknowledge clear debts. The Notes at the ends of chapters,

[2]This has become chapter 10 in the current edition.

[3]These have become chapter 11 in the current edition.

and referential material in the text have the aim of giving an impression of what the subject is about, of how mathematics is an activity involving people, and that it is a subject in a state of active development, even in its foundations.

I would like to thank the following for helpful comments and criticisms: Philip Higgins, Alan Pears, Guy Hirsch, Peter May, Terry Wall, Frank Adams, Jim Dugundji, R. E. Mosher. I also thank all those who wrote and pointed out misprints and obscurities.

I would like to conclude this preface on a personal note, which I hope may be of use to readers starting on, or aspiring to, a career in mathematical research, and wondering what that might entail. There is little in print on the methodology of mathematical research. There is material on problem solving, but there is little on evaluation, on problem choice, or on problem and concept formulation. There is material on the psychology of invention, but not so much on the training and development of invention, nor so much on the ends to which one should harness whatever invention one has. There are autobiographies available, but some perhaps give the impression that to do research in mathematics it is helpful to be a genius in the first place.

At the time of writing the first edition of this book, I was not clear as to what direction I wanted to take in my research, and to some extent writing the book was a displacement activity, distracting attention from the necessity of decision. However, as the book progressed and I tried to make the exposition clear, difficulties in the subject began to emerge. As draft succeeded draft, I became clearer about what I did not know, and a new range of possibilities began to arise. The original intention was a standard exposition of known, but scattered, material. This turned into an idiosyncratic treatment which itself suggested a new research line which has kept me busy ever since.

So I would like to commend to readers the idea that writing and rewriting mathematics with an intention to make things clear, well organised and comprehensible, and perhaps with some particular formal changes in mind, may in itself be a stimulus to further mathematical activity. These ideas are confirmed by some remarks of the composer Maurice Ravel, who argued for copying from other composers. If you have something to contribute, then this will appear of itself, and if you have no new ideas in this area, then at the least you will have made things clear to yourself. Thus the quality of your understanding is improved. There is also the simple joy of a well-crafted work.

There is also an argument for the *quality of misunderstanding*. It may be easy for you to feel that you do not understand because you are stupid. But your lack of understanding may be a reaction to a lack of clarity in the current expositions, or a feeling that the current expositions require something more to carry conviction. These expositions may prove results, but not explain them in a way you find agreeable. The only way to tell is to try and write an account for yourself, and see if you can make the matter clearer. For example, the material on the gluing theorem in chapter 7 was my response to a lack of understanding of the complicated proofs in the literature of results such as 7.5.3 (Corollary 2). By the time of the final draft, I was clear why these results were true.

The account of groupoids which has dominated both editions of this book came about in the following way. It was annoying to have a Van Kampen theorem which did not compute the fundamental group of a circle. I found that the account of the Van Kampen theorem given by [Olu58] could be generalised to yield this computation (see [Bro64]). So I started an exposition of Olum's non-abelian cohomology, with the laudable motivation that this would also introduce cohomological ideas to the reader. Unfortunately, when I looked at the 30 pages of my draft, I had to admit that they were pretty boring.

At the same period I had been pursuing some references on free groups. These lead me to the article [Hig63], which introduced free products with amalgamation of groupoids. So it seemed reasonable to set an exercise on the fundamental groupoid $\pi(U \cup V)$ of a union of spaces. Since the result was not in the literature as such, it seemed reasonable to write out a solution. The solution turned out to have the qualities of elegance, concision and clarity which I had been hoping for, but had not obtained, in my previous account. This suggested that the exposition should be turned round to give groupoids a central rather than peripheral rôle, particularly in view of a warm reception to a seminar I gave on the topic to the London Algebra Seminar in 1965. In 1967, I met Professor G. W. Mackey of Harvard University who told me of his work using groupoids in ergodic theory. This suggested that the groupoid concept had a wider application than I had envisaged, and so I also worked up chapter 9 on covering spaces, emphasising the groupoid viewpoint. It is perhaps only now possible to see the many disparate strands of work in which the notion of groupoid is usefully involved (cf. the survey article [Bro87]).

I would like to acknowledge here the help of two people who started me on a mathematical career. I have a great debt to the late Professor J. H. C. Whitehead, who was patient with my hesitancies and confusions. His many successes in exposing the formalities underlying geometric phenomena are a background to this text, and one part of my aims was to give an exposition of his lemmas on the homotopy type of adjunction spaces. From him I also absorbed the attitude of not giving up a mathematical idea until its essentials had been extracted, whatever the apparent relevance or otherwise to current fashions.

My second debt is to Professor M. G. Barratt, who initiated me into the practicalities and impracticalities of mathematical research. As to the impracticalities, I once got a ten-page letter from him which ended: 'Dawn breaks; I hope nothing else does!' As to the practicalities, I remember thinking to myself after a long session with Michael: 'If Michael Barratt can try out one damn fool thing after another, why can't I?' This has seemed a reasonable way of proceeding ever since. What is not so clear is why the really foolish projects (such as higher homotopy groupoids, based on flimsy evidence, and counter to current traditions) have turned out the most fun.

My thanks also go to Ellis Horwood and his staff. The production of this new edition has been greatly assisted by their expertise and enthusiasm.

RONALD BROWN
March, 1988

Preface to the third edition

The second edition of this book went out of print in 1994 or so, partly due to the publisher Ellis Horwood Ltd being sold. The final impetus for a new version came with funding from a Leverhulme Emeritus Fellowship, 2002–2004.

Background to the approach in this book

The republishing of this text is intended to make available material unavailable in other texts. The retitling of the book and some chapters is intended to show a major emphasis of this and previous editions: *the modelling of geometry, principally topology, by the algebra of groupoids.* The way this modelling works illustrates some important general principles, which are worth explaining.

The process of understanding some geometry may, over many years, go through the stages

$$\text{geometry} \longrightarrow \begin{array}{c}\text{underlying}\\ \text{processes}\end{array} \longrightarrow \text{algebra} \longrightarrow \text{algorithms} \longrightarrow \text{computation}$$

each of which has its own interest, character and problems. The algebra gives a precise, but possibly not total, expression of what we envision as the underlying processes. The end product, some computations, will hopefully give some specific answers to geometrical questions. The computations may be by hand, or if sufficiently complex, in a computer implementation. The success of the computations in answering geometrical questions 'proves' the methodology, in the old sense of 'prove' as 'test'.

In choosing algebra to model geometry there is a tendency to take the algebraic structures which are to hand—groups, rings, fields, and so on—but a certain eclecticism is necessary, and one should allow the geometry, an intuition for the underlying processes, and a feeling for mathematical structures, to formulate the algebra so that it expresses our intuitions. Experience shows that in this way understanding and modes of computation are increased. Possibly more importantly, we may in this way find new forms of algebra, algorithms and modes of computation, of general significance. In this way, mathematics develops language for description, verification, deduction and calculation, and that is a major reason for its necessity in science and technology.

This book shows the use of groupoids for this aim and in this context of basic algebraic topology.

The chief difference between groupoids and groups is that in groupoids there is a *partial multiplication* which is defined under geometric conditions: two arrows compose if and only if the end point of one is the initial point of the other. The arrows a, b in the diagram

$$x \xrightarrow{a} y \xrightarrow{b} z$$

'compose' to give an arrow $ba : x \to z$. This corresponds to the composition of journeys. Conversely, the analysis of a journey through various places requires precisely this notion of partial composition. So, to express our intuitions, we are, if we so allow ourselves, forced into considering groupoids rather than just groups. The theory of groupoids has added to group theory a spatial component, coming from the geography of the places we visit in a journey. For this reason, groupoids can model more of the geometry than groups alone. This leads not only to more powerful theorems with simpler and more natural proofs, but also to new theorems and new landscapes.

The process of modelling needs more explanation. We prove that the fundamental groupoid takes certain constructions on spaces to analogous constructions on groupoids. The use of analogy is fundamental to mathematics, though not always acknowledged publicly. When we note that $2+3 = 3+2$ and $2 \times 3 = 3 \times 2$ are examples of the commutative law, we are making an analogy between addition and multiplication of numbers. We then note that the commutative law applies to many situations, for example also to the addition of vectors. But numbers are not analogous to vectors, so how does this analogy come about? This seems to be an anomaly.

We mentioned above an analogy between certain constructions on spaces and constructions on groupoids. Yet spaces are very different from groupoids: they are different types of mathematical structure, without many analogies between the way they are defined. How can we make analogies between constructions on them?

The answer is a very important principle. In mathematics, and in many areas, analogies are not between objects themselves, but between the relations between objects. We will define many constructions by their relations to all other objects of the same type—this is called a 'universal property'. A certain construction on topological spaces which we call 'pushout' is defined by its relation to all topological spaces; a certain construction on groupoids which we also call 'pushout' is defined by its relation to all groupoids. Further, relations between topological spaces are given by the continuous functions between them. Relations between groupoids are defined by the morphisms between groupoids. Once the term pushout is defined in each context, the analogy between the two definitions will be completely clear. However the *construction* of a pushout in spaces is necessarily different in detail from the *construction* of a pushout of groupoids. Thus without the universal properties, the analogy might not be seen. All this is the essence of the 'categorical approach', and explains why category theory, which we use explicitly or implicitly, has been a major unifying force in the mathematics of the 20th century.

As an instance, the fact that the fundamental group of the circle is an infinite cyclic group is seen as an analogy between the following two diagrams (which are both pushouts!)

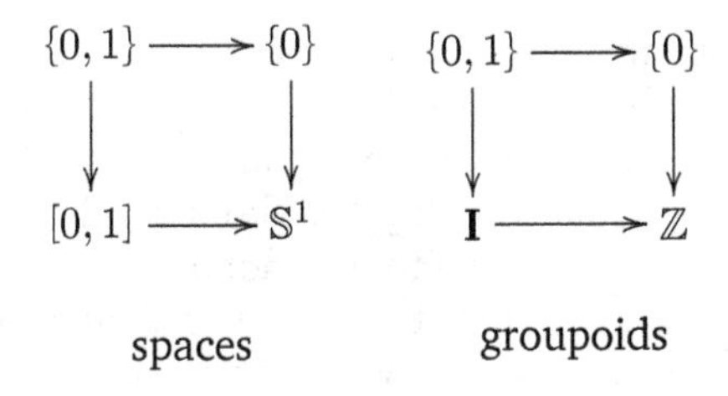

The left hand diagram shows the circle as obtained from the unit interval $[0, 1]$ by identifying, in the category of spaces, the two end points $0, 1$. The right hand diagram shows the infinite group of integers as obtained from the finite groupoid $\mathbf{I}$, again by identifying $0, 1$, but this time in the category of groupoids. The right hand diagram is not available for groups.

The groupoid $\mathbf{I}$ with its special arrow $\iota : 0 \to 1$ has the following property: if g is an arrow of a groupoid G then there is a unique morphism $\hat{g} : \mathbf{I} \to G$ whose value on ι is g. Thus the groupoid $\mathbf{I}$ with ι plays for groupoids the same role as does for groups the infinite cyclic group $\mathbb{Z}$ with the element 1: they are each free on one generator in their respective category. However we can draw a complete diagram of the elements of $\mathbf{I}$ as follows:

$$\circlearrowright 0 \underset{\iota^{-1}}{\overset{\iota}{\rightleftarrows}} 1 \circlearrowleft$$

whereas we cannot draw a complete picture of the elements of $\mathbb{Z}$.

The analogy between the above pushouts of spaces and groupoids generalises to the analogy between a space Y obtained from a space X by identifying various points of a discrete subspace, and a groupoid H obtained from a groupoid G by identifying various objects of G. The necessary algebra of groupoids for this is developed in Chapter 8, and the relation with topology in Chapter 9.

A necessary background to this method of universal properties is the notion of a category, which we give in Chapter 6, developing just that amount of theory needed for our results. We also develop some simple set theoretic examples of universal properties in the Appendix.

Other analogies we make are in Chapter 10, between covering maps of spaces, and covering morphisms of groupoids; and in Chapter 11, between orbit spaces under the action of a group, and orbit groupoids under the action of a group. Here also we prove the main result by verifying the required universal property of an orbit groupoid, and go on to examine ways of constructing and calculating orbit groupoids. This approach is not only easier to follow and more efficient, but also can suggest wider possibilities for investigation.

The types of problems in this book which we translate from topology to algebra using groupoids come largely under the heading of *local-to-global problems*. These problems are an important feature of mathematics and science. Much of algebraic topology is about methods for solving such problems. The distinctive nature of the fundamental group in these methods of algebraic topology is that it gives a nonabelian invariant, and the use of groupoids gives an extra power to this aspect. This power is also shown by the increasing use of groupoids in many fields of mathematics and physics, partly through allowing not only a more flexible approach to symmetry than groups alone, but also allowing for an algebra corresponding to transitions. Forms of multiple groupoids yield higher dimensional nonabelian methods for local-to-global problems.

With regard to the index, we take the line as in previous editions that it is better to be over- rather under-indexed, so that the index can possibly act as a guide, particularly in the e-version, with hyperref. Thus 'topological group' appears in the index under 'topological' and under 'group'. We may have failed on this principle in a number of instances, for which apologies in advance.

The changes for this edition are modest but significant.

1) We have already mentioned the retitling of the book and some chapters, and there has been a division of each the previous last two chapters into two, in order to give a clearer structure to the content.

2) Corrections have been made where known and found, and the diagrams have been improved where possible.

Finally, even if this is obvious to you, we must mention a major social change in freedom of mathematics information since the last edition of this book, namely the use since the 1980s of TEX and LATEX, and of course the growth of the Internet. These together have led in mathematics to: free electronic journals; free independent Internet encyclopaedias; search engines; the notion of e-book; the preprint arXiv; and many others. This means that a reader with access to the Internet can follow up and evaluate topics and concepts in a way not dreamt of in the 1980s. Hence if you want to follow up any reference to wider reading given in this book, you are well advised also to check out for yourself using Internet resources.

Acknowledgements

I have been greatly helped by a number of organisations and people.

The Leverhulme Foundation awarded me a Leverhulme Emeritus Fellowship for 2002–2004, part of whose aim was to publish this new edition. This book will also be a foundation for a book again supported in that period by the Fellowship, authored with Philip Higgins and Rafael Sivera, entitled *Nonabelian algebraic topology*. It will include an exposition of work of the author and Philip Higgins over the period 1974–2005 on applications of crossed complexes and related higher order groupoids in algebraic topology and the cohomology of groups. I would like to thank Tony Bak and Peter May for the strong support they gave to my Fellowship proposal.

Booksurge LLC have the organisation which makes this book readily available, in terms of printing and price.

La Monte Yarrol made a first draft of the LaTeX for the whole book in 1994. Peilang Wu revised this draft in 2003, and Gareth Evans has revised and overseen the final production version, during 2005-6. They both greatly improved the diagrams.

John Robinson and Ben Dickins designed the cover. For more information on the origin of the sculpture 'Journey', see `www.popmath.org.uk`.

I have been fortunate in many happy collaborations on the 'groupoid project'.

Philip Higgins introduced me to groupoids. This resulted in a long collaboration, fully employing his algebraic mastery and expository skills. His 1971 book 'Categories and groupoids', which overlaps this one, is available as a downloadable reprint.

Jean-Louis Loday showed how to take nonabelian methods in a new direction, yielding in particular a nonabelian tensor product of groups with homotopical and algebraic applications.

Jean Pradines and Kirill Mackenzie introduced me to Lie groupoids.

Tim Porter and Chris Wensley at Bangor have helped me enormously over the years: Tim particularly in algebraic and categorical background, and Chris in computational methods. Many research students have ventured into this groupoid territory, made important contributions and often became collaborators: Phil Heath (Hull), Lew Hardy, Tony Seda, Razak

Salleh, Keith Dakin, Nick Ashley, Andy Tonks, Ghafar Mosa, Fahd Al-Agl, Osman Mucuk, Ilhan İçen, Anne Heyworth, Emma Moore.

I also thank Alexander Grothendieck for an exuberant correspondence in the years 1982–1991: he described this as *a baton rompu*, which roughly means 'ranging over this and that', and indeed it dealt with many matters of mathematics and life.

RONALD BROWN
Deganwy, January 2006

Does the pursuit of truth give you as much pleasure as before? Surely it is not the knowing, but the learning, not the possessing but the acquiring, not the being there but the getting there, that afford the greatest satisfaction. If I have clarified and exhausted something, I leave it in order to go again into the dark. Thus is that insatiable man so strange; when he has completed a structure it is not in order to dwell in it comfortably, but to start another.

Karl Friedrich Gauss

I am a part of all that I have met;
Yet all experience is an arch wherethro'
Gleams that untravelled world, whose margin fades
For ever and for ever when I move.
How dull it is to pause, to make an end,
. . . .
And this gray spirit yearning in desire
To follow knowledge like a shining star,
Beyond the utmost bounds of human thought.
. . . .
Come, my friends,
'Tis not too late to seek a newer world.
. . . .
Made weak by time and fate, but strong in will
To strive, to seek, to find, and not to yield.

Alfred Lord Tennyson, *Ulysses*

Alice laughed. 'There's no use trying,' she said, 'one *can't* believe impossible things.'

'I daresay you haven't had much practice,' said the Queen. 'When I was your age, I always did it for half-an-hour a day. Why, sometimes I've believed as many as six impossible things before breakfast.'

Lewis Carroll, *Through the looking glass*

Chapter 1

Some topology on the real line

In this chapter, we introduce the system of neighbourhoods of points on the real line $\mathbb{R}$; this presents $\mathbb{R}$ with structure additional to its usual structures of addition, multiplication, and order. This additional structure makes $\mathbb{R}$ a *topological space*, and the study of this and other topological spaces is the subject matter of this book.

There are several reasons for giving this example of a topological space before a definition of such an object. First of all, we hope in this way to familiarise the reader with some of the techniques and terminology which will recur constantly. Secondly, the real line is central to mathematics, in the way that atoms are central to physics, and cells to biology. So time devoted to this example is well spent.

1.1 Neighbourhoods in $\mathbb{R}$

Definition Let N be a subset of $\mathbb{R}$, and let $a \in \mathbb{R}$. Then N is a *neighbourhood* of a if there is a real number $\delta > 0$ such that the open interval about a of radius δ is contained in N; that is, if there is a $\delta > 0$ such that

$$]a - \delta, a + \delta[\subseteq N.$$

EXAMPLES

1. $\mathbb{R}$ itself is a neighbourhood of any a in $\mathbb{R}$.
2. The closed interval $[0, 1]$ is a neighbourhood of $\frac{1}{2}$, but not of 0, nor of 1.
3. The set $\{0\}$ consisting of 0 alone is not a neighbourhood of 0.

4. $[0,1[\,\cup\,]1,2]$ is not a neighbourhood of 1.
5. $[0,1] \cup [2,3] \cup]3\frac{6}{7}, 8[$ is a neighbourhood of 2π.
6. The set $\mathbb{Q}$ of rationals is not a neighbourhood of any $a \in \mathbb{R}$: for any interval $]a-\delta, a+\delta[$ (where $\delta > 0$) must contain irrational as well as rational points, and so $]a-\delta, a+\delta[$ cannot be contained in $\mathbb{Q}$.
7. This is a mildly pathological example. For each integer $n > 0$, let

$$X_n = \{x \in \mathbb{R} : (2n+1)^{-1} < |x| < (2n)^{-1}\}.$$

Let X be the union of $\{0\}$ and the sets X_n for all $n > 0$ [cf. Fig. 1.1]. Then X is not a neighbourhood of 0, since any interval about 0 contains points not in X (in addition, of course, to points of X).

$-\frac{1}{2}$ $-\frac{1}{3}$ 0 $\frac{1}{5}$ $\frac{1}{4}$ $\frac{1}{3}$ $\frac{1}{2}$

Fig. 1.1

The notion of neighbourhood is useful in formulating the definition of something being true 'near' a given point. In fact, let P be a property which applies to real numbers, and may or may not hold for any given real number. Let a be in $\mathbb{R}$. We say P *holds near* a, or *is valid near* a, if P holds for all points in some neighbourhood of a.

For example, let $f : \mathbb{R} \to \mathbb{R}$ be the function $x \mapsto x^2 + x^3$. Then f is positive near the point 1 (since $f(x) > 0$ for all $x > 0$); but it is not true that f is positive near 0, or near -1.

There is still no notion of absolute nearness, that is, of a point x being 'near a'. This is to be expected; the only definition of x being near a that makes sense is that x is near a if x is in some neighbourhood of a. But any x in $\mathbb{R}$ is then near a.

At this stage we could still dispense with the arbitrary neighbourhoods and work entirely with intervals. However, the elegance and flexibility of the general notion will appear as we proceed.

We now derive some simple properties of neighbourhoods.

1.1.1 *Let $a \in \mathbb{R}$, and let M, N be neighbourhoods of a. Then $M \cap N$ is a neighbourhood of a.*

Proof Since M, N are neighbourhoods of a, there are real numbers $\delta, \delta' > 0$ such that

$$]a-\delta,\ a+\delta[\ \subseteq M, \qquad]a-\delta',\ a+\delta'[\ \subseteq N.$$

Let $\delta'' = \min(\delta, \delta')$. Then

$$]a-\delta'',\ a+\delta''[\ \subseteq M \cap N$$

and so $M \cap N$ is a neighbourhood of a. □

1.1.2 *Let* $a \in M \subseteq N \subseteq \mathbb{R}$. *If* M *is a neighbourhood of* a, *then so also is* N.

Proof If M is a neighbourhood of a, then there is a $\delta > 0$ such that $]a - \delta,\ a+\delta[\ \subseteq M$. Hence $]a-\delta,\ a+\delta[\ \subseteq N$, and the result follows. □

1.1.3 *An open interval is a neighbourhood of any of its points.*

Proof Suppose first that I is an open interval of the form $]a, b[$ where $a, b \in \mathbb{R}$. Let $x \in I$, and let $\delta = \min(x-a, b-x)$. Then δ is positive and $]x-\delta,\ x+\delta[\ \subseteq I$, whence I is a neighbourhood of x.

The proofs for the other kinds of open intervals are also simple. □

Clearly 1.1.3 is false if the word 'open' is removed.

A point a in $\mathbb{R}$ determines the sets N which are neighbourhoods of a. Also a set A determines the set of points of which A is a neighbourhood; this set is called the *interior of* A, and is written Int A; thus $x \in \operatorname{Int} A$ if and only if A is a neighbourhood of x. Since x belongs to any of its neighbourhoods, Int A is a subset of A.

EXAMPLES

8. If I is an open interval, then $\operatorname{Int} I = I$.
9. If A is finite then Int A is empty.
10. If $A = [a, b]$ then $\operatorname{Int} A =\ ]a, b[$.
11. If $A = \mathbb{Q}$ the set of rationales, then $\operatorname{Int} A = \varnothing$.
12. If $A = \mathbb{R} \setminus \mathbb{Q}$, then $\operatorname{Int} A = \varnothing$.

1.1.4 *If* $A \subseteq B$, *then* $\operatorname{Int} A \subseteq \operatorname{Int} B$.

Proof If A is a neighbourhood of x, then so also, by 1.1.2, is B. □

1.1.5 *If* N *is a neighbourhood of* a, *then so also is* Int N.

Proof Let N be a neighbourhood of a, and let $\delta > 0$ be such that $]a-\delta,\ a+\delta[\ \subseteq N$. Then $]a-\delta,\ a+\delta[\ \subseteq \operatorname{Int} N$, by 1.1.4 and since the interior of an open interval is the same interval. So Int N is a neighbourhood of a. □

EXERCISES

1. Let $a \in \mathbb{R}$ and let N be a subset of $\mathbb{R}$. Prove that the following conditions are equivalent.
(a) N is a neighbourhood of a.
(b) There is a $\delta > 0$ such that $[a-\delta,\ a+\delta]\ \subseteq N$.
(c) There is an integer $n > 0$ such that $[a - n^{-1},\ a+n^{-1}]\ \subseteq N$.

2. Prove that if A is a countable set of real numbers, then $\operatorname{Int} A = \varnothing$.
3. Prove that $\operatorname{Int}(A \cap B) = \operatorname{Int} A \cap \operatorname{Int} B$.
4. Does $\operatorname{Int}(A \cup B) = \operatorname{Int} A \cup \operatorname{Int} B$?
5. Let (A_n) be a sequence of subsets of $\mathbb{R}$. Does

$$\operatorname{Int}\left(\bigcap_n A_n\right) = \bigcap_n \operatorname{Int} A_n?$$

6. Let C be a neighbourhood of $c \in \mathbb{R}$, and let $a + b = c$. Prove that there are neighbourhoods A of a, B of b such that $x \in A$ and $y \in B$ implies $x + y \in C$.
7. Write down and prove a similar result to that of Exercise 6, but with $c = ab$.
8. Let C be a neighbourhood of c, where $c \neq 0$. Prove that there is a neighbourhood C' of c^{-1} such that if x is in C', then x^{-1} is in C.
9. Prove that $\operatorname{Int}(\operatorname{Int} A) = \operatorname{Int} A$.
10. Let $A_1 = [-1, 1[\setminus\{0\}$, $A_2 = \{2\}$, $A_3 = \mathbb{Q} \cap [3, 4]$, and let $A = A_1 \cup A_2 \cup A_3$. Show that exactly fourteen distinct subsets of $\mathbb{R}$ may be constructed from A by means of the operations Int and complementation with respect to $\mathbb{R}$.

1.2 Continuity

In this section, we define continuity of real functions (that is, functions whose domain and range are subsets of $\mathbb{R}$). This definition is entirely in terms of neighbourhoods. In this section, we shall usually take the range of a real function to be $\mathbb{R}$ itself, but the definitions and results are the same as for the general case, when the range is any subset of $\mathbb{R}$.

Let $f : A \to \mathbb{R}$ be a function, where A is a subset of $\mathbb{R}$. Let $a \in A$.

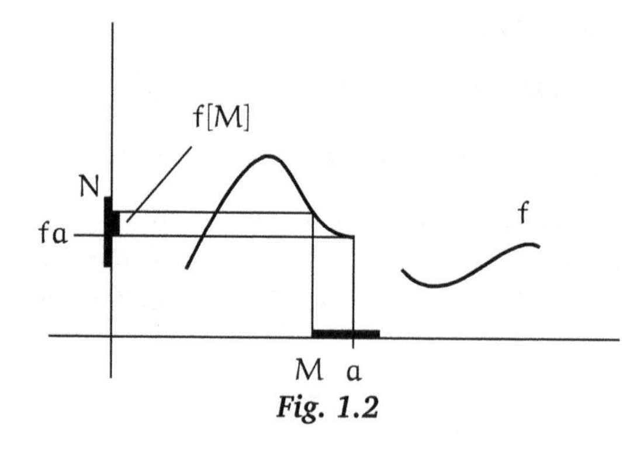

Fig. 1.2

Definition C The function f is *continuous* at a if for each neighbourhood N of $f(a)$ there is a neighbourhood M of a such that $f[M] \subseteq N$. Further, f is *continuous* if f is continuous at each a in A.

In this definition, $f[M]$ is the image[‡] of the set M by f. So the definition can be restated:

Definition C′ The function f is *continuous at* a if, for each neighbourhood N of $f(a)$, there is a neighbourhood M of a such that for all $x \in \mathbb{R}$

$$x \in M \cap A \Rightarrow f(x) \in N.$$

The statement $f[M] \subseteq N$ is equivalent to $M \cap A \subseteq f^{-1}[N]$. Suppose now that A is a neighbourhood of a. Then $M \cap A$ is a neighbourhood of a, so that $f[M] \subseteq N$ implies that $f^{-1}[N]$ is a neighbourhood of a. On the other hand, we always have

$$ff^{-1}[N] \subseteq N.$$

So if $f^{-1}[N]$ is a neighbourhood of a, then $f^{-1}[N]$ is itself a neighbourhood M of a such that $f[M] \subseteq N$. This shows that if A is a neighbourhood of a, we can restate Definition C as:

Definition C″ The function f is *continuous* at a if, for every neighbourhood N of $f(a)$, $f^{-1}[N]$ is a neighbourhood of a.

This last definition has only one quantifier, whereas Definition C has two, and Definition C′ has *three*. Thus Definition C″ is the easiest to understand, but we emphasise that it applies only to the case when A, the domain of f, is a neighbourhood of a.

Another advantage of Definition C″ is that it is easy to negate. We suppose A is a neighbourhood of a: then f is not continuous at a if, for some neighbourhood N of $f(a)$, $f^{-1}[N]$ is not a neighbourhood of a. This is illustrated in the examples which follow.

EXAMPLES

1. Let $l \in \mathbb{R}$ and let $f : \mathbb{R} \to \mathbb{R}$ be the constant function $x \mapsto l$. Let $a \in \mathbb{R}$. The domain of f is $\mathbb{R}$, which is a neighbourhood of a. If N is a neighbourhood of l, then $f^{-1}[N] = \mathbb{R}$, which is a neighbourhood of a. Therefore, f is continuous at a, and since a is arbitrary, f is a continuous function.
2. Let $f : \mathbb{R} \to \mathbb{R}$ be the identity function $x \mapsto x$. Let $a \in \mathbb{R}$. The domain of f is a neighbourhood of a, and $f(a) = a$. If N' is a neighbourhood of $f(a)$, then $f^{-1}[N] = N$, so that $f^{-1}[N]$ is a neighbourhood of a. Thus f is a continuous function.

[‡]See A.1.5 of the Appendix.

3. Consider the function

$$f : \mathbb{R} \to \mathbb{R}$$

$$x \mapsto \begin{cases} 0, & x < 1 \\ 2, & x \geqslant 1. \end{cases}$$

Here again, $\mathbb{R}$ is a neighbourhood of 1. In this case, f is not continuous at 1. For let $N = [1, \to[$. Then N is a neighbourhood of $f(1) = 2$, but $f^{-1}[N] = [1, \to[$ which is not a neighbourhood of 1.

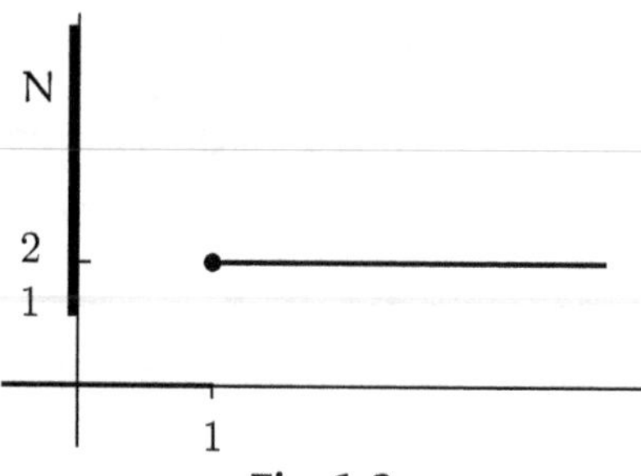

Fig. 1.3

4. Consider the function

$$f : [1, \to[\ \to \mathbb{R}$$
$$x \mapsto 2.$$

Here the domain of f is $[1, \to[$, which is not a neighbourhood of 1; so to prove continuity at 1, we must use Definition C. Let N be a neighbourhood of 2. Then

$$f[\mathbb{R}] = \{2\}$$

which is a subset of N. Since $\mathbb{R}$ is a neighbourhood of 1, we have proved continuity at 1. This example should be compared carefully with Example 3.

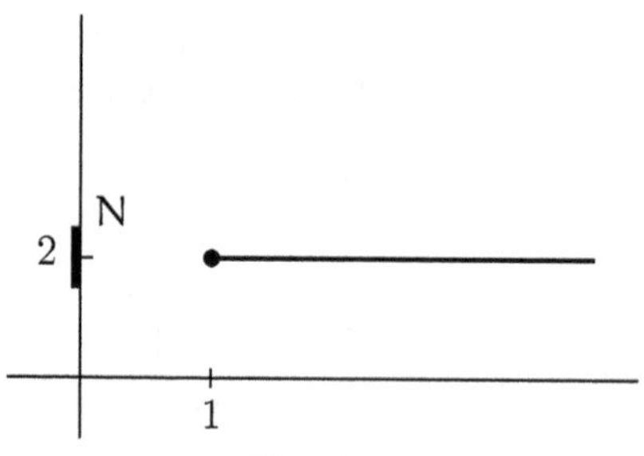

Fig. 1.4

5. Let $f : \mathbb{R} \to \mathbb{R}$ be defined as follows. If α is a rational number and is, in its lowest terms, p/q where p, q are integers such that $q > 0$, then $f(\alpha) = 1/q$ (in particular, $f(0) = 1$). If β is irrational, then $f(\beta) = 0$. We prove that f is not continuous at any rational number, but is continuous at any irrational number.

Let $\alpha = p/q$ be such that $f(\alpha) = 1/q$. Let $N = \,]0, \to[$. Then N is a neighbourhood of $f(\alpha)$, but $f^{-1}[N]$ contains no irrational numbers and so is not a neighbourhood of α. Therefore f is not continuous at α.

In order to prove that f is continuous at an irrational number β, we use the fact that rational numbers of high denominators are needed to approximate closely an irrational number.

Let N be a neighbourhood of $0 = f(\beta)$. We prove that $M = f^{-1}[N]$ is a neighbourhood of β. Certainly, M contains all irrational numbers, since these are all sent to 0 by f, and we show that all rational numbers close enough to β are also contained in M.

Let n be a positive integer such that

$$[-n^{-1}, n^{-1}] \subseteq N;$$

such an integer exists since N is a neighbourhood of 0 [cf. Exercise 1 of Section 1.1]. Let m be an integer such that $\beta \in [m, m+1]$. There are only a finite number of rational numbers p/q which are in $[m, m+1]$ and are such that $0 < q \leqslant n$. One of these, say r, will be closest to β. Let

$$|\beta - r| = \delta, \qquad I = \,]\beta - \delta, \beta + \delta[.$$

Since β is irrational, δ is positive. Also, all rational numbers p/q in I satisfy $q > n$. Hence

$$f[I] \subseteq \,]-n^{-1}, n^{-1}[, \text{ whence } I \subseteq M.$$

Therefore M is a neighbourhood of β, and f is continuous at β.

Examples 1, 2, and 5 show that some functions are proved continuous by working directly with the definition. More usually, though, we construct continuous functions by taking a basic stock of continuous functions and giving rules for making more complicated continuous functions from those of the basic stock.

We take from analysis the definitions and continuity of the functions sin and log. These, with the identity function and the constant function, form our basic stock. Let $f : A \to \mathbb{R}$, $g : B \to \mathbb{R}$ be functions; then $f + g$, $f.g$, f/g are respectively the functions

$$x \mapsto f(x) + g(x), \qquad x \mapsto f(x).g(x), \qquad x \mapsto f(x)/g(x)$$

with respective domains $A \cap B$, $A \cap B$, $(A \cap B) \setminus \{x \in B : g(x) = 0\}$. Let A, B be subsets of $\mathbb{R}$, and let $a \in A \cap B$.

1.2.1 *(Sum, product, and quotient rule.) If* f, g *are continuous at* a, *so also are* $f + g$, $f.g$; *if* f/g *is defined at* a, *then it is continuous at* a.

We omit the proof, which is given, for example, in many analysis books [cf. also Exercise 1 following, and Sections 2.5, 2.8].

1.2.2 *(Restriction rule) Let* $f : A \to \mathbb{R}$ *be a real function continuous at* $a \in A$. *Let* B *be a subset of* A *containing* a. *Then* $f \mid B$ *is continuous at* a.

Proof Let $g = f \mid B$. Let N be a neighbourhood of $g(a) = f(a)$. Since f is continuous at a, there is a neighbourhood M of a such that $f[M] \subseteq N$. Then

$$g[M] = g[M \cap B] = f[M \cap B] \subseteq f[M] \subseteq N.$$

So $g[M] \subseteq N$, and g is continuous at a. □

The restriction rule can be used to deduce Example 4 from Example 1.

Let $f : A \to \mathbb{R}$, $g : B \to \mathbb{R}$ be real functions, and let $h : A \cap f^{-1}[B] \to \mathbb{R}$ be the composite function $x \mapsto g(f(x))$.

1.2.3 *(Composite rule) If* $a \in A \cap f^{-1}[B]$, f *is continuous at* a *and* g *is continuous at* $f(a)$, *then* h *is continuous at* a.

Proof Let N be a neighbourhood of $h(a)$. Since g is continuous at $f(a)$, there is a neighbourhood M of $f(a)$ such that $g[M] \subseteq N$. Since f is continuous at a, there is a neighbourhood L of a such that $f[L] \subseteq M$. It is easy to verify that

$$h[L] = g[f[L]],$$

whence $h[L] \subseteq N$. Therefore h is continuous at a. □

1.2.4 *(Inverse rule.) Let* $f : A \to \mathbb{R}$ *be a real function which is injective, so that* f *has an inverse* $f^{-1} : f[A] \to A$. *If* A *is an interval and* f *is continuous, then* f^{-1} *is continuous.*

For a proof see many books on a first course in analysis. This result is also a consequence of general theorems on connectivity and compactness [cf. Chapter 3]. The assumption that A is an interval is essential [cf. Exercise 4 of Section 1.2].

EXAMPLES

6. By repeated application of 1.2.1 to the identity function and constant functions, we can prove in turn the continuity of $x \mapsto x^n$ $(n \geqslant 0)$, polynomial functions, and rational functions.

7. The continuity of cos, tan, sec, cosec follows easily from 1.2.1 and 1.2.3. For example, cos is the composite of sin and $x \mapsto \pi/2 - x$, and tan is the quotient sin / cos.

By 1.2.2, $\sin \mid [-\pi/2, \pi/2]$ is continuous; but this function is injective, so its inverse $\sin^{-1}$ is continuous by 1.2.4. Similarly, we derive the continuity of $\cos^{-1}$, and $\tan^{-1}$. The function log is injective and continuous; therefore, its inverse exp is continuous. The function $\sqrt[n]{\ }$ is continuous, since it is the inverse of $x \mapsto x^n$ if n is odd, and of $x \mapsto x^n$ $(x \geqslant 0)$ if n is even. So we can prove continuity of functions such as

$$x \mapsto (\sin x)^{1/n} + \log x + x^{17}.$$

EXERCISES

1. Prove 1.2.1 by use of Exercises 6, 7 of Section 1.1.
2. Prove the following 'Sandwich Rule' (also called the Squeeze Rule). Let $\lambda, \mu : A \to \mathbb{R}$ be two functions continuous at $a \in A$ and such that $\lambda(a) = \mu(a)$. Let $f : A \to \mathbb{R}$ be a function such that for some neighbourhood M of a

$$x \in M \cap A \Rightarrow \lambda(x) \leqslant f(x) \leqslant \mu(x).$$

Then f is continuous at a.

Use this rule to prove the continuity of the function

$$x \longmapsto \begin{cases} x \sin x^{-1} & x \neq 0 \\ 0 & x = 0. \end{cases}$$

3. Let $f : A \to \mathbb{R}$ be a function, let $a \in A$, and let N be a neighbourhood of a. Prove that if $f \mid N \cap A$ is continuous at a, then so also is f.
4. Let f be the function

$$x \longmapsto \begin{cases} x & 0 \leqslant x \leqslant 1 \\ x - 1 & 2 < x \leqslant 3. \end{cases}$$

Prove that f is continuous and injective, but that f^{-1} is not continuous at 1.
5. Let $f : [a, b] \to [c, d]$ be a monotonic bijection. Prove that f is continuous.
6. Let $f : \mathbb{R} \to \mathbb{R}$. Prove that f is continuous if and only if for every subset A of $\mathbb{R}$

$$f^{-1}[\operatorname{Int} A] \subseteq \operatorname{Int} f^{-1}[A].$$

7. Let $f : A \to \mathbb{R}$ be a real function and let $a \in A$. Prove the equivalence of the following statements:
(a) f is continuous at a.
(b) for all $\varepsilon > 0$, there is a $\delta > 0$ such that

$$f\,]a - \delta, a + \delta[\ \subseteq\ ]a - \varepsilon, a + \varepsilon[.$$

(c) For all positive integers m there is a positive integer n such that

$$f\,]a - n^{-1}, a + n^{-1}[\ \subseteq\ ]a - m^{-1}, a + m^{-1}[.$$

8. Prove the following 'gluing rule'. Let $A = A_1 \cup A_2 \subseteq \mathbb{R}$, $a \in A_1 \cap A_2$. Let $f : A \to \mathbb{R}$ be a function such that $f \mid A_1$, $f \mid A_2$ are continuous at a. Then f is continuous at a. Prove also that f is continuous if $f \mid A_1$, $f \mid A_2$ are continuous and $A_1 \setminus A_2 \subseteq \operatorname{Int} A_1$, $A_2 \setminus A_1 \subseteq \operatorname{Int} A_2$.
9. Prove that the set of all continuous functions $[0,1] \to \mathbb{R}$ is uncountable.

1.3 Open sets, closed sets, closure

In this section, we introduce some more topological concepts on the real line $\mathbb{R}$.

First we consider the open sets. A subset U of $\mathbb{R}$ is *open* if U is a neighbourhood of each of its points. Now for any set U, $\operatorname{Int} U$ is the set of points of which U is a neighbourhood. So U is open if and only if $U = \operatorname{Int} U$.

Examples of open sets are the empty set (which has no points and so is a neighbourhood of each of them) and, by 1.1.3, any open interval. Other examples may be constructed by means of the following results.

1.3.1 *The union of any family of open sets is open.*

Proof Let $(U_i)_{i \in J}$ be a family of subsets of $\mathbb{R}$ such that each U_i is open, and let $U = \bigcup_{i \in J} U_i$. If U is empty, it is open, if not, let $u \in U$; we prove that U is a neighbourhood of u.

First, $u \in U_i$ for some i. Since U_i is open, it is a neighbourhood of u. Therefore U, which contains U_i, is also a neighbourhood of u. □

1.3.2 *A subset of $\mathbb{R}$ is open if and only if it is the union of a countable set of disjoint open intervals.*

Proof The union of any family of open intervals is open by 1.1.3 and 1.3.1.

To prove the converse, let U be an open set. If U is empty the result is true since U is the union of the empty family of intervals.

Suppose U is not empty. Two points x, y of U are called equivalent, written $x \sim y$, if the closed interval with end points x, y is contained in U. It is easily verified that $\sim$ is an equivalence relation on U. By the definition of $\sim$, the equivalence classes are intervals of $\mathbb{R}$ (that is, if x, y belong to an equivalence class E, and $x < y$, then any point z such that $x < z \leqslant y$ also belongs to E). These equivalence classes are also disjoint and cover U.

Now the open intervals of $\mathbb{R}$ are exactly those intervals of $\mathbb{R}$ which do not contain any of their end points—this is a non-trivial fact about $\mathbb{R}$ being a consequence of the completeness of the order relation (see the Glossary under *bounded* and *interval*). So to prove the theorem let E be one of the above intervals and let a be an end point of E. If $a \in U$ then, for some

$\delta > 0$, $]a - \delta, a + \delta[$ is contained in U and hence also in E, and this is absurd. Therefore $a \notin U$ and so $a \notin E$. Thus E is an open interval.

Let φ be the function which sends each element of $\mathbb{Q} \cap E$ to its equivalence class. Then φ is a surjection to the set of equivalence classes since each non-empty open interval of $\mathbb{R}$ contains a rational number. Since $\mathbb{Q} \cap U$ is countable it follows that the number of equivalence classes is countable. □

The simple criterion 1.3.2 allows some pathological and complicated examples. For example, in Fig. 1.1 the union of all the sets X_n, that is, the set $X \setminus \{0\}$, is an open set.

EXAMPLE *The Cantor Set.* This is a subset K of $\mathbb{I} = [0, 1]$ such that $\mathbb{I} \setminus K$ is open.

The *middle-third* of a closed interval $[a, b]$ is the open interval,

$$]a + (b - a)/3, b - (b - a)/3[\ .$$

If (I_α) is a family of disjoint closed intervals, and $U = \bigcup I_\alpha$, then the *middle-third* of U is the union of the middle-thirds of each I_α.

Now let $\mathbb{I} = [0, 1]$; we define sets X_n, I_n by induction. First, X_1 is the middle third of $\mathbb{I}$ and

$$I_1 = \mathbb{I} \setminus X_1.$$

Suppose X_n, I_n have been defined, and I_n is a finite union of disjoint closed intervals. Then X_{n+1} is defined to be the union of X_n and the middle-third of I_n, and we set $I_{n+1} = \mathbb{I} \setminus X_{n+1}$. The construction is illustrated in Fig. 1.5.

It is easy to prove by induction that I_n is the union of 2^n closed intervals each of length 3^{-n}, that X_n is a union of disjoint open intervals, and that

$$\mathbb{I} \supseteq I_1 \supseteq I_2 \supseteq \cdots; \quad X_1 \subseteq X_2 \subseteq X_3 \subseteq \cdots.$$

Let $X = \bigcup_{n \geqslant 1} X_n$ so that X is open. The *Cantor set* is

$$K = \mathbb{I} \setminus X = \bigcap_{n \geqslant 1} I_n.$$

There is a convenient representation of the points of K by ternary decimals. We recall (see any book on analysis) that each point of $\mathbb{I}$ can be represented as

$$.a_1 a_2 a_3 \ldots = \sum_{n=1}^{\infty} a_n 3^{-n}, \quad a_n = 0, 1, \text{ or } 2.$$

The points of X_1 have $a_1 = 1$, and those of I_1 of $a_1 = 1$ or 2. The points of X_2 have $a_1 = 1$ or $a_2 = 1$, while for I_2 neither a_1 nor a_2 can be 1. In fact, it is not hard to prove by induction that the points of I_n are exactly those real numbers whose representation as ternary decimals $.a_1a_2\ldots$ have no 1's in the first n places. So the points of K are represented by ternary decimals $.a_1a_2\ldots$ with $a_n \neq 1$ for all $n \geqslant 1$; and the points of X are represented by ternary decimals $.a_1a_2\ldots$ in which at least one of the a_n is 1.

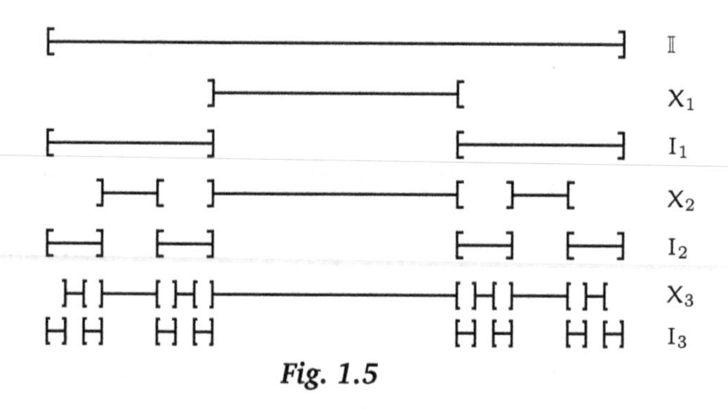

Fig. 1.5

This completes our account of the Cantor set.

The open sets in $\mathbb{R}$ generalise the open intervals. The corresponding generalisations of the closed intervals are the closed sets: a subset C of $\mathbb{R}$ is *closed* if $\mathbb{R} \setminus C$ is open. Thus a closed interval $[a, b]$ is closed since its complement is $]\leftarrow, a[\ \cup \ ]b, \rightarrow[$ which is open. Corresponding to 1.3.1, we have,

1.3.3 *The intersection of any family of closed sets is closed.*

Proof Let (C_α) be a family of closed sets, and let $C = \bigcap C_\alpha$. We must prove that $\mathbb{R} \setminus C$ is open.

By the De Morgan laws,

$$\mathbb{R} \setminus C = \bigcup (\mathbb{R} \setminus C_\alpha).$$

But C_α is closed, so that $\mathbb{R} \setminus C_\alpha$ is open, and $\bigcup(\mathbb{R} \setminus C_\alpha)$ is open by 1.3.1. □

A corollary of 1.3.3 is that the Cantor set K is closed: for $K = \bigcap_{n \geqslant 1} I_n$, and $\mathbb{R} \setminus I_n$ is open, since it is the union of open intervals.

There are subsets of $\mathbb{R}$ which are neither open nor closed, for example, the half-open interval $[0, 1[$. A natural question is: which subsets of $\mathbb{R}$ are both open and closed?

1.3.4 *The only subsets of* $\mathbb{R}$ *which are both open and closed are* $\varnothing$ *and* $\mathbb{R}$.

Proof Let U be a non-empty, open, proper subset of $\mathbb{R}$.

By 1.3.2, U is the union of a family of disjoint open intervals. Since U is neither $\varnothing$ nor $\mathbb{R}$, one of these intervals has an end point, say a. Now $a \in \mathbb{R} \setminus \mathrm{U}$, but $\mathbb{R} \setminus \mathrm{U}$ cannot be a neighbourhood of a since a is the end point of an interval contained in U. Therefore $\mathbb{R} \setminus \mathrm{U}$ is not open, so U is not closed. □

The open sets U and $\mathbb{R}$ are characterised by the property: $\mathrm{U} = \operatorname{Int} \mathrm{U}$. There is another operation on subsets of $\mathbb{R}$ which characterises the closed sets.

Let A be a subset of $\mathbb{R}$. We divide the points of R into three sets. First, we have the interior points of A; these form a set $\operatorname{Int} A$ which has already been discussed. Second, we have the *exterior* points of A, which form a set $\operatorname{Ext} A$; these are the points which are interior to $\mathbb{R} \setminus A$, so that $x \in \operatorname{Ext} A$ if and only if $\mathbb{R} \setminus A$ is a neighbourhood of x. Finally, the points which remain form the *frontier* $\operatorname{Fr} A$.

For example, let $A = [0, 1[$. The interior of A is $]0, 1[$, the exterior of A is $]\leftarrow, 0[\ \cup\ ]1, \rightarrow[$, and the frontier of A consists of the points 0 and 1.

The *closure* $\overline{A}$ of a set A is obtained by adding to $\operatorname{Int} A$ the points of the frontier of A; that is,

$$\overline{A} = \operatorname{Int} A \cup \operatorname{Fr} A.$$

In fact $A \subseteq \overline{A}$, so that $\overline{A} = A \cup \operatorname{Fr} A$. We prove $A \subseteq \overline{A}$ as follows: Let $x \in A$. If $x \in \operatorname{Int} A$, then certainly $x \in \overline{A}$. Suppose $x \in A \setminus \operatorname{Int} A$: then A is not a neighbourhood of x, and neither is $\mathbb{R} \setminus A$; hence $x \in \operatorname{Fr} A$, and so $x \in \overline{A}$.

We set out the definitions of these operators in terms of neighbourhoods:

(a) $x \in \operatorname{Int} A \Leftrightarrow A$ is a neighbourhood of $x \Leftrightarrow$ some neighbourhood of x does not meet $\mathbb{R} \setminus A$.
(b) $x \in \operatorname{Ext} A \Leftrightarrow \mathbb{R} \setminus A$ is a neighbourhood of $x \Leftrightarrow$ some neighbourhood of x does not meet A.
(c) $x \in \operatorname{Fr} A \Leftrightarrow$ every neighbourhood of x meets both A and $\mathbb{R} \setminus A$.
(d) $x \in \overline{A} \Leftrightarrow$ every neighbourhood of x meets A.

It is necessary to explain why (d) defines $\overline{A}$. Suppose first that $x \in \overline{A}$, and that N is a neighbourhood of x. If $x \in A$, then $x \in N \cap A$ and so N meets A; if $x \in \operatorname{Fr} A$, then N meets A by (c). So if $x \in \overline{A}$ then every neighbourhood of x meets A.

Conversely, suppose every neighbourhood of x meets A. If $x \in A$, then $x \in \overline{A}$; and if $x \in \mathbb{R} \setminus A$, then every neighbourhood of x meets both A and $\mathbb{R} \setminus A$, so that $x \in \operatorname{Fr} A$. In either case, $x \in \overline{A}$.

The closure operation is probably the most important of these topological operators—when we define closure for subsets of a general topological space, we shall take (d) above as the definition.

We conclude this section by proving:

1.3.5 *A subset* A *of* $\mathbb{R}$ *is closed if and only if* $A = \overline{A}$.

Proof Suppose A is closed. Then $\mathbb{R} \setminus A$ is open, so that $\operatorname{Ext} A = \mathbb{R} \setminus A$. Hence

$$\overline{A} = \operatorname{Int} A \cup \operatorname{Fr} A = \mathbb{R} \setminus \operatorname{Ext} A = A.$$

Conversely, suppose $A = \overline{A}$. Then

$$\mathbb{R} \setminus A = \operatorname{Ext} A = \operatorname{Int}(\mathbb{R} \setminus A).$$

Therefore $\mathbb{R} \setminus A$ is open, and A is closed. □

EXERCISES

1. Prove that the Cantor set K is uncountable.
2. Prove that the function

$$\begin{aligned} f : K &\to \mathbb{I} \\ \sum_1^{\infty} a_n 3^{-n} &\mapsto \sum_1^{\infty} a_n 2^{-n} \end{aligned}$$

is continuous, increasing, and surjective.

3. Let X_n be as in the construction of the Cantor set K. Prove that $X_n \setminus X_{n-1}$ is the union of 2^{n-1} open intervals $X_{n,p}$ each of length 3^{-n}. Prove that the function f of Exercise 2 takes the same value at the end points of each $X_{n,p}$. Deduce that f extends to a continuous function $g : \mathbb{I} \to \mathbb{I}$ which is constant on each $X_{n,p}$. Sketch the graph of g.
4. Determine the exterior, frontier, and closure of the following subsets of $\mathbb{R}$: (i) $\mathbb{Q}$, (ii) $\mathbb{R} \setminus \mathbb{Q}$, (iii) $\{0\}$, (iv) $\{n^{-1} : n \text{ a positive integer}\}$, (v) $\mathbb{Z}$.
5. Let A be a subset of $\mathbb{R}$. Prove that

$$\mathbb{R} \setminus \overline{A} = \operatorname{Int}(\mathbb{R} \setminus A).$$

6. Let A be a subset of $\mathbb{R}$. Prove that if C is a closed set containing A, then C contains $\overline{A}$.
7. Let A be an open subset of $\mathbb{R}$ and $f : A \to \mathbb{R}$ a function. Prove that f is continuous if and only if $f^{-1}[U]$ is open for each open set U of $\mathbb{R}$.
8. Prove that any closed set of $\mathbb{R}$ is the intersection of countably many open sets. [Use 1.3.2.]

1.4 Some generalisations

The idea of neighbourhood on which much of the previous sections was based, is applicable to other situations and this will of course be discussed in detail in later chapters. Here we wish to prepare the reader for the full scale axiomatics of the next chapter by showing some other examples of neighbourhoods.

There is an easy generalisation to the Euclidean plane $\mathbb{R}^2 = \mathbb{R} \times \mathbb{R}$. Let us identify $\mathbb{R}^2$ with $\mathbb{C}$, the set of complex numbers. Then, we can define in $\mathbb{C}$ a neighbourhood of a complex number a to be any subset N of $\mathbb{C}$ which contains a set

$$B(a, \delta) = \{z \in \mathbb{C} : |z - a| < \delta\}$$

for some $\delta > 0$. Here the 'open ball' $B(a, \delta)$ replaces what in $\mathbb{R}$ was the open interval $]a - \delta, a + \delta[$. As will be clear later, most of the previous discussions and definitions go through without change. In particular, we can define continuity for functions with domain and range subsets of either $\mathbb{R}$ or $\mathbb{C}$, since all that is needed for the definition of continuity is the notion of neighbourhood.

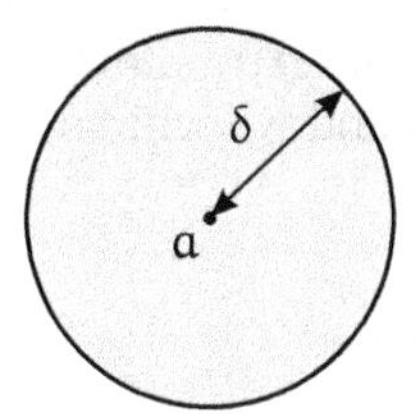

Fig. 1.6

Similarly, there is the notion of an open ball $B(a, \delta)$ in $\mathbb{R}^3$, namely, the set of points z whose Euclidean distance from a is less than δ. Given the notion of open ball, we can again define a neighbourhood of a in $\mathbb{R}^3$ to be any subset of $\mathbb{R}^3$ containing an open ball $B(a, \delta)$ for some $\delta > 0$. These definitions of neighbourhood find their proper place in the definition of neighbourhoods in metric spaces and in normed vector spaces.

These ideas also lead to definitions of neighbourhoods for subsets A of $\mathbb{R}^3$; vis., if $a \in A$, then a *neighbourhood in* A of a is a set $N \cap A$ where N is a neighbourhood of a in the above sense of neighbourhoods for points of $\mathbb{R}^3$. For example, in Fig. 1.7 which pictures the *2-sphere* (a) and *Möbius band* (b), a neighbourhood in A of a is any subset of A containing a 'disc' about a such as that shown.

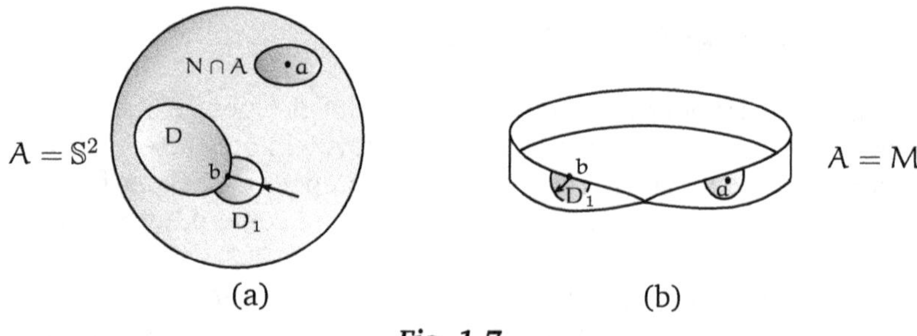

Fig. 1.7

These examples are simple. But we can also give examples of neighbourhoods where it is not easy to visualise the whole set.

Consider the Möbius band M. This has only one edge, and this edge E can be considered as a (somewhat twisted) circle. If we cut a disc D out of the 2-sphere $\mathbb{S}^2$, then the edge of this disc is again a circle, E′ say. Let us suppose that these are models (in cloth perhaps) in $\mathbb{R}^3$ and that we have arranged the models so that E and E′ have the same length. It would then seem reasonable to produce a new model by stitching E to E′ and so joining the two models. Unfortunately, as experiment will show, the whole thing gets hopelessly tangled. The point is, that this sort of model making is impossible in $\mathbb{R}^3$—an extra dimension is needed. (The *proof* of this assertion is very difficult and we will not give it.)

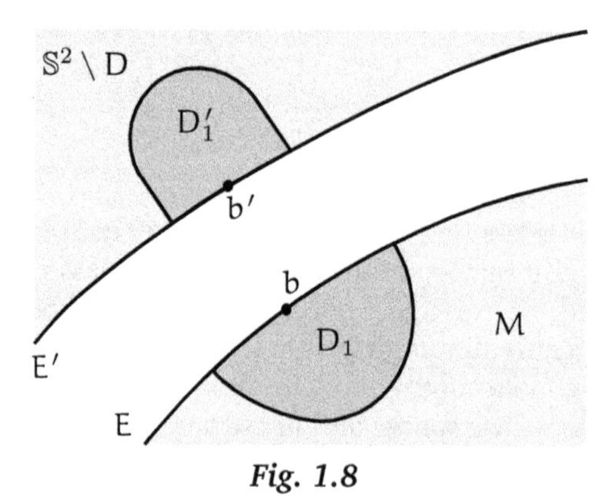

Fig. 1.8

The question is: can we anyway say what we mean by this stitching process without having to produce the result as a subset of $\mathbb{R}^3$? One of the properties of the model we should like is that if in Fig. 1.7 the point b′ of

$\mathbb{S}^2$ is identified with b in M, then the curve shown should be continuous. This can be arranged by defining neighbourhoods suitably. First, the model K should, as a set, be the union of M and $\mathbb{S}^2 \setminus D$ (we suppose $\mathbb{S}^2$ and M are disjoint). If $a \in K \setminus E$, then $a \in \mathbb{S}^2 \setminus D$, or $a \in M \setminus E$; about a we can find discs contained in $\mathbb{S}^2 \setminus D$ or $M \setminus E$ as the case may be, and a neighbourhood of a in K shall be any subset of K containing some such disc. For a point b on E the situation is different. Suppose b is identified with b' on E'. About b and b' consider 'half-discs' D_1 and D_1' (Fig. 1.8). Then any subset of K containing $D_1 \cup (D_1' \setminus E)$ is to be a neighbourhood in K of b, and all neighbourhoods of b in K are to be obtained by this construction for some half-discs D_1, D_1'. It is easy to see that this definition gives the required continuity for curves through b and b'.

We must now leave the particular examples and go forward to the general theory. Examples of the type of the last one will be discussed again in chapter 4.

NOTES

The Cantor set is not used elsewhere in this book, but in many branches of topology it furnishes a source of examples and constructions. For example, it can be used in the construction of fractals and space filing curves ([Man77]).

Chapter 2

Topological spaces

In section 2.1 of this chapter, we give the axioms for a topological space: the topological spaces are the objects of study of the rest of this book.

A topological space consists of a set X and a 'topology' on X. The importance for science of the notion of a topology is that it gives a precise but general sense to the intuitive ideas of nearness and continuity. As we saw in chapter 1 the basis of both these ideas is that of neighbourhood, and so it is this idea of neighbourhood which we axiomatise at first. It will appear quickly that there are equivalent ways of defining a topology, for example by means of open sets, or of closed sets. These definitions do not have the same intuitive appeal as the neighbourhood definition, but they are logically simpler and in some cases give the best method of defining a topology.

The type of structure of the real line $\mathbb{R}$ with its neighbourhoods of points is that for each r in $\mathbb{R}$ we have a set of subsets of $\mathbb{R}$ called neighbourhoods of r. More precisely, we have a function $r \mapsto \mathcal{N}(r)$ where $\mathcal{N}(r)$ is the set of neighbourhoods of r.

The way to generalise this structure is apparent. We consider a set X and a function $\mathcal{N} : x \mapsto \mathcal{N}(x)$ assigning to each x in X a set $\mathcal{N}(x)$ of subsets of X called neighbourhoods of x. The function $\mathcal{N}$ will be a topology on X if it satisfies suitable axioms.

The question of what axioms are suitable can not be decided on *a priori* grounds. The decision rests on the character of (a) the theory derived from the axioms, and (b) the examples of objects which satisfy the axioms. We shall see in later chapters how the theory presented here gives a language which is essential for certain aspects (in fact the topological aspects!) of geometry.

2.1 Axioms for neighbourhoods

Let X be a set and $\mathcal{N}$ a function assigning to each x in X a non-empty set $\mathcal{N}(x)$ of subsets of X. The elements of $\mathcal{N}(x)$ will be called *neighbourhoods of x with respect to $\mathcal{N}$* (or, simply, *neighbourhoods of x*). The function $\mathcal{N}$ is called a *neighbourhood topology* if Axioms N1–N4 below are satisfied; and then X with $\mathcal{N}$ is called a *topological space*.

The following axioms must hold for each x in X.

N1 *If N is a neighbourhood of x, then $x \in N$.*

N2 *If N is a subset of X containing a neighbourhood of x, then N is a neighbourhood of x.*

N3 *The intersection of two neighbourhoods of x is again a neighbourhood of x.*

N4 *Any neighbourhood N of x contains a neighbourhood M of x such that N is a neighbourhood of each point of M.*

We allow the *empty* topological space in which $X = \varnothing$ and $\mathcal{N}$ is the empty function from $\varnothing$ to $\{\varnothing\}$, the set of subsets of $\varnothing$.

EXAMPLE The real line $\mathbb{R}$ with its neighbourhoods of points is a topological space. The first three axioms have been verified in section 1.1, and we now verify N4. Let N be a neighbourhood of $x \in \mathbb{R}$; there is a $\delta > 0$ such that $]x-\delta, x+\delta[\ \subseteq N$. Let $M = \ ]x-\delta, x+\delta[$. Then M is a neighbourhood of each of its points, and so N is a neighbourhood of each point of M.

This neighbourhood topology on $\mathbb{R}$ is called the *usual topology* on $\mathbb{R}$.

The first three axioms for neighbourhoods have a clear meaning. The fourth axiom has a very important use in the structure of the theory, that of linking together the neighbourhoods of different points of X.

Intuitively, Axiom N4 can be expressed as follows: *a neighbourhood of x is also a neighbourhood of all points sufficiently close to x*. Another way of expressing the axiom is that *each point is inside any of its neighbourhoods*. To explain what is meant by this, we introduce the notion of the interior or a subset of X.

Any point x in X determines its neighbourhoods, which are subsets of X. On the other hand, given a subset A of X we can find those points of which A is a neighbourhood—the set of all such points is called the *interior* of A and is written $\operatorname{Int} A$.

2.1.1 $\operatorname{Int} A \subseteq A$.

2.1.2 *If* $A \subseteq B \subseteq X$, *then* $\operatorname{Int} A \subseteq \operatorname{Int} B$.

2.1.3 $\operatorname{Int} A$ *is a neighbourhood of any of its points.*

2.1.4 $\operatorname{Int}(\operatorname{Int} A) = \operatorname{Int} A$.

Proof If $\operatorname{Int} A = \emptyset$, then 2.1.1–2.1.3 (and so 2.1.4) are trivially satisfied. Suppose then $x \in \operatorname{Int} A$, so that A is a neighbourhood of x.

By Axiom N1, $x \in A$. Hence $\operatorname{Int} A \subseteq A$. Further, if $A \subseteq B$ then B is also a neighbourhood of x (Axiom N2), so that $x \in \operatorname{Int} B$. Hence $\operatorname{Int} A \subseteq \operatorname{Int} B$.

Let M be a neighbourhood of x such that $M \subseteq A$ and A is a neighbourhood of each point of M (such an M exists by Axiom N4). Then $M \subseteq \operatorname{Int} A$, and so $\operatorname{Int} A$ is a neighbourhood of x, by Axiom N2. This proves 2.1.3.

By 2.1.3, $\operatorname{Int} A \subseteq \operatorname{Int}(\operatorname{Int} A)$. By 2.1.1 and 2.1.2 $\operatorname{Int}(\operatorname{Int} A) \subseteq \operatorname{Int} A$. □

The sentence 'each point is inside any of its neighbourhoods' means that if N is a neighbourhood of x, then x belongs to $\operatorname{Int} N$. Also, $\operatorname{Int} N$ is again a neighbourhood of x. Thus Axioms N1–N4 imply:

N4′ *If* N *is a neighbourhood of* x*, then so also is* $\operatorname{Int} N$.

This axiom implies Axiom N4, since, by definition N is a neighbourhood of each point of $\operatorname{Int} N$.

EXERCISES

1. Let X be a set with a neighbourhood topology. Prove that if A, B are subsets of X then,

$$\operatorname{Int} A \cap \operatorname{Int} B = \operatorname{Int}(A \cap B)$$
$$\operatorname{Int} A \cup \operatorname{Int} B \subseteq \operatorname{Int}(A \cup B).$$

Prove also that if $(X_\lambda)_{\lambda \in L}$ is a family of subsets of X then,

$$\operatorname{Int} \bigcap_{\lambda \in L} X_\lambda \subseteq \bigcap_{\lambda \in L} \operatorname{Int} X_\lambda.$$

2. Let $\leqslant$ be an order relation on the set X. Let $x \in X$ and $N \subseteq X$. We say that N is a neighbourhood of x if there is an open interval I of X such that

$$x \in I \subseteq N.$$

Prove that these neighbourhoods of points of X form a neighbourhood topology on X. This topology is called the *order topology* on X. What is the order topology on $\mathbb{R}$?

3. Prove that the following are neighbourhood topologies on a set X.
(a) The *discrete topology*: N is a neighbourhood of x if and only if $x \in N \subseteq X$.
(b) The *indiscrete topology*: N is a neighbourhood of $x \in X$ if and only if $N = X$.
For what X do these topologies coincide? Let $x \in X$. What is $\operatorname{Int}\{x\}$ if X has the discrete topology?, the indiscrete topology?

4. Prove that the order topology on $\mathbb{Z}$ is the discrete topology, but that the order topology on $\mathbb{Q}$ is not discrete.

5. Let $\leqslant$ be a partial order on the set X. Discuss the possibility, or impossibility, of using $\leqslant$ to define a neighbourhood topology on X.

6. Let X be an uncountable set. Prove that the following define distinct neighbourhood topologies on X.
(a) N is a neighbourhood of $x \in X$ if $x \in N \subseteq X$ and $X \setminus N$ is finite.
(b) N is a neighbourhood of $x \in X$ if $x \in N \subseteq X$ and $X \setminus N$ is countable.
Can either of these topologies be the discrete topology?
7. Let $X = \mathbb{Z}$ and let p be a fixed integer. A set $N \subseteq \mathbb{Z}$ is a p-*adic neighbourhood* of $n \in \mathbb{Z}$ if N contains the integers $n + mp^r$ for some r and all $m = 0, \pm 1, \pm 2, \dots$ (so that in a given neighbourhood r is fixed but m varies). Prove that the p-adic neighbourhoods form a neighbourhood topology on $\mathbb{Z}$, the p-*adic topology*. Is this topology the same as the order topology?, the discrete topology?, the indiscrete topology?

The reader familiar with ring theory should develop two generalisations of the p-adic topology on $\mathbb{Z}$. First, replace the ring $\mathbb{Z}$ by an arbitrary ring R, so that now $p \in R$. Second, replace the element p by any ideal P of R. What is the P-adic topology in R if (a) $P = R$, (b) $P = \{0\}$?
8. Prove that Axioms N1–N4 are independent.

2.2 Open sets

Let $\mathcal{N}$ be a neighbourhood topology on the set X. A subset U of X is *open* (with respect to $\mathcal{N}$) if U is a neighbourhood of each of its points. Thus U is open if and only if $U = \operatorname{Int} U$.

2.2.1 *Let $x \in X$ and $N \subseteq X$; N is a neighbourhood of x if and only if there is an open set U such that*

$$x \in U \subseteq N.$$

Proof If N is a neighbourhood of x, then $\operatorname{Int} N$ is an open set such that $x \in \operatorname{Int} N \subseteq N$. Conversely, if U is an open set such that $x \in U \subseteq N$, then U is a neighbourhood of x and hence so also is N. □

The most important properties of open sets are given by:

2.2.2 *The open sets of X satisfy*
O1 *X and $\varnothing$ are open sets.*
O2 *If U, V are open sets, then $U \cap V$ is open.*
O3 *If $(U_\lambda)_{\lambda \in L}$ is any family of open sets, then $\bigcup_{\lambda \in L} U_\lambda$ is open.*

Proof The relation $\operatorname{Int} \varnothing \subseteq \varnothing$ implies that $\operatorname{Int} \varnothing = \varnothing$; thus $\varnothing$ is open. If $x \in X$, then x has at least one neighbourhood N; but $N \subset X$ and so X is a neighbourhood of x. Thus X is open.

If $U \cap V$ is empty, then it is open. If it is not empty, let $x \in U \cap V$. Then U and V are both neighbourhoods of x, and hence $U \cap V$ is a neighbourhood of x. Thus $U \cap V$ is open.

Let $U = \bigcup_{\lambda \in L} U_\lambda$. If U is empty, it is open. If not, let $x \in U$. Then $x \in U_\lambda$ for some $\lambda \in L$, and U_λ, being open, is a neighbourhood of x. But $U_\lambda \subseteq U$. So U also is a neighbourhood of x. □

We now show that the innocent seeming properties O1, O2, O3 suffice to axiomatise topological spaces in terms of open sets.

Let $\mathcal{U}$ be a set of subsets of X, called open sets, satisfying O1, O2, O3.

For each $x \in X$ a set $\mathcal{M}(x)$ of *$\mathcal{U}$-neighbourhoods* of x is defined by:

$$N \in \mathcal{M}(x) \Leftrightarrow N \subseteq X \textit{ and there is a } U \in \mathcal{U} \textit{ such that } x \in U \subseteq N.$$

(Compare 2.2.1.) The function $x \mapsto \mathcal{M}(x)$ is said to be *associated with* $\mathcal{U}$.

$\mathcal{M}$ *is* a neighbourhood topology on X. The Axioms N1, N2 are immediately verified while Axiom N3 follows from O2. Also, if $x \in U \in \mathcal{U}$, then $U \in \mathcal{M}(x)$, and this implies Axiom N4.

We now prove:

2.2.3 *$\mathcal{U}$ is the set of open sets of $\mathcal{M}$.*

Proof Certainly, each $U \in \mathcal{U}$ is open with respect to $\mathcal{M}$. Suppose, conversely, that $U \subseteq X$ and U is open with respect to $\mathcal{M}$. If $U = \emptyset$, then $U \in \mathcal{U}$. If $U \neq \emptyset$, then for each $x \in U$ there is a set U_x in $\mathcal{U}$ such that $x \in U_x \subseteq U$. Let U' be the union of these U_x for all x in U. Then $U' \subseteq U$ since each $U_x \subseteq U$; and $U \subseteq U'$ since each x in U belongs to U_x. So $U = U'$. Hence, $U = \bigcup_{x \in U} U_x$ belongs to $\mathcal{U}$ by O3. □

Suppose $\mathcal{N}$ is a neighbourhood topology on X, and $\mathcal{U}$ is the set of open sets of $\mathcal{N}$. It is immediate from 2.2.1 that $\mathcal{N}$ is the neighbourhood topology associated with $\mathcal{U}$. Since 2.2.1 is a consequence of Axiom N4, we have shown another use for this axiom—it ties together the neighbourhoods and the open sets.

A set of subsets of X called open sets and satisfying O1, O2, O3 is called an *open set topology* on X. We have proved that the structures of open set topology and neighbourhood topology determine one another: so topology may be developed using either as a starting point.

As we shall see later, topological spaces may be axiomatised in terms of other structures, for example, closed sets, closure, interior, or the relation $A \subseteq \operatorname{Int} B$. We shall use the word *topology* to denote the set of these equivalent structures, and shall be more specific when necessary. A topological space will be a set with a topology; thus a topological space carries all these structures, and may be defined by any one of them.

A topological space is really a pair $(X, \mathcal{T})$ where $\mathcal{T}$ is a topology on X. It is often convenient to use the symbol X to denote this pair. Such a notation causes confusion only when we are considering two topologies on the same set X. In such case, we shall write $X_\mathcal{T}$ for the topological space consisting of the set X and the topology $\mathcal{T}$. We call X the *underlying set* of $X_\mathcal{T}$.

Closed sets, closure

Let X be a topological space. A subset C of X is *closed* if $X \setminus C$ is open.

2.2.4 *The closed sets of X satisfy*
C1 *X and $\varnothing$ are closed sets.*
C2 *If C, D are closed sets, then $C \cup D$ is closed.*
C3 *If $(C_\lambda)_{\lambda \in L}$ is a family of closed sets of X, then $\bigcap_{\lambda \in L} C_\lambda$ is a closed set.*

This is immediate from O1, O2, O3 and the De Morgan laws.

If $\mathcal{C}$ is a set of subsets of a set X, called closed sets, which satisfy C1, C2, C3, then the set $\mathcal{U}$ of complements with respect to X of the elements of $\mathcal{C}$ is a set of open sets satisfying O1, O2, O3; thus $\mathcal{C}$ determines a topology on X, and $\mathcal{C}$ is exactly the set of closed sets of this topology. Thus topological spaces may be axiomatised in terms of closed sets.

Let X be a topological space and let $A \subseteq X$. The *closure* of A is the set $\overline{A}$ of points x in X such that every neighbourhood of x meets A.

2.2.5 $X \setminus \overline{A} = \operatorname{Int}(X \setminus A)$.

Proof Each of the following statements is obviously equivalent to its successor.

(a) $x \in X \setminus \overline{A}$.
(b) There is a neighbourhood N of x not meeting A.
(c) There is a neighbourhood N of x such that $N \subseteq X \setminus A$.
(d) $X \setminus A$ is a neighbourhood of x.
(e) $x \in \operatorname{Int}(X \setminus A)$. □

2.2.6 $A \subseteq \overline{A}$.
2.2.7 *If $A \subseteq B$, then $\overline{A} \subseteq \overline{B}$.*
2.2.8 *$\overline{A}$ is a closed set.*
2.2.9 *If A is a closed set, then $A = \overline{A}$.*
2.2.10 $\overline{\overline{A}} = \overline{A}$.

Proof Let $x \in A$. Then any neighbourhood N of x meets A (since $x \in N$). So $x \in \overline{A}$, and 2.2.6 is proved.

2.2.7 is obvious. For 2.2.8, $X \setminus \overline{A} = \operatorname{Int}(X \setminus A)$, and $\operatorname{Int}(X \setminus A)$ is open by 2.1.4. Hence $\overline{A}$ is closed.

Suppose A is a closed set and $x \notin \overline{A}$. Then $X \setminus A$ is a neighbourhood of x not meeting A. So $x \notin \overline{A}$. Thus $\overline{A} \subseteq A$ and so $A = \overline{A}$.

Finally, 2.2.10 follows from 2.2.8 and 2.2.9. □

2.2.11 *If $A, B \subseteq X$, then $\overline{A \cup B} = \overline{A} \cup \overline{B}$.*

Proof This can be deduced from 2.2.5 and Exercise 1 of Section 2.1. Alternatively, we argue as follows.

$\overline{A} \cup \overline{B}$ is closed and contains $A \cup B$. Hence $\overline{A \cup B} \subset \overline{A} \cup \overline{B}$. On the other hand, $A \subseteq A \cup B$ implies $\overline{A} \subseteq \overline{A \cup B}$; similarly, $\overline{B} \subseteq \overline{A \cup B}$, whence $\overline{A} \cup \overline{B} \subseteq \overline{A \cup B}$. □

EXERCISES

1. What are the open sets of X when X is discrete, that is, has the discrete topology?, is indiscrete, that is, has the indiscrete topology? What is the closure of $\{x\}$, $x \in X$, in these cases?
2. Let X be a topological space and let $A \subseteq X$. Prove that $\operatorname{Int} A$ is the union of all open sets U such that $U \subseteq A$, and $\overline{A}$ is the intersection of all closed sets C such that $A \subseteq C$.
3. Let X be a topological space, and let $A \subseteq X$. A point x in X is called a *limit point* of A if each neighbourhood of x contains points of A other than x. The set of limit points of A is written $\widehat{A}$. Prove that $\overline{A} = A \cup \widehat{A}$, and that A is closed $\Leftrightarrow \widehat{A} \subseteq A$. Give examples of non-empty subsets A of $\mathbb{R}$ such that (i) $\widehat{A} = \varnothing$, (ii) $\widehat{A} \neq \varnothing$ and $\widehat{A} \subseteq A$, (iii) A is a proper subset of $\widehat{A}$, (iv) $\widehat{A} \neq \varnothing$ but $A \cap \widehat{A} = \varnothing$.
4. Let X be a topological space and let $A \subseteq B \subseteq X$. We say that A is *dense in B* if $B \subseteq \overline{A}$, and A is *dense* if $\overline{A} = X$. Prove that if A is dense in X and U is open then

$$U \subseteq \overline{A \cap U}.$$

5. Let $\mathbb{I} = [0, 1]$. Define an order relation $\leqslant$ on $\mathbb{I}^2 = \mathbb{I} \times \mathbb{I}$ by

$$(x, y) \leqslant (x', y') \Leftrightarrow y < y' \text{ or } (y = y' \text{ and } x \leqslant x').$$

The *television topology* on $\mathbb{I}^2$ is the order topology with respect to $\leqslant$ (the name is due to E. C. Zeemann). Let A be the set of points $(2^{-1}, 1 - n^{-1})$ for positive integral n. Prove that in the television topology on $\mathbb{I}^2$

$$\overline{A} = A \cup \{(0, 1)\}.$$

6. A topological space is *separable* if it contains a countable, dense subset. Which of the following topological spaces are separable? (i) $\mathbb{Q}$ with the order topology, (ii) $\mathbb{R}$ with the usual topology, (iii) $\mathbb{I}^2$ with the television topology, (iv) an uncountable set with the indiscrete topology, (v) the spaces defined in Exercise 6 of Section 2.1.
7. Prove that if A is the closure of an open set, then $A = \overline{\operatorname{Int} A}$. Prove that at most fourteen distinct sets can be constructed from A by the operations of closure and complementation.
8. For any subset A of a topological space X, define $\operatorname{Ext} A$ (the *exterior* of A), $\operatorname{Bd} A$ (the *boundary* of A) and $\operatorname{Fr} A$ (the *frontier* of A) as follows:

$$\begin{aligned} \operatorname{Ext} A &= \operatorname{Int}(X \setminus A), \\ \operatorname{Bd} A &= A \setminus \operatorname{Int} A, \\ \operatorname{Fr} A &= \operatorname{Bd} A \cup \operatorname{Bd}(X \setminus A). \end{aligned}$$

Prove that the following relations hold:

(i) $\overline{A} = \text{Int}\,A \cup \text{Fr}\,A = A \cup \text{Fr}\,A = A \cup \text{Bd}(X \setminus A)$.
(ii) $\text{Int}(\text{Bd}\,A) = \varnothing$.
(iii) $\text{Bd}(\text{Int}\,A) = \varnothing$.
(iv) $\text{Bd}(\text{Bd}\,A) = \text{Bd}\,A$.
(v) $\text{Fr}\,A = \overline{A} \cap \overline{(X \setminus A)}$.
(vi) $\text{Fr}\,A$ is closed. If A is closed then $\text{Bd}\,A = \text{Fr}\,A$.
(vii) $\text{Fr}(\text{Fr}(\text{Fr}\,A)) = \text{Fr}(\text{Fr}\,A) \subset \text{Fr}\,A$.
(viii) $\text{Fr}\,A = \varnothing \Leftrightarrow A$ is both open and closed.
(ix) $\text{Bd}\,A = A \cap \overline{(X \setminus A)}$.
(x) $\text{Bd}\,A$ is closed $\Leftrightarrow A$ is the union of a closed and an open set.
(xi) $\text{Ext}(\text{Ext}\,A) = \text{Int}\,\overline{A}$.
(xii) $\text{Ext}\,\text{Ext}\,\text{Ext}\,\text{Ext}\,A = \text{Ext}\,\text{Ext}\,A$.

9. A topological space H is defined as follows. The underlying set of H is $\mathbb{R}$, and for each $x \in H$ and $N \subseteq H$, N is a neighbourhood of $x \Leftrightarrow$ there are real numbers x', x'' such that

$$x \in [x', x''[\ \subseteq N.$$

Prove that H is a topological space and that (i) each interval $[a, b[$ is both open and closed, (ii) H is separable, (iii) if $A \subseteq H$, then $A \setminus \widehat{A}$ is countable. (This topology on $\mathbb{R}$ is the *half-open topology*.)

10. Let X be a non-empty set and $i : \mathcal{P}(X) \to \mathcal{P}(X)$ a function such that for all $A, B \in \mathcal{P}(X)$

(i) $i(A) \subseteq A$
(ii) $i(i(A)) = i(A)$
(iii) $i(X) = X$
(iv) $i(A \cap B) = i(A) \cap i(B)$.

For each x in X, define A to be a neighbourhood of x if $x \in i(A)$. Prove that these neighbourhoods form a topology $\mathcal{N}$ on X. Which of the axioms (i)–(iv) are essential in the proof?

*11. With the notation of Exercise 10, prove that i is the interior operator for the topology $\mathcal{N}$.

*12. Let X be a non-empty set and $\lhd$ a relation on subsets of X such that

(i) $\varnothing \lhd \varnothing$, $X \lhd X$
(ii) $A \subset A'$, $A' \lhd B'$ and $B' \subset B$ imply $A \lhd B$
(iii) $A \lhd B$ implies $A \subset B$.
(iv) $A \lhd B$ and $A' \lhd B'$ imply $A \cap A' \lhd B \cap B'$
(v) $A_i \lhd B_i$ for all $i \in I$ implies $\cup_{i \in I} A_i \lhd \cup_{i \in I} B_i$.

For each $x \in X$, $A \subseteq X$ define A to be a neighbourhood of x if $\{x\} \lhd A$. Prove that these neighbourhoods define a topology on X for which $A \lhd B \Leftrightarrow A \subseteq \text{Int}\,B$.

*13. Show how to axiomatise topologies using the closure operator. [Use 2.2.5 and Exercise 10.]

2.3 Product spaces

Let X, Y be topological spaces. We consider the problem of defining a reasonable topology on the set $X \times Y$. For example, if $X = Y = \mathbb{R}$, this is the problem of finding the notion of 'nearness' in the Euclidean plane $\mathbb{R}^2 = \mathbb{R} \times \mathbb{R}$.

We consider the abstract situation. Let $x \in X$, $y \in Y$; we wish to define neighbourhoods of (x, y) in $X \times Y$ in such a way that Axioms N1–N4 are satisfied.

An obvious first attempt is to say that the neighbourhoods of (x, y) shall be the sets $M \times N$ for M, N neighbourhoods of x, y respectively. This would correspond to the intuitive idea '(x, y) is near to (x', y') if x is near to x' and y is near to y''. However, with this definition Axiom N2 is not satisfied: the set P in the following figure is not of the form $M \times N$ for any sets $M \subseteq X$, $N \subseteq Y$.

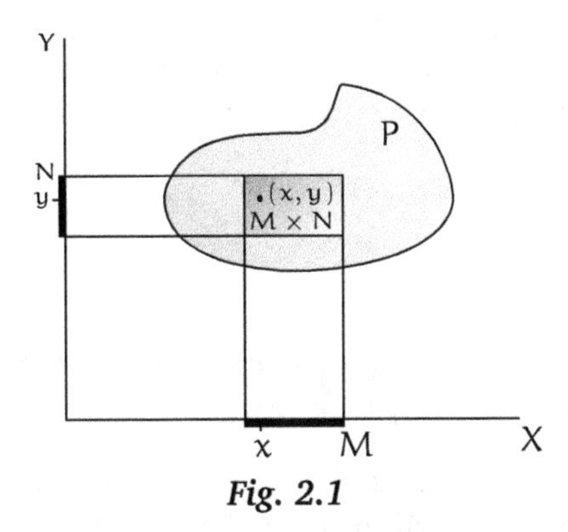

Fig. 2.1

We therefore make a virtue of necessity. The sets $M \times N$ as above we call the *basic neighbourhoods* of (x, y), and we define a *neighbourhood of* (x, y) *in* $X \times Y$ to be any subset of $X \times Y$ containing a basic neighbourhood of (x, y). We show that with these neighbourhoods $X \times Y$ is a topological space.

Let P be a neighbourhood of (x, y), and $M \times N$ a basic neighbourhood of (x, y) contained in P. Clearly, $(x, y) \in P$, and if $P \subseteq Q$ then Q is a neighbourhood of (x, y). This verifies Axioms N1 and N2.

Let $M^0 = \operatorname{Int} M$, $N^0 = \operatorname{Int} N$. Then $M^0 \times N^0$ is a basic neighbourhood of (x, y) and of any $(x', y') \in M^0 \times N^0$. Hence, P is a neighbourhood of any $(x', y') \in M^0 \times N^0$. This verifies Axiom N4.

Finally, let P' be a neighbourhood of (x, y) containing the basic neighbourhood $M' \times N'$ of (x, y). Then $(M \cap M') \times (N \cap N') = (M \times N) \cap (M' \times$

$N') \subseteq P \cap P'$, and so $P \cap P'$ is a neighbourhood of (x, y). This verifies Axiom N3 and completes the proof that $X \times Y$ is a topological space.

The product $X_1 \times \cdots \times X_n$ of n topological spaces $X_1, \ldots, X_n$ is defined inductively by

$$X_1 \times \cdots \times X_n = (X_1 \times \cdots \times X_{n-1}) \times X_n.$$

It is easily shown that a set $P \subseteq X_1 \times \cdots \times X_n$ is a neighbourhood of $(x_1, \ldots x_n)$ if and only if there are neighbourhoods M_i of $x, i = 1, \ldots, n$, such that $M_1 \times \cdots \times M_n \subseteq P$. In particular, the product topology of $\mathbb{R}^n = \mathbb{R} \times \cdots \times \mathbb{R}$ is called the *usual topology* on $\mathbb{R}^n$.

Let X, Y be topological spaces.

2.3.1 *If* U, V *are open in* X, Y *respectively, then* $U \times V$ *is open in* $X \times Y$.

2.3.2 *If* C, D *are closed in* X, Y *respectively, then* $C \times D$ *is closed in* $X \times Y$.

Proofs U is a neighbourhood of each x in U, V is a neighbourhood of each y in V. So $U \times V$ is a neighbourhood of each (x, y) in $U \times V$.

That $C \times D$ is closed is immediate from the formula

$$(X \times Y) \setminus (C \times D) = (X \times (Y \setminus D)) \cup ((X \setminus C) \times Y).$$

□

2.3.3 *A set* U *is open in* $X \times Y \Leftrightarrow$ *there are sets* U_λ, V_λ $(\lambda \in L)$ *open in* X, Y *respectively such that* $U = \bigcup_{\lambda \in L} U_\lambda \times V_\lambda$.

Proof The implication $\Leftarrow$ is clear from 2.3.1 and property O3 of open sets.

In order to prove the implication $\Rightarrow$, let U be open in $X \times Y$. For each $\lambda \in U$ there is a basic neighbourhood $M \times N$ of λ such that $M \times N \subseteq U$. Let $U_\lambda = \operatorname{Int} M$, $V_\lambda = \operatorname{Int} N$. Then U_λ, V_λ are open and $U = \bigcup_{\lambda \in U} U_\lambda \times V_\lambda$. □

EXAMPLE Let $\alpha = (a, b) \in \mathbb{R}^2$, and let $r > 0$. The *open ball about* α *of radius* r is the set

$$B(\alpha, r) = \{(x, y) \in \mathbb{R}^2 : (x - a)^2 + (y - b)^2 < r^2\}.$$

This open ball is an open set: For, let $\alpha' = (a', b') \in B(\alpha, r)$ and let $s = \sqrt{(a' - a)^2 + (b' - b)^2}$. Then $s < r$. Let $0 < \delta < (r - s)/\sqrt{2}$, $M =]a' - \delta, a' + \delta[$, $N =]b' - \delta, b' + \delta[$. Then $M \times N \subseteq B(\alpha, r)$ and so $B(\alpha, r)$ is a neighbourhood of α'.

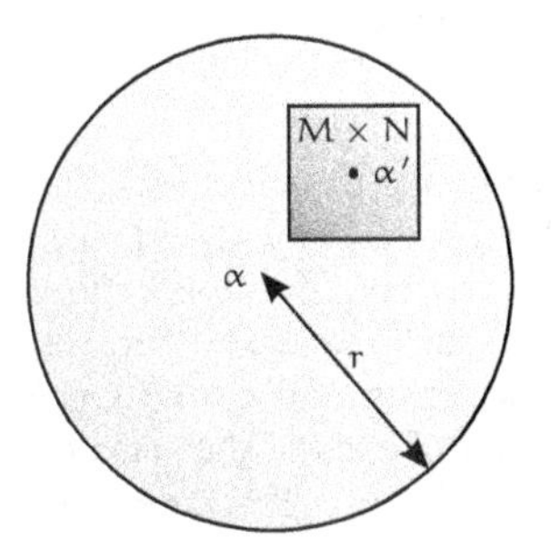

Fig. 2.2

2.3.4 *Let* $A \subseteq X$, $B \subseteq Y$. *Then*

$$\operatorname{Int}(A \times B) = \operatorname{Int} A \times \operatorname{Int} B,$$
$$\overline{A \times B} = \overline{A} \times \overline{B}.$$

Proof The relation $\operatorname{Int} A \times \operatorname{Int} B \subseteq \operatorname{Int}(A \times B)$ follows from the fact that $\operatorname{Int} A \times \operatorname{Int} B$ is an open set contained in $A \times B$. On the other hand, suppose $(x, y) \in \operatorname{Int}(A \times B)$. Then $A \times B$ is a neighbourhood of (x, y) and so contains a basic neighbourhood $M \times N$ of (x, y). Thus, $M \subseteq A$, $N \subseteq B$ and so A, B are neighbourhoods of x, y respectively. Hence, $(x, y) \in \operatorname{Int} A \times \operatorname{Int} B$.

The proof of the second relation is left as an exercise to the reader. □

We recall that the projections $p_1 : X \times Y \to X$, $p_2 : X \times Y \to Y$ are the functions $(x, y) \mapsto x$, $(x, y) \mapsto y$.

2.3.5 *If* U *is open in* $X \times Y$, *then* $p_1[U]$ *is open in* X, $p_2[U]$ *is open in* Y.

Proof We prove only that $p_1[U]$ is open in X. This is clear if $p_1[U]$ (and so U) is empty.

Suppose $x \in p_1[U]$. Then there is a y in Y such that $(x, y) \in U$. Since U is open, it contains a basic neighbourhood $M \times N$ of (x, y). So $M = p_1[M \times N] \subseteq p_1[U]$. Hence, $p_1[U]$ is a neighbourhood of x. □

It is not true that the projection of a closed set of $X \times Y$ is closed [Exercise 3].

EXERCISES

1. State which of the following sets of points (x, y) of $\mathbb{R}^2$ are (a) open, (b) closed, (c) neither open nor closed. (i) $|x| < 1$ and $|y| < 1$, (ii) $|x| + |y| \leqslant 1$, (iii) $x^2 + y^2 > 1$, (iv) $|xy| < 1$, (v) $\sin \pi x = 0$, (vi) one of x, y is rational, (vii) $y = \sin(1/x)$, $x \neq 0$, (viii) there is an integer $n > 0$ such that $xy = 1/n$, (ix) there are integers p, q such that $q > 0$, p/q is in its lowest terms, and $x = p/q$, $y = 1/q$.
2. Find the closure and interior of each of the sets of Exercise 1.
3. Prove that the set $\{(x, y) \in \mathbb{R}^2 : xy = 1\}$ is closed in $\mathbb{R}^2$ but that its projections in $\mathbb{R}$ are not closed.
4. Prove the second relation in 2.3.4.
5. Let $A \subseteq X \times Y$, $x \in X$, and let $A_x = \{y \in Y : (x, y) \in A\}$. Prove that A_x is open (closed) in Y if A is open (closed) in $X \times Y$. Give examples of subsets A of $\mathbb{R}^2$ such that (i) A is not open but A_x is open for each x in $\mathbb{R}$, (ii) A is not closed but A_x is closed for each x in $\mathbb{R}$.
6. Let $A \subseteq \mathbb{R}^3$ be the set of points

$$\{((2 + \cos \alpha t) \cos t, (2 + \cos \alpha t) \sin t, \sin \alpha t) : t \in \mathbb{R}\}.$$

Prove that A is closed if and only if α is a rational multiple of π, and that if A is not closed then $\overline{A}$ is the *anchor ring*

$$\{(x, y, z) \in \mathbb{R}^3 : ((x^2 + y^2)^{\frac{1}{2}} - 2)^2 + z^2 = 1\}.$$

7. Prove directly that the open sets given by 2.3.3 satisfy the Axioms O1, O2, O3 for an open set topology on $X \times Y$.
8. Prove that if X, Y are separable spaces, then so also is $X \times Y$.
9. Prove that $X \times Y$ is discrete (indiscrete) if and only if both X and Y are discrete (indiscrete).

2.4 Relative topologies and subspaces

Let X be a topological space and A a subset of X. We consider the problem of defining a topology on A so that A becomes a topological space. The theory here works out slightly simpler if the topology is defined in terms of open sets rather than in terms of neighbourhoods.

The *induced*, or *relative*, *topology* on A (with respect to X) is that in which the open sets are the sets $U \cap A$ where U is an open set of X. These sets $U \cap A$ are called *open in* A. (Thus *open in* X means the same as open.)

We must verify that this does define a topology.

(a) Axiom O1 is trivially verified, since $A \cap \varnothing = \varnothing$, $A \cap X = A$.
(b) Let V, V' be open in A. Then $V = U \cap A$, $V' = U' \cap A$ where U, U' are open in X. Hence, $V \cap V' = U \cap U' \cap A$ is open in A.

(c) Let $(V_\lambda)_{\lambda \in L}$ be a family of sets open in A, and let V be the union of this family. For each $\lambda \in L$, there is an open set U_λ such that $V_\lambda = U_\lambda \cap A$. So

$$V = \bigcup_{\lambda \in L} (U_\lambda \cap A) = \left(\bigcup_{\lambda \in L} U_\lambda \right) \cap A$$

is open in A.

EXAMPLE Let $X = \mathbb{R}$, $A = [0, 1[$. Then A itself is open in A, and $[0, \frac{1}{2}[$ is open in A; neither of these sets are open in X. On the other hand $]\frac{1}{2}, 1[$ is open in A and open in X.

Let X be a topological space. A topological space A which is a subset of X and whose topology is the relative topology as a subset of X is called a *subspace* of X. A subset of X is usually assumed to have the relative topology (if it has a topology at all) and so to be a subspace of X. The relative topologies on $\mathbb{N}$, $\mathbb{Z}$, $\mathbb{Q}$ as subsets of $\mathbb{R}$ are called the *usual topologies* on these sets. In the case of $\mathbb{N}$, $\mathbb{Z}$, the usual topologies are the discrete topologies: for, if $n \in \mathbb{Z}$, then

$$\{n\} = \mathbb{Z} \cap \left]n - \frac{1}{2}, n + \frac{1}{2}\right[;$$

hence $\{n\}$ is open in $\mathbb{Z}$ and so any subset of $\mathbb{Z}$ is open in $\mathbb{Z}$. A similar argument applies to $\mathbb{N}$.

For the rest of this section, we suppose that A is a subspace of the topological space X.

Let $a \in A$. The neighbourhoods of a for the topology of A are called *neighbourhoods in* A of a.

2.4.1 *Let $a \in A$, A set $N \subseteq A$ is a neighbourhood in A of $a \Leftrightarrow$ there exists M, a neighbourhood in X of a, such that $N = M \cap A$.*

Proof $\Leftarrow$ Let $N = M \cap A$, where M is a neighbourhood in X of a. Then $\operatorname{Int} M$ is open in X and so $(\operatorname{Int} M) \cap A$ is open in A. Hence N is a neighbourhood in A of a since $a \in (\operatorname{Int} M) \cap A \subseteq N$.

$\Rightarrow$ Let N be a neighbourhood in A of a. Then there is a set V open in A such that $a \in V \subset N$. Also, $V = U \cap A$ where U is open in X. Then, $M = U \cup N$ is a neighbourhood in X of a such that $M \cap A = N$. □

For example, if $X = \mathbb{R}$, $A = [0, 1]$, then $[0, \frac{1}{2}[$ is a neighbourhood in A of 0; and if $A = \{0\}$ then A itself is a neighbourhood in A of 0.

Again, let $X = \mathbb{R}$, $A = \{(x, y) \in \mathbb{R}^2 : x^2 + y^2 = 1\}$. The thickened part of A in Fig. 2.3 is a neighbourhood in A of P, but is not a neighbourhood in A of Q.

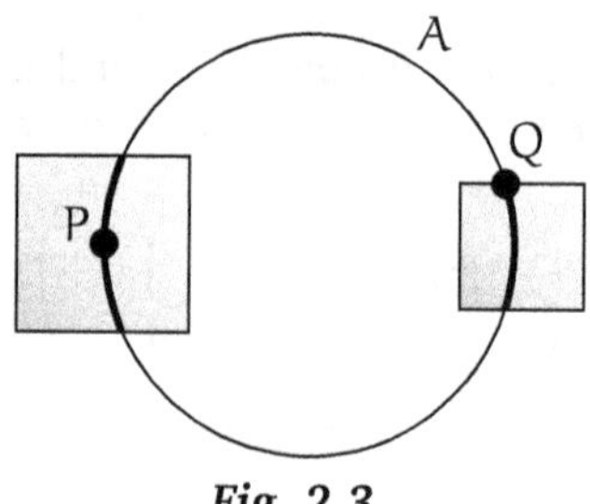

Fig. 2.3

A subset C of A is *closed in* A if C is closed in the topology of A, that is, if $A \setminus C$ is open in A. Also, if $C \subseteq A$, then we denote by

$$\mathrm{Cl}_A\ C$$

the closure of C with respect to the relative topology of A. This operation is very simply related to the closure $\overline{C}$ of C in X.

2.4.2 *If* $C \subseteq A$, *then*

$$\mathrm{Cl}_A\ C = \overline{C} \cap A.$$

Proof Suppose M is a subset of X. Since C is contained in A, we have M meets C if and only if $M \cap A$ meets C. It follows from this and 2.4.1 that if $x \in A$, then all neighbourhoods in A of x meet C if and only if all neighbourhoods in X of x meet C. □

2.4.3 *If* $C \subseteq A$, *then* C *is closed in* A *if and only if there is a set* D *closed in* X *such that* $C = D \cap A$.

Proof If C is closed in A, then $C = \mathrm{Cl}_A\ C$ and so $C = \overline{C} \cap A$. The result follows (with $D = \overline{C}$).

Conversely, if $C = D \cap A$ where D is closed in X, then $C \subseteq D$ and so $\overline{C} \subseteq D$. It follows easily that $C = \overline{C} \cap A$ and so $C = \mathrm{Cl}_A\ C$. This shows that C is closed in A. □

2.4.4 *If* A *is closed in* X, *then any set closed in* A *is also closed in* X. *The same holds with the word closed replaced by open.*

Proof The first part follows from 2.4.3 and the fact that the intersection of two closed sets is closed. The second part is similar. □

EXERCISES

1. Prove that the relation 'X is a subspace of Y' is a partial order relation for topological spaces.
2. Prove that the set $\{x \in \mathbb{Q} : -\sqrt{2} \leqslant x \leqslant \sqrt{2}\}$ is both open and closed in $\mathbb{Q}$.
3. Let A be the subspace of $\mathbb{R}$ of points $1/n$ for $n \in \mathbb{Z}\setminus\{0\}$. Prove that A is discrete, but that the subspace of $A \cup \{0\}$ of $\mathbb{R}$ is not discrete.
4. Prove that a subspace of a discrete space is discrete, and a subspace of an indiscrete space is indiscrete.
5. Let A be the subspace $[0, 2] \setminus \{1\}$ of $\mathbb{R}$. Prove that $[0, 1[$ is both open and closed in A.
6. Let $x \in X$ and let A be a neighbourhood (in X) of x. Prove that the neighbourhoods in A of x are exactly the neighbourhoods in X of x which are contained in A.
7. Let $\leqslant$ be an order relation on the set X. If $A \subseteq X$ then the restriction of $\leqslant$ is an order relation on A. Show that it is not necessarily true that if A, X have the order topologies, then A is a subspace of X. What is the order topology on $\mathbb{Q}$?
8. Let A, B be subspaces of X, Y respectively. Prove that $A \times B$ is a subspace of $X \times Y$.
9. Let A be a subspace of X, and let Int, Int_A denote respectively the interior operators for X, A. Prove that if $B \subseteq X$, then

$$(\text{Int}\, B) \cap A \subseteq \text{Int}_A(B \cap A).$$

Give an example for which $(\text{Int}\, B) \cap A \neq \text{Int}_A(B \cap A)$.
10. Let A be a subspace of X, and let Cl, Cl_A denote respectively the closure operators for X, A. Prove that if $B \subseteq X$, then

$$\text{Cl}_A(B \cap A) \subseteq (\text{Cl}\, B) \cap A.$$

Give an example for which $\text{Cl}_A(B \cap A) \neq (\text{Cl}\, B) \cap A$.
*11. A set $A \subseteq X$ is *locally closed* if, for each a in A, there exists N, a neighbourhood in X of a, such that $N \cap A$ is closed in N. Prove that A is locally closed $\Leftrightarrow$ A is the intersection of a closed set and an open set of X.
12. Write an account of relative topologies inverting the order of the present section; that is, define neighbourhoods in A using 2.4.1, prove that these neighbourhoods form a topology on A, and show that the sets open in A are as defined here.
*13. Let H be the real line with the half-open topology [Exercise 9 of Section 2.2]. Prove that the subspace $\{(x, x) \in H \times H : x \in H\}$ of $H \times H$ is not separable.

2.5 Continuity

One of the main reasons for studying the concept of a topological space is that it provides the most natural context for dealing with continuity. In this section, we define continuity for functions $X \to Y$, where X, Y are topological spaces, and we give also a number of important rules for constructing continuous functions.

Let $X_{\mathcal{S}}$, $Y_{\mathcal{T}}$ be topological spaces with underlying sets X, Y and topologies $\mathcal{S}$, $\mathcal{T}$ respectively. By a *function* $X_{\mathcal{S}} \to Y_{\mathcal{T}}$ is meant the triple $(f, \mathcal{S}, \mathcal{T})$ consisting of a function $f : X \to Y$ and the two topologies $\mathcal{S}$, $\mathcal{T}$. The purpose of this notation is that two functions $(f, \mathcal{S}, \mathcal{T})$, $(f', \mathcal{S}', \mathcal{T}')$ are equal if and only if $f = f'$, $\mathcal{S} = \mathcal{S}'$, $\mathcal{T} = \mathcal{T}'$. However, we make an abuse of language and denote such a function $(f, \mathcal{S}, \mathcal{T})$ also by f.

Let X, Y be topological spaces, and $f : X \to Y$ a function. We say f is *continuous* if, for all x in X, N is a neighbourhood of $f(x)$ implies $f^{-1}[N]$ is a neighbourhood of x. This condition is obviously equivalent to: for all x in X, if N is a neighbourhood of $f(x)$, then there is a neighbourhood M of x such that $f[M] \subseteq N$.

A *map* $X \to Y$ is simply a continuous function $X \to Y$.

2.5.1 *Let X, Y be topological spaces. Any constant function* $X \to Y$ *is continuous.*

Proof Let $f : X \to Y$ be a constant function and let $x \in X$. If N is a neighbourhood of $f(x)$, then $f^{-1}[N] = X$ which is a neighbourhood of x. □

2.5.2 *Let X be a topological space. The identity* $1_X : X \to X$ *is continuous.*

This follows easily from the rule $1_X^{-1}[N] = N$ for $N \subseteq X$.

2.5.3 *Let* $f : X \to Y$ *be a map and let* $A \subseteq X$, $B \subseteq Y$ *be such that* $f[A] \subseteq B$. *Then* $f \mid A, B$, *the restriction*‡ *of* f, *is a map.*

Proof Let $g = f \mid A, B$, let $a \in A$ and let M be a neighbourhood in B of $f(a)$. Then there exists N, a neighbourhood in Y of $f(a)$, such that $M = N \cap B$. Since f is a map, $f^{-1}[N]$ is a neighbourhood in X of a. Hence $g^{-1}[M] = f^{-1}[N] \cap A$ is a neighbourhood in A of a. □

A corollary of 2.5.3 is that, if $A \subset X$, then the inclusion function $i : A \to X$ is a map: for $i = 1_X \mid A$.

2.5.4 *If* $f : X \to Y$, $g : Y \to Z$ *are maps, so also is* $gf : X \to Z$.

Proof Let $x \in X$ and let N be a neighbourhood of $gf(x)$. Then

$$(gf)^{-1}[N] = f^{-1}g^{-1}[N]$$

and so $(gf)^{-1}[N]$ is a neighbourhood of x.

‡See A.1.4 of the Appendix.

2.5.5 *If X, Y are topological spaces, then the projections* $p_1 : X \times Y \to X$, $p_2 : X \times Y \to Y$ *are maps.*

Proof Let N be a neighbourhood of $x = p_1(x, y)$. Then $p_1^{-1}[N] = N \times Y$ is a neighbourhood of (x, y). □

Now, functions $f : Z \to X$, $g : Z \to Y$ determine uniquely a function $(f, g) : Z \to X \times Y$ whose components are f, g, that is, (f, g) is $z \mapsto (fz, gz)$.

2.5.6 *Let* $f : Z \to X$, $g : Z \to Y$ *be maps. Then* $(f, g) : Z \to X \times Y$ *is a map.*

Proof Let $h = (f, g)$, so that h sends $z \mapsto (f(z), g(z))$. Let P be a neighbourhood of $h(z)$, and let $M \times N$ be a basic neighbourhood of $h(z)$ contained in P. Then $h^{-1}[P]$ contains the set

$$\begin{aligned} h^{-1}[M \times N] &= \{z \in Z : f(z) \in M,\ g(z) \in N\} \\ &= f^{-1}[M] \cap g^{-1}[N]. \end{aligned}$$

It follows that $h^{-1}[P]$ is a neighbourhood of z. □

This result can also be expressed: *a function* $h : Z \to X \times Y$ *is continuous* $\Leftrightarrow$ $p_1 h, p_2 h$ *are continuous.* The implication $\Rightarrow$ follows from 2.5.4 and 2.5.5, while the converse implication follows from 2.5.6 since $p_1 h, p_2 h$ are the components f, g of h.

There are a number of useful corollaries of 2.5.6.

2.5.7 *The diagonal map* $\Delta : X \to X \times X$ *is continuous.*

2.5.8 *The twisting function* $T : X \times Y \to Y \times X$ *which sends* $(x, y) \mapsto (y, x)$ *is continuous.*

2.5.9 *If* $f : X \to X'$, $g : Y \to Y'$ *are continuous, then so is* $f \times g : X \times Y \to X' \times Y'$.

Proofs The diagonal map Δ is simply $(1_X, 1_X)$, and so is continuous. The twisting map T is (p_2, p_1), where p_1, p_2 are the projections of $X \times Y$. Finally, $f \times g = (fp_1, gp_2)$, and so $f \times g$ is continuous. □

EXAMPLES

1. We use some of these results to prove a sum and product rule for maps $X \to \mathbb{R}$. Let $f, g : X \to \mathbb{R}$ be maps. Then $f + g$, $f.g$ are the functions

$$X \xrightarrow{(f,g)} \mathbb{R} \times \mathbb{R} \xrightarrow{+} \mathbb{R}, \qquad X \xrightarrow{(f,g)} \mathbb{R} \times \mathbb{R} \xrightarrow{\cdot} \mathbb{R}$$

respectively where +, . denote the addition and multiplication functions $(x, y) \mapsto x + y$, $(x, y) \mapsto xy$. We shall prove later (see p. 53) that these latter functions are continuous—the continuity of $f + g$, $f.g$ follows.

2. The following type of example will occur frequently in later chapters. Suppose $F : X \times X \to X$ is a map, and consider the function

$$\begin{aligned} G : X \times X &\to X \\ (x, y) &\mapsto F(y, F(x, y)). \end{aligned}$$

Then G is a map since it is the composite of the maps

$$\begin{array}{ccccccccc} X \times X & \xrightarrow{1 \times \Delta} & X \times X \times X & \xrightarrow{T \times 1} & X \times X \times X & \xrightarrow{1 \times F} & X \times X & \xrightarrow{F} & X \\ (x, y) & \mapsto & (x, y, y) & \mapsto & (y, x, y) & \mapsto & (y, F(x, y)) & \mapsto & G(x, y). \end{array}$$

3. Let $\mathbb{S}^1 = \{(x, y) \in \mathbb{R}^2 : x^2 + y^2 = 1\}$, and consider the function

$$\begin{aligned} f : [0, 1[&\to \mathbb{S}^1 \\ t &\mapsto (\cos 2\pi t, \sin 2\pi t). \end{aligned}$$

Then f is a bijection, and it is continuous since its components are continuous. However f^{-1} is not continuous since $M = [0, \frac{1}{2}]$ is a neighbourhood in $[0, 1[$ of 0, but $f[M]$ is not a neighbourhood in $\mathbb{S}^1$ of $f(0) = (1, 0)$. This confirms the intuitive idea that breaking a loop of string is a non-continuous process.

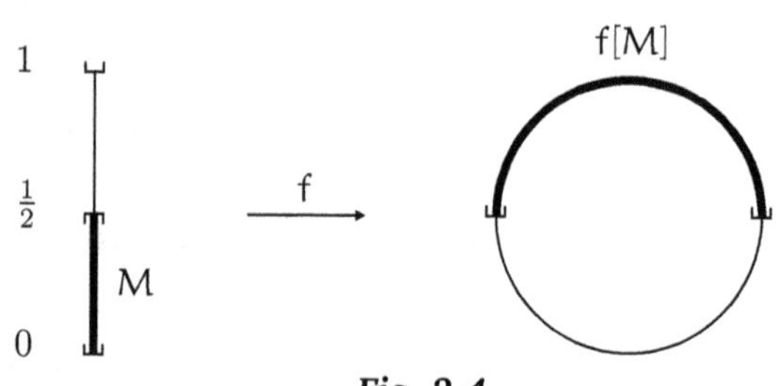

Fig. 2.4

Continuity of a function is a 'local' property in the following sense.

2.5.10 *Let X, Y be topological spaces and $f : X \to Y$ a function such that each $x \in X$ has a neighbourhood N such that $f \mid N$ is continuous. Then f is continuous.*

Proof Let $x \in X$ and let M be a neighbourhood of $f(x)$. Let N be a neighbourhood of x such that $f \mid N$ is continuous. Then

$$(f \mid N)^{-1}[M] = f^{-1}[M] \cap N \subseteq f^{-1}[M].$$

Since $f \mid N$ is continuous, $(f \mid N)^{-1}[M]$ is a neighbourhood in N of x, and so is also a neighbourhood in X of x (since N is a neighbourhood of x). Therefore, $f^{-1}[M]$ is a neighbourhood in X of x. □

Finally, we prove two versions of the 'gluing rule'.

2.5.11 *Let* X, Y *be topological spaces and* $f : X \to Y$ *a function. Let* $X = A \cup B$ *where* $A \setminus B \subseteq \operatorname{Int} A$, $B \setminus A \subseteq \operatorname{Int} B$. *If* $f \mid A$, $f \mid B$ *are continuous, then* f *is continuous.*

Proof Let $x \in X$ and let P be a neighbourhood of $f(x)$.

Suppose first $x \in A \cap B$. By continuity of $f \mid A$, $f \mid B$ and 2.4.1 there are neighbourhoods M, N of x in X such that

$$f^{-1}[P] \cap A = (f \mid A)^{-1}[P] = M \cap A,$$
$$f^{-1}[P] \cap B = (f \mid B)^{-1}[P] = N \cap B.$$

So $M \cap N \subseteq (M \cap A) \cup (N \cap B) = f^{-1}[P]$, and hence $f^{-1}[P]$ is a neighbourhood of x.

Suppose next $x \in A \setminus B$. Then A is a neighbourhood of x and hence so also is $M \cap A$ where M is constructed as above. *A fortiori*, $f^{-1}[P]$ is a neighbourhood of x. A similar argument applies if $x \in B \setminus A$. □

2.5.12 *Let* X, Y *be topological spaces and* $f : X \to Y$ *a function. Let* A, B *be closed subsets of* X *such that* $X = A \cup B$. *If* $f \mid A$, $f \mid B$ *are continuous, so also is* f.

Proof Since A, B are closed and have union X, the set $A \setminus B = X \setminus B$ is open, and so $A \setminus B \subseteq \operatorname{Int} A$. Similarly, $B \setminus A \subseteq \operatorname{Int} B$. So the result follows from 2.5.11. □

EXERCISES

1. Prove the continuity of the following functions $\mathbb{R}^3 \to \mathbb{R}$.

   (i) $(x, y, z) \mapsto P(x, y, z)$ where P is a polynomial.
   (ii) $(x, y, z) \mapsto \sin xz + \cos(x + y + z)$.

2. Let $F, G : X \times X \to X$ be maps. Prove the continuity of the functions

   (i) $X \times X \times X \to X$, $(x, y, z) \mapsto F(G(y, z), x)$.
   (ii) $X \times X \times X \to X \times X$, $(x, y, z) \mapsto (F(x, z), G(y, y))$.

3. Let $X = A \cup B$, Y be topological spaces and let $f : X \to Y$ be a function such that $f \mid A$, $f \mid B$ are continuous. Prove that f is continuous if

$$\overline{(A \setminus B)} \cap (B \setminus A) = \emptyset, \quad (A \setminus B) \cap \overline{(B \setminus A)} = \emptyset.$$

4. Let X be a topological space and let $f, g : X \to \mathbb{R}$ be maps. Prove that the following functions $X \to \mathbb{R}$ are maps.

(i) $x \mapsto |f(x)|$

(ii) $x \mapsto f(x)/g(x)$ (if $g(x)$ is never 0)

(iii) $x \mapsto \max\{f(x), g(x)\}$, $x \mapsto \min\{f(x), g(x)\}$.

5. Prove that X has the discrete topology $\Leftrightarrow$ for all spaces Y any function $X \to Y$ is continuous. Find a similar characterisation of the indiscrete topology.

6. Let $x_0 \in X$, $y_0 \in Y$ and let $X \vee Y$ be the subspace $X \times \{y_0\} \cup \{x_0\} \times Y$ of $X \times Y$. Let $i_1 : X \to X \vee Y$, $i_2 : Y \to X \vee Y$ be the functions $x \mapsto (x, y_0)$, $y \mapsto (x_0, y)$ respectively. Prove that a function $f : X \vee Y \to Z$ is continuous $\Leftrightarrow$ fi_1, fi_2 are continuous.

2.6 Other conditions for continuity

The set of neighbourhoods at a point x in a topological space X contains 'large' neighbourhoods, for example X itself. However, for many purposes, such as deciding continuity, it is only necessary to look at 'small' neighbourhoods of x, or at neighbourhoods of a particular type. For example, in $X \times Y$ it is the basic neighbourhoods $M \times N$ which are important. The precise way of expressing these notions is in terms of a *base* for the neighbourhoods of x.

A *base for the neighbourhood* at $x \in X$ is a set $\mathcal{B}(x)$ of neighbourhoods of x such that if N is a neighbourhood of x then N contains some B of $\mathcal{B}(x)$.

EXAMPLES

1. The set of all neighbourhoods of x is a base for the neighbourhoods at x. So also is the set of all open neighbourhoods of x.

2. The intervals $]-1/n, 1/n[$ for positive integral n form a base for the neighbourhoods of 0 in $\mathbb{R}$. So also do the closed intervals $[-1/n, 1/n]$.

3. The basic neighbourhoods $M \times N$ of (x, y) in $X \times Y$ form a base for the neighbourhoods of (x, y).

4. Let M be a neighbourhood of x. The neighbourhoods N of x such that $N \subseteq M$ form a base for the neighbourhoods of x.

If we have such a base $\mathcal{B}(x)$ for each $x \in X$, then the function $\mathcal{B} : x \mapsto \mathcal{B}(x)$ is called a *base for the neighbourhoods of* X. Our main result on bases is the following.

Let $\mathcal{B}$, $\mathcal{B}'$ be bases for the neighbourhoods of X, X' respectively. Let $f : X \to X'$ be a function.

2.6.1 f *is continuous* $\Leftrightarrow$ *for each* x *in* X *and* $N \in \mathcal{B}'(f(x))$, *there is an* $M \in \mathcal{B}(x)$ *such that* $f[M] \subseteq N$.

Proof The proof is simple.

$\Rightarrow$ Let $x \in X$, $N \in \mathcal{B}'(f(x))$. Then $f^{-1}[N]$ is a neighbourhood of x and so there is an $M \in \mathcal{B}(x)$ such that $M \subseteq f^{-1}[N]$. This implies $f[M] \subseteq N$.

$\Leftarrow$ Let $x \in X$ and let P be a neighbourhood of $f(x)$. Then there exists $N \in \mathcal{B}'(f(x))$ such that $N \subseteq P$. By assumption there is an $M \in \mathcal{B}(x)$ such that $M \subseteq f^{-1}[N]$. So $f^{-1}[P]$, which contains $f^{-1}[N]$, is a neighbourhood of x. □

The argument here is similar to that of 2.5.6.

The continuity of a function can also be described in terms of open sets, closed sets, or closure. This fact, which is of vital importance later, is contained in the following omnibus theorem (in which (a)–(d) are the important conditions).

Let X, Y be topological spaces and $f : X \to Y$ a function.

2.6.2 *The following conditions are equivalent.*
(a) f *is continuous.*
(b) *If* U *is open in* Y, *then* $f^{-1}[U]$ *is open in* X.
(c) *If* C *is closed in* Y, *then* $f^{-1}[C]$ *is closed in* X.
(d) *If* A *is a subset of* X, *then*

$$f[\overline{A}] \subseteq \overline{f[A]}.$$

(e) *If* B *is a subset of* Y, *then*

$$\overline{f^{-1}[B]} \subseteq f^{-1}[\overline{B}].$$

(f) *If* D *is a subset of* Y, *then*

$$f^{-1}[\operatorname{Int} D] \subseteq \operatorname{Int} f^{-1}[D].$$

Proof (a) $\Rightarrow$ (b) If $f^{-1}[U]$ is empty, then it is open. Otherwise, let $x \in f^{-1}[U]$. Then $f(x) \in U$ and so U is a neighbourhood of $f(x)$. Hence $f^{-1}[U]$ is a neighbourhood of x.
(b) $\Rightarrow$ (a) This follows easily from the fact that if N is a neighbourhood of $f(x)$, then $\operatorname{Int} N$ is open.
(b) $\Leftrightarrow$ (c) This is a simple consequence of

$$f^{-1}[Y \setminus C] = X \setminus f^{-1}[C].$$

(a) $\Rightarrow$ (d) Let $y \in f[\overline{A}]$ so that $y = f(x)$ where $x \in \overline{A}$. Let N be a neighbourhood of y. Then $f^{-1}[N]$ meets A since $f^{-1}[N]$ is a neighbourhood of x. Hence N meets $f[A]$, and so $y \in \overline{f[A]}$.
(d) $\Rightarrow$ (e) Let $A = f^{-1}[B]$ so that $f[A] \subseteq B$. Then

$$f[\overline{A}] \subseteq \overline{f[A]} \subseteq \overline{B}$$

whence $\overline{f^{-1}[B]} = \overline{A} \subseteq f^{-1}[\overline{B}]$.

(e) $\Leftrightarrow$ (f) This is an immediate consequence of the rules

$$f^{-1}[Y \setminus D] = X \setminus f^{-1}[D],$$
$$X \setminus \overline{f^{-1}[D]} = \mathrm{Int}(X \setminus f^{-1}[D]).$$

(f) $\Rightarrow$ (b) Let U be open in Y, so that $U = \mathrm{Int}\, U$. Then

$$f^{-1}[U] = f^{-1}[\mathrm{Int}\, U] \subseteq \mathrm{Int}\, f^{-1}[U].$$

So $f^{-1}[U] = \mathrm{Int}\, f^{-1}[U]$ and $f^{-1}[U]$ is open. □

It should be confessed that 2.6.2 is useful to solve rigorously some earlier exercises, particularly those of section 2.3. For example, to prove that the set $A = \{(x, y) \in \mathbb{R}^2 : |xy| < 1\}$ is open in $\mathbb{R}^2$, we consider the function $f : \mathbb{R}^2 \to \mathbb{R}$ which sends $(x, y) \mapsto |xy|$. Then f is continuous and $A = f^{-1}]-1, 1[$. Since $]-1, 1[$ is open in $\mathbb{R}$, A is open in $\mathbb{R}^2$. The reader should work again through the Exercises in Section 2.3 to show how 2.6.2 can be used.

EXAMPLES

5. Let $f, g : X \to \mathbb{R}^n$ be maps. Then the set A of points on which f, g agree is closed in X. For let $h = f - g : X \to \mathbb{R}^n$; then h is continuous (as we shall prove later), $\{0\}$ is closed in $\mathbb{R}^n$ and so, by 2.6.2(c), $A = h^{-1}[\{0\}]$ is closed in X.

6. Let $f : X \to \mathbb{R}$ be a map and let $A \subseteq X \times \mathbb{R}$ be the graph of f. Then A is closed in $X \times \mathbb{R}$ since A is the set of points of $X \times \mathbb{R}$ on which the maps p_2, fp_1 agree.

For example, the set $\{(x, y) : y = x^2 + e^x + \sin x\}$ is closed in $\mathbb{R}^2$.

7. The function $x \mapsto x/|x|$ is a continuous function $\mathbb{R}^{\neq 0} \to \mathbb{R}$. The graph of this function is closed in $\mathbb{R}^{\neq 0} \times \mathbb{R}$, but not in $\mathbb{R}^2$.

8. Let X be the graph of the function $x \mapsto \sin \pi/x$ $(x \neq 0)$. Then X is closed in $Z = \mathbb{R}^{\neq 0} \times \mathbb{R}$, but not in $\mathbb{R}^2$. In fact, let $J = \{(0, y) \in \mathbb{R}^2 : -1 \leqslant y \leqslant 1\}$; we prove that $\overline{X}$, the closure of X in $\mathbb{R}^2$, is $X \cup J$.

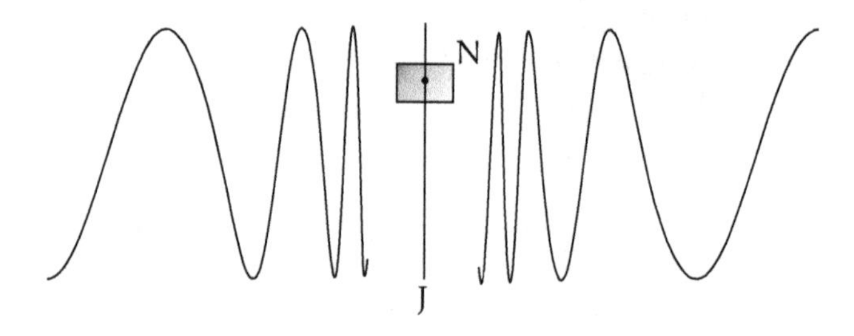

Fig. 2.5

First, $\sin \pi/x$ assumes all values in $[-1,1]$ for x in any interval $[1/(n+1), 1/n]$ (n is a positive integer); therefore, if $z \in J$, then any neighbourhood N of z meets X. That is, $X \cup J \subseteq \overline{X}$.

Second, $|\sin \pi/x| \leqslant 1$ $(x \neq 0)$; therefore, if $(x,y) \in \overline{X}$, then $|y| \leqslant 1$.

Third, X is closed in Z and hence $X = Z \cap \overline{X}$ [2.4.2]. So the only points of $\overline{X}$ which are not in X are the points of J.

EXERCISES

1. Let A be subspace of X and let $\mathcal{B}$ be a base for the neighbourhoods of X. Construct from $\mathcal{B}$ a base of the neighbourhoods of A.
2. Let $\mathcal{B}(x)$, $\mathcal{B}'(x')$, be bases for the neighbourhoods of $x \in X$, $x' \in X'$ respectively. Prove that the sets $M \times N$ for $M \in \mathcal{B}(x)$, $N \in \mathcal{B}'(x')$, form a base for the neighbourhoods of $(x,x') \in X \times X'$, and that the sets $M \times M$ for $M \in \mathcal{B}(x)$ form a base for the neighbourhoods of $(x,x) \in X \times X$.
3. A topological space X is said to satisfy the *first axiom of countability* if there is a base $\mathcal{B}$ for the neighbourhoods of X such that $\mathcal{B}(x)$ is countable for each x in X. Prove that the following satisfy the first axiom of countability: $\mathbb{R}$, $\mathbb{Q}$, a discrete space, a space with a countable number of open sets.
4. Prove that subspaces and (finite) products of spaces satisfying the first axiom of countability also satisfy the first axiom of countability.
5. A topological space X has a countable base for the neighbourhoods at x. Prove that there is a base for the neighbourhoods of x of sets B_n, $n \in \mathbb{N}$, such that $B_n \supseteq B_{n+1}$, $n \in \mathbb{N}$.
6. Deduce Exercises 9 and 10 of Section 2.4 from 2.6.2(e), (f).
7. Prove that the continuity of $f : X \to Y$ is not equivalent to the condition: if $A \subseteq X$, then $\operatorname{Int} f[A] \subseteq f[\operatorname{Int} A]$.
8. Give an example of X, Y, a map $f : X \to Y$ and subsets A of X, and B, D of Y such that $f[\overline{A}] \neq \overline{f[A]}, \overline{f^{-1}[B]} \neq f^{-1}[\overline{B}]$, $f^{-1}[\operatorname{Int} D] \neq \operatorname{Int} f^{-1}[D]$. [Let X have the discrete topology.]
9. Let $f, g : X \to \mathbb{R}$ be maps. Prove that the sets

$$\{x \in X : f(x) \geqslant g(x)\}, \quad \{x \in X : f(x) \leqslant g(x)\}$$

are closed in X.

10. Deduce the version 2.5.12 of the gluing rule from 2.6.2(c). Generalise this rule to the case when X is the union of n closed sets $C_1, \ldots, C_n$.

    Is this rule true if the sets C_i need not be closed? if their number is not finite?
11. Let $f : X \to \mathbb{R}$ be a map and let $A = \{(x, 1/(f(x)) \in X \times \mathbb{R} : f(x) \neq 0\}$. Prove that A is closed in $X \times \mathbb{R}$.
12. A map $f : X \to Y$ is *closed* if f maps closed sets of X to closed sets of Y, and is *open* if f maps open sets of X to open sets of Y. Give examples of X, Y, f for which (i) f is neither open nor closed, (ii) f is open but not closed, (iii) f is closed by not open, (iv) f is both open and closed.

2.7 Comparison of topologies, homeomorphism

There is partial order on the set of topologies on a set X defined as follows. Let $\mathcal{S}$, $\mathcal{T}$ be topologies in X. We say $\mathcal{S}$ is *finer* than $\mathcal{T}$ (and $\mathcal{T}$ is *coarser* than $\mathcal{S}$) if the identity function $i : X_{\mathcal{S}} \to X_{\mathcal{T}}$ is continuous. In such a case we write $\mathcal{S} \geqslant \mathcal{T}$ (or $\mathcal{T} \leqslant \mathcal{S}$). Clearly, this relation is transitive and reflexive.

Let $\mathcal{S}$, $\mathcal{T}$ be topologies on X. If $N \subseteq X$, then $i^{-1}[N] = N$. So we have: *$\mathcal{S}$ is finer then $\mathcal{T}$ if and only if for each x in X each $\mathcal{T}$-neighbourhood of x is also an $\mathcal{S}$-neighbourhood of x.* It follows easily that $\leqslant$ is also an anti-symmetric relation, which is to say that if $\mathcal{S} \geqslant \mathcal{T}$ and $\mathcal{T} \geqslant \mathcal{S}$ then $\mathcal{S} = \mathcal{T}$, and the neighbourhoods of any x in X are the same for $\mathcal{S}$ as for $\mathcal{T}$.

By 2.6.2 this relation can also be described in terms of open, or of closed, sets: $\mathcal{S} \geqslant \mathcal{T} \Leftrightarrow$ each set open for $\mathcal{T}$ is also open for $\mathcal{S} \Leftrightarrow$ each set closed for $\mathcal{T}$ is also closed for $\mathcal{S}$.

The finest topology on X is the discrete topology and the coarsest topology is the indiscrete topology.

Two topologies $\mathcal{S}$, $\mathcal{T}$ on X may be *incomparable*, in which case $\mathcal{S}$ is neither finer nor coarser then $\mathcal{T}$. For example, let $X = \{x, y\}$, let $\mathcal{S}$ be the topology whose open sets are $\varnothing$, $\{x\}$, X and let $\mathcal{T}$ be the topology whose open sets are $\varnothing$, $\{y\}$, X. Clearly, $\mathcal{S}$ and $\mathcal{T}$ are incomparable.

Let X, Y be topological spaces. A function $f : X \to Y$ is a *homeomorphism* if (i) f is a bijection, and (ii) for all x in X, M is a neighbourhood of $x \Leftrightarrow f[M]$ is a neighbourhood of $f(x)$. Clearly (i) and (ii) are equivalent to (i) and (ii′): for all x in X, N is a neighbourhood of $f(x) \Leftrightarrow f^{-1}[N]$ is a neighbourhood of x.

This definition, though possibly the most intuitive, is not the most elegant or useful. Better conditions for homeomorphism are given in 2.7.1: let X, Y be topological spaces and $f : X \to Y$ a function.

2.7.1 *The following conditions are equivalent.*
(a) f *is a homeomorphism.*
(b) f *is continuous, a bijection, and* $f^{-1} : Y \to X$ *is continuous.*
(c) f *is continuous and there is a continuous function* $g : Y \to X$ *such that* $gf = 1_X$, $fg = 1_Y$.

Proof (b) is obviously equivalent to (a). Given (b) then $g = f^{-1}$ satisfies $gf = 1_X$, $fg = 1_Y$. Conversely, if g exists as in (c), then f must be a bijection and g must be f^{-1}. So (c) implies (b). □

This shows that we may replace (ii) in the definition of homeomorphism by (ii′): U *is open in* X $\Leftrightarrow$ f[U] *is open in* Y, or similar conditions involving f^{-1}, or closed sets.

It is important to note that a continuous bijection need not be a homeomorphism. For example, let $\mathcal{S}$, $\mathcal{T}$ be topologies on a set X. The identity

$i : X_{\mathcal{S}} \to X_{\mathcal{T}}$ is a homeomorphism $\Leftrightarrow \mathcal{S} = \mathcal{T}$; and i is continuous $\Leftrightarrow \mathcal{S} \geqslant \mathcal{T}$. The relation $\mathcal{S} \geqslant \mathcal{T}$ does not imply $\mathcal{S} = \mathcal{T}$.

If $f : X \to Y$ is a homeomorphism we write $f : X \approx Y$, and if a homeomorphism $f : X \approx Y$ exists we say that X is homeomorphic to Y or X is of the same homeomorphism type as Y and write $X \approx Y$. We leave the reader to check that the relation $X \approx Y$ is an equivalence relation.

2.7.2 *Any two open intervals of $\mathbb{R}$ are homeomorphic.*

Proof First let $a, b \in \mathbb{R}$ $(a < b)$ and consider the function

$$\begin{aligned} f :\ ]0,1[&\to\ ]a,b[\\ t &\mapsto a(1-t) + bt. \end{aligned}$$

Then f is continuous, a bijection, and with continuous inverse $s \mapsto (s - a)/(b - a)$. Thus, f is a homeomorphism. It follows that all bounded open intervals in $\mathbb{R}$ are homeomorphic.

We now show that any interval in $\mathbb{R}$ is homeomorphic to a bounded interval. Consider the function

$$\begin{aligned} g : \mathbb{R} &\to\ ]-1,+1[\\ r &\mapsto r/(1+|r|). \end{aligned}$$

Then g is continuous, a bijection and with inverse $s \mapsto s/(1-|s|)$, which is continuous. Let I be an interval of $\mathbb{R}$. Then $g[I]$ is a bounded interval and $g \mid I, g[I]$ is a homeomorphism. □

This example illustrates that to prove two spaces are homeomorphic, one simply constructs a homeomorphism from one to the other. It is usually more difficult to prove that two given spaces are not homeomorphic.

Topology is often characterised as the study of those properties of spaces which are not changed under homeomorphism. For this reason, homeomorphic spaces are also called *topologically equivalent*.

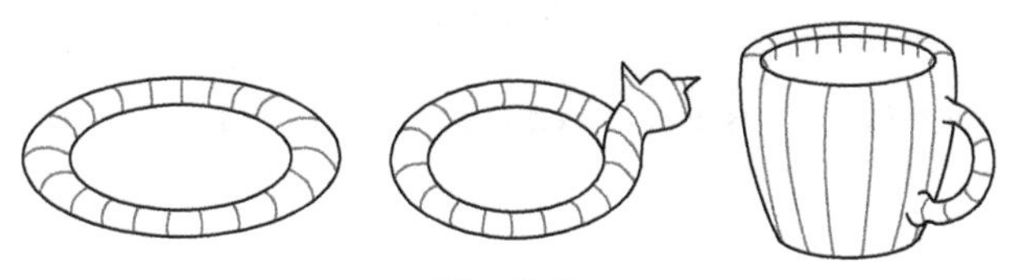

Fig. 2.6

Topology has also been called 'rubber sheet geometry', because if a surface X is constructed from sheets of rubber, then elastic deformations such as pulling and squashing, do not change the homeomorphism type of X. As an example, Fig. 2.6 illustrates surfaces in $\mathbb{R}^3$ which are all topologically equivalent.

One final definition will be needed later. Let X, Y be topological spaces. A function $f : X \to Y$ is an *embedding* if its restriction $f \mid X, f[X]$ is a homeomorphism. For example, any inclusion mapping of a subspace into a total space is an embedding and, in general, an embedding is the composite of a homeomorphism and an inclusion map.

EXAMPLE Let X Y be topological spaces, and let $y \in Y$. The function $f : X \to X \times Y$, $x \mapsto (x, y)$, is an embedding. First, f is continuous, since its components are the identity map and the constant map $x \mapsto y$. Second, the inverse of $f \mid X, f[X]$ is simply the restriction of the projection $p_1 : X \times Y \to Y$ and so this inverse is continuous.

The following gluing rule is often useful for constructing homeomorphisms (cf. Section 4 of Chapter 4).

2.7.3 *Let $X = X_1 \cup X_2$, $Y = Y_1 \cup Y_2$ be topological spaces such that X_1, X_2 are closed in X and Y_1, Y_2 are closed in Y. Let $f_1 : X_1 \to Y_1$, $f_2 : X_2 \to Y_2$ be homeomorphisms which restrict to the same homeomorphism $f_0 : X_1 \cap X_2 \to Y_1 \cap Y_2$. Then the function*

$$f : X \to Y$$

$$x \mapsto \begin{cases} f_1 x, & x \in X_1 \\ f_2 x, & x \in X_2 \end{cases}$$

is well-defined and is a homeomorphism.

Proof The function f is well-defined since f_1, f_2 agree on $X_1 \cap X_2$. The continuity of f follows from 2.5.12. A similar argument shows that f^{-1} is defined and continuous. □

EXERCISES

1. Construct homeomorphisms between the subsets A, B of $\mathbb{R}^2$ in each of the following cases.

(i) $A = \mathbb{R}^2$, $B = \{(x, y) : y > 0\}$,
(ii) $A = \{(x, y) : x^2 + y^2 \leqslant 1\}$, $B = \{(x, y) : |x| \leqslant 1, |y| \leqslant 1\}$,
(iii) $A = \{(x, y) : y \geqslant 0\}$, $B = \{(x, y) : y \geqslant x^2\}$,
(iv) $A = \mathbb{R}^2$, $B = \{(x, y) : x^2 + y^2 < 1\}$,
(v) $A = \{(x, y) : y \geqslant 0 \text{ and } |x| \leqslant 1\}$, $B = \{(x, y) : y \geqslant 0 \text{ and } |x| \leqslant (1 + |y|)^{-1}\}$,
(vi) $A = \{(x, y) : y \neq 0\}$, $B = \{(x, y) : y < 0 \text{ or } y > 1\}$.

2. Let $\mathbb{S}^1 = \{z \in \mathbb{C} : |z| = 1\}$, let $\alpha \in \mathbb{R}$ be irrational. Prove that the function $f : \mathbb{R} \to \mathbb{S}^1 \times \mathbb{S}^1, t \mapsto (e^{2\pi i\alpha t}, e^{2\pi i t})$, is not an embedding.
3. Let $f : [0,1] \to [a,b]$ be an order preserving bijection. Prove that f is a homeomorphism.

[In Exercises 4–7 it should be assumed that if $f : [a,b] \to \mathbb{R}$ is a map, then Im f is a closed, bounded interval.]
4. Let $f : [0,1] \to [0,1]$ be a homeomorphism. Prove that $f(0)$ is 0 or 1, and that $f]0,1[\ = \ ]0,1[$.
5. Prove that $[0,2]$ is not homeomorphic to $[-1,0[\ \cup\ ]1,2]$.
6. Prove that there is no continuous surjection $[0,1] \to\ ]0,1[$; construct a continuous surjection $]0,1[\ \to [0,1]$.
7. Let $X \subseteq \mathbb{R}$ be the union of the open intervals $]3n, 3n+1[$ and the points $3n+2$ for $n = 0, 1, 2, \ldots$. Let $Y = (X \setminus \{2\}) \cup \{1\}$. Prove that there are continuous bijections $f : X \to Y$, $g : Y \to X$, but that X, Y are not homeomorphic.
8. Is the half-open topology on $\mathbb{R}$ finer, coarser, or incomparable to the usual topology on $\mathbb{R}$?
9. Let A, the *anchor ring*, be the set of points in $\mathbb{R}^3$

$$(\cos\theta(2+\cos\varphi), \sin\theta(2+\cos\varphi), \sin\varphi), \quad \theta, \varphi \in \mathbb{R}.$$

Construct a homeomorphism from A to the *torus* $\mathbb{S}^1 \times \mathbb{S}^1$.
10. Let $\mathbb{E}^2 = \{(x,y) \in \mathbb{R}^2 : x^2 + y^2 \leqslant 1\}$. The space $\mathbb{S}^1 \times \mathbb{E}^2$ is called the *solid torus*. Prove that the *3-sphere*

$$\mathbb{S}^3 = \{(x_1, \ldots, x_4) \in \mathbb{R}^4 : x_1^2 + \cdots + x_4^2 = 1\}$$

is the union of two spaces each homeomorphic to a solid torus and with intersection homeomorphic to a torus. [Consider the subspaces of $\mathbb{S}^3$ given by $x_1^2 + x_2^2 \leqslant x_3^2 + x_4^2$ and by $x_1^2 + x_2^2 \geqslant x_3^2 + x_4^2$.]
11. Construct the homeomorphism $f : \mathbb{I}^2 \to \mathbb{I}^2$ (where $\mathbb{I}^2 = \mathbb{I} \times \mathbb{I}$) such that f maps $\mathbb{I} \times \{0,1\} \cup \{0\} \times \mathbb{I}$ onto $\{0\} \times \mathbb{I}$. [We use this in chapter 7.]

2.8 Metric spaces and normed vector spaces

Let $\mathbb{K}$ denote either the real numbers, the complex numbers or the quaternions. The reader not familiar with quaternions may simply not consider them at this stage—in fact the only property of $\mathbb{K}$ we use is that $\mathbb{K}$ is a field (with non-commutative multiplication if $\mathbb{K}$ is the quaternions) and that for each element $\alpha \in \mathbb{K}$ there is defined an *absolute value,* or *modulus,* $|\alpha| \in \mathbb{R}$ with the properties:

(a) $|\alpha| > 0$ if $\alpha \neq 0$, and $|0| = 0$,
(b) $|\alpha\beta| = |\alpha|\,|\beta|$,

(c) $|\alpha + \beta| \leqslant |\alpha| + |\beta|$, for all $\alpha, \beta \in \mathbb{K}$.

We shall be considering vector spaces V over $\mathbb{K}$. Now, it is usual in elementary work to write αx for the multiple of the vector x in V by the scalar α in $\mathbb{K}$—this is expressed by saying that V is considered as a *left* vector space over $\mathbb{K}$. However, it turns out that in the case $\mathbb{K}$ is non-commutative it is more convenient to write $x\alpha$ instead of αx—that is, to consider V as a *right* vector space over $\mathbb{K}$, with scalar multiplication a function $V \times \mathbb{K} \to V$. Given such a right vector space, we can always define left scalar multiplication by $\alpha x = x\alpha, v \in V, \alpha \in \mathbb{K}$. However, if $\alpha, \beta \in \mathbb{K}, x \in V$, then

$$\beta(\alpha x) = (\alpha x)\beta = (x\alpha)\beta = x(\alpha\beta) = (\alpha\beta)x.$$

So we obtain the usual associativity rule $\beta(\alpha x) = (\beta\alpha)x$ if and only if $\mathbb{K}$ is commutative. Thus a vector space over $\mathbb{R}$ or $\mathbb{C}$ can, and will, be considered as both a left and a right vector space, while a vector space over the quaternions $\mathbb{H}$ will be considered only as a right vector space. To cover all cases, we frame our axioms for normed spaces in terms of right vector spaces.

Let V be any right vector space over $\mathbb{K}$. A *norm* on V is a function

$$\| \; \| : V \to \mathbb{R}$$

such that for any x, y in V and α in $\mathbb{K}$:

NVS1 $\|x\| > 0$ if $x \neq 0$,

NVS2 $\|x\alpha\| = |\alpha|\|x\|$,

NVS3 $\|x + y\| \leqslant \|x\| + \|y\|$.

Then V with such a norm is called a *normed vector space*. Intuitively, a norm gives a measure of the size of elements of V. From the formal viewpoint, the three axioms tie in the norm with the addition and scalar multiplication of the vector space structure on V.

The following examples show the importance of this concept.

EXAMPLES

1. The field $\mathbb{K}$ itself is a normed vector space over $\mathbb{K}$ with norm $\|x\| = |x|, x \in \mathbb{K}$.

2. More generally, the n-dimensional vector space $\mathbb{K}^n$ over $\mathbb{K}$ has many norms of which two are particularly important. (i) The *Euclidean norm* or *modulus* on $\mathbb{K}^n$ is written $|x|$ and is defined by

$$|(x_1, \ldots, x_n)| = \left\{ \sum_{i=1}^{n} |x_i|^2 \right\}^{\frac{1}{2}}, \quad x_i \in \mathbb{K}.$$

The verification of Axioms NVS1 and NVS2 is trivial, while Axiom NVS3 follows from the well-known Cauchy-Schwarz inequality [cf. Section 5.4].

The resulting normed vector space is called n-*dimensional Euclidean space* if $\mathbb{K} = \mathbb{R}$, n-*dimensional unitary space* if $\mathbb{K} = \mathbb{C}$, and *n-dimensional symplectic space* if $\mathbb{K} = \mathbb{H}$, the field of quaternions. (ii) The *Cartesian norm* on $\mathbb{K}^n$ is defined by

$$\|(x_1, \ldots, x_n)\| = \max\{|x_1|, \ldots, |x_n|\}, \quad x_i \in \mathbb{K}.$$

The verification that this is a norm is very simple and is left to the reader.

As a further example, let p be a real number such that $p \geqslant 1$, and let

$$\|(x_1, \ldots, x_n)\| = \left\{ \sum_{i=1}^{n} |x_i|^p \right\}^{\frac{1}{p}}, \quad x_i \in \mathbb{K}.$$

This defines a norm on $\mathbb{K}^n$, Axiom NVS3 being the generalised Minkowski inequality proved in many books on analysis.

3. Let $\mathcal{C}$ be a vector space over $\mathbb{R}$ of all maps $[0,1] \to \mathbb{R}$. There are two norms on $\mathcal{C}$ which are important in analysis, the *sup norm* defined by

$$\|f\|_S = \sup\{|f(x)| : x \in [0,1]\}$$

and the *integral norm* defined by

$$\|f\|_I = \int_0^1 |f(x)|\, dx.$$

The verification that the sup norm is a norm is entirely trivial—the continuity of any f in $\mathcal{C}$ is used only to show that $\|f\|_S$ is well defined. On the other hand, the continuity of any f in $\mathcal{C}$ is used in an essential way in showing that for the integral norm, $\|f\|_I > 0$ if $f \neq 0$.

As we shall see in the next section, a norm on a vector space V induces in a natural way a topology on V, and also on subsets of V. However, a subset of V need not be a vector space, and so not a normed vector space. For this reason, we widen our outlook and consider the more general concept of a metric space.

Let X be any set. A metric on X is a function $d : X \times X \to \mathbb{R}^{\geqslant 0}$ with the following properties.

M1 $d(x,y) = 0 \Leftrightarrow x = y$

M2 $d(x,y) = d(y,x)$

M3 $d(x,z) \leqslant d(x,y) + d(y,x)$.

A set X with a metric d is called a *metric space*, and is denoted by X_d or simply X.

EXAMPLES

4. Let V be a normed vector space and X a subset of V. Then the function

$$d(x, y) = \|x - y\|, \quad x, y \in X$$

is a metric on X. In particular, V itself will always be taken to have this metric.

5. Let X be a set, and define $d(x, y)$ to be 0 if $x = y$ and 1 otherwise. This is the *discrete* metric on X.

6. Let X_d be a metric space. Then we can define a new metric on X by

$$d'(x, y) = \min\{1, d(x, y)\}.$$

We can in a metric space define generalisations of open and of closed intervals in $\mathbb{R}$. Let X be a metric space with metric d. Let $a \in X$, $r \geqslant 0$. The *open ball about* a *of radius* r is the set

$$B(a, r) = \{x \in X : d(x, a) < r\}.$$

The *closed ball about* a *of radius* r is the set

$$E(a, r) = \{x \in X : d(x, a) \leqslant r\}$$

and the *sphere about* a *of radius* r is

$$S(a, r) = \{x \in X : d(x, a) = r\}.$$

The closed ball is sometimes called a *cell*, or *disc*.

In a normed vector space V, the sets $B(0, 1)$, $E(0, 1)$, $S(0, 1)$ are called the *standard ball, cell* and *sphere* and are written $B(V)$, $E(V)$, $S(V)$. In particular, if $V = \mathbb{R}^n$ with the Euclidean norm, these are denoted by $\mathbb{B}^n$, $\mathbb{E}^n$, $\mathbb{S}^{n-1}$. The standard 1-cell $\mathbb{E}^1$ is of course the interval $[-1, 1]$. If $V = \mathbb{R}^n$ with the Cartesian norm, then $E(V)$ is the n-fold product of $[-1, 1]$ with itself, and is written $\mathbb{J}^n$.

The following diagram illustrates $E(V)$ when $V = \mathbb{R}^2$ with respectively the Cartesian, Euclidean, $\|\ \ \|_1$, and $\|\ \ \|_{3/2}$, norms.

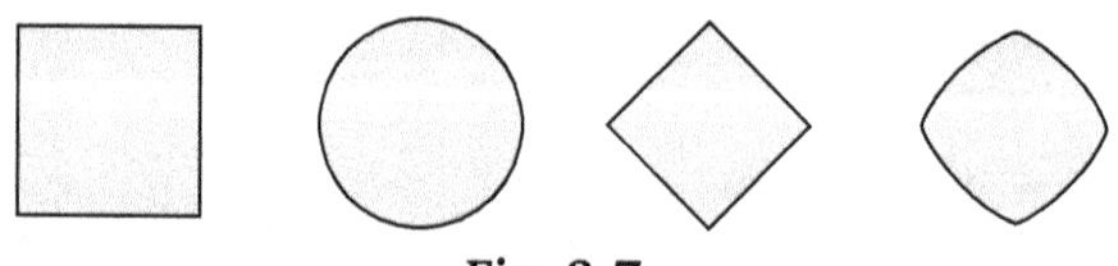

Fig. 2.7

Let X be a metric space and let $a \in X$, $r > 0$.

2.8.1 (a) *If* $a' \in B(a, r)$ *then there is a* $\delta > 0$ *such that*

$$B(a', \delta) \subseteq B(a, r).$$

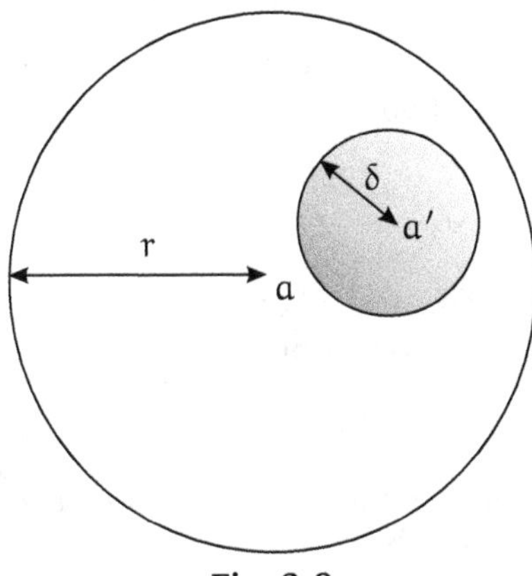

Fig. 2.8

(b) *If* $a' \in X \setminus E(a, r)$*, then there is a* $\delta > 0$ *such that*

$$B(a', \delta) \subseteq X \setminus E(a, r).$$

Proof Let $a' \in B(a, r)$. Then $\delta = r - d(a', a)$ is positive. Also if $x \in B(a', \delta)$ then $d(x, a) \leqslant d(x, a') + d(a', a) < \delta + r - \delta = r$. This proves (a). The proof of (b) is similar and is left to the reader. □

Metric topologies

We now show how a metric d on X defines a topology on X.

Definition A set $N \subseteq X$ is a *neighbourhood of* $a \in X$ if there is a real number $r > 0$ such that $B(a, r) \subseteq N$.

We verify Axioms N1–N4. The relation $d(a, a) = 0$ implies $a \in B(a, r)$; this verifies Axiom N1. The verification of Axioms N2 and N3 is simple and is left to the reader. By 2.8.1 an open ball is a neighbourhood of each of its points; so the verification of Axiom N4 is immediate.

Thus the neighbourhoods on X form a topology on X called the topology *induced by* d or the *metric topology*. Clearly, the open balls $B(a, r)$, all $r > 0$, form a base for the neighbourhoods of a; so also do the open balls $B(a, r)$ for r rational and positive, or for r of the form $1/n$, n a positive integer. This proves that X has a countable base for the neighbourhoods at each point. Notice also that $E(a, r/2) \subseteq B(a, r)$, so the closed balls of radius $r > 0$ also form a base for the neighbourhoods of r.

EXAMPLES

7. The *usual metric* on $\mathbb{R}$ is the metric $d(x, y) = |x - y|$. The topology induced by d is clearly the usual topology on $\mathbb{R}$. Similarly, the fields $\mathbb{C}$, $\mathbb{H}$ of complex numbers and quaternions have topologies induced by $d(x, y) = |x - y|$.

8. The Cartesian norm on $\mathbb{K}^n$ induces the product topology. This is proved as follows. If N_i is a neighbourhood in $\mathbb{K}$ of a_i, then $N_i \supset B(a_i, r_i)$ for some $r_i > 0$. Let $a = (a_1, \dots, a_n)$, $\delta = \min\{r_1, \dots, r_n\}$. Then

$$N_1 \times \cdots \times N_n \supseteq B(a, \delta).$$

This shows that each product neighbourhood of a is a metric neighbourhood of a. The converse is trivial since

$$B(a, r) = B(a_1, r) \times \cdots \times B(a_n, r).$$

9. Let X be a set and let d be the discrete metric on X. Then, if $a \in X$, the open balls about a are given by

$$B(a, r) = \begin{cases} X, & r > 1 \\ \{a\}, & 0 < r \leqslant 1. \end{cases}$$

Hence, any set containing a is a neighbourhood of a. That is, the discrete metric induces the discrete topology.

10. There is no metric on X inducing the indiscrete topology (unless $\#X \leqslant 1$). However, let us define a *pseudo-metric* to be a function $d : X \times X \to \mathbb{R}^{\geqslant 0}$ satisfying Axioms M2, M3 and the following weak form of Axiom M1.

M1′ $d(x, x) = 0$.

A pseudo-metric induces a topology on X in the same way as does a metric, since Axiom M1′ is enough to show that $a \in B(a, r)$. The indiscrete topology on X is induced by the *indiscrete pseudo-metric* given by $d(x, y) = 0$ for all $x, y \in X$.

It is clear from 2.8.1(a) that the open balls of X are open sets, and from 2.8.1(b) that the closed balls of X are closed sets. However, the closed ball is not necessarily the closure of the open ball, since in Example 9 above

$$B(a, 1) = \{a\}, \quad E(a, 1) = X.$$

Let X, Y be metric spaces and $f : X \to Y$ a function. We can use 2.6.1 to give another description of the continuity of f, namely, f is continuous $\Leftrightarrow$ for each $a \in X$ and each $\varepsilon > 0$, there is a $\delta > 0$ such that

$$fB(a, \delta) \subseteq B(fa, \varepsilon).$$

Further, we may replace here any open ball by a closed ball, and also restrict any or both of ε, δ to rational numbers $1/n$ for n a positive integer.

Let d, e be metrics on X. The metrics are *equivalent* if they induce the same topology on X or, what is the same thing, if the identity $1 : X_d \to X_e$ is a homeomorphism. Let subscripts (e.g., B_d, B_e) be used to distinguish the balls for the two metrics. Then d, e *are equivalent if and only if for each* $a \in X$

(a) *for each* $\varepsilon > 0$ *there is a* $\delta > 0$ *such that* $B_d(a, \delta) \subseteq B_e(a, \varepsilon)$, *and*
(b) *for each* $\varepsilon > 0$ *there is a* $\delta > 0$ *such that* $B_e(a, \delta) \subseteq B_d(a, \varepsilon)$.

EXAMPLE

11. Let $d : \mathbb{R} \times \mathbb{R} \to \mathbb{R}^{\geqslant 0}$ be defined by

$$d(x, y) = \frac{|x - y|}{1 + |x - y|}.$$

Then d is a metric on $\mathbb{R}$; the only verification which is non-trivial is of the triangle inequality M3, and this is proved as follows:

Let $x \neq z$. Then

$$\begin{aligned} d(x, z) &= (1 + |x - z|^{-1})^{-1} \\ &\leqslant (1 + (|x - y| + |y - z|)^{-1})^{-1} \\ &= |x - y|(1 + |x - y| + |y - z|)^{-1} + |y - z|(1 + |x - y| + |y - z|)^{-1} \\ &\leqslant d(x, y) + d(y, z). \end{aligned}$$

This metric d is equivalent to the usual metric. For it is easily seen that

$$|x - y| < \varepsilon \Leftrightarrow d(x, y) < \varepsilon(1 + \varepsilon)^{-1} \Rightarrow d(x, y) < \varepsilon.$$

Let $B(x, \varepsilon)$ be the open ball for the metric d. Then

$$]x - \varepsilon, x + \varepsilon[\ = \ B(x, \varepsilon(1 + \varepsilon)^{-1}) \subset B(x, \varepsilon)$$

and this implies the equivalence of the two metrics. □

A metric d on a set X is *bounded* if there is a real number r such that $d(x, y) \leqslant r$ for all x, y in X. The last example shows that $\mathbb{R}$ admits an equivalent bounded metric, and a similar argument applies to any metric space X. It is also easy to check that the metric d' of Example 6 of p. 48 is a bounded metric equivalent to the given one.

There is a very useful continuity criterion for additive and bi-additive maps on normed vector spaces. Let U, V, W be normed vector spaces. A function $f : V \to W$ is *additive* if $f(x+y) = f(x) + f(y)$ for all $x, y \in V$; and $g : U \times V \to W$ is *bi-additive* if

$$g(x, y + y') = g(x, y) + g(x, y'),$$
$$g(x + x', y) = g(x, y) + g(x', y)$$

for all $x, x' \in U$, $y, y' \in V$.

2.8.2 (a) *An additive function* $f : V \to W$ *is continuous if there is a real number* $r > 0$ *such that*

$$\|f(x)\| \leqslant r\|x\| \text{ for all } x \in V.$$

(b) *A bi-additive function* $g : U \times V \to W$ *is continuous if there is a real number* $r > 0$ *such that*

$$\|g(x, y)\| \leqslant r\|x\|\|y\| \text{ for all } x \in U, y \in V.$$

Proof We prove (b) first. Let $\varepsilon > 0$, $(a, b) \in U \times V$. Let $\|X\|, \|y\| < \delta$ where $\delta < 1$. Then

$$\begin{aligned}\|g(a + x, b + y) - g(a, b)\| &\leqslant \|g(x, y)\| + \|g(a, y)\| + \|g(x, b)\| \\ &\leqslant r\{\|x\|\|y\| + \|a\|\|y\| + \|b\|\|x\|\} \\ &< r\delta\{1 + \|a\| + \|b\|\} \text{ since } \delta < 1 \\ &= k\delta \text{ say, where } k > 0.\end{aligned}$$

So $g[B(a, \delta) \times B(b, \delta)] \subseteq B(g(a, b), \varepsilon)$ if $\delta < \varepsilon/k$. This proves continuity of g.

We use (b) to prove (a). If $f : V \to W$ is additive and satisfies $\|f(x)\| \leqslant r\|x\|$, then $g : V \times \mathbb{K} \to W$ defined by $g(x, \lambda) = f(x)\lambda$ is bi-additive and satisfies $\|g(x, \lambda)\| \leqslant r|\lambda|\|x\|$. Since $f = gi$ where $i : V \to V \times \mathbb{K}$ is $x \mapsto (x, 1)$, the continuity of g implies that of f. □

It is easy to give a direct proof of (a), and we leave this to the reader.

Remark For f linear and g bilinear the converses of 2.8.2(a) and (b) are true. However we do not need this fact and so leave its proof as an exercise.

EXAMPLES

12. Let $V \times V$ be given the norm

$$\|(x, y)\| = \max\{\|x\|, \|y\|\}.$$

The axioms for a norm are easily verified. The addition map $V \times V \to V$ given by $(x, y) \mapsto x + y$ is additive and satisfies

$$\|x+y\| \leqslant \|x\| + \|y\| \leqslant 2\|(x,y)\|.$$

Therefore addition is continuous.

13. Let $f : \mathbb{R}^n \times \mathbb{R}^m \to V$ be bilinear, that is, f is additive and also

$$f(\lambda x, y) = f(x, \lambda y) = \lambda f(x, y) \text{ for } \lambda \in \mathbb{R}.$$

Let $e_j, j = 1, \ldots, n$ denote the standard basis elements of $\mathbb{R}^n$, and let $\mathbb{R}^n, \mathbb{R}^m, \mathbb{R}^l$ have Euclidean or Cartesian norm. This implies that if $x = \sum \lambda_j e_j$, then $|\lambda_j| \leqslant \|x\|$. Hence if $x = \sum_{j=1}^{n} \lambda_j e_j$, $y = \sum_{i=1}^{m} \mu_i e_i$, then

$$\begin{aligned} \|f(x,y)\| &\leqslant \sum_{i,j} |\lambda_j||\mu_i|\|f(e_j, e_i)\| \\ &\leqslant \|x\|\|y\| \sum_{i,j} \|f(e_j, e_i)\|. \end{aligned}$$

It follows that f is continuous.

14. A similar, and simpler, argument to that of the last example shows that if $\mathbb{R}^n$ has the Euclidean or Cartesian norms, then any linear function $\mathbb{R}^n \to V$ is continuous. This implies that the Euclidean and Cartesian norms are equivalent (that is, define the same metric topology), since the identity $\mathbb{R}^n \to \mathbb{R}^n$ is continuous whichever of these norms we put on each $\mathbb{R}^n$. Actually it can be proved, as an application of compactness, that *on a finite dimensional normed vector space any two norms are equivalent.* [Exercise 11 of Section 3.5.]

15. The sup norm $\| \ \|_S$ and the integral norm $\| \ \|_I$ on the space $\mathcal{C}$ of continuous functions $[0, 1] \to \mathbb{R}$ are not equivalent. For let $0 < r \leqslant 1$ and let $f_r : [0, 1] \to \mathbb{R}$ be the function whose graph is shown in Fig. 2.9. Then $\|f_r\|_S = 1$, but $\|f_r\|_I = r/2$, Hence $f_r \in E_I(0, r/2)$ but $f_r \notin E_S(0, \frac{1}{2})$. Thus $E_I(0, r/2) \nsubseteq E_S(0, \frac{1}{2})$ for any $0 < r \leqslant 1$. So the two norms are not equivalent.

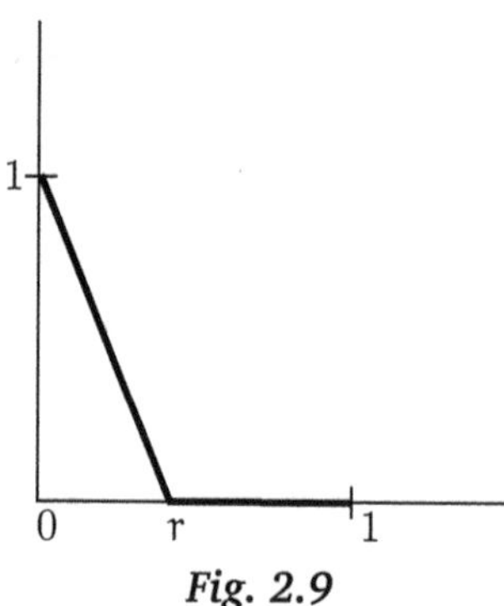

Fig. 2.9

16. Let V_1, V_2 be the same vector space V but with norms $\| \|_1, \| \|_2$ respectively. If we apply 2.8.2(b) to the two identity maps $V_1 \to V_2$, $V_2 \to V_1$ we see that these norms are equivalent if there are real numbers $r, s > 0$ such that

$$\|x\|_1 \leqslant r\|x\|_2, \qquad \|x\|_2 \leqslant s\|x\|_1.$$

(This sufficient condition is also necessary—cf. the Remark on p. 52). This gives another proof that in $\mathbb{R}^n$ the Euclidean norm $\| \|_2$ and the Cartesian norm $\| \|_\infty$ are equivalent, since for any x in $\mathbb{R}^n$

$$\|x\|_\infty \leqslant \|x\|_2 \leqslant \sqrt{n}\|x\|_\infty.$$

In this section, we have shown that every metric on a set X induces a topology on X. On the other hand, not every topology on X is induced by a metric. One example of this, the indiscrete topology, has been given already, and other examples will be given later. The characterisation of metric topologies has been completely solved (for an account of this see [Kel55], [Eng68]), but this kind of problem is outside the scope of this book.

Products and subspaces

Let X, Y be metric spaces. On $X \times Y$ we can define a metric by

$$d((x, y), (x', y')) = \max\{d(x, x'), d(y, y')\}.$$

The verification of the axioms for a metric is simple, and is left to the reader. The topology on $X \times Y$ induced by this metric is simply the product topology, as is easily seen from the formula

$$B(a, r) \times B(b, r) = B((a, b), r) \subseteq B(a, s) \times B(b, t)$$

where $r = \min\{s, t\}$.

Let X be a metric space with metric d and let A be a subset of X. Let $d_A = d \mid A \times A$. It is easily verified that d_A is a metric on A. The open balls in X we write $B(a, r)$ and in A, $B_A(a, r)$. Clearly

$$B_A(a, r) = B(a, r) \cap A.$$

Fig. 2.10 gives a picture of a subset A of $\mathbb{R}^2$ (with the Euclidean metric) and various open balls in A.

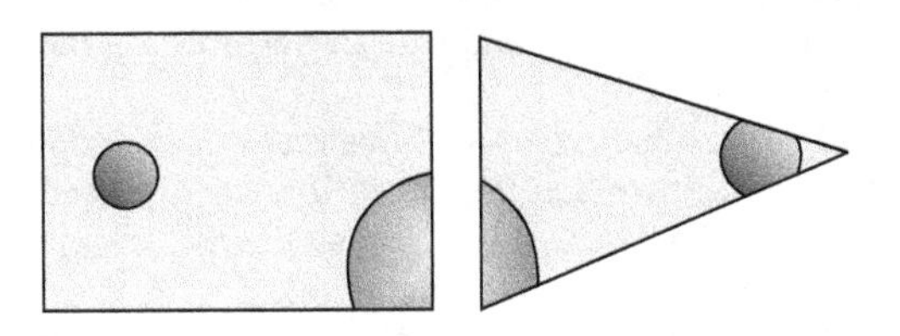

Fig. 2.10

2.8.3 *The metric topology on* A *is its relative topology as a subset on* X.

Proof Let $a \in A$. If N is a neighbourhood of a in X, then $N \supseteq B(a, r)$ for some $r > 0$ and so $N \cap A \supseteq B_A(a, r)$. Thus each subspace neighbourhood of a is also a metric neighbourhood.

Conversely, suppose M is a neighbourhood of a for the metric d_A. Then $M \supseteq B_A(a, r)$ for some $r > 0$. Let

$$N = B(a, r) \cup M.$$

Then N is a neighbourhood of a in X and $N \cap A = M$. Therefore, M is also a subspace neighbourhood of a. □

Because of this proposition, a subspace A of a metric space X will mean a subset with the metric d_A and the metric topology.

Exercises

1. Let X be a metric space. Prove that the topology of X is discrete if and only if for each x in X there is an $r > 0$ such that $B(x, r) = \{x\}$.
2. Let $f : \mathbb{R}^{\geqslant 0} \to \mathbb{R}^{\geqslant 0}$ be a continuous function such that

   (i) $f(x) = 0 \Leftrightarrow x = 0$,
   (ii) $x \leqslant x' \Rightarrow f(x) \leqslant f(x')$,
   (iii) $f(x + x') \leqslant f(x) + f(x')$.

Let d be a metric on X. Show that the composition $e = fd$ is a metric on X equivalent to d. Show also that the function $x \mapsto x/(1+x)$ satisfies conditions (i), (ii), and (iii) above.

3. Let X, Y be metric spaces. Show that the following formulae define metrics on $X \times Y$ whose metric topology is the product topology.

(a) $d((x,y),(x',y')) = d(x,x') + d(y,y')$,
(b) $d((x,y),(x',y')) = [\{d(x,x')\}^2 + \{d(y,y')\}^2]^{\frac{1}{2}}$.

4. In Euclidean space two open balls meet if the distance between their centres is less than the sum of their radii. Is this true in an arbitrary metric space? in an arbitrary normed vector space?

5. Let $d : X \times X \to \mathbb{R}^{\geqslant 0}$ be a metric on X. Prove that d is continuous.

6. Let d be a metric on X. Prove that the metric topology is the coarsest topology $\mathcal{T}$ on X such that each function $d_x : X_{\mathcal{T}} \to \mathbb{R}^{\geqslant 0}$, $y \mapsto d(x,y)$, is continuous.

7. Let V be a normed vector space over $\mathbb{K}$, and let $a \in V$, $\alpha \in \mathbb{K}$ ($\alpha \neq 0$). Prove that the functions $V \to V$ given by $x \mapsto x + a$, $x \mapsto x\alpha$ are homeomorphisms. Prove also that any open ball in V is homeomorphic to V.

8. Let V be a normed vector space and let $a \in V$, $r > 0$. Prove that $B(a,r)$ is the interior of $E(a,r)$, and $E(a,r)$ is the closure of $B(a,r)$.

9. Let A, B be subsets of the normed vector space V. Prove that if one of A, B is open, then so also is

$$A + B = \{a + b : a \in A, b \in B\}.$$

10. Let $V_1, \dots, V_n, V$ be normed vector spaces over $\mathbb{R}$. We say $u : V_1 \times \cdots \times V_n \to V$ is *multilinear* if, for any $\lambda, \mu \in \mathbb{R}$, and $x_i, x_i' \in V_i$ $(i = 1, \dots, n)$

$$u(x_1, \dots, \lambda x_i + \mu x_i', \dots, x_n) = \lambda u(x_1, \dots, x_i, \dots, x_n) + \mu u(x_1, \dots, x_i', \dots, x_n).$$

Prove that such a multilinear map is continuous $\Leftrightarrow$ there is a real number $r > 0$ such that for all $x_i \in V_i$

$$\|u(x_1, \dots, x_n)\| \leqslant r\|x_1\| \cdots \|x_n\|.$$

(This applies to normed vector spaces over $\mathbb{K}$, since such objects are also, by restriction of the field, normed vector spaces over $\mathbb{R}$.)

[Exercise 10 implies the necessity of the condition for equivalent norms given in Example 16, p. 54.]

11. Let V_1, V_2 be the same vector space V with distinct, but equivalent, norms $\| \|_1, \| \|_2$ respectively. Construct homeomorphisms

$$B(V_1) \to B(V_2),\ E(V_1) \to E(V_2),\ S(V_1) \to S(V_2).$$

12. Let $f : \mathbb{E}^n \to \mathbb{E}^n$ be a map such that $f \mid \mathbb{B}^n, \mathbb{B}^n$ is a homeomorphism. Prove that $f[\mathbb{S}^{n-1}] \subseteq \mathbb{S}^{n-1}$.

13. Brouwer has proved the following theorem known as the *Invariance of Domain* [cf. [Nag65], [Spa66]]. *Let* A, B *be subsets of* $\mathbb{R}^n$ *and* $f : A \to B$ *a homeomorphism. Then* $f[\text{Int}\, A] \subseteq \text{Int}\, B$.

Use the Invariance of Domain to prove (i) (*Invariance of Dimension*). If $f : \mathbb{R}^m \to \mathbb{R}^n$ is a homeomorphism, then $m = n$. (ii) If $f : \mathbb{E}^n \to \mathbb{E}^n$ is a homeomorphism, then $f \mid \mathbb{B}^n, \mathbb{B}^n$ and $f \mid \mathbb{S}^{n-1}, \mathbb{S}^{n-1}$ are defined and are homeomorphisms.

14. Let p be a prime number. For each $n \in \mathbb{N}$ define $\nu_p(n)$ to be the exponent of p in the decomposition of n into prime numbers. If $x = \pm m/n$ is any non-zero rational number, with $m, n \in \mathbb{N}$, define

$$\nu_p(x) = \nu_p(m) - \nu_p(n).$$

Finally, if x, y are rational numbers define

$$d(x,y) = \begin{cases} p^{-\nu_p(x-y)}, & x \neq y \\ 0, & x = y. \end{cases}$$

(i) Prove that d is a metric on $\mathbb{Q}$ and that d satisfies the following strong form of the triangle inequality

$$d(x,z) \leqslant \max\{d(x,y), d(y,z)\}.$$

The topology induced by d is called the *p-adic topology*.

(ii) Prove that the topology induced by d on $\mathbb{Z}$ is the p-adic topology of Exercise 7 of Section 2.1.

(iii) Justify the following statement: in the p-adic topology on $\mathbb{Q}$, small rational numbers are those which are multiples of large powers of p.

*15. A subset A of $\mathbb{R}^m$ is *convex* if for any x, y in A the line segment joining x to y (that is, the set of points $(1-t)x + ty$, $0 \leqslant t \leqslant 1$) is contained in A. This exercise outlines a proof that any two open convex subsets A, B of $\mathbb{R}^m$ are homeomorphic. The steps are as follows:

(i) There is a homeomorphism $f : \mathbb{R}^m \to \mathbb{B}^m$ such that $f[A]$ is convex. So we may suppose A is bounded. Let $a \in A$. Then, there is a real number $\delta > 0$ such that $B(a, \delta) \subseteq A$.

(ii) For each x in A, $x \neq 0$, let

$$r(x) = \sup\{\lambda \in \mathbb{R} : \lambda(x - a) \in A\}.$$

Then $r(x)$ is well-defined and non-zero, and the function $x \mapsto r(x)$ is continuous.

(iii) The function $x \mapsto a + (\delta/r(x))(x - a)$ is a homeomorphism $A \to B(a, \delta)$.

(iv) A, B are homeomorphic.

2.9 Distance from a subset

Let X be a metric space with metric d, and let A be a (non-empty) subset of X. For each $x \in X$ we define the *distance of* x *from* A to be

$$\operatorname{dist}(x, A) = \inf\{d(x,a) : a \in A\}.$$

2.9.1 *The function* $x \mapsto \operatorname{dist}(x, A)$ *is a continuous function* $X \to \mathbb{R}^{\geqslant 0}$.

Proof We prove that for any $\varepsilon > 0$ there is a $\delta > 0$ such that

$$d(x,y) \leqslant \delta \Rightarrow |\operatorname{dist}(x,A) - \operatorname{dist}(y,A)| \leqslant \varepsilon. \qquad (*)$$

For any a in A

$$\begin{aligned} d(x,a) &\leqslant d(x,y) + d(y,a), \\ d(y,a) &\leqslant d(y,x) + d(x,a). \end{aligned}$$

We apply $\inf_{a \in A}$ to each of these inequalities to obtain

$$\begin{aligned} \operatorname{dist}(x,A) &\leqslant d(x,y) + \operatorname{dist}(y,A), \\ \operatorname{dist}(y,A) &\leqslant d(y,x) + \operatorname{dist}(x,A). \end{aligned}$$

whence

$$|\operatorname{dist}(x,A) - \operatorname{dist}(y,A)| \leqslant d(x,y).$$

This proves (*) with $\delta = \varepsilon$. □

2.9.2 $x \in \overline{A} \Leftrightarrow \operatorname{dist}(x,A) = 0$.

Proof The inverse image of $\{0\}$ under $x \mapsto (x,A)$ is a closed set containing A, and so containing $\overline{A}$. Thus $x \in \overline{A} \Rightarrow d(x,A) = 0$.

On the other hand, if $x \notin \overline{A}$, then there is a closed ball $E(x,r)$, $r > 0$, not meeting A. Hence $\operatorname{dist}(x,A) \geqslant r$. □

2.9.3 *Let A, B be disjoint closed sets in X. There are disjoint open sets U, V in X such that $A \subseteq U$, $B \subseteq V$.*

Proof Let $f : X \to \mathbb{R}$ be the function $x \mapsto \operatorname{dist}(x,A) - \operatorname{dist}(x,B)$, and let $U = f^{-1}[\mathbb{R}^{<0}]$, $V = f^{-1}[\mathbb{R}^{>0}]$. By 2.9.1, U and V are open, and they are clearly disjoint. If $x \in A$, then $\operatorname{dist}(x,B) > 0$ since B is closed and A, B are disjoint. Therefore, $f(x) < 0$ and so $x \in U$. Thus $A \subseteq U$ and, similarly, $B \subseteq V$. □

In a topological space, we say a subset N is a *neighbourhood* of a subset A if there is an open set U such that $A \subseteq U \subseteq N$. We can express 2.9.3 succinctly as: in a metric space, disjoint closed sets have disjoint neighbourhoods. A topological space with this property is called *normal*—examples of non-normal spaces are given in the Exercises.

Exercises

1. A topological space X is called T_1 if, for each x in X, the set $\{x\}$ is closed; and X is called *Hausdorff* if distinct points of X have disjoint neighbourhoods. Prove that a metric space is Hausdorff, and that a Hausdorff space is T_1.

2. A subset A of a topological space X is called a G_δ-set if A is the intersection of a countable number of open sets of X. Prove that a closed subset of a metric space is a G_δ-set.
3. Give examples of (i) a metric space X and a subset A of X which is not a G_δ-set, (ii) a topological space X and a closed subset A of X which is not a G_δ-set.
4. Let X be the unit interval $[0,1]$ with the following topology. The neighbourhoods of t for $0 < t \leqslant 1$ are the usual ones. The neighbourhoods of 0 are the usual ones and also the sets $N \setminus A$ where N is a usual neighbourhood of 0 and A is a set $\{x_1, x_2, x_3, \ldots\}$ of points x_n such that $x_n \neq 0$ for any n, and $x_n \to 0$ as $n \to \infty$. Prove that this defines a topology on X, and that X is Hausdorff but not normal.
5. Prove that a topological space X is $T_1 \Leftrightarrow$ for each x, y in X there is a neighbourhood of x not containing y.
6. Let X be a metric space, A a subset of X, and $r > 0$. We define

$$B(A, r) = \{x \in X : \operatorname{dist}(x, A) < r\}.$$

If $B(A,r) \subseteq N \subseteq X$ for some $r > 0$, then N is a neighbourhood of A. Prove that the converse of this implication is false.
7. Generalise the notion of a base for the neighbourhoods of a point to the notion of a base for the neighbourhoods of a set. Give an example of a subset A of $\mathbb{R}$ such that A does not have a countable base for its neighbourhoods.

2.10 Hausdorff spaces

We recall [Exercise 1 of Section 2.9] that a topological space X is Hausdorff if distinct points of X have disjoint neighbourhoods.‡ The following characterisation of this property is more aesthetic, and often more useful.

2.10.1 *A topological space X is Hausdorff if and only if the diagonal*

$$\Delta(X) = \{(x, x) \in X \times X : x \in X\}$$

is closed in $X \times X$.

Proof Let $\Delta = \Delta(X)$. The following statements are each equivalent to their successors (since $x \neq x' \Leftrightarrow (x, x') \notin \Delta$).

(a) X is Hausdorff.
(b) if $x \neq x'$, then there exist neighbourhoods M, M' of x, x' such that $M \cap M' = \varnothing$.
(c) if $x \neq x'$, then there exist neighbourhoods M, M' of x, x' such that $(M \times M') \cap \Delta = \varnothing$.
(d) if $x \neq x'$, then $(X \times X) \setminus \Delta$ is a neighbourhood of (x, x').
(e) Δ is closed in $X \times X$.

‡A standard joke is that X is Hausdorff if any two points can be housed off from each other.

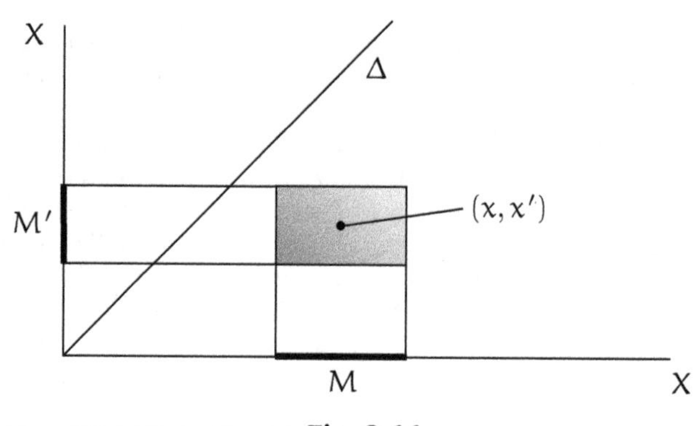

Fig. 2.11

□

2.10.2 *Let* $f, g : Y \to X$ *be maps of the topological space* Y *to the Hausdorff space* X. *Then the set* A *of points on which* f, g *agree is closed in* Y.

Proof $A = (f, g)^{-1}[\Delta(X)]$. □

2.10.3 *Subspaces, and products, of Hausdorff spaces are Hausdorff.*

Proof Let A be a subspace of the Hausdorff space X. Then

$$\Delta(A) = \Delta(X) \cap (A \times A).$$

Therefore, $\Delta(A)$ is closed in $A \times A$, and so A is Hausdorff.

Let X, Y be Hausdorff spaces and let

$$\begin{aligned} T : X \times X \times Y \times Y &\to X \times Y \times X \times Y \\ (x, x', y, y') &\mapsto (x, y, x', y'). \end{aligned}$$

Then T is a homeomorphism and

$$T[\Delta(X) \times \Delta(Y)] = \Delta(X \times Y).$$

Therefore, $\Delta(X \times Y)$ is closed in $X \times Y \times X \times Y$. □

2.10.4 *Let* $f : Y \to X$ *be a continuous injection and let* X *be Hausdorff. Then* Y *is Hausdorff.*

Proof Since f is an injection

$$\Delta(Y) = (f \times f)^{-1}[\Delta(X)].$$

Therefore, $\Delta(Y)$ is closed in $Y \times Y$. □

This result has two important special cases.

(a) A finer topology than a Hausdorff topology is also Hausdorff.
(b) If X and Y are homeomorphic and X is Hausdorff, then so also is Y.

In this book, the spaces of main interest will all be Hausdorff. But the reader should beware of thinking that non-Hausdorff spaces are of little importance. In the Exercises we sketch the construction of two important classes of non-Hausdorff spaces, firstly the Zariski topology and, secondly, the sheaf of germs of functions.

EXERCISES

1. Write down proofs of 2.10.2, 2.10.3, and 2.10.4 using directly the definition of Hausdorff spaces.
2. The *Zariski topology* on $\mathbb{R}^n$ (or on $\mathbb{C}^n$) is that in which C is closed if and only if there is a set of polynomials in n variables such that C is the set of points on which all these polynomials vanish. In the case $n = 1$, this is the topology in which C is closed $\Leftrightarrow$ $C = \mathbb{R}$ or C is finite. The Zariski topology on $\mathbb{R}$ is not Hausdorff. (The proof that the Zariski topology on $\mathbb{R}^n$ is not Hausdorff requires knowledge of the ideal theoretic properties of polynomial rings—cf. [ZS60, Ch. VII §3].)
3. *The sheaf of germs of functions.* Let X, Y be topological spaces. For each $x \in X$ let $F(x)$ denote the set of continuous functions from some neighbourhood of x to Y. An equivalence relation $\sim$ is defined in $F(x)$ by $f \sim g \Leftrightarrow f, g$ agree on some neighbourhood of x. The set of equivalence classes is written $G(x)$, and $G = \bigcup_{x \in X} G(x)$.

An element of $G(x)$ is called a *function-germ* or *germ* at x, and G is the *sheaf of germs of continuous functions*. If $f \in F(x)$, the germ of f (that is the equivalence class of f) is written f^x. The value of f^x at x is well-defined by $f^x(x) = f(x)$. Let U be an open set containing x, and let $f : U \to Y$ be continuous. Then f defines a germ f^y for each $y \in U$ and the set $f^U = \{f^y : y \in U\}$ is defined to be a basic neighbourhood of f^x. The topology on G is that in which a neighbourhood of f^x is any set containing a basic neighbourhood.

Prove that G is in fact a topological space, and that G is non-Hausdorff even if $X = Y = \mathbb{R}$. (If $f, g : \mathbb{R} \to \mathbb{R}$ are such that $f(x) = g(x)$ for $x \leqslant a$, $f(x) \neq g(x)$ for $x > a$, then f^a, g^a are germs which do not have disjoint neighbourhoods.)

Prove that if $X = Y = \mathbb{R}$ the above construction can be varied by replacing the word continuous by (i) integrable, (ii) differentiable, (iii) of class C^∞, (i.e., with derivatives of all orders), (iv) polynomial, (v) analytic (i.e., expressible locally by power series). Prove that in the last two cases the corresponding sheaf of germs is a Hausdorff space.
4. Let X be a topological space. Let $(x_n)_{n>0}$ be a sequence of points of X. If $Y \subseteq X$, we say (x_n) is *eventually in* Y if there is a number n_0 such that $n \geqslant n_0$ implies $x_n \in Y$. If $x \in X$, we say $x_n \to x$ *as* $n \to \infty$, or (x_n) *has limit* x, (or, briefly, $(x_n) \to x$) if for any neighbourhood N of x, (x_n) is eventually in N. Let

$\mathbb{L} = \{0\} \cup \{n^{-1} : n \in \mathbb{N}^{>0}\}$ have it relative topology as a subset of $\mathbb{R}$. Prove that $(x_n) \to x$ if and only if the function $g : \mathbb{L} \to X$ which sends $n^{-1} \mapsto x_n$, $0 \mapsto x$ is continuous. Prove also that if X is Hausdorff, then the conditions $(x_n) \to x$, $(x_n) \to y$, imply $x = y$.

5. Consider the following conditions on a space X: (a) X is *Fréchet*, that is, if $x \in X$ and $A \subseteq X$, then $x \in \overline{A}$ if and only if there is a sequence of points (x_n) of A such that $(x_n) \to x$. (b) X is *sequential*, that is a subset U of X is open if and only if every sequence converging to a point of U is eventually in U. (c) A subset U of X is open if and only if $U \cap A$ is open in A for every countable subset A of X. Prove that (a) is satisfied if X satisfies the first axiom of countability [Exercise 3 of Section 2.6], and that (a) $\Rightarrow$ (b) $\Rightarrow$ (c). [Use Exercise 5 of Section 2.6 for the proof that first countable implies Fréchet.]

6. Let X be a sequential space, let Y be a space and $f : X \to Y$ a function. Prove that the following conditions are equivalent, (a) f is continuous, (b) $fg : \mathbb{L} \to Y$ is continuous for all continuous functions $g : \mathbb{L} \to X$, (c) for all x in X and sequences (x_n) in X, $(x_n) \to x$ implies $(fx_n) \to fx$.

7. Let X be the space of Exercise 4 of Section 2.9 but defined using $\mathbb{Q} \cap [0, 1]$ instead of $[0, 1]$. Prove that X is not sequential, but that X satisfies (c) of Exercise 5.

8. Let X be the set $[0, 1]$ retopologised as follows. The neighbourhoods of t in $]0, 1]$ are the usual neighbourhoods. The neighbourhoods of 0 are the usual neighbourhoods and also any set containing $\{0\} \cup U$ where U is the usual open neighbourhood of $\mathbb{L}^* = \mathbb{L} \setminus \{0\}$. Prove that X is sequential but not Fréchet. Prove also that $X \setminus \mathbb{L}^*$ is not sequential.

9. Give an example of a topological space X which is not indiscrete, and in which limits of sequences are not unique.

10. Let X be an uncountable set with the topology that $C \subseteq X$ is closed if C is countable or if $C = X$. Let $g : \mathbb{L} \to X$ be continuous. Prove that there is an integer n_0 such that $n \geqslant n_0 \Rightarrow g(n^{-1}) = g(0)$. Prove also that in X limits of sequences are unique. Let Y be the underlying set of X with the discrete topology. Let $f : X \to Y$ be the identity function. Prove that f is not continuous, but fg is continuous for all continuous functions $g : \mathbb{L} \to X$.

11. Prove that the space X of Exercise 10 is not a pseudometric space. [A pseudometric space satisfies the first axiom of countability.]

12. Let X be an uncountable set and let $x_0 \in X$. Let X have the topology in which a subset C of X is closed if C is countable or if $x_0 \in C$. Prove that this is a topology and that X with this topology is a Hausdorff, non-metric space.

13. Let $\triangle \subseteq \mathbb{R}^2$ be the set of points inside and on the right-angled triangle ABC, which we suppose has a right-angle at A and satisfies $AC > AB$. This exercise outlines the construction of a continuous surjection $f : [0, 1] \to \triangle$. Let D on BC be such that AD is perpendicular to BC. Let $a = .a_1a_2a_3\ldots$ be a binary decimal, so that each a_n is 0 or 1. Then we construct a sequence (D_n) of points of $\triangle$ as follows: D_1 is the foot of the perpendicular from D onto the hypotenuse of the larger or smaller of the triangles ADB, ADC according as $a_1 = 1$ or 0 respectively. This construction is now repeated using D_1 in place of D and the appropriate triangle of ADB, ADC in place of ABC. For example, Fig. 2.12 illustrates the points D_1 to D_5 for the binary decimal $.10110\ldots$.

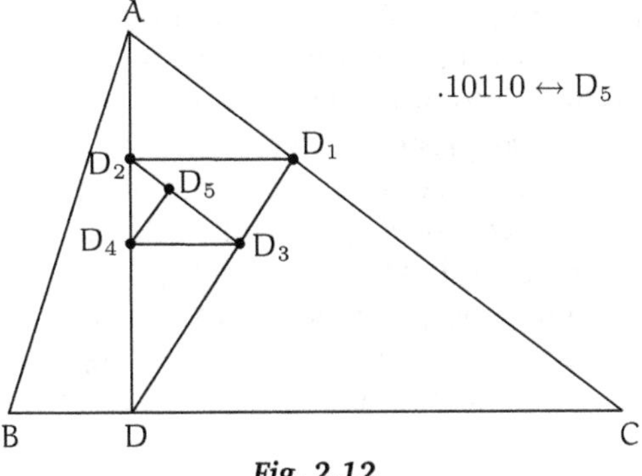

Fig. 2.12

The reader should give a decent inductive definition of the sequence (D_n) and prove in turn: (i) the sequence (D_n) tends to a limit $D(a)$ in $\triangle$; (ii) if $\lambda \in [0,1]$ is represented by distinct binary decimals a, a' then $D(a) = D(a')$; hence, the point $D(\lambda)$ in $\triangle$ is uniquely defined; (iii) if $f : [0,1] \to \triangle$ is the function $\lambda \mapsto D(\lambda)$ then f is surjective. (iv) f is continuous.

14. Using the previous exercise, prove the existence of continuous surjections from $[0,1]$ to the sets (i) $\mathbb{E}^2$, (ii) $\mathbb{I}^2$, (iii) $\mathbb{I}^n$.

15. Prove that there is a continuous surjection $\mathbb{R} \to \mathbb{R}^m$.

NOTES

Most books on topology start with the axioms for open sets. However the idea of neighbourhood seems more intuitive, and was found earlier. You would find it helpful to look up some of the history of set theory and of topology, in order to see the roots of our subject in analysis and in geometry. (See for example [Man64], the historical notes in [Bou66], and [Wil49].)

The modern theory of *locales* axiomatises the main properties of the lattice of open sets of a space, and gives a broader context in which to discuss the notion of continuity. The theory of locales has been called *pointless topology*. (See [Joh83], [Joh82]).

A variety of axiomatisations for topological spaces are listed in the Exercises of [Vai60]. Topological spaces are not adequate to deal with the notion of uniform continuity—for this, there is needed the concept of uniform space which may be found for example in [DK70], [Bou66]. A general account of the type of axiom system of which uniform spaces and topological spaces form a particular example is given in [Csa63]. The relation between convergence and topologies is best shown by means of the filters of [Bou66]. Spaces whose topology can be defined by sequences are discussed, for example, by [Fra65]. For an account of the general theory of sheaves see [God58]; applications of sheaves to algebraic topology are given in [Swa64] and to algebraic geometry in [Hir66] (but none of these books on sheaves is for the beginner in topology).

The results on maps into products given on p. 35 are extended in later sections by using *universal properties* in section A.4.

The idea of a sheaf has lead to an important generalisation of a topology, namely a *Grothendieck topology*, in which the inclusions of open sets are replaced by more general maps (cf. [Sch72], [Joh02]). This illustrates the maxim that a good concept turns up in various disguises, generalisations, and ramifications. However, an understanding of this particular notion requires a grounding in category theory (see chapter 6, and [Mac71], [HS79], [Sch72]).

Chapter 3

Connected spaces, compact spaces

3.1 The sum of topological spaces

Let X_1, X_2 be disjoint subspaces of a topological space X, and suppose $X = X_1 \cup X_2$. In general, it is not possible to recover the topology of X from the topologies of X_1, X_2. For example, if $a \neq b$, then the set $\{a, b\}$ has four distinct topologies, while the sets $\{a\}, \{b\}$ have each only one topology.

A case when the topology of X is determined by the topologies of X_1, X_2 is when U is open in X if both $U \cap X_1$ is open in X_1 and $U \cap X_2$ is open in X_2. In this case, we say X is a *topological sum* of X_1, X_2 and we write

$$X = X_1 \sqcup X_2.$$

The open sets of $X_1 \sqcup X_2$ are then simply the unions $U_1 \cup U_2$ for U_1 open in X_1, U_2 open in X_2.

EXAMPLE Let $X = [0, 2] \setminus \{1\}$ with its usual topology as a subspace of $\mathbb{R}$. Then $X = [0, 1[\sqcup]1, 2]$. On the other hand, $[0, 2]$ itself is not $[0, 1] \sqcup]1, 2]$.

Intuitively, a sum should be thought of as a space which is in two pieces. But one should bear in mind that any X is $X \sqcup \varnothing$.

3.1.1 *Let X_1, X_2 be disjoint subspaces of the topological space X such that $X = X_1 \cup X_2$. The follows conditions are equivalent.*

(a) $X = X_1 \sqcup X_2$,
(b) X_1, X_2 *are both open in* X,

(c) X_1 *is both open and closed in* X,
(d) $\overline{X_1} \cap X_2 = \varnothing$ *and* $X_1 \cap \overline{X_2} = \varnothing$.

Proof That (a) $\Rightarrow$ (b) is immediate from the definition of $X_1 \sqcup X_2$, while (b) $\Leftrightarrow$ (c) follows from the fact that $X_1 = X \setminus X_2$.

If X_1 is open and closed in X, then so also is X_2. Hence, $\overline{X_1} = X_1$, $\overline{X_2} = X_2$, and so (c) $\Rightarrow$ (d). Conversely, $\overline{X_1} \cap X_2 = \varnothing$ implies X_2 is open, so that X_1 is closed, while $X_1 \cap \overline{X_2} = \varnothing$ implies X_1 is open. Thus (d) $\Rightarrow$ (c).

Finally we prove that (c) $\Rightarrow$ (a). Let U_α be open in X_α, $\alpha = 1, 2$. Then U_α is open in X and so $U_1 \cup U_2$ is open in X. □

The most useful property of the sum $X_1 \sqcup X_2$ is concerned with functions $X_1 \sqcup X_2 \to Y$. Let $i_1 : X_1 \to X_1 \sqcup X_2$, $i_2 : X_2 \to X_1 \sqcup X_2$ be the two inclusion functions.

3.1.2 *If* $f_1 : X_1 \to Y, f_2 : X_2 \to Y$ *are maps, then there is a unique map* $f : X_1 \sqcup X_2 \to Y$ *such that* $f i_1 = f_1$, $f i_2 = f_2$.

Proof We suppose f_1, f_2 given. Then $f : X_1 \sqcup X_2 \to Y$ given by $x \mapsto f_\alpha(x)$ for $x \in X_\alpha$ ($\alpha = 1, 2$) is the only function $X_1 \sqcup X_2 \to Y$ such that $f i_1 = f_1$, $f i_2 = f_2$. We prove that f is continuous.

Let U be open in Y. Then

$$\begin{aligned} f^{-1}[U] &= (f^{-1}[U] \cap X_1) \cup (f^{-1}[U] \cap X_2) \\ &= f_1^{-1}[U] \cup f_2^{-1}[U]. \end{aligned}$$

Therefore, $f^{-1}[U]$ is open in $X_1 \sqcup X_2$. □

The situation of 3.1.2 is summed up in the diagram

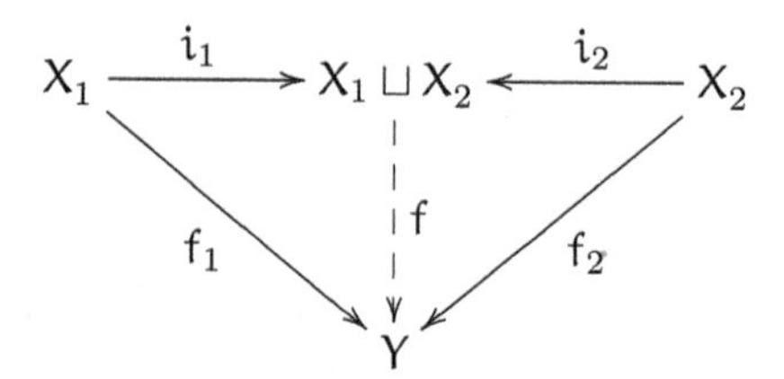

in which the dotted arrow indicates a function to be constructed.

A consequence of 3.1.2 is that a function $f : X_1 \sqcup X_2 \to Y$ is continuous if $f i_1, f i_2$ are continuous. This fact is used in the proof of the next result, in which $\{1, 2\}$ is a discrete space.

3.1.3 *Let* X_1, X_2 *be subspaces of* X. *Then* $X = X_1 \sqcup X_2$ *if and only if there is a map* $f : X \to \{1, 2\}$ *such that* $X_\alpha = f^{-1}[\alpha]$, $\alpha = 1, 2$.

Proof Suppose $X = X_1 \sqcup X_2$. Define $f : X \to \{1,2\}$ by $f[X_\alpha] = \alpha$, $\alpha = 1,2$. Then for $\alpha = 1,2$ $fi_\alpha : X_\alpha \to \{1,2\}$ is a constant function, and hence continuous. Therefore f is continuous.

Now suppose that $f : X \to \{1,2\}$ is continuous, and $X_\alpha = f^{-1}[\alpha]$, $\alpha = 1,2$. Then X_1, X_2 are disjoint and have union X. Also $\{1\}$ is open and closed in $\{1,2\}$ and so X_1 is open and closed in X. □

3.1.4 *Let Y be any space with underlying set $X_1 \cup X_2$ such that $X_1 \cap X_2 = \varnothing$ and the inclusions $j_\alpha : X_\alpha \to Y$, $\alpha = 1,2$ are continuous. Then the topology of $X_1 \sqcup X_2$ is finer than that of Y.*

Proof Let $f : X_1 \sqcup X_2 \to Y$ be the identity function, so that $fi_\alpha = j_\alpha$, $\alpha = 1,2$. By 3.1.2, f is continuous. □

This result is expressed roughly by: $X_1 \sqcup X_2$ has the finest topology such that the inclusions $i_\alpha : X_\alpha \to X_1 \sqcup X_2$ are continuous.

We shall need in chapter 4 to form a sum of spaces X_1, X_2 which are not disjoint. For this we take the universal property 3.1.2 as a definition.

Definition A *sum* of topological spaces X_1, X_2 is a pair of maps $i_1 : X_1 \to X$, $i_2 : X_2 \to X$ which is φ-universal: that is, if $f_1 : X_1 \to Y$, $f_2 : X_2 \to Y$ are any maps, then there is a unique map $f : X \to Y$ such that $fi_1 = f_1$, $fi_2 = f_2$.

The usual universal argument [Appendix §4] shows that the space X is then uniquely defined up to a homeomorphism. For this reason, we denote a sum simply by $X_1 \sqcup X_2$.

A sum $X_1 \sqcup X_2$ can always be constructed: Its underlying set is to be a sum of the underlying sets of X_1, X_2, so that we have injections $i_1 : X_1 \to X_1 \sqcup X_2$, $i_2 : X_2 \to X_1 \sqcup X_2$. The open sets of $X_1 \sqcup X_2$ are to be the sets

$$i_1[U_1] \cup i_2[U_2] = U_1 \sqcup U_2$$

for U_1 open in X_1, U_2 open in X_2. We leave the reader to verify that this does define a sum.

If $f_\alpha : X_\alpha \to Y$ are maps ($\alpha = 1,2$) then the map $X_1 \sqcup X_2 \to Y$ defined by f_1, f_2 is written $(f_1, f_2)^t$.

Exercises

1. Prove that if X_1, X_2 are metrisable, then $X_1 \sqcup X_2$ is metrisable.
2. Let X_1, X_2 be subspaces of X such that $X = X_1 \sqcup X_2$. Let $f : Y \to X$ be a map. Prove that Y is the sum of $f^{-1}[X_1]$ and $f^{-1}[X_2]$.

3. Let X_1, X_2, X_3 be topological spaces. Prove that there are homeomorphisms,

$$X_1 \sqcup X_2 \to X_2 \sqcup X_1,$$
$$X_1 \sqcup (X_2 \sqcup X_3) \to (X_1 \sqcup X_2) \sqcup X_3,$$
$$X_1 \times (X_2 \sqcup X_3) \to (X_1 \times X_2) \sqcup (X_1 \times X_3).$$

4. Prove that the following properties hold for $X_1 \sqcup X_2$ if and only if they hold for both X_1 and X_2: separable, first axiom of countability, Hausdorff.
5. Let X_0, X_1, X_2 be subspaces of X such that $X = X_1 \cup X_2$, $X_0 = X_1 \cap X_2$ and

$$X \setminus X_0 = (X_1 \setminus X_2) \sqcup (X_2 \setminus X_1).$$

Prove that a function $f : X \to Y$ is continuous if $f \mid X_1$, $f \mid X_2$ are continuous.

3.2 Connected spaces

Let X be a topological space. A pair $\{X_1, X_2\}$ of subspaces of X is called a *partition* of X if X_1, X_2 are non-empty, disjoint, and $X = X_1 \sqcup X_2$. Intuitively, X has a partition if it falls into two bits. This leads to the definition; X is *connected* if it has no partition, and otherwise is *disconnected*. We have immediately from 3.1.1 and 3.1.3:

3.2.1 *Let X be a topological space. The following conditions are equivalent.*

(a) X *is connected.*
(b) *If a subset A of X is open and closed in X, then $A = \varnothing$ or $A = X$.*
(c) *If $X = A \dot\cup B$ where $\overline{A} \cap B = \varnothing, A \cap \overline{B} = \varnothing$ then $A = \varnothing$ or $B = \varnothing$.*
(d) *If $f : X \to \{1, 2\}$ is continuous, then f is constant.*

The last condition is probably the most useful.

A subset Y of a topological space X is connected if Y with its induced topology is a connected space. The connectedness of Y can be described in terms of the closure operator in X: Let $A, B \subseteq Y$, and let $\overline{B}, \mathrm{Cl}_Y B$ be the closures of B in X, Y respectively. Then $\mathrm{Cl}_Y B = Y \cap \overline{B}$ [2.4.2], so that $A \cap \overline{B} = A \cap \mathrm{Cl}_Y B$; and, similarly, $\overline{A} \cap B = \mathrm{Cl}_Y A \cap B$. So Y is connected if and only if the conditions $Y = A \cup B$, $\overline{A} \cap B = A \cap \overline{B} = \varnothing$ imply $A = \varnothing$ or $B = \varnothing$.

The connectedness of Y can also be described in terms of the open sets of X [cf. Exercises 6, 7, 8].

It is to be expected from the fact that connectedness involves the open sets of X that connectedness is a *topological invariant*: that is, if X is homeomorphic to Y, then X is connected if and only if Y is connected. In fact, we prove a stronger result.

3.2.2 *If* X *is connected, and* $f : X \to Y$ *is continuous, then* $\operatorname{Im} f$ *is connected.*

Proof Let $f' = f \mid X, \operatorname{Im} f$. If $\operatorname{Im} f$ is disconnected, then there is continuous surjection $g : \operatorname{Im} f \to \{1, 2\}$, and $gf' : X \to \{1, 2\}$ is a continuous surjection. This implies that X is disconnected. □

A discrete space with more than one point is disconnected, while $\varnothing$ and $\{a\}$ are connected spaces. By 1.3.4, the real line $\mathbb{R}$ is connected. Any open interval of $\mathbb{R}$ is homeomorphic to $\mathbb{R}$ and hence is connected by 3.2.2.

3.2.3 *If* A *is dense in a topological space* X*, and* A *is connected, then* X *is connected.*

Proof Let A be connected and dense in X. Let $f : X \to \{1, 2\}$ be continuous. Then $f \mid A$ is continuous and hence constant, say with value 1. By 2.6.2(d)

$$f[\overline{A}] \subseteq \overline{f[A]} = \overline{\{1\}} = \{1\}.$$

Since $\overline{A} = X$, this implies that f is constant. □

A corollary of 3.2.3 is that if $A \subseteq B \subseteq \overline{A} \subseteq X$ and A is connected, then B is connected— for the proof replace X in 3.2.3 by B with its relative topology.

3.2.4 *A subset* X *of* $\mathbb{R}$ *is connected if and only if* X *is an interval.*

Proof If X is empty or a singleton, it is both an interval and is connected. So suppose X has more than one point.

If X is an interval of $\mathbb{R}$, then X is contained in the closure of an open interval, and so is connected by the remark following 3.2.3.

Suppose, conversely, that X is connected. If X contains at most one point, then it is an interval. If X contains points a, b with $a < b$, let $x \in\ ]a, b[$ and suppose $x \notin X$. Then the set

$$X \cap\ ]\leftarrow, x] = X \cap\ ]\leftarrow, x[$$

is a non-empty proper subset of X both open and closed in X. This contradicts the assumption that X is connected. Hence, $x \in X$, and so X is an interval. □

A very direct way of proving that an interval X of $\mathbb{R}$ is connected is to show that any map $f : X \to \{0, 1\}$ is constant—an outline proof is as follows. Let $x, y \in X$ and suppose for example $x < y$ and $fx = 0$. Let $s = \sup\{z : x \leqslant z \leqslant y \text{ and } fz = 0\}$. It is easy to prove that $fs = 0$ and to derive a contradiction from the assumption $s < y$. Hence $fy = 0$, and so $fx = fy$.

EXAMPLES

1. The interval $\mathbb{I} = [0, 1]$ is connected. The function $t \mapsto e^{2\pi i t}$ is a continuous surjection $\mathbb{I} \to \mathbb{S}^1$. Hence $\mathbb{S}^1$ is connected.
2. Let $f : A \to \mathbb{R}$ be a map where A is an interval of $\mathbb{R}$. By the last two results, $\operatorname{Im} f$ is an interval. Hence f takes any value between two given values.
3. A space is *totally disconnected* if its only (non-empty) connected subsets consist of single points. Examples of totally disconnected spaces are discrete spaces, $\mathbb{Q}$, $\mathbb{R} \setminus \mathbb{Q}$ and also

$$\mathbb{L} = \{0\} \cup \{n^{-1} : n \text{ a positive integer}\}.$$

(The notation $\mathbb{L}$ will be standard for this space.) The proof that the last three space are totally disconnected is easy using 3.2.4.

3.2.5 *Let $(A_\lambda)_{\lambda \in \Lambda}$ be a family of connected subspaces of X, whose intersection is non-empty. Then*

$$A = \bigcup_{\lambda \in \Lambda} A_\lambda$$

is connected.

Proof Let $f : A \to \{1, 2\}$ be continuous. Then $f \mid A_\lambda$ is constant (since A_λ is connected) and so f is constant (since $\bigcap_{\lambda \in \Lambda} A_\lambda \neq \varnothing$). Therefore, A is connected. □

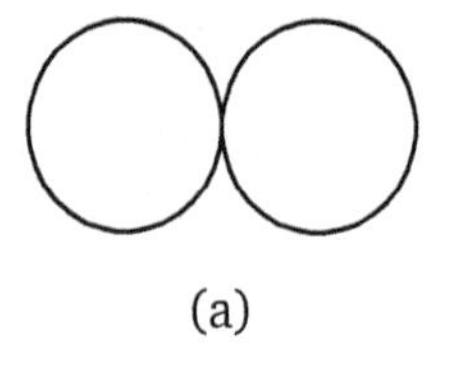

(a)

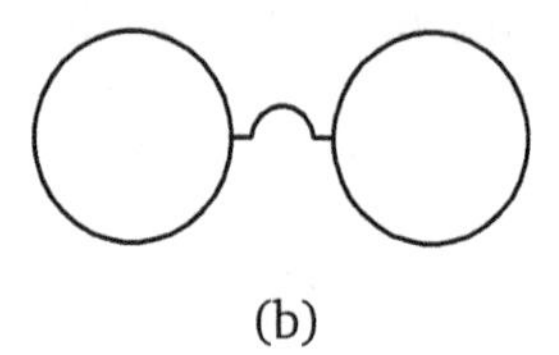

(b)

Fig. 3.1

For example, the space illustrated in (a) of Fig. 3.1 is connected, being the union of two spaces, homeomorphic to $\mathbb{S}^1$ and meeting in a single point. By two applications of 3.2.5, the space illustrated in (b) of Fig. 3.1 is connected.

3.2.6 *If X, Y are connected, then so also is $X \times Y$.*

Proof Let X, Y be connected and let $f : X \times Y \to \{1, 2\}$ be continuous. We prove that f is constant.

Let (x, y), $(x', y') \in X \times Y$. The space $\{x\} \times Y$ is homeomorphic to Y and hence is connected. Therefore, f is constant on $\{x\} \times Y$ and, in particular, $f(x, y) = f(x, y')$. Similarly, $f(x, y') = f(x', y')$. Therefore, $f(x, y) = f(x', y')$, and f is constant. □

EXAMPLES

4. Since $\mathbb{R}$ is connected, so also is $\mathbb{R}^n$.
5. $\mathbb{I}^n$, the n-fold product of $\mathbb{I} = [0, 1]$, is connected.
6. Let $X = \{(x, \sin \pi/x) : 0 \neq x \in \mathbb{R}\}$, let $J = \{0\} \times [-1, 1]$ and let $Y = X \cup J$ (cf. Fig. 2.5, p. 41). We prove that Y is connected.

Let $X_+ = \{(x, y) \in X : x > 0\}$, $X_- = \{(x, y) \in X : x < 0\}$. The function $\mathbb{R}^{>0} \to X_+$ sending $x \mapsto (x, \sin \pi/x)$ is continuous and surjective. Hence X_+ is connected. By 3.2.3, $X_+ \cup J = \overline{X_+}$ is connected. Similarly, $X_- \cup J$ is connected. By 3.2.5, $X \cup J$ is connected.

EXERCISES

1. Let X be a connected metric space with unbounded metric. Prove that every sphere $S(a, r)$ in X is non-empty. Is this true for X disconnected?
2. Let $X \subset \mathbb{R}^2$ be the subspace of points (x, y) such that either (i) x is irrational and $0 \leqslant y \leqslant 1$, or (ii) x is rational and $-1 \leqslant y < 0$. Prove that X is connected. Prove also that if $f : [0, 1] \to X$ is continuous, and $p_1 : X \to \mathbb{R}$ is the projection on the first coordinate, then $p_1 f$ is constant.
3. Let A be a connected subset of the connected space X. Let B be open and closed in $X \setminus A$. Prove that $A \cup B$ is connected. [Use Exercise 5 of Section 3.1 to extend a map $A \cup B \to \{1, 2\}$ over X.]
4. Prove 3.2.3, 3.2.5 by using directly condition 3.2.1(b) for connectedness.
5. Let A be a connected subset of the topological space X. Is $\operatorname{Int} A$ necessarily connected?
6. Let A be a subset of the metric space X. Prove that A is disconnected if and only if there are sets U, V open in X such that (i) $U \cap V = \varnothing$, (ii) $A \subset U \cup V$, and (iii) $A \cap U$, $A \cap V$ are non-empty.
7. Show that Exercise 6 is false for arbitrary spaces by considering the space $X = \{0, 1, 2\}$ in which $\varnothing$, X, $\{0\}$, $\{0, 1\}$, $\{0, 2\}$ are the only open sets.
8. Prove that a subset A of a topological space X is disconnected if and only if there are sets U, V open in X such that (i) $U \cap V \subseteq X \setminus A$, (ii) $A \subseteq U \cup V$, and (iii) $A \cap U, A \cap V$ are non-empty.
9. If $A, B \subseteq X, A$ is open, $\overline{A} \subseteq B$, and B and $\operatorname{Fr} A$ are connected, then $B \setminus A$ is connected.
10. Let $\leqslant$ be an order relation on X. This order is *without gaps* if $]x, y[$ is non-empty for each x, y in X such that $x < y$. Prove that X with its order topology is connected if and only if the order is complete and without gaps.

3.3 Components and locally connected spaces

Let X be a topological space and let $x \in X$. The *component of* x *in* X is $C(x)$, the union of all connected sets containing x. By 3.2.5, $C(x)$ is connected; therefore, $C(x)$ is the largest connected set containing x. But $\overline{C(x)}$ is also connected [3.2.3]. Therefore, $C(x) = \overline{C(x)}$. This proves:

3.3.1 *The component of* x *in* X *is a closed subset of* X.

A component need not be open. For example, in the space $\mathbb{L}$ the component of 0 is $\{0\}$, which is closed but not open in $\mathbb{L}$. Again, the components of points of $\mathbb{Q}$ are not open in $\mathbb{Q}$. An obvious question is therefore: under what conditions are the components always open?

Definition A space X is *locally connected at a point* x in X if the connected neighbourhoods of x form a base for the neighbourhoods at x. (This is sometimes expressed as: x has a base of connected neighbourhoods.) The space X is *locally connected* if it is locally connected at each x in X. Thus X is locally connected if, for each x in X, each neighbourhood of x contains a connected neighbourhood of x.

One more definition: if $A \subseteq X$, the *components of* A are the components of the points of the subspace A. So the components of A are subsets of A, except that the empty set $\varnothing$ has no components.

3.3.2 X *is locally connected if and only if the components of each open set of* X *are open sets of* X.

Proof Suppose X is locally connected, V is open in X, C is a component of V, and $x \in C$. Since V is a neighbourhood of x, and X is locally connected, there is a connected neighbourhood U of x such that $U \subseteq V$. Therefore, $U \subseteq C$, and C is a neighbourhood of x. Therefore, C is open.

For the converse, we start with a neighbourhood V of x which we may suppose to be open (otherwise we replace V by $\operatorname{Int} V$). The component of V which contains x is open in X (by assumption), and so is a connected neighbourhood of x contained in V. □

A special case of 3.3.2 is that, if X is locally connected, then each component of X is open.

EXAMPLES

1. The following spaces are not locally connected: the rationals, $\mathbb{Q}$; the irrationals, $\mathbb{R} \setminus \mathbb{Q}$; and $\mathbb{L} = \{0\} \cup \{1/n : n \in \mathbb{N}, n \neq 0\}$.
2. A connected set need not be locally connected: the space $Y = X \cup J$ of Example 6, p. 71, is not locally connected, since points of J have no 'small' connected neighbourhoods [cf. Fig. 3.2].

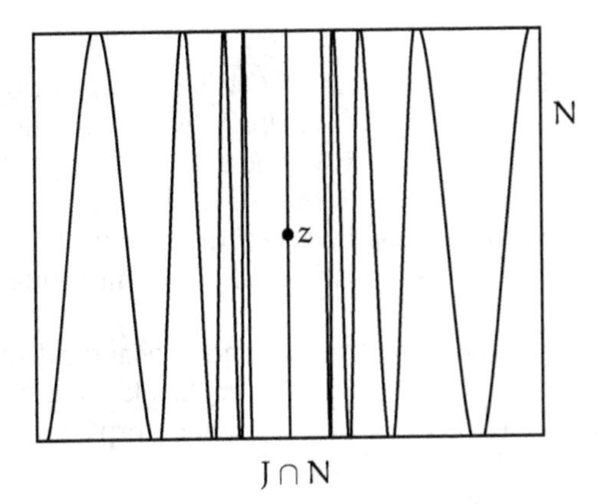

Fig. 3.2

In fact we leave as an exercise for the reader the detailed proof of the following statement: if $z \in J$, there is in Y a neighbourhood N of z such that the component of z in N is $N \cap J$.

3. Let X consist of the line segment joining $(1,1)$ in $\mathbb{R}^2$ to the points of $\mathbb{L} \times \{0\}$ [Fig. 3.3]. Then X is connected, but not locally connected, since $(0,0)$ has no 'small' connected neighbourhoods. (Here also, we leave a detailed proof to the reader.) Is $X \setminus \{(1,1)\}$ connected?

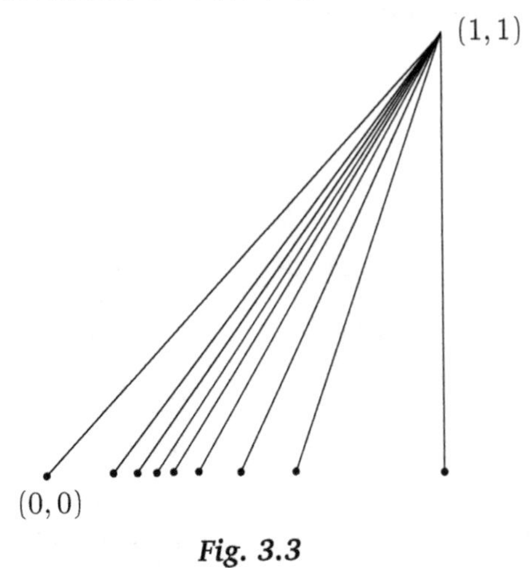

Fig. 3.3

4. A discrete space is locally connected and also totally disconnected.

Cut points

Let $f : X \to Y$ be a homeomorphism. If C is a component of a point x of X, then $f[C]$ is a component of $f(x)$ in Y: this follows from the fact that $A \subseteq X$ is connected if and only if $f[A]$ is connected. Consequently, f induces a bijection of the components of X to the components of Y. So the number of components of X is a topological invariant of X. However, this is not a very subtle invariant, since it fails to distinguish different connected spaces.

Definition Let X be connected space, and k a natural number or $\mathbb{N}$. A point x in X is a *cut point of order* k if $X \setminus \{x\}$ has k components.

Let X be connected, and $f : X \to Y$ a homeomorphism. A point x in X is a cut point of order k in X if and only if $f(x)$ is a cut point of order k in Y. Therefore, the number of cut points of order k is a topological invariant of X.

EXAMPLES

5. The closed interval $[0, 1]$ has two cut points of order 1; the half-open interval $[0, 1[$ has one cut point of order 1; the open interval $]0, 1[$ has no cut points of order 1. Therefore, no two of the spaces $[0, 1]$, $[0, 1[$ and $]0, 1[$ are homeomorphic.

6. The following 1-dimensional‡ spaces can be distinguished by the numbers of cut points of various orders:

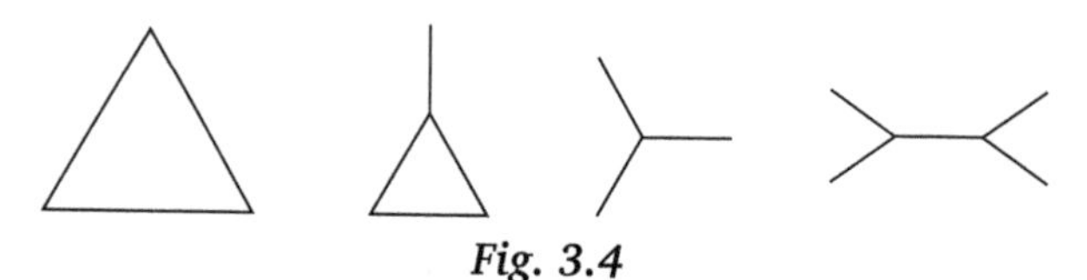

Fig. 3.4

7. These methods however fail to distinguish between the following spaces:

Fig. 3.5

However, if we remove two points from the first space, two components are left, while with the second space we can get one, two, or three components by removing different pairs of points. Hence, the two spaces are not homeomorphic.

8. The space $\mathbb{R}^n$ has cut points of order 1 only if $n > 1$. That is, if $a \in \mathbb{R}^n$, then $\mathbb{R}^n \setminus \{a\}$ is connected if $n > 1$. For let $x, y \in \mathbb{R}^n \setminus \{a\}$, and let $z \in \mathbb{R}^n \setminus \{a\}$

‡The term 1-dimensional is to be understood at this stage only intuitively.

be such that the lines L, M joining z to x, y respectively do not pass through a (this is possible since $n > 1$). Then $L \cup M$ is a connected set containing x and y, so that x and y belongs to the same component of $\mathbb{R}^n \setminus \{a\}$.

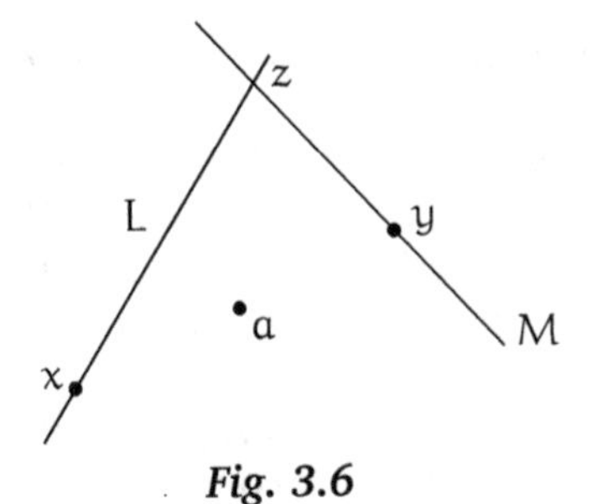

Fig. 3.6

This proves that $\mathbb{R}^n$ is not homeomorphic to $\mathbb{R}^1$ if $n \neq 1$. It is much more difficult to prove the *Invariance of Dimension:* if $\mathbb{R}^m$ is homeomorphic to $\mathbb{R}^n$, then $m = n$. All the present proofs of this theorem, as of the Invariance of Domain, use techniques of homology theory or of subdivisions of simplicial complexes.

Actually, a stronger result than the Invariance of Dimension is true: if $f : \mathbb{R}^m \to \mathbb{R}^n$ is a continuous bijection, then $m = n$ and f is a homeomorphism.

9. Another problem is to distinguish between surfaces, for example the 2-sphere $\mathbb{S}^2$ and the torus $T^2 = \mathbb{S}^1 \times \mathbb{S}^1$. This latter space is homeomorphic to the anchor ring [Fig. 3.7(ii)].

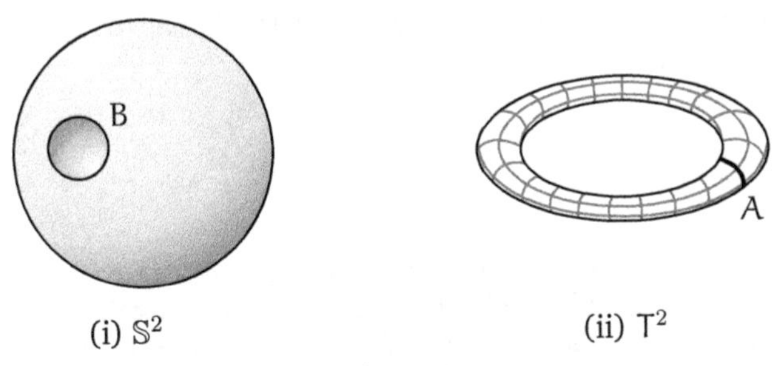

(i) $\mathbb{S}^2$ (ii) T^2

Fig. 3.7

Since removing points is a useful method for distinguishing 1-dimensional spaces, it seems reasonable that to distinguish 2-dimensional spaces we need to remove 1-dimensional subspaces (supposing also that we know what the words 1-dimensional and 2-dimensional mean). For example, let A be the meridian circle on T^2; then $T^2 \setminus A$ has only one component. On the other hand, it seems likely that if B is any subspace of $\mathbb{S}^2$ homeomorphic to $\mathbb{S}^1$, then $\mathbb{S}^2 \setminus B$ has two components. This is equivalent to the *Jordan Curve Theorem*: if B is a subspace of $\mathbb{R}^2$ homeomorphic to $\mathbb{S}^1$ (i.e., B is a simple closed curve in $\mathbb{R}^2$) then $\mathbb{R}^2 \setminus B$ has two components. We prove this in chapter 9.

As we shall see in chapter 9, $\mathbb{S}^2$ and T^2 can be distinguished by their fundamental groups.

10. If two spaces are homeomorphic, they must have the same 'local' properties. Let X be the graph of $x \mapsto \sin(\pi/x)$, and let $Y = X \cup \{(0,0)\}$. Then Y is connected, and any point of Y is a cut point of order 2. But Y is not homeomorphic to $\mathbb{R}$, since $\mathbb{R}$ is locally connected, and Y is not locally connected at $(0,0)$.

11. We can also define *local cut points*. A point x in X is a local cut point of order k if each neighbourhood V of x contains a connected neighbourhood U of x such that $U \setminus \{x\}$ has k components. If X is homeomorphic to Y, then X and Y must have the same number of local cut points of order k for each $k = 1, 2, \ldots$. The spaces of Fig. 3.8 are distinguished by the fact that one has a local cut point of order 4, and the other does not.

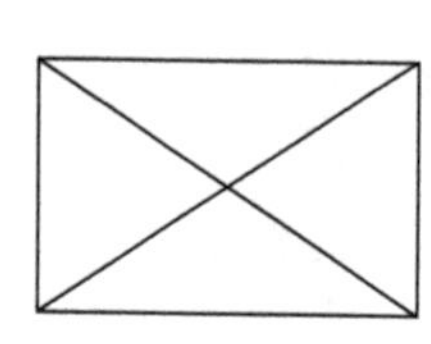
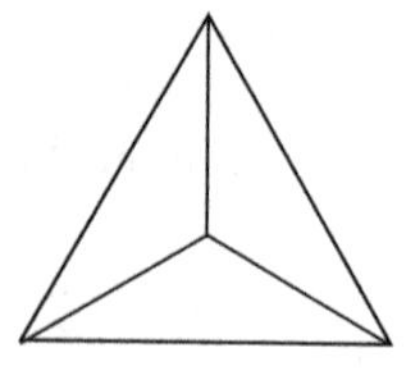

Fig. 3.8

By the use of homology theory these methods can be generalised to higher dimensional spaces.

EXERCISES

1. Prove that if X has a finite number of components, then each component is open.
2. Prove that the space X of Exercise 2 of Section 3.2 is not locally connected.
3. Decide whether or not the following 1-dimensional spaces are homeomorphic

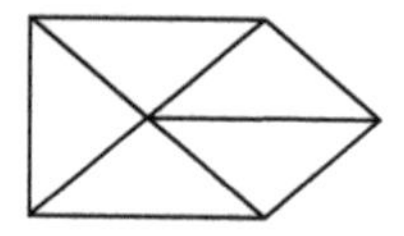
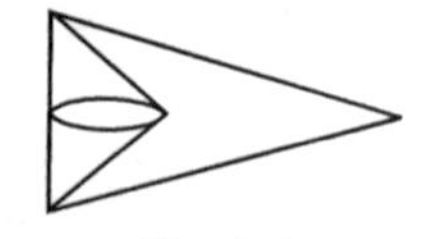
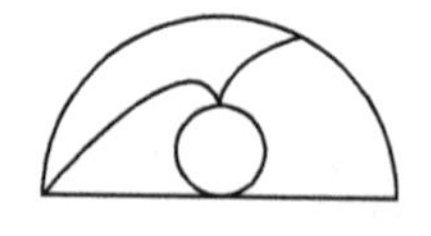

Fig. 3.9

4. Construct a locally connected subspace X of $\mathbb{R}^2$ in which for each $r > 0$ there is an x in X such that $B(x, r)$ (the open ball in X) is not connected.
5. Prove that $\mathbb{R}^n$ is locally connected.

6. Let $X = (\mathbb{L} \times \mathbb{I}) \setminus (\{0\} \times]0,1[)$. Prove that the components of $(0,0)$ and $(0,1)$ in X are single points. Let $f : X \to \{1,2\}$ be continuous. Prove that $f(0,0) = f(0,1)$.
7. Let X be the subspace of $\mathbb{R}^2$

$$\{0\} \cup \{(x, x \sin \pi/x) : x \neq 0\}$$

with the Euclidean metric. Prove that if $r > 0$ is sufficiently small then $B(0,r)$ (the open ball in X) is not connected.
*8. Let A be a non-empty subset of a metric space X. We say A is *bounded* if $\sup\{d(x,y) : x, y \in A\}$ exists, and then this number is called the *diameter* of A. Let $x, y \in X$. If there is a connected set containing x, y and of diameter < 1, let $\sigma(x,y)$ be the infinum of the diameters of such sets. If no such set exists, let $\sigma(x,y) = 1$. Prove that $(x,y) \mapsto \sigma(x,y)$ is a metric on X.

Let $L \subseteq X$ be the subset of X of points at which X is locally connected. Prove that σ induces the discrete topology on $X \setminus L$ (if $X \setminus L \neq \varnothing$) and that σ induces the same topology on L as does d.

Prove that in L, each open ball for the metric σ is connected if of radius < 1.
*9. Let X be connected and locally connected, and $f : X \to Y$ continuous. Prove that $\operatorname{Im} f$ is locally connected if f maps closed sets of X to closed sets of Y, but not in general.
*10. Let C be a component of the open set U of the locally connected space X. Prove that $\operatorname{Fr} C \subseteq \operatorname{Fr} U \subseteq X \setminus U$.

3.4 Path-connectedness

In this section, we discuss a type of connectedness which is stronger (in the precise sense given by 3.4.4 below) than that of section 3.2, and which is to some extent more intuitive. For 'nice' spaces, for example, the cell-complexes of chapter 4, the two notions of connectedness are equivalent.

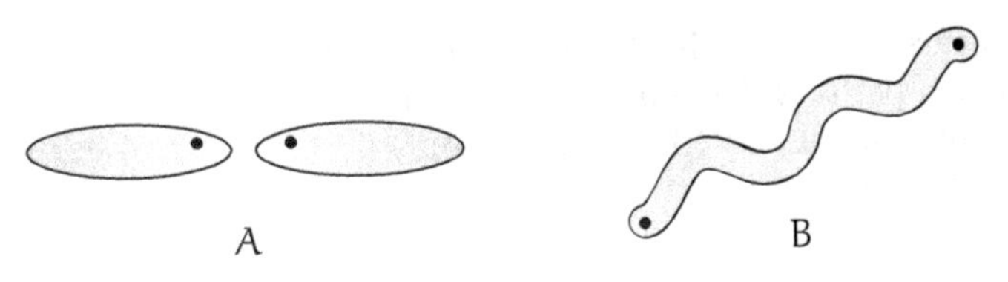

Fig. 3.10

Consider the spaces A and B of Fig. 3.10. It is intuitively clear that any two points of B (such as those shown by dots) can be joined by a continuous curve lying wholly in B. But this is false for the space A. The best general expression of these ideas is in terms of paths and path-connectedness.

Let X be a topological space and let $r \in \mathbb{R}^{\geqslant 0}$.

Definition A *path in* X *of length* r is a continuous function $a : [0, r] \to X$. We write $|a|$ for r. Then $a(0)$, the *source* of a is written $\sigma(a)$, and $a(r)$ the *target* of a is written $\tau(a)$, and we say a *joins* $\sigma(a)$ to $\tau(a)$. We call $\sigma(a)$ and $\tau(a)$ the *end points of* a.

A point x in X determines a unique constant path of length r with value x. If $r = 0$, this path is called the *zero path at* x.

It is important to note that a path in X is not just a set of points, but is a function. For example, the two paths $[0, 1] \to \mathbb{R}$ given by $t \mapsto t$ and $t \mapsto t^2$ are distinct paths in $\mathbb{R}$ joining 0 to 1. Our illustrative figures should then show the graph of a path—but it is usually more convenient to illustrate the image of the path.

We now consider two simple operations on paths. The *reverse* of a path $a : [0, r] \to X$ is the path

$$\begin{aligned} -a : [0, r] &\to X \\ t &\mapsto a(r - t). \end{aligned}$$

Thus $|-a| = |a|$ and $-a$ joins $\tau(a)$ to $\sigma(a)$.

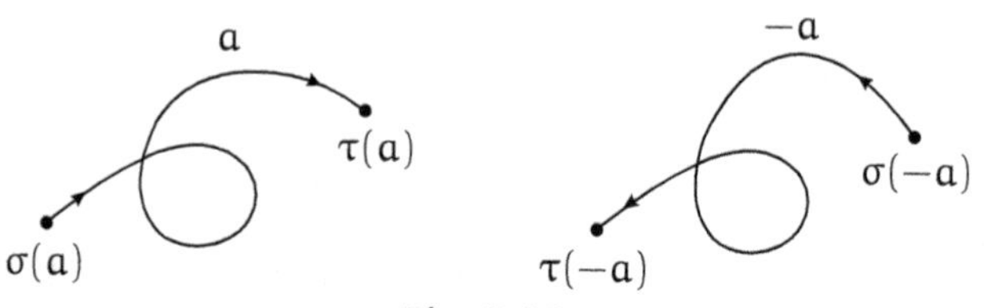

Fig. 3.11

The reverse of a path is always defined. On the other hand, the *sum* $b + a$ of two paths is defined if and only if the final point of a coincides with the initial point of b (that is, if and only if $\tau(a) = \sigma(b)$). In such case, $b + a$ is the path

$$\begin{aligned} b + a : [0, |b| + |a|] &\to X \\ t &\mapsto \begin{cases} a(t), & 0 \leqslant t \leqslant |a| \\ b(t - |a|), & |a| \leqslant t \leqslant |a| + |b|. \end{cases} \end{aligned}$$

Clearly the condition $\tau(a) = \sigma(b)$ is essential for $b + a$ to be defined; and with this condition, $b + a$ is continuous by the gluing rule [2.5.12].

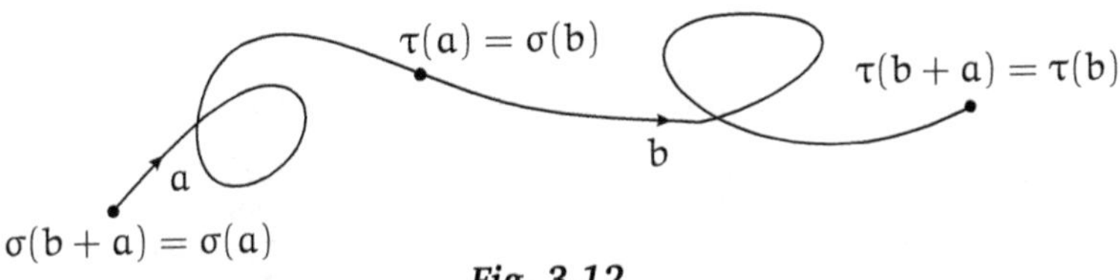

Fig. 3.12

We use the additive notation because of its convenience in dealing with differences ($b - c$ is a good abbreviation of $b + (-c)$). Our convention that $b + a$ means first a then b [Fig. 3.12] is related to the convention that the composition gf of functions means first apply f, then apply g.

The reader should be warned that addition of paths is not commutative: if $b + a$ is defined, then $a + b$ need not be defined. And even if both $b + a$ and $a + b$ are defined, they are in general unequal.

Definition A topological space X is *path-connected* if for any x, y in X, there is a path in X joining x to y.

EXAMPLE

1. We recall that a subset A of a normed vector space V is *convex* if, for any x, y in A, the *line segment*

$$[x, y] = \{(1 - t)x + ty : 0 \leqslant t \leqslant 1\}$$

is contained in A. A convex set A is path-connected, since if $x, y \in A$ then $t \mapsto (1-t)x + ty$ is a path in A from x to y of length 1. Examples of convex sets are $B(x, r)$, $E(x, r)$. For example, if $y, z \in B(x, r)$ and $0 \leqslant t \leqslant 1$, then

$$\begin{aligned} \|(1 - t)y + tz - x\| &\leqslant \|(1 - t)(y - x)\| + \|t(z - x)\| \\ &< (1 - t)r + tr = r \end{aligned}$$

whence $(1 - t)y + tz \in B(x, r)$. Again, any interval of $\mathbb{R}$ is convex.

Other examples of path-connected sets may be constructed from the following results.

3.4.1 *Let $f : X \to Y$ be continuous and surjective. If X is path-connected, then so also is Y.*

Proof Let $y, y' \in Y$; then there are points x, x' in X such that $f(x) = y$, $f(x') = y'$. Since X is path-connected, there is a path a joining x to x'. Then fa joins y to y'. □

3.4.1 *(Corollary 1) Let X be homeomorphic to Y. Then X is path-connected if and only if Y is path-connected.*

3.4.2 *Let $(A_\lambda)_{\lambda \in \Lambda}$ be a family of path-connected subspaces of X such that $\bigcap_{\lambda \in \Lambda} A_\lambda$ is non-empty. Then $A = \bigcup_{\lambda \in \Lambda} A_\lambda$ is path-connected.*

Proof Let $x, y \in A$, $z \in \bigcap_{\lambda \in \Lambda} A_\lambda$, and suppose $x \in A_\lambda$, $y \in A_\mu$. Since A_λ is path-connected, there is a path a joining x to z. Since A_μ is path-connected, there is a path b joining z to y. Then $b + a$ is a path joining x to y. □

3.4.3 *If X, Y are path-connected, then $X \times Y$ is path-connected.*

Proof Let $(x, y), (x', y') \in X \times Y$. Let a be a path in X joining x to x', b be a path of Y joining y to y'. Then $t \mapsto (a(t), y)$ joins (x, y) to (x', y), and $t \mapsto (x', b(t))$ joins (x', y) to (x', y'). The sum of these paths joins (x, y) to (x', y'). □

We can use 3.4.2 to define path-components of X: if $x \in X$, then the *path-component of* x is the union of all path-connected subsets P of X such that P contains x. This union will be a path-connected by 3.4.2, and will contain x. Hence it is the largest path-connected subset of X containing x.

A path-component need not be closed. To show this, we consider a little the relationship between the two kinds of connectedness.

3.4.4 *If X is path-connected, then X is connected.*

Proof Let $x, y \in X$, and let a be a path joining x to y. Then Im a is by 3.2.2 a connected set containing x and y. Hence x and y belong to the same component of X. Thus X has only one component, and must be connected. □

EXAMPLE

2. Let $Y = X \cup J$ be the connected space of Example 2, p. 72 and Example 6, p. 71. Then Y is not path-connected.

Proof Let $g : Y \to \{0, 1\}$ be the function which sends points of J to 0 and points of X to 1. Of course, g is not continuous, but we prove that for any path $f : [0, r] \to Y$, the composite gf is continuous.

Let $x \in [0, r]$. If $fx \in X$, then fx has a neighbourhood N which does not meet J; hence there is a neighbourhood M of x such that $f[M] \subset N$, whence $gf[M] = \{1\}$; continuity of gf at x follows easily.

On the other hand, suppose $fx \in J$. As stated on p. 72, there is a neighbourhood (in Y) of fx such that the component of fx in N is $N \cap J$. Therefore, there is a connected neighbourhood M of x such that $f[M] \subseteq N \cap J$. Hence, $gf[M] = \{0\}$, and continuity of gf at x follows easily.

We have now proved that gf is continuous. Since $[0, r]$ is connected it follows that gf is constant. So $\operatorname{Im} f$ is contained either in X or in J, and Y is not path-connected. □

The philosophy of local path-connectedness is different from that of local connectedness. The definition of the former concept that first suggests itself is that a space X is locally path-connected if each point x in X has a base of path-connected neighbourhoods. However, because of extensions of this property to higher-dimensional kinds of connectedness, we take a different definition which, it turns out, is equivalent to the above.

Definition A space X is *locally path-connected* if, for each point x in X, any neighbourhood of U of x contains a neighbourhood V of x such that any two points of V can be joined by a path in U.

3.4.5 *A space X is locally path-connected ⇔ each point of X has a base of open path-connected neighbourhoods.*

Proof The implication ⇐ is trivial, and so we prove the implication ⇒. Let $x \in X$ and let U be an open neighbourhood of x. Let U' be the path-component of U containing x.

Let $y \in U'$. Then U is a neighbourhood of y, and so U contains a neighbourhood V of y such that any two points of V can be joined by a path in U. This implies that V is contained in U'. Hence U' is a neighbourhood of y. Therefore (since $x \in U'$) U' is an open, path-connected neighbourhood of x. □

EXERCISES

1. In the following, X is a subspace of $\mathbb{R}^2$ and x_0, x_1 are points of X. Write down, if possible, explicit paths in X joining x_0 to x_1.

(i) $X = \{(x, y) \in \mathbb{R}^2 : |x| \geqslant 1 \text{ or } |y| \geqslant 1\}$, $x_0 = (-2, 0)$, $x_1 = (2, 0)$.
(ii) $X = \{(x, y) \in \mathbb{R}^2 : x + y \neq 1\}$, $x_0 = (4, -5)$, $x_1 = (-6, 8)$.
(iii) $X = \{(x, y) \in \mathbb{R}^2 : [x] = [y]\}$, $x_0 = (-\frac{1}{2}, -1)$, $x_1 = (1, \frac{3}{2})$.
(iv) $X = \{(x, y) \in \mathbb{R}^2 : [x] + [y] = 1\}$, $x_0 = (\frac{3}{2}, 0)$, $x_1 = (0, \frac{3}{2})$.

2. Let V be a normed vector space over $\mathbb{R}$. A 'bent line' in V is the union of a finite number of line segments $[u_i, u_{i+1}]$, $i = 1, \ldots, n-1$, and such a bent line is said to *join* u_1 to u_n. A subset A of V of 'polygonally connected' if, for any two points u, v of A, there is a bent line joining u to v and lying wholly in A.

Prove that any open, connected subset of V is polygonally connected. Give examples of subsets of $\mathbb{R}^2$ which are path-connected but not polygonally connected.

3. Let $X = \{a, b, c\}$ with the topology whose open sets are $\varnothing, \{c\}, \{a, c\}, \{b, c\}, X$. Prove that X is path-connected.

4. Prove that X is locally path-connected if and only if the path-components of each open set of X are open. Give an example of a space which is path-connected but not locally path-connected.
5. Prove that two points x, y in X lie in the same path-component if and only if they can be joined by a path of length 1.
6. Let V be a normed vector space over $\mathbb{R}$ of dimension > 1. Prove that $S(V)$ is path-connected.
7. Prove that if X is a countable subset of $\mathbb{R}^n$ $(n > 1)$, then $\mathbb{R}^n \setminus X$ is path-connected.
8. Let X_1, X_2 be subsets of X such that $X = \operatorname{Int} X_1 \cup \operatorname{Int} X_2$. Prove that, if X is path-connected, then each path-component of X_1 meets X_2.

3.5 Compactness

The reader will certainly be aware of the importance for mathematics of the distinction between finite and infinite sets. As examples, consider the statements (a) the sum of elements of a set of A of real numbers is well-defined, (b) a set A of real numbers has a least element, (c) the intersection of the elements of a set A of open sets is open. Each of these is true if A is finite but may be false if A is infinite. This wide range of techniques applicable to finite sets but not infinite ones is the reason for the importance of the notion of a compact space in topology.

In order to define compactness we need some preliminary definitions.

Let X be a topological space. A *cover* of X is a set $\mathcal{A}$ of sets such that the union of the elements of $\mathcal{A}$ contains X. A *subcover* of $\mathcal{A}$ is a subset $\mathcal{B}$ of $\mathcal{A}$ such that $\mathcal{B}$ covers X. A cover $\mathcal{A}$ of X is *open* if each set of $\mathcal{A}$ is open in X.

EXAMPLES

1. Let $k \in \mathbb{R}^{>0}$, and let $\mathcal{A}$ be the set of intervals $]x-k, x+k[$ for each $x \in \mathbb{R}$. Then $\mathcal{A}$ is an open cover of $\mathbb{R}$. Similarly, in any metric space X, the set of all open balls $B(x, k)$, $x \in X$, is an open cover of X.
2. Let X be a metric space and let $x \in X$. The set of open balls $B(x, n)$ for all positive n in $\mathbb{N}$ is an open cover of X.
3. The set of intervals $]1/n, 1]$ for n a positive integer, is an open cover of $]0, 1]$.
4. If $\mathcal{A}$ is an open cover of Y, and $f : X \to Y$ is continuous, then the set of $f^{-1}[A]$ for all A in $\mathcal{A}$ is an open cover of X.
5. For any topological space X, the set $\{X\}$ is an open cover of X as is $\{X, \varnothing\}$.

Definition A topological space X is *compact* if every open cover of X has a finite subcover.

This means of course, that to prove a space X is compact we have to start with any open cover $\mathcal{A}$ of X and construct a finite subcover of $\mathcal{A}$. To prove X non-compact, we have to produce an open cover $\mathcal{A}$ of X without finite subcover.

EXAMPLES

6. An infinite discrete space X is not compact, since the set of singletons $\{x\}$ for each x in X is an open cover of X without finite subcover.
7. Any finite space X is compact, since any open cover of X is a finite set.
8. The interval $]0, 1]$ is not compact—the open cover of Example 3 has no finite subcover.
9. If X is a metric space with unbounded metric, then X is not compact. To prove this, let $x \in X$ and consider the open cover of Example 2. This cover has a finite subcover if and only if $X = B(x, n)$ for some n, in which case X has bounded metric.
10. The previous example has a converse, namely, that if X is a non-compact metric space, then X admits an equivalent unbounded metric. The proof of this theorem is not as simple as that of its converse.

Compactness is a topological invariant. In fact, we have the stronger result.

3.5.1 *Let X be compact and* $f : X \to Y$ *continuous and surjective. Then Y is compact.*

Proof Let $\mathcal{A}$ be an open cover of Y. Since f is continuous the set $\mathcal{B}$ of sets $f^{-1}[A]$ for all A in $\mathcal{A}$ is an open cover of X. Since X is compact, $\mathcal{B}$ has a finite subcover $\mathcal{C}$. The set of A in $\mathcal{A}$ for which $f^{-1}[A]$ is in $\mathcal{C}$ is a finite subcover of $\mathcal{A}$. □

3.5.2 *Remark*

Let C be a subspace of the topological space X. Then we have covers of C by sets open in X, and by sets open in C. We distinguish these by calling them open covers of the *set* C, and of the *space* C, respectively. An open cover of the space C clearly consists of sets $A \cap C$ for A in an open cover of the set C. So the statements (a) every open cover of the space C has a finite subcover, and (b) every open cover of the set C has finite subcover, are equivalent, and either may be used as a criterion for compactness of C.

3.5.3 *Remark*

Let $\mathcal{A}$ be a cover of the space X. A *refinement* of $\mathcal{A}$ is a cover $\mathcal{B}$ of X such that each set of $\mathcal{B}$ is contained in some set of $\mathcal{A}$. Suppose $\mathcal{B}$ is an open cover which refines an open cover $\mathcal{A}$. Then, if $\mathcal{B}$ has a finite subcover, so also does $\mathcal{A}$. Thus, when trying to construct finite subcovers of an open cover $\mathcal{A}$ we may at will replace $\mathcal{A}$ by an open refinement.

The next theorem gives the simplest non-trivial example of a compact space. (The proof given here is due to R. M. F. Moss and G. Roberts.)

3.5.4 *The unit interval $\mathbb{I}$ is compact.*

Proof Let $\mathcal{A}$ be an open cover of $\mathbb{I}$. For each x in $\mathbb{I}$ we choose an interval U_x, open in $\mathbb{I}$, such that U_x contains x and is contained in some set of $\mathcal{A}$. The set $\mathcal{B}$ of these intervals U_x is an open cover of $\mathbb{I}$ which refines $\mathcal{A}$. By Remark 3.5.3, we may assume from that start that each element of $\mathcal{A}$ is an interval.

Let $f : \mathbb{I} \to \{0, 1\}$ be the function defined by $fx = 0$ if $[0, x]$ can be covered by a finite number of sets of $\mathcal{A}$, and $fx = 1$ otherwise. We shall prove that f is constant on each set of $\mathcal{A}$.

Let $U \in \mathcal{A}$, let $x \in U$ and suppose $fx = 0$. Then $[0, x]$ is covered by a finite subset $\mathcal{B}$ of $\mathcal{A}$ and so for any y in U, since U is an interval $[0, y]$ is covered by the finite set $\mathcal{B} \cup \{U\}$. Thus we have shown, as required, that f is 0 either on all or none of U.

It follows immediately that f is continuous; therefore, f is constant (since $\mathbb{I}$ is connected) and the image of f is $\{0\}$ (since $f0 = 0$). Hence $[0, 1]$ can be covered by a finite number of sets of $\mathcal{A}$. □

It is instructive to examine the failure of similar attempts to prove that the intervals $]0, 1]$ and $[0, 1[$ are compact (they are non-compact by Example 8 and a similar example for $[0, 1[$). The proof for $]0, 1]$ breaks down because we cannot prove that the unique value of f is 0. The proof for $[0, 1[$ breaks down because $f1$ is not defined.

3.5.5 *A closed subset of a compact space is compact.*

Proof Let C be a closed subset of the compact space X and, applying 3.5.2, let $\mathcal{A}$ be an open cover of the set C. Since $X \setminus C$ is open,

$$\mathcal{A}' = \mathcal{A} \cup \{X \setminus C\}$$

is an open cover of X. By compactness of X, $\mathcal{A}'$ has a finite subcover $\mathcal{B}$ say.

Here $\mathcal{B}$ is an open cover of X, and so an open cover of the set C. If $\mathcal{B}$ does not contain $X \setminus C$, then $\mathcal{B}$ is a finite subcover of $\mathcal{A}$. In any case, $\mathcal{B} \setminus \{X \setminus C\}$ is certainly a finite subcover of $\mathcal{A}$. □

The next theorem has a slightly complicated formulation—this is due to the fact that we wish to include in one theorem (and one proof) a number of highly important special cases.

3.5.6 *Let B, C be compact subsets of X, Y respectively, and let $\mathcal{W}$ be a cover of $B \times C$ by sets open in $X \times Y$. Then B, C have open neighbourhoods U, V respectively such that $U \times V$ is covered by a finite number of sets of $\mathcal{W}$.*

Proof The proof is carried out in two steps, first when B has a single point b, and next for B arbitrary.

Step 1 – $B = \{b\}$

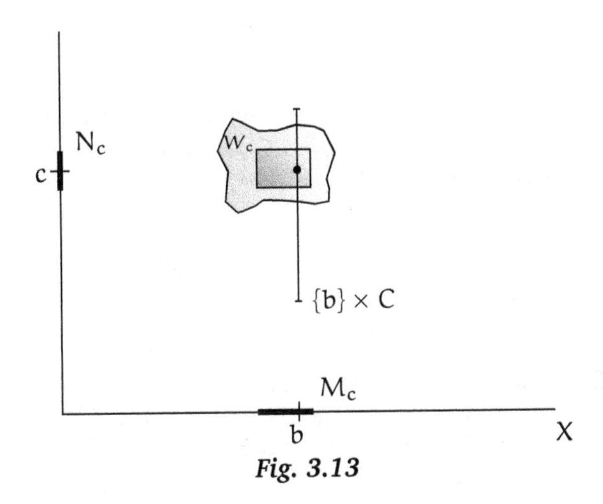

Fig. 3.13

For each c in C there are open neighbourhoods of M_c of b (in X), N_c of c (in Y) such that $M_c \times N_c$ is contained in some set W_c of $\mathcal{W}$. The set $\{N_c : c \in C\}$ is an open cover of the set C which, by compactness, has a finite subcover $\{N_f : f \in F\}$. Let

$$U = \bigcap_{f \in F} M_f, \qquad V = \bigcup_{f \in F} N_f.$$

Since F is finite, U is an open neighbourhood of b. Clearly,

$$\{b\} \times C \subseteq U \times V \subseteq \bigcup_{f \in F} W_f.$$

Step 2 – B *arbitrary*

By step 1, for each b in B there are open neighbourhoods U_b of b (in X), V_b of C (in Y), such that $U_b \times V_b$ is contained in a finite union, say $\bigcup\{W_f : f \in F_b\}$, of sets of $\mathcal{W}$. The set $\{U_b : b \in B\}$ is an open cover of B which, by compactness, has a finite subcover $\{U_g : g \in G\}$. Let

$$U = \bigcup_{g \in G} U_g, \qquad V = \bigcap_{g \in G} V_g.$$

Then U, V are open neighbourhoods of B, C respectively and $U \times V$ is contained in the union of the finite number of sets $W_f, f \in F_g, g \in G$. □

3.5.6 *(Corollary 1) The product of compact spaces is compact.*

Proof This follows from 3.5.6 by taking $B = X$, $C = Y$ (and so $U = X$, $V = Y$). □

3.5.6 *(Corollary 2) Let B, C be compact subsets of X, Y respectively, and let W be an open subset of $X \times Y$ containing $B \times C$. Then B, C have open neighbourhoods U, V respectively such that $U \times V \subseteq W$.*

Proof This is the case of 3.5.6 when $\mathcal{W} = \{W\}$. □

3.5.6 *(Corollary 3) If B, C are disjoint compact subsets of the Hausdorff space X, then B, C have disjoint open neighbourhoods.*

Proof In 3.5.6 (Corollary 2), let $X = Y$ and $W = (X \times X) \setminus \Delta[X]$ where Δ is the diagonal of $X \times X$; W is open since X is Hausdorff. Since B and C are disjoint, $B \times C$ is contained in W. By 3.5.6 (Corollary 2), B, C have open neighbourhoods U, V such that $U \times V \subseteq W$. Hence $U \cap V = \varnothing$. □

3.5.6 *(Corollary 4) A compact subset of a Hausdorff space is closed.*

Proof Let C be a compact subset of the Hausdorff space X, and let $x \in X \setminus C$. By 3.5.6 (Corollary 3), x and C have disjoint neighbourhoods. Hence $X \setminus C$ is open. □

As an application of these rules we prove

3.5.7 *A subset of Euclidean n-space $\mathbb{R}^n$ is compact if and only if it is closed and bounded.*

Let A be a closed, bounded subset of $\mathbb{R}^n$. Let J_r be the subspace of $\mathbb{R}^n$ of points $(x_1, \ldots, x_n)$ such that $|x_i| \leqslant r$—thus J_r is the product of the interval $[-r, r]$ with itself n times [cf. Exercise 8 of Section 2.4]. The interval $[-r, r]$ is homeomorphic to $[0, 1]$ (if $r > 0$) and so is compact. By 3.5.6 (Corollary 1), J_r is compact. Since A is bounded, it is contained in J_r for some r. Thus A is a closed subset of a compact space and so is compact.

The converse follows from Example 9 and 3.5.6 (Corollary 4). □

EXAMPLES

11. The Cantor set K [The Example of Section 1.3] is a closed, bounded subset of $\mathbb{R}$ and so is compact.

12. The subsets of $\mathbb{R}^n$ (with the Euclidean norm)

$$\mathbb{S}^{n-1} = \{x \in \mathbb{R}^n : \|x\| = 1\},$$
$$\mathbb{E}^n = \{x \in \mathbb{R}^n : \|x\| \leqslant 1\}$$

are both closed, bounded subsets of $\mathbb{R}^n$ and so are compact.
13. If X is indiscrete, then any subset A of X is compact, but A will be closed in X only if $A = \varnothing$ or $A = X$.
14. Let W be the subspace of $\mathbb{R}^2$ of points (x, y) such that $y > |x|$. Let $B = \{0\}$, $C = \,]0, 1]$. Then W is an open set of $\mathbb{R}^2$ containing $B \times C$ but, if U is a neighbourhood of 0, then W does not contain even $U \times C$.

A map $f : X \to Y$ of spaces is called *closed* if $f[C]$ is closed in Y for each closed set C of X. For example, a continuous bijection $f : X \to Y$ is closed if and only if f is a homeomorphism.

3.5.8 *Any map from a compact space to a Hausdorff space is closed.*

Proof Let $f : X \to Y$ be a map where X is compact and Y is Hausdorff. Let C be closed in X. Then C is compact [3.5.5] whence $f[C]$ is compact [3.5.1] and so $f[C]$ is closed [3.5.6 (Corollary 4)]. □

3.5.8 *(Corollary 1) A continuous bijection from a compact space to a Hausdorff space is a homeomorphism.*

Proof If f is a closed bijection, then f^{-1} is continuous. □

The following proposition is required later; the simple proof is left as an exercise.

3.5.9 *A topological space which is a finite union of compact spaces is itself compact.*

There is a characterisation of compactness by means of closed sets which is of great importance in some contexts but not, as it turns out, in this book. For this reason it has been left as an exercise [Exercise 10].

There is a generalisation of 3.5.6 which uses the product topology defined in section 5.7:

3.5.10 *(Tychonoff's theorem) The topological product of any family of compact spaces is compact.*

This theorem is of great importance in functional analysis and in a further study of some of the topics in this book. However the theorem is not essential to our present purposes, and so we refer the reader to [Kel55] or [Dug68] for a proof.

EXERCISES

1. Give an example of a space with two points in which not all compact sets are closed.
2. Prove that $\mathbb{R}$ with the Zariski topology is compact.
3. Prove that a discrete space is compact if and only if it is finite.
4. Use the result of sections 3.2 and 3.5 to prove that if $a, b \in \mathbb{R}$ $(a < b)$ and $f : [a, b] \to \mathbb{R}$ is continuous, then $\operatorname{Im} f$ is a closed, bounded interval.
5. Let X be a compact topological space, and $f : X \to \mathbb{R}$ a continuous function. Prove that there are elements a, b in X such that $f(a) = \inf(\operatorname{Im} F)$, $f(b) = \sup(\operatorname{Im} f)$. Deduce that if $f(x) > 0$ for all x in X, then there is a positive real number r such that $f(x) > r$ for all x in X.
6. Let $X = \mathbb{R} \times Y$ where Y is an indiscrete space with two elements a, b. Prove that in X the sets $A = [0, 1[\times\{a\} \cup [1, 2] \times \{b\}$, $B = [0, 1] \times \{a\} \cup]1, 2] \times \{b\}$ are both compact, but $A \cap B$ is non-compact.
7. Let $(C_i)_{i \in I}$ be a family of closed, compact subsets of X. Prove that $\bigcap_{i \in I} C_i$ is compact.
8. Let A, B be non-empty subsets of the metric space X. Let

$$\operatorname{dist}(A, B) = \inf\{(d(a, b) : a \in A, b \in B\}.$$

Prove that if A, B are disjoint and closed, and A is compact, then there is an element a of A such that

$$\operatorname{dist}(A, B) = \operatorname{dist}(a, B) > 0.$$

9. Prove that $\mathbb{I}^2$ with the television topology is compact, connected and Hausdorff.
10. Prove that the following conditions on a topological space X are equivalent. (i) X is compact. (ii) If $(C_i)_{i \in I}$ is a family of closed subsets of X such that $\bigcap_{i \in I} C_i = \emptyset$, then $\bigcap_{a \in A} C_a = \emptyset$ for some finite subset A of I. (iii) If $(C_i)_{i \in I}$ is a family of closed subsets of X such that $\bigcap_{a \in A} C_a \neq \emptyset$ for all finite subsets A of I, then $\bigcap_{i \in I} C_i \neq \emptyset$.
11. Let $F : \mathbb{R}^n \to V$ be a linear isomorphism, where $\mathbb{R}^n$ is Euclidean space and V is a normed vector space over $\mathbb{R}$. Prove that the function $h : \mathbb{S}^{n-1} \to \mathbb{R}$ which sends $x \mapsto \|f(x)\|$ is continuous, and that there is a positive real number δ such that $h(x) \geqslant \delta$ for all x in $\mathbb{S}^{n-1}$. Show that for all y in V, $|f^{-1}(y)| \leqslant \delta^{-1}\|y\|$. Finally, prove that f is a homeomorphism.
12. Prove that n-dimensional complex space $\mathbb{C}^n$ is linearly homeomorphic to $\mathbb{R}^{2n}$. Prove that, if V is a finite dimensional normed vector space over $\mathbb{R}$ or $\mathbb{C}$, then all norms on V are equivalent. [This theorem is proved in [Die60] by completeness methods. An advantage of such methods is that they also prove that any finite dimensional subspace of a normed vector space V is closed in V.]
13. Let X_n be the subset $\{n^{-1}\} \times [-n, n]$ of $\mathbb{R}^2$ and let $Y = \mathbb{R}^2 \setminus \bigcup_{n \geqslant 1} X_n$. Prove that Y is connected but not path-connected.

*14. Let $\varphi : X \times C \to \mathbb{R}$ be a map where C is compact. Prove that the function $x \mapsto \sup_{c \in C} \varphi(x, c)$ is a continuous function $X \to \mathbb{R}$.

3.6 Further properties of compactness

The basic results on compactness are given in the last section. The more technical results of this section will be used in later chapters, but the study of these results can be omitted till they are needed.

Locally compact spaces, normal spaces

Definition A topological space X is *locally compact* if each x in X has a base of compact neighbourhoods.

EXAMPLES
1. Euclidean n-space $\mathbb{R}^n$ is locally compact since, if $x \in \mathbb{R}^n$, then the closed balls $E(x, r)$ for $r > 0$ are compact and form a base for the neighbourhoods of x.
2. The space $\mathbb{Q}$ of rational numbers is not locally compact since a neighbourhood of 0 in $\mathbb{Q}$ cannot be closed in $\mathbb{R}$ and so cannot be compact.

In the literature, it is common to define a space X to be locally compact if each point of X has a compact neighbourhood. We have not adopted this definition for two reasons:
(i) It would be contrary to the general spirit of local properties. If P is a property of topological spaces, it is usual to say X is locally P if each point of X has a base of neighbourhoods with property P.
(ii) The property of locally compact spaces needed later is exactly the one we have taken for a definition.

For Hausdorff spaces the two definitions are equivalent. This is an easy consequence of the following result.

3.6.1 *A compact Hausdorff space is locally compact.*

Proof Let X be compact and Hausdorff, let $x \in X$ and let W be an open neighbourhood of x. We must find a compact neighbourhood of x contained in W.

Let $C = X \backslash W$. Then C is closed in X and so is compact. By 3.5.6 (Corollary 3), x and C have disjoint open neighbourhoods M, N say. The closure $\overline{M}$ of M is contained in $X \setminus N$ which is itself contained in W. Also $\overline{M}$ is compact (since it is closed in X) and is a neighbourhood of x. □

Definition A topological space X is *normal* if disjoint closed sets of X have disjoint neighbourhoods.

It is immediate from 3.5.5 and 3.5.6 (Corollary 3) that any compact, Hausdorff space is normal. We showed in section 2.9 that any metric space is normal.

Normal spaces have another property important in many parts of topology, for example in metrisation theorems and in the theory of ANRs. The following theorem will be found in many texts.

3.6.2 *(Tietze extension theorem) A space X is normal if and only if for any closed subspace C of X any map $f : C \to \mathbb{I}$ extends over X (i.e., is the restriction of a map $X \to \mathbb{I}$).*

The proof that the extension condition implies normality is easy since let $C = C_1 \cup C_2$ where C_1, C_2 are disjoint, non-empty closed subsets of X and let $f : C \to \mathbb{I}$ be 0 on C_1 and 1 on C_2. Let $g : X \to \mathbb{I}$ be an extension of f over X. Then, for each r in $]0, 1[$, the sets

$$g^{-1}[0, r[, \qquad g^{-1}]r, 1]$$

are disjoint open sets containing C_1, C_2 respectively; in fact for various r these sets form a kind of 'continuous family' of open sets between C_1 and C_2. One word of warning—it is not always possible to find g such that $C_1 = g^{-1}[0]$, $C_2 = g^{-1}[1]$; conditions for this will be mentioned in the Exercises.

Proper maps

Definition Let $f : X \to Y$ be a map of topological spaces. Then f is *proper* if, for all spaces Z,

$$f \times 1 : X \times Z \to Y \times Z$$

is a closed map.

By taking Z to consist of a single point, we see that a proper map is always closed. Similarly, to say that a constant map $X \to \{y\}$ is proper is equivalent to saying that for all Z the projection $X \times Z \to Z$ is closed. The result on proper maps that we shall need (in section 5.8) is the following and its corollary.

3.6.3 *If $f : X \to Y$ is a closed map such that $f^{-1}[y]$ is compact for each y in Y, then f is proper.*

Proof Let $h = f \times 1 : X \times Z \to Y \times Z$, let C be a closed subset of $X \times Z$, and let $D = h[C]$—we must prove that D is closed, i.e., that the set $D' = (Y \times Z) \setminus D$ is open.

Let $(y, z) \in D'$. Since the complement of $f[X] \times Z$ is open, we may assume $y \in f[X]$. Let $C' = (X \times Z) \setminus C$, so that C' is open. It is easily verified that

$$f^{-1}[y] \times \{z\} \subseteq C'$$

and so, by our assumptions and 3.5.6 (Corollary 2), there are open sets U, V such that

$$f^{-1}[y] \times \{z\} \subseteq U \times V \subseteq C'.$$

Let $U' = X \setminus U$, $V' = Z \setminus V$. Then $C \subseteq (U' \times Z) \cup (X \times V')$ and so

$$D = h[C] \subseteq (f[U'] \times Z) \cup (f[X] \times V') = Q \text{ say.}$$

Since f is a closed map, Q is a closed set. Therefore, $Q' = (Y \times Z) \setminus Q$ is an open set contained in D'. But $y \notin f[U']$, nor does $z \in V'$, so it follows that $(y, z) \in Q'$. Hence D' is open. □

The converse of 3.6.3 is true [see the Exercises], but will not be needed here.

3.6.3 *(Corollary 1) Any map from a compact space to a Hausdorff space is proper.*

Proof Let $f : X \to Y$ be a map where X is compact and Y is Hausdorff. Then f is closed by 3.5.8. If $y \in Y$, then $\{y\}$ is closed in Y; hence $f^{-1}[y]$ is closed in X and so $f^{-1}[y]$ is compact. □

Lebesgue covering lemma

Let X be a metric space and $\mathcal{A}$ an open cover of X. We consider the following question: is there a real number $r > 0$ such that the open cover $\mathcal{B}_r = \{B(x, r) : x \in X\}$ refines $\mathcal{A}$? Clearly the set of all r for which this is so is an interval L of $\mathbb{R}$, and L may be empty. If L is non-empty, then the real number $l = \sup L$ is called the *Lebesgue number* of the cover, $\mathcal{A}$ (we allow the rather boring case $l = \infty$, which, intuitively, means $\mathcal{A}$ has lots of large sets).

EXAMPLES

3. If $\mathcal{A} = \mathcal{B}_r$ then the Lebesgue number of $\mathcal{A}$ is r.
4. Let $X = \mathbb{R}$, and let $\mathcal{A}$ consist of the open intervals $]n, n+2[$ for each $n \in \mathbb{Z}$. Then the Lebesgue number of $\mathcal{A}$ is $\frac{1}{2}$.
5. Let $X = \;]0, 1[$ and let $\mathcal{A}$ consist of the open intervals $]n^{-1}, 1[$ for all positive integral n. Then $\mathcal{A}$ has no Lebesgue number.
6. Let $X = [0, 2] \setminus \{1\}$, and let $\mathcal{A}$ consist of the intervals $[0, 1[$ and $]1, 2]$. Then $\mathcal{A}$ has no Lebesgue number.

3.6.4 *(Lebesgue covering lemma) If X is a compact metric space, then any open cover of X has a Lebesgue number.*

Proof Let $\mathcal{A}$ be an open cover of X. Since X is compact, $\mathcal{A}$ has a finite subcover and any refinement of this is a refinement of $\mathcal{A}$. So we may assume $\mathcal{A}$ finite.

For each A in $\mathcal{A}$ and x in X let

$$f_A(x) = \text{dist}(x, X \setminus A),$$
$$f(x) = \max\{f_A(x) : A \in \mathcal{A}\}.$$

Each f_A is continuous and hence f is continuous.

If $x \in X$ then $x \in A$ for some A in $\mathcal{A}$ and for that A, $f_A(x) > 0$ (since $X \setminus A$ is closed). Hence $f(x) > 0$ for all x in X.

The set $f[X]$ is a compact and hence closed subset of $\mathbb{R}$. Therefore, $r = \inf f[X]$ belongs to $f[X]$, and so $r > 0$.

If $x \in X$ then $f(x) \geqslant r$ whence $f_A(x) \geqslant r$ for some A in $\mathcal{A}$ and so, by definition of f_A, $B(x, r) \subseteq A$. This proves that the cover $\mathcal{B}_r$ refines $\mathcal{A}$. □

We now illustrate the main type of application of 3.6.4. Let $a : [0, r] \to Z$ be a path in a topological space Z. A *subdivision* of a is a sequence $a_1, \ldots, a_n$ (for some n) of paths in Z such that

$$a = a_n + \cdots + a_1. \qquad (*)$$

Such a subdivision is usually denoted by the expression (*).

3.6.4 *(Corollary 1) Let $\mathcal{U}$ be an open cover of Z and $a : [0, r] \to Z$ a path in Z. Then there is a subdivision $a = a_n + \cdots + a_1$ such that for each $i = 1, \ldots, n$, $\text{Im}\, a_i$ is contained in some set of $\mathcal{U}$.*

Proof Let δ be the Lebesgue number of the covering $\{a^{-1}[U] : U \in \mathcal{U}\}$ of the compact metric space $[0, r]$. Let n be an integer such that $0 < r/n < \frac{1}{2}\delta$, and let a_{i+1} be the path $t \mapsto a(t + ir/n)$, $i = 0, \ldots, n-1$ of length r/n. Then clearly $a = a_n + \cdots + a_1$. Further, for each $i = 0, \ldots, n-1$, the interval $[ir/n, (i+1)r/n]$ is contained in some set $a^{-1}[U]$, $U \in \mathcal{U}$; hence

$$\text{Im}\, a_i = a[ir/n, (i+1)r/n] \subseteq U.$$

□

Compactifications

If K is a compact space, then removing points from K is liable to produce a non-compact space. The idea of *compactification* is the reverse procedure—given a space X, can points be added to X to produce a compact space? There are a variety of such compactifications. They will not be used elsewhere in this book, but their importance is such that a brief survey should be given of the ramifications.

The simplest compactification is the *Alexandroff one-point compactification* X^+ of X. Here $X^+ = X \cup \{\omega\}$, where ω is a point not in X, with the topology in which X is a subspace and the neighbourhoods of ω are the complements in X^+ of closed compact subsets of X. A precise definition is given in Exercise 6, and the principal properties are summarised in Exercises 6–17. From these properties it can be shown that the one-point compactification of $\mathbb{R}^n$ is homeomorphic to the n-sphere $\mathbb{S}^n$.

Another important compactification, but for which it is difficult to find a textbook exposition, is the *Freudenthal compactification* $X^\wedge$. The intuitive idea is that the open interval $]0,1[$ should be compactified by adding two *ends*, for example by adding 0 and 1 to give the closed interval $[0,1]$. The plane should be compactified by adding one end, as in the Alexandroff compactification. On the other hand, the following infinitely branched space is compactified with an infinite number of ends.

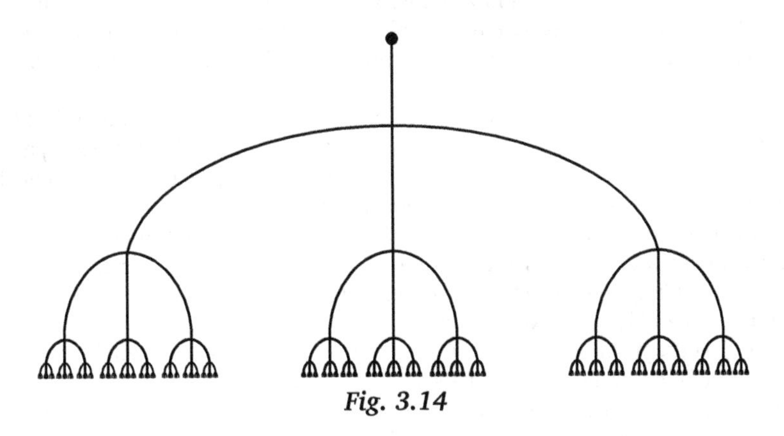

Fig. 3.14

The definition of the *ends* of a space involves the number of components remaining when compact subsets are removed. The Freudenthal compactification $X^\wedge$ is defined for a Hausdorff space X which has a base of open sets (chapter 5) with compact boundary. Then $X^\wedge \setminus X$ is a compact space containing X as a subspace and such that $X^\wedge \setminus X$ is zero-dimensional. For a good account in English, see [Hou74]. The number of ends of a covering space of a cell complex (see chapter 10) has important relations to group theory (see the articles [Hou74], [SW79] and the introductory article [Sch79]). The most striking recent result in the area is [Dun85].

A very large compactification of a space X is the Stone-Čech compactification βX. There is a map $i : X \to \beta X$ with the property that (i) βX is compact and Hausdorff, (ii) if $g : X \to Y$ is any map to a compact Hausdorff space Y, then there is a unique map $g' : \beta X \to Y$ such that $g'i = g$. It is this universal property which gives this compactification its importance, particularly in analysis. The map $i : X \to \beta X$ is an embedding if and only if X is *completely regular*, that is, X is T_0 and for each point x of X and closed subset A of X not containing x, there is a continuous function $f : X \to [0, 1]$ such that $f(x) = 0$ and $f[A] = \{1\}$. For more information, see [Wal74] and [Joh82]. The reader should be warned that β applied to even a simple space such as the positive integers $\mathbb{N}$ yields a horrendously complicated space which has been the subject of many papers.

For information on other compactifications, follow up the references in [IK87].

EXERCISES

1. Prove that the product of two locally compact spaces is locally compact.
2. Prove that a closed subspace of a locally compact space is locally compact, but that an arbitrary subspace of a locally compact space need not be locally compact.
3. Let A be a compact subspace of a locally compact space, and let W be a neighbourhood of A. Prove that there is a compact neighbourhood M of A such that $M \subseteq W$.
4. A space X is a *k-space* if a subset A of X is closed in X if and only if $A \cap C$ is closed in C for every compact subset C of X. Prove that a locally compact space if a k-space, as is any sequential space [Exercise 5 of Section 2.10]. Prove that the space X of Exercise 4 of Section 2.9 is not a k-space.
5. Show that the following statement is true if f is open but not in general: if X is locally compact and $f : X \to Y$ is a continuous surjection, then Y is locally compact.
*6. *The Alexandroff 1-point compactification.* Let X be a topological space, let ω be a point not in X and let $X^* = X \cup \{\omega\}$. Define a neighbourhood topology on X^* by (i) if $x \in X$ and M is a neighbourhood in X of x, then M and $M \cup \{\omega\}$ are neighbourhoods in X^* of x; (ii) if A is a closed compact subset of X, then $X^* \setminus A$ is a neighbourhood of ω. Prove (a) X^* is compact and X is a subspace of X^*, (b) X is locally compact and Hausdorff $\Leftrightarrow$ X^* is Hausdorff, (c) if X^* is Hausdorff and $i_1 : X \to X_1^*$ is any homeomorphism into a compact Hausdorff space X_1^* such that the image of i_1 is the complement of a single point of X_1^*, then there is a unique homeomorphism $g : X^* \to X_1^*$ such that $gi = i_1$, where $i : X \to X^*$ is the inclusion. Prove also that, if X is compact, then $X^* = X \sqcup \{\omega\}$. [The point ω is called the *point at infinity* of X^*.]
7. Given an open cover $\mathcal{U}$ of X, prove that the following prescription defines a topological space X'. (i) $X' = X \cup \{\omega\}$ where ω is a point not belonging to X, (ii) if $x \in X$ then any subset N of X' such that $x \in N$ is a neighbourhood of x, (iii) a subset M of X' is a neighbourhood of ω if and only if $\omega \in M$ and $X' \setminus M$ is contained in a finite union of sets of $\mathcal{U}$. Prove that with this topology on X', the set $\{\omega\}$ is open in X' if and only if $\mathcal{U}$ has a finite subcover.
8. Let X' be the space defined by an open cover of X as in the previous exercise. Prove that the projection $X \times X' \to X'$ is closed if and only if $\{\omega\}$ is open in X'. Deduce that if a constant map $X \to \{y\}$ is proper, then X is compact.
9. Let $f : X \to Y$ be continuous and injective. Then the following are equivalent: (a) f is proper; (b) f is closed; (c) f is a homeomorphism onto a closed subspace of Y.

10. Let $f : X \to Y$ be continuous, let $B \subseteq Y$ and let $A = f^{-1}[B]$. Prove that, if f is proper, then so also is $f \mid A, B$.
11. Prove that, if $f : X \to Y$ is proper, then f is closed and $f^{-1}[y]$ is compact for each y in Y.
12. Prove that if $f : X \to X'$, $g : Y \to Y'$ are proper then so also if $f \times g$. Deduce that if $f : X \to X'$ is proper and X is Hausdorff, then $\operatorname{Im} f$ is Hausdorff.
13. Let $f : X \to Y$, $g : Y \to Z$ be continuous. Prove that (a) if f, g are proper and $\operatorname{Im} f$ is closed, then gf is proper; (b) if gf is proper and f is surjective, then g is proper; (c) if gf is proper and g is injective then f is proper; (d) if gf is proper and Y is Hausdorff then f is proper. Deduce that (e) if f is proper, so also is $f \mid A$ for any closed subset A of X, (f) if X is Hausdorff and $f : X \to Y$, $g : X \to Z$ are proper, then $(f, g) : X \to Y \times Z$ is proper.
14. Prove that if $f : X \to Y$ is continuous, $\operatorname{Im} f$ is a Hausdorff k-space and $f^{-1}[K]$ is compact for every compact subset K of $\operatorname{Im} f$, then f is proper.
15. Let X, Y be locally compact, Hausdorff spaces and X^*, Y^* their Alexandroff compactifications. Let $f : X \to Y$ be continuous and let $f^* : X^* \to Y^*$ be the extension of f which sends the point at infinity to the point at infinity. Prove that f is proper and has closed image if and only if f^* is continuous.
16. Prove the following (i) if C is closed in the normal space X then there is a map $f : X \to \mathbb{I}$ such that $C = f^{-1}[0]$ if and only if C is a G_δ-set; (ii) if C_1, C_2 are disjoint, closed G_δ-sets in the normal space X, then there is a map $f : X \to \mathbb{I}$ such that $C_1 = f^{-1}[0]$, $C_2 = f^{-1}[1]$. [You should assume the Tietze extension theorem.]
17. A *continuum* is a compact, connected space. Read the proof of the following theorem in [HY61] (Theorem 2.9): If a, b are two points of a compact Hausdorff space X, and if X is not the union of two disjoint open sets one containing a and the other containing b, then X contains a continuum containing a and b.
18. A subset Q of X is a *quasicomponent* of X if for any partition $\{X_1, X_2\}$ of X, Q is contained in X_1 or in X_2, and Q is maximal with respect to this property. Prove that (i) the quasicomponents of X cover X, (ii) each quasicomponent of X is closed, (iii) every component of X is contained in a quasicomponent, (iv) in a compact, Hausdorff space the components and quasicomponents coincide.

NOTES

The first three chapters of [HY61] form excellent supplementary reading to the topics we have discussed, particularly for the results on continua (i.e., compact, connected spaces). Most other books on general topology are biased more towards analysis. For more results on proper maps and, in particular, for solutions of some of the exercises, see [Bou66]; however, the proofs there use filters in an essential way. The Tietze extension theorem has led to a large theory of retracts which is surveyed in [Hu65]. This theorem is also used in the metrisation problem—the problem of finding necessary and sufficient conditions of a topological character for a space to be metrisable. An account of the solution of this is given in [Kel55]. In this context, an important role is played by the paracompact spaces, although the use of these spaces in topology is beginning to be taken over by the use of partitions of unity, cf. [Mok64], [Dol63], [Eng68].

For a review of many properties of topological spaces, and for examples, counterexamples, and tables of relationships, see [SS78].

The exercises on proper maps and the 1-point compactification cover important material. I owe a number of these to W. F. Newns. For more information and hints, see [Bou66], [Str76] and [Jam84]. The paper [Why42] suggests the excellent term *compact map* instead of proper map. For the notion of compactifying a map, see [Why42], the references there, [Her71], [Fir74], [Dyc72], [Dyc76], [Dyc84] and [Jam84]. Indeed, tracing back the references in [Dyc84] will show the large amount of study that has gone into the notion of proper or, equivalently, perfect, map. A sequential version of proper maps is given in [Bro73].

Chapter 4

Identification spaces and cell complexes

4.1 Introduction

In chapter 1, we considered briefly some examples of topological spaces obtained by identifications. In this chapter, we shall discuss this process in full generality. But first we shall consider some examples in order to clarify the set-theoretic processes involved.

EXAMPLES

1. The interval $[0, 2]$ can be thought of as obtained by joining two intervals of length 1. The circle $\mathbb{S}^1$ is obtained from $[0, 1]$ by identifying 0 and 1 [Fig. 4.1].

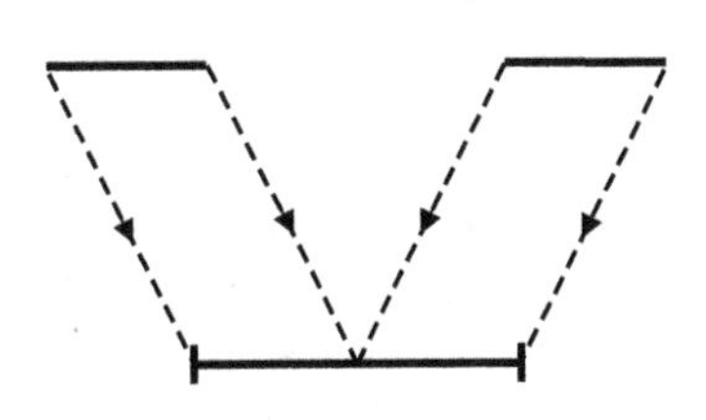

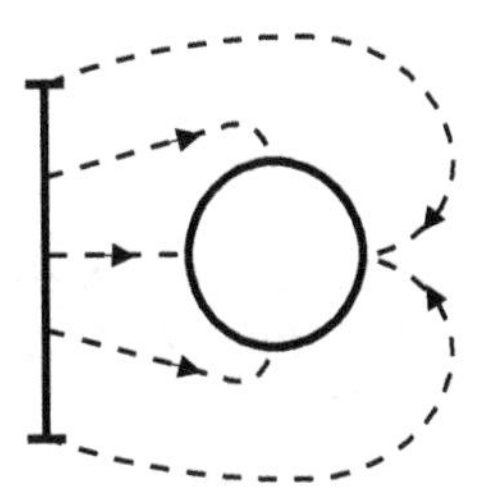

Fig. 4.1

2. From the square $\mathbb{I}^2 = \mathbb{I} \times \mathbb{I}$ we can obtain two spaces by identifying two opposite sides according to the schemes shown in Fig. 4.2 (the figures have

used topological license in stretching and bending).

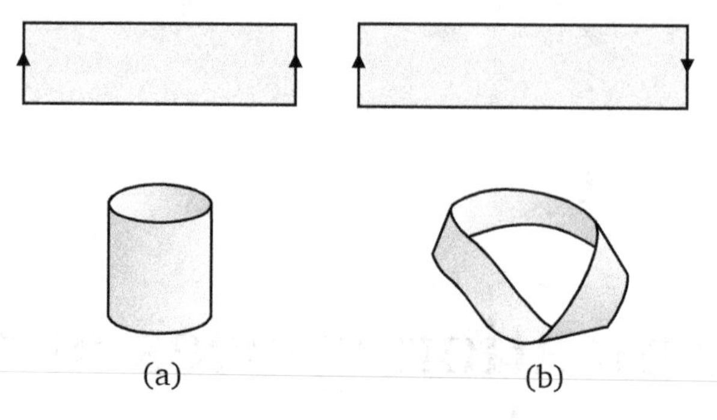

Fig. 4.2

The arrows on the sides indicate whether the sides should be stuck together in the same or in opposite senses. In (a) the sides are stuck together in the same sense and the result is the cylinder $\mathbb{S}^1 \times \mathbb{I}$. In (b), the sides are stuck together in the opposite sense, and the result is the Möbius band.

3. By identifying two pairs of sides of $\mathbb{I}^2$ we can obtain the *torus* (or *anchor ring*), which is simply $\mathbb{S}^1 \times \mathbb{S}^1$ [cf. Exercise 9 of Section 2.7].

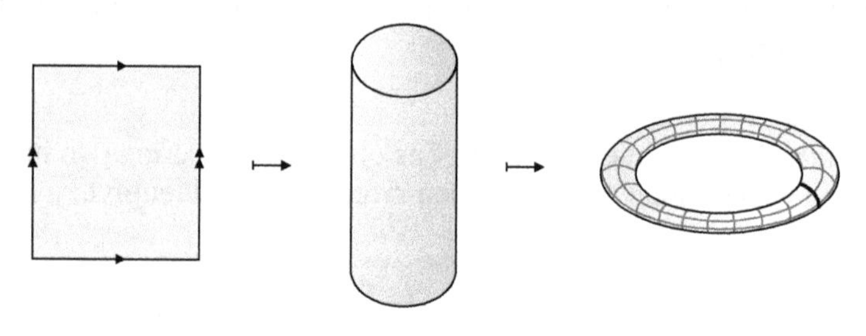

Fig. 4.3

4. If we try and identify the sides of $\mathbb{I}^2$ according to the scheme of Fig. 4.4 then, by making one identification, we obtain a cylinder but the final identification cannot be represented properly in three dimensions. We do in fact obtain a space as we shall show in detail later; this space is called the *Klein bottle* and is homeomorphic to the space constructed at the end of chapter 1. There is a 3-dimensional model of the Klein bottle [Fig. 4.5], but this model has 'self-intersections' which the 'real' Klein bottle does not have.

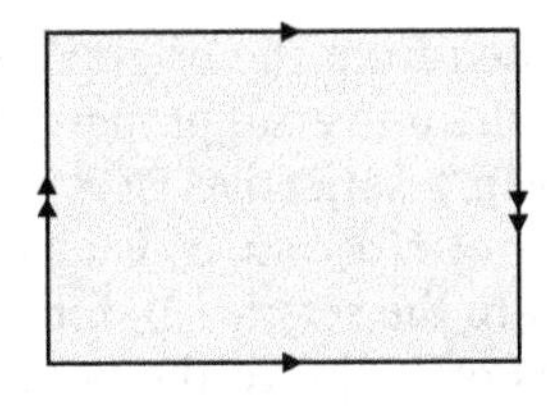

Fig. 4.4

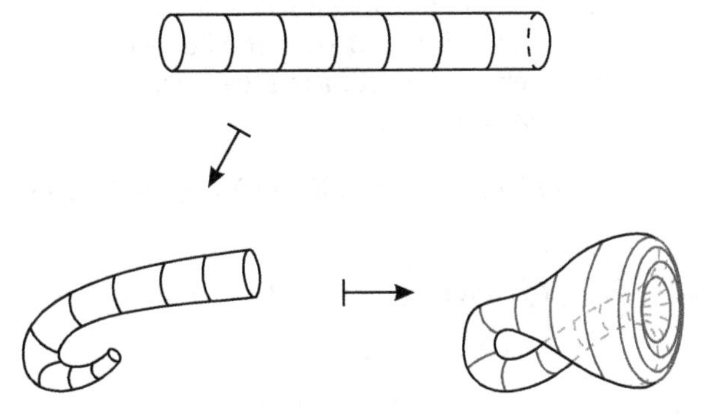

Fig. 4.5

This last example illustrates a general principle, namely, that it may be possible to describe a space Y as obtained by identifications from a simple space X, and this description may be adequate to show us the properties of Y even though Y cannot be visualised.

We have to formulate in general the notion of constructing a space Y by identifications from a given space X. For the moment, let us forget about the topologies on X and Y.

In X, we suppose, given a relation R, where the interpretation of aRb is that a shall be identified with b. But if a is to be identified with b then b must be identified with a; and if, further, b is to be identified with c then a must also be identified with c. Also a is 'identified' with itself. This shows that we must consider not only R but also the equivalence relation E generated by R [Appendix A.4], and that the totality of points of X identified with a given point a of X are the points b such that bEa; the set of such b is, of course, the equivalence class $\operatorname{cls} a$.

The equivalence classes of E form a set of disjoint, non-empty, subsets of X whose union is X. This set X/E of equivalence classes shall be the set Y. There is a function $f : X \to Y$ which sends $a \mapsto \operatorname{cls} a$. Clearly, f is a surjection. So we have passed from the original relation R to a surjection $f : X \to Y$, characterised by the universal property A.4.6 of the Appendix.

Conversely, suppose given a surjective $f : X \to Y$, where Y is now any set. Then the relation $aE_f b \Leftrightarrow f(a) = f(b)$ is an equivalence relation in X whose equivalence classes are the sets $f^{-1}[y]$ for each y in Y. The function f identifies all the elements of $f^{-1}[y]$ to the point y. This shows that the notions of *set with identifications* and *surjective function* are closely related.

We shall need a generalisation of the above identifications. Suppose there is given a family $(X_\alpha)_{\alpha \in A}$ of sets and for each α, β in A a relation $R_{\alpha\beta}$ from X_α to X_β; that is $R_{\alpha\beta}$ is a subset of $X_\alpha \times X_\beta$, and $aR_{\alpha\beta}b$ means $a \in X_\alpha$, and $(a, b) \in R_{\alpha\beta}$. The following result shows that there is a set Y obtained from the family (X_α) by identifying a with b whenever $a \in X_\alpha$, $b \in X_\beta$, and $aR_{\alpha\beta}b$. The set Y is characterised, as in the case of identifications in a single set, by a φ-universal property.

4.1.1 *There is a set* Y *and a family* $(f_\alpha : X_\alpha \to Y)_{\alpha \in A}$ *of functions such that*

(a) *for all* a, b, $aR_{\alpha\beta}b \Rightarrow f_\alpha a = f_\beta b$,
(b) *if* $(g_\alpha : X_\alpha \to Z)_{\alpha \in A}$ *is a family such that for all* a, b

$$aR_{\alpha\beta}b \Rightarrow g_\alpha a = g_\beta b,$$

then there is a unique function $g^* : Y \to Z$ *such that*

$$g^* f_\alpha = g_\alpha \textit{ for all } \alpha \textit{ in } A.$$

Proof Let $X = \bigsqcup_{\alpha \in A} X_\alpha$ be the sum of the family, and let $i_\alpha : X_\alpha \to X$ be the injections. In X we define a relation R by

$$i_\alpha(a) R i_\beta(b) \Leftrightarrow aR_{\alpha\beta}b.$$

Let E be the equivalence relation generated by R, let $Y = X/E$ and let $f : X \to Y$ be the projection. Let $f_\alpha = f i_\alpha : X_\alpha \to Y$.

The functions g_α define a function $g : X \to Z$ such that $g i_\alpha = g_\alpha$, α in A. By (b), for any a, b in X, $aRb \Rightarrow ga = gb$. So g defines $g^* : Y \to Z$ such that $g^* f = g$, whence $g^* f_\alpha = g^* f i_\alpha = g i_\alpha = g_\alpha$, α in A.

If $g' : Y \to Z$ satisfies $g' f_\alpha = g_\alpha$, then $g' f i_\alpha = g i_\alpha$. Hence $g' f = g$ and so $g' = g^*$. □

These considerations may, initially, seem abstract. However we shall see that the main point of spaces with identifications is that they give a means of constructing functions, and this is both a necessary and a sufficient condition for their utility. Put in another way, we have moved from a local consideration—what happens in a given space—to a global consideration—the relation of this space to other spaces. This widening of the point of view has proved very fruitful in mathematics.

4.2 Final topologies, identification topologies

We have now treated the set theoretic part of the notion of identification space, and shown that it can be subsumed under the general notion of a family $(X_\alpha)_{\alpha \in A}$ of sets and a family

$$(f_\alpha : X_\alpha \to Y)_{\alpha \in A}$$

of functions. In specific situations, (f_α) is often the universal family constructed in 4.1.1, but for the moment this fact is irrelevant.

We now suppose that each X_α is a topological space and we construct from (f_α) a reasonable topology on Y. The problem centres of course on the word reasonable. In the case considered in 4.1.1, the specific virtue of Y was that we could construct functions from Y. So the topology on Y which we choose is that which enables us to decide whether or not functions from Y are continuous.

Definition A topology $\mathcal{F}$ on Y is said to be *final* with respect to the functions (f_α) if, for any topological space Z and function $g : Y_{\mathcal{F}} \to Z$, we have g is continuous if and only if $gf_\alpha : X_\alpha \to Z$ is continuous for each α in A.

EXAMPLE

1. Suppose, for example, that we are in the situation of 4.1.1 and $(f_\alpha : X_\alpha \to Y)$ is a family satisfying the φ-universal property considered there. Let each X_α be a topological space and let Y have the final topology. Then we have: *if* $(g_\alpha : X_\alpha \to Z)_{\alpha \in A}$ *is any family of continuous functions such that*

$$aR_{\alpha\beta}b \Rightarrow g_\alpha a = g_\beta b$$

then there is a unique continuous function $g^* : Y \to Z$ *such that* $g^* f_\alpha = g_\alpha, \alpha \in A$. In fact, the universal property ensures that there exists $g^* : Y \to Z$, such that $g^* f_\alpha = g_\alpha$, $\alpha \in A$. The continuity of g^* follows from the fact that Y has the final topology.

We shall show that a final topology always exists, but we first point out some simple consequences.

4.2.1 *If $\mathcal{F}$ is the final topology on Y with respect to* (f_α) *then*

(a) *each* $f_\alpha : X_\alpha \to Y_{\mathcal{F}}$ *is continuous,*
(b) *if $\mathcal{T}$ is any topology on Y such that each* $f_\alpha : X_\alpha \to Y_{\mathcal{T}}$ *is continuous, then $\mathcal{F}$ is finer than $\mathcal{T}$.*

Proof (a) The identity $1 : Y_{\mathcal{F}} \to Y_{\mathcal{F}}$ is continuous and hence $1f_\alpha : X_\alpha \to Y$ is continuous for each α in A. Since $1f_\alpha = f_\alpha$, it follows that f_α is continuous.

(b) Let $g : Y_{\mathcal{F}} \to Y_{\mathcal{T}}$ be the identity function. Then $gf_\alpha = f_\alpha : X_\alpha \to Y_{\mathcal{T}}$ is continuous for each α in A. Hence g is continuous, and so $\mathcal{F}$ is finer than $\mathcal{T}$. □

We now show that the final topology exists. We suppose given (X_α) and (f_α) as before.

4.2.2 *The final topology $\mathcal{F}$ on Y with respect to (f_α) exists and is characterised by either of the following conditions:*

(a) *If $U \subseteq Y$, then U is open in $\mathcal{F}$ if and only if $f_\alpha^{-1}[U]$ is open in X_α for each α in A.*
(b) *The same as* (a), *but with 'open' replaced by 'closed'.*

Proof First we show that (a) does define a topology. Clearly $\varnothing$, Y are open in $\mathcal{F}$. If U, V are open in $\mathcal{F}$, then

$$f_\alpha^{-1}[U \cap V] = f_\alpha^{-1}[U] \cap f_\alpha^{-1}[V]$$

which is open in X_α, and so $U \cap V$ belongs to $\mathcal{F}$. Similarly, the formula $f_\alpha^{-1}[\bigcup U_i] = \bigcup f_\alpha^{-1}[U_i]$, shows that the union of any family of sets open in $\mathcal{F}$ is again open in $\mathcal{F}$. Therefore (a) does define a topology. The proof that (b) also defines a topology in terms of closed sets is similar, and the relation $f_\alpha^{-1}[Y \setminus U] = X_\alpha \setminus f_\alpha^{-1}[U]$ shows that these topologies are the same.

We now prove that this topology is the final topology. Clearly, each $f_\alpha : X_\alpha \to Y_{\mathcal{F}}$ is continuous. Suppose $g : Y_{\mathcal{F}} \to Z$ is a function where Z is a topological space.

If g is continuous, then so also is each composite gf_α. Suppose, conversely, that each gf_α is continuous. Let U be open in Z. Then

$$f_\alpha^{-1} g^{-1}[U] = (gf_\alpha)^{-1}[U]$$

and, therefore, $f_\alpha^{-1} g^{-1}[U]$ is open in X_α. It follows that $g^{-1}[U]$ is open in $Y_{\mathcal{F}}$. Hence g is continuous, as we were required to prove. □

EXAMPLE

2. Let $X = \bigsqcup_{\alpha \in A} X_\alpha$ be the sum of the underlying sets of the family $(X_\alpha)_{\alpha \in A}$ of topological spaces, and let $i_\alpha : X_\alpha \to X$ be the injections. The final topology on X with respect to (i_α) is called the *sum topology*. Such a topology was defined in chapter 3 when the indexing set was finite—clearly the definitions coincide in this case.

By means of the topological sum we can reduce final topologies with respect to a family (f_α) to final topologies with respect to a single function f. In fact, with the assumptions as for 4.2.1, 4.2.2 we have:

4.2.3 *Let X be the sum of spaces (X_α) and let $f : X \to Y$ be the function determined by (f_α). Then the final topologies on Y with respect to f and with respect to (f_α) coincide.*

Proof Let $i_\alpha : X_\alpha \to X$ be the injection. Let $g : Y \to Z$ be a function where Z is a topological space. Then

$$gf_\alpha = gfi_\alpha$$

since $fi_\alpha = f_\alpha$. Also gf is continuous if and only if each gfi_α is continuous. Thus the conditions (a) g is continuous if and only if gf is continuous, and (b) g is continuous if and only if each gf_α is continuous, are equivalent. □

We now concentrate attention on the case of a single function $f : X \to Y$. Let Y have the final topology with respect to f. Let $Y_1 = Y \setminus f[X]$. If $y \in Y_1$ then $f^{-1}[y]$ is empty, and so $\{y\}$ is both open and closed in Y. Also, $f[X]$ is open and closed in Y. Therefore, Y is the topological sum of $f[X]$ and the discrete space Y_1. This shows that the case of major interest is when f is a surjection.

Let X, Y be topological spaces and $f : X \to Y$ a function. We say f is an *identification map* if f is a surjection and Y has the final topology with respect to f. This topology on Y is also called the *identification topology* with respect to f, and we say Y is an *identification space* of f.

There is a useful characterisation of identification topologies in addition to those given by the definition and by 4.2.2. A subset A of X is *saturated* (more precisely, *saturated with respect to* f, or f-*saturated*) if $f^{-1}f[A] = A$. For example, any set $f^{-1}[B]$ is saturated.

Let $f : X \to Y$ be an identification map. If B is a subset of Y then $ff^{-1}[B] = B$ (since f is surjective). Also, if A is saturated, then $f[A] = ff^{-1}f[A]$. Hence the *open sets of* Y *are the sets* $f[V]$ *for all saturated open sets* V *of* X; and the same statement holds with open replaced by closed.

There is a difficulty in the description of neighbourhoods in Y. Let $A \subseteq Y$ and suppose N is a neighbourhood of A. Then there is an open set U such that $A \subseteq U \subseteq N$ whence

$$f^{-1}[A] \subseteq f^{-1}[U] \subseteq f^{-1}[N].$$

So N is a neighbourhood of A implies $f^{-1}[N]$ is a neighbourhood of $f^{-1}[A]$. The converse of this last implication is false: for suppose $f^{-1}[N]$ is a neighbourhood of $f^{-1}[A]$. Then we can find an open set V such that $f^{-1}[A] \subseteq V \subseteq f^{-1}[N]$, but it may be impossible to find such a V which is saturated with respect to f [Exercise 1 of Section 4.3].

The following result gives a useful class of identification maps.

4.2.4 *Let $f : X \to Y$ be a continuous surjection. If f is an open map, or a closed map, then f is an identification map.*

Proof Suppose that f is an open map. Let U be a subset of Y. By continuity, if U is open in Y then $f^{-1}[U]$ is open in X. On the other hand f is a surjection, so $ff^{-1}[U] = U$. Hence $f^{-1}[U]$ is open if and only if U is open.

A similar proof applies with open replaced by closed. □

4.2.4 *(Corollary 1) A continuous surjection from a compact space to a Hausdorff space is an identification map.*

Proof Such a function is a closed map, by 3.5.8 (Corollary 1). □

A consequence of 4.2.4 (Corollary 1) is that if X is compact and $f : X \to Y$ is a surjection, then there is at most one Hausdorff topology on Y such that f is continuous. If such a Hausdorff topology exists, it is clearly the most 'reasonable' topology on Y. Unfortunately, the identification topology need not be Hausdorff, nor need it be, when X is not compact, the most 'reasonable' Hausdorff topology.

Since identification topologies are special cases of final topologies, we can apply to identification topologies the results of 4.2.2. We illustrate this in the proof of our next result, which can also be proved directly from the definition.

4.2.5 *The composite of identification maps is an identification map.*

Proof Let $f : X \to Y$, $g : Y \to Z$ be identification maps. Then $gf : X \to Z$ is certainly continuous and surjective. We show that if $h : Z \to W$ and hgf are continuous, then h is continuous.

Suppose then $hgf : X \to W$ is continuous. Then hg is continuous (since f is an identification map) and h is continuous (since g is an identification map). □

EXERCISES

1. Let X, Y be topological spaces and $f : X \to Y$ an injection. Prove that f is an identification map if and only if f is a homeomorphism.
2. Let $f : X \to Y$ be an identification map. What is the topology of Y if X is discrete? indiscrete?
3. Prove that the following maps are identification maps

   (i) The projections $X \times Y \to X$, $X \times Y \to Y$.
   (ii) $\{(x, y) \in \mathbb{R}^2 : xy = 0\} \to \mathbb{R}, \quad (x, y) \mapsto x$.
   (iii) $\{(x, y) \in \mathbb{R}^2 : x^2y^2 = 0\} \to \mathbb{R}^{\neq 0}, \quad (x, y) \mapsto x$.
   (iv) $\mathbb{I} \to \mathbb{S}^1, \quad t \mapsto e^{2\pi i t}$.

4. Let Y have the final topology with respect to $f : X \to Y$. Prove that $f \mid X, f[X]$ is an identification map.
5. Let X, Y be topological spaces and $f : X \to Y$ a continuous surjection. Suppose that each point y in Y has a neighbourhood N such that $f \mid f^{-1}[N], N$ is an identification map. Prove that f is an identification map. Deduce that the *covering map*

$$p : \mathbb{R} \to \mathbb{S}^1, \quad t \mapsto e^{2\pi i t}$$

is an identification map.
6. Let A be a subspace of X. A *retraction of* X *onto* A is a map $r : X \to A$ such that $r \mid A$ is the identity. Prove that a retraction of X onto A is an identification map. Deduce that the map

$$\mathbb{R}^{n+1} \setminus \{0\} \to \mathbb{S}^n, \quad x \mapsto x/|x|$$

is an identification map.
7. Prove that if $f : X \to Y$, $f' : X' \to Y'$, $g : Y \to Z$ are open surjections, then so also are $gf : X \to Z$ and $f \times f' : X \times X' \to Y \times Y'$.
8. Let $f : X \to Y$ be an identification map. For each $A \subset X$, let $f^{\dagger}[A] = \{a \in A : f^{-1}f[a] \subseteq A\}$. Prove that the following conditions are equivalent. (i) f is a closed map. (ii) If A is closed in X, then also is $f^{-1}f[A]$. (iii) If A is open in X, then so also is $f^{\dagger}[A]$. (iv) For each y in Y, every neighbourhood N of $f^{-1}[y]$ contains a saturated neighbourhood of $f^{-1}[y]$.
9. Let $f : X \to Y$ be an identification map, and let $R_f = \{(x, x') \in X \times X : fx = fx'\}$. Prove that, if Y is Hausdorff, then R_f is closed in $X \times X$. Prove that if f is an open map, and R_f is closed in $X \times X$, then Y is Hausdorff.
10. Prove the following 'transitive law' for final topologies. Suppose there are given functions $f_{\alpha\lambda} : X_{\alpha\lambda} \to Y_{\lambda}$ for each λ in Λ and α in A_{λ}, and functions $g_{\lambda} : Y_{\lambda} \to Z$ for each λ in Λ, and a topology for each $X_{\alpha\lambda}$. Prove that if Y_{λ} has the final topology with respect to $(f_{\alpha\lambda})_{\alpha \in A_{\lambda}}$, then the final topology on Z with respect to $(g_{\lambda})_{\lambda \in \Lambda}$ coincides with the final topology with respect to the family of composites $(g_{\lambda} f_{\alpha\lambda})_{\lambda \in \Lambda, \alpha \in A_{\lambda}}$. Show that 4.2.3 and 4.2.5 are corollaries of this transitive law.
11. Let Σ be a set of subspaces of X and for each $S \in \Sigma$ let $i_S : S \to X$ be the inclusion. The final topology on X with respect to $(i_S)_{S \in \Sigma}$ is called the *fine topology with respect to* Σ (or the *weak topology* with respect to Σ); and the set X with this topology is written X_{Σ}. Prove that: (i) Each inclusion $i_S : S \to X_{\Sigma}$ is a homeomorphism into. (ii) If $X \in \Sigma$ then $X = X_{\Sigma}$. (iii) Σ is a set of subspaces of X_{Σ}, and $(X_{\Sigma})_{\Sigma} = X_{\Sigma}$. (iv) The identity $X_{\Sigma} \to X$ is continuous.
12. Let Σ, Σ' be sets of subspaces of X, X' respectively. Let $f : X \to X'$ be a function such that for each S of Σ (i) there is an S' of Σ' such that $f[S] \subset S'$, (ii) $f \mid S$ is continuous. Prove that $f : X_{\Sigma} \to X'_{\Sigma}$ is continuous. Suppose now that $X = X'$: prove that $X_{\Sigma} = X_{\Sigma'}$ if both Σ is a refinement of Σ' and Σ' is a refinement of Σ.

4.3 Subspaces, products, and identification maps

The following result is important in itself and will also help the discussion of examples of identification spaces.

Let $f : X \to Y$ be an identification map, and let $A \subseteq X$. Then $A, f[A]$ are subspaces of X, Y respectively, and we ask: is $g = f \mid A, f[A]$ an identification map?

4.3.1 *The following conditions are equivalent.*

(a) $g = f \mid A, f[A]$ *is an identification map,*
(b) *each* g*-saturated set which is open in* A *is the intersection of* A *with an* f*-saturated set open in* X,
(c) *as for* (b) *but with 'open' replaced by 'closed'.*

Proof We state an elementary exercise in set theory [Exercise 5 of Section A.1 of the Appendix]: if $U \subseteq f[A]$ and $V \subseteq Y$, then

$$U = f[A] \cap V \Leftrightarrow g^{-1}[U] = A \cap f^{-1}[V].$$

(a) $\Rightarrow$ (b) Let U' be open in A and g-saturated. Then $U = g[U']$ is open in $f[A]$, whence $U = f[A] \cap V$ where V is open in Y. So

$$U' = g^{-1}g[U'] = g^{-1}[U] = A \cap f^{-1}[V].$$

Clearly, $f^{-1}[V]$ is open in X and f-saturated.
(b) $\Rightarrow$ (a) Let U be a subset of $f[A]$ such that $g^{-1}[U]$ is open in A. By condition (b), $g^{-1}[U] = A \cap f^{-1}[V]$ for some V open in Y. Hence $U = f[A] \cap V$, and so U is open in $f[A]$.

The proof of (a) $\Leftrightarrow$ (c) is the same as the above but with the word 'open' replaced by 'closed'. □

4.3.1 *(Corollary 1) Each of the following conditions implies that* $g = f \mid A, f[A]$ *is an identification map.*

(a) *For all* U, *if* U *is* g*-saturated and open in* A, *then* $f^{-1}f[U]$ *is open in* X.
(b) *As for* (a), *but with 'open' replaced by 'closed'.*
(c) *The set* A *is* f*-saturated and open in* X.
(d) *As for* (c), *but with 'open' replaced by 'closed'.*

Proof We give the proof only for cases (a) and (c).
(a) For any subset U of A

$$g^{-1}g[U] = A \cap f^{-1}f[U].$$

So condition (b) of 4.3.1 is satisfied.
(c) We reduce this to case (a). Let U be g-saturated and open in A. Then

$$f^{-1}f[U] \subset f^{-1}f[A] = A$$

whence

$$f^{-1}f[U] = g^{-1}g[U] = U.$$

Also U is open in X since A is open in X. □

EXAMPLES

1. Consider the subspace of $\mathbb{R}^2$

$$X = \mathbb{I} \times \{0\} \cup \mathbb{I} \times \{1\}.$$

Let $Y = [0,1[\ \cup \{y_0, y_1\}$ where y_0, y_1 are distinct points not in $[0,1[$.

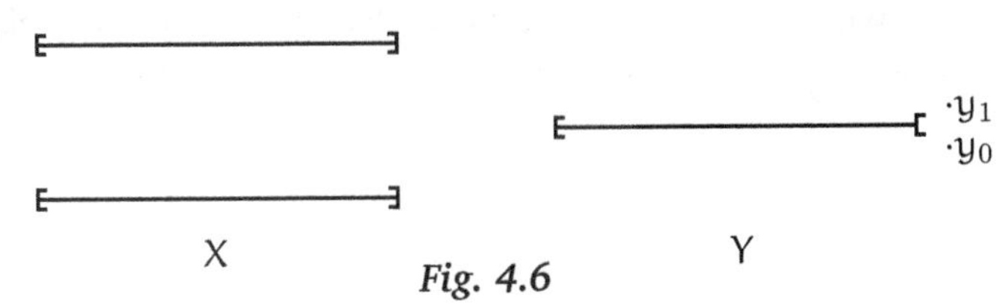

Fig. 4.6

Let $f : X \to Y$ be the function

$$(t, \varepsilon) \mapsto \begin{cases} t, & t \neq 1 \\ y_0, & t = 1, \varepsilon = 0 \\ y_1, & t = 1, \varepsilon = 1. \end{cases}$$

Thus f identifies $(t, 0)$ with $(t, 1)$ for each t in $[0,1[$. Let

$$I_0 = [0,1[\ \cup \{y_0\}, \quad I_1 = [0,1[\ \cup \{y_1\}.$$

Let $f_0 : \mathbb{I} \times \{0\} \to I_0$ be the restriction of f. It is an easy deduction from 4.3.1 (Corollary 1a) that f_0 is an identification map. Since f_0 is a bijection, it is therefore a homeomorphism [Exercise 1 of Section 4.2]. Thus I_0 is (both as a set and topologically) essentially $\mathbb{I}$ with 1 renamed y_0. Similar remarks apply to I_1.

From this we have that the sets

$$]t, 1[\ \cup\ \{y_0\}, \quad t \in [0,1[$$

form a base for the neighbourhood of y_0, and there is a similar base for the neighbourhood of y_1. Hence every neighbourhood of y_0 meets every neighbourhood of y_1; that is, Y is not Hausdorff.

2. Let A be a subset of the topological space X. Then X *with* A *shrunk to a point* is a topological space, written X/A, which is obtained from X by identifying all of A to a single point. More precisely, the elements of X/A

are the equivalence classes in X under the equivalence relation generated by

$$x \sim y \Leftrightarrow x \in A \text{ and } y \in A.$$

These equivalence classes are therefore the sets $\{x\}$ for x in $X \setminus A$ and also, when A is non-empty, the set A.

It is convenient to identify the point x of $X \setminus A$ with the point $\{x\}$ of X/A, and so to regard $X \setminus A$ as a subset of X/A. This causes confusion only when $A \in X \setminus A$, in which case we can always replace the equivalence class A by some point not in $X \setminus A$. Let $f : X \to X/A$ be the projection

$$x \mapsto \begin{cases} x, & x \in X \setminus A \\ A, & x \in A. \end{cases}$$

We give X/A the final topology with respect to f.

Now f is surjective and so an identification map. If A is empty, or consists of a single point, then X/A can be identified with X.

The set $X \setminus A$ is f-saturated and f is the identity on $X \setminus A$. By 4.3.1 (Corollary 1), $X \setminus A$ is a subspace of X/A if A is open or is closed in X.

The application of Example 1 of Section 4.2 is important: *if* $g : X \to Z$ *is any map such that* $g[A]$ *consists of a single point of Z, then there is a unique map* $g^* : X/A \to Z$ *such that* $g^*f = g$.

The following terminology is convenient. A function $g : X \to Z$ is *constant on* A if $g \mid A$ is a constant map; also g *shrinks* A *to a point* if g is constant on A, is surjective, and $g \mid X \setminus A$ is injective. Thus, in the latter case, the function $g^* : X/A \to Z$ defined by g is a bijection. The universal property of 4.1.1 and Example 1 above shows that if $g : X \to Z$, $g' : X \to Z'$ are two identification maps which shrink A to a point, then there is a unique homeomorphism $h : Z \to Z'$ such that $hg = g'$.

3. We consider a special case of the last example. Let Y be any space. The *cone on* Y is

$$CY = Y \times \mathbb{I}/Y \times \{0\}.$$

The point v of CY which is the set $Y \times \{0\}$ is called the *vertex* of CY.

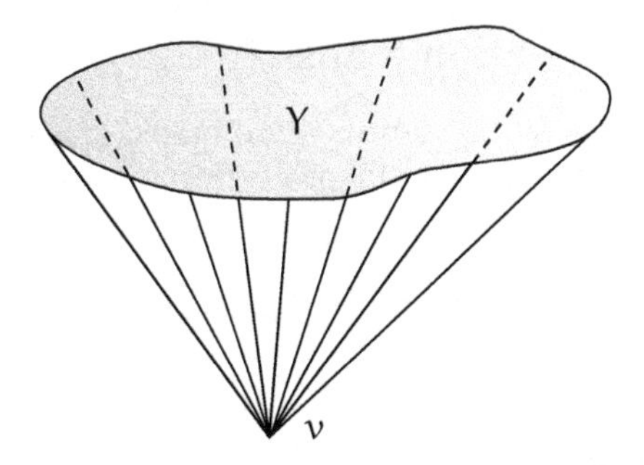

Fig. 4.7

Let B be a subspace of Y, let $i : B \times \mathbb{I} \to Y \times \mathbb{I}$ be the inclusion function, and let $p' : B \times \mathbb{I} \to CB$, $P : Y \times \mathbb{I} \to CY$ be the identification maps. Then $pi[B \times \{0\}]$ is the vertex of CY and so there is a unique map $g : CB \to CY$ such that $gp' = pi$. Also, g is injective. We now show that g need not be an embedding.

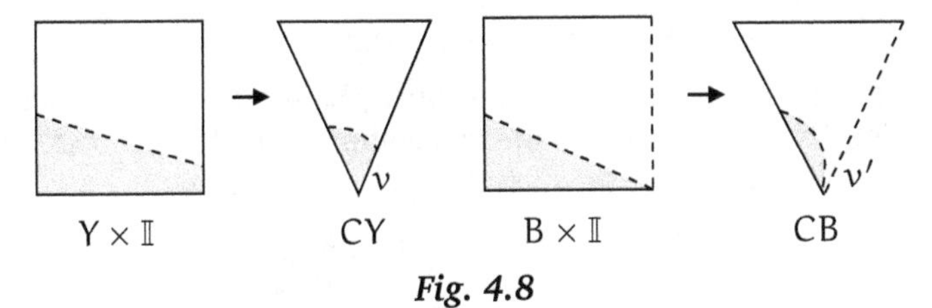

Fig. 4.8

Let $Y = \mathbb{I}$, $B = [0, 1[$, and let v, v' be the vertices of CY, CB respectively. The difference between the topologies of CY and CB is illustrated in Fig. 4.8. The neighbourhoods in CY of v look as expected, but the shaded set in CB is a neighbourhood of v' since it is the image of a saturated open neighbourhood of $B \times \{0\}$. (A detailed justification of these pictures is left as an exercise.)

In this case, it might be considered more reasonable to give CB the topology which makes $g : CB \to CY$ an embedding. Actually, in chapter 5 we shall discuss the *coarse topology* which behaves better with regard to subspaces than does the identification topology. Until chapter 5 we shall need a topology on CY for which we can construct continuous functions *from* CY *to* some other space. For these purposes the identification topology is the best.

We shall see in section 4.6, that when Y is compact Hausdorff so also is CY; in this case, the coarse and identification topologies on CY coincide.

Products of identification maps[‡]

Let $f : X \to Y$, $f' : X' \to Y'$ be identification maps. It is not true in general that $f \times f' : X \times X' \to Y \times Y'$ is an identification map—an example is given below. However we can prove:

4.3.2 *Let* $f : X \to Y$ *be an identification map and let* B *be locally compact. Then*

$$f \times 1 : X \times B \to Y \times B$$

is an identification map.

Proof Let $W \subseteq X \times B$ be open and saturated with respect to

$$h = f \times 1.$$

We must prove that $h[W]$ is open in $Y \times B$.

To this end, let $(y_0, b_0) \in h[W]$ and suppose $y_0 = fx_0$. Then $(x_0, b_0) \in W$ since W is saturated. Let B_0 be the subset of B such that

$$\{x_0\} \times B_0 = (\{x_0\} \times B) \cap W;$$

this is the 'x_0-section' of W. Since W is open, B_0 is a neighbourhood of b_0. Since B is locally compact, B_0 contains a compact neighbourhood C of b_0.

Let $U \subseteq X$ be the largest set such that $U \times C \subseteq W$, that is,

$$U = \{x \in X : \{x\} \times C \subseteq W\}.$$

Then

$$(y_0, b_0) \in f[U] \times C \subseteq h[W].$$

So to prove that $h[W]$ is a neighbourhood of (y_0, b_0), it is sufficient to prove that $f[U]$ is a neighbourhood of y_0.

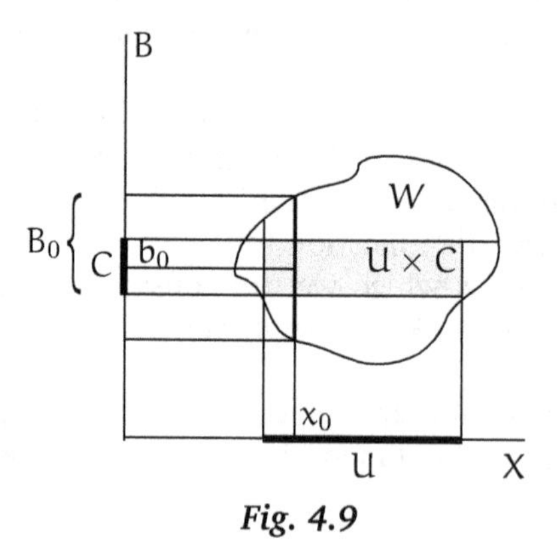

Fig. 4.9

[‡]The main result here and its application in 4.6.6 are not needed until chapter 7.

U *is open in* X Let $x \in U$ so that $\{x\} \times C \subseteq W$. Since C is compact and W is open, there is a neighbourhood M of x such that $M \times C \subset W$ [3.5.6 (Corollary 2)]. This implies that $M \subseteq U$, by definition of U. So U is open.

U *is* f*-saturated* We have $U \subseteq f^{-1}f[U]$ and

$$f^{-1}f[U] \times C = h^{-1}h[U \times C] \subseteq h^{-1}h[W] = W.$$

So $f^{-1}f[U] \subseteq U$, by definition of U. Hence $f^{-1}f[U] = U$.

It follows that $f[U]$ is a neighbourhood of y_0. □

EXAMPLE

4. We now show that 4.3.2 is false without some assumptions on B. Let $f : \mathbb{Q} \to \mathbb{Q}/\mathbb{Z}$ be the identification map and let $h = f \times 1 : \mathbb{Q} \times \mathbb{Q} \to (\mathbb{Q}/\mathbb{Z}) \times \mathbb{Q}$. Then h is not an identification map.

Proof Let $r_0 = 1$ and for each non-zero n in $\mathbb{Z}$ let $r_n = \sqrt{2}/|n|$—thus r_n is irrational and $r_n \to 0$ as $n \to \infty$. Let A_n be any open subset of $[n, n+1] \times \mathbb{R}$ such that the closure of A_n meets $\{n, n+1\} \times \mathbb{R}$ in the two points (n, r_n) and (n, r_{n+1}) (such a set A_n is shaded in Fig. 4.10.) Let A be the union of these sets A_n, $n \in \mathbb{Z}$, and let $B = \overline{A} \cap (\mathbb{Q} \times \mathbb{Q})$. Then B is closed in $\mathbb{Q} \times \mathbb{Q}$ and saturated with respect to h. We leave it as an exercise to the reader to prove that in $(\mathbb{Q}/\mathbb{Z}) \times \mathbb{Q}$ the point $(f0, 0)$ belongs to the closure of $h[B]$. □

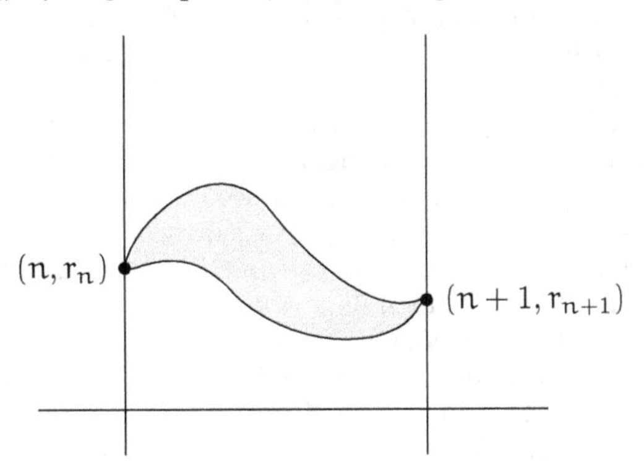

Fig. 4.10

The fact that 4.3.2 is false in general has led to suggestions for changing the maps used in topology, or for changing the product topology. [cf. Section 5.9].

EXERCISES

1. Let A be the subset $\mathbb{L} \setminus \{0\}$ of $\mathbb{I}$ and let $p : \mathbb{I} \to \mathbb{I}/A$ be the identification map. Let $N = [0, \frac{1}{2}] \cup \{1\}$. Prove that N is a saturated neighbourhood of 0, but $p[N]$ is not a neighbourhood of $p0$.
2. Let $f : \mathbb{R} \to [-1, 1]$ be the function

$$x \mapsto \begin{cases} \sin 1/x & x \neq 0 \\ 0 & x = 0 \end{cases}$$

and let $[-1, 1]$ have the identification topology with respect to f. Prove that the subspace $[-1, 1] \setminus \{0\}$ has its usual topology but the only neighbourhood of 0 is $[-1, 1]$.
3. Let A, B be subsets of X such that A is closed and $A \subseteq B$. Prove that B/A is a subspace of X/A.
4. Let $A \subseteq X$. Prove that $X \setminus A$ is a subspace of X/A if and only if the following condition holds: a subset U of $X \setminus A$ is open in $X \setminus A$ if and only if $U = V \cap (X \setminus A)$ where V is open in X and V either contains A or does not meet A. Give an example of X, A for which this condition holds, yet A is neither open nor closed in X. Give an example of X, A for which this does not hold.
5. Let A be a subspace of X. Prove that X/A is Hausdorff if (i) $X \setminus A$ is Hausdorff, (ii) $X \setminus A$ is a subspace of X/A, and (iii) if $x \in X \setminus A$ then x and A have disjoint neighbourhoods in X. Prove that if $X = [0, 2]$, $A =]1, 2]$, then X/A is not Hausdorff.
6. Let A be a closed subset of X and let B be a proper subset of A. Let $X' = X \setminus B$, $A' = A \setminus B$. Prove that X'/A' is homeomorphism to X/A.
7. Let $f : X \to Y$ be an identification map and let $A \subseteq X$. Suppose that there is a map $u : X \to A$ such that for all $x \in X$, $fux = fx$. Prove that $f[A] = Y$ and $g = f \mid A$ is an identification map.
8. [This and the following exercises use the notation of Exercises 11 and 12 of Section 4.2.] Let Σ be a set of subspaces of X and let $A \subseteq X$. The *restriction* of Σ to A is the set $\Sigma \mid A = \{S \cap A : S \in \Sigma\}$. So we can form the space $A_{\Sigma|A}$. On the other hand, A with its relative topology as a subset of X_Σ determines a space A_Σ say. Prove that the identity function $A_{\Sigma|A} \to A_\Sigma$ is continuous. Prove also that if A and each S of Σ is closed in X, then $A_{\Sigma|A} = A_\Sigma$ and A_Σ is a closed subspace of X_Σ.
9‡. For any space X, let $\mathcal{C}X$ denote the set of compact subspaces of X, and write kX for $X_{\mathcal{C}X}$. Refer to the definition of a k-space in Exercise 4 of Section 3.6 in order to prove that X is a k-space if and only if $X = kX$. Prove also that (i) X, kX have the same compact subspaces, (ii) kX is a k-space, (iii) the topology of kX is the finest topology $\mathcal{T}$ on the underlying set of X such that X and $X_\mathcal{T}$ have the same compact subspaces.
10. Prove that a closed subspace of a k-space is a k-space.
11. Prove that an identification space of a k-space is a k-space.
12. The *weak product* of spaces X, Y is the space $X \times_W Y = k(X \times Y)$. Prove that $X \times_W Y$ has the fine topology with respect to the sets $A \times B$ for A compact in X,

‡In this and the following exercises, all spaces are assumed to be Hausdorff.

B compact in Y. Prove also that (i) $X \times_W Y = kX \times_W kY$, (ii) the weak product is associative and commutative, (iii) the projections from $X \times_W Y$ to X and Y are continuous, (iv) if $X \times Y = X \times_W Y$, then X and Y are k-spaces, (v) the diagonal map $X \to X \times_W X$ is continuous if and only if X is a k-space, (vi) $X \times_W Y = X \times Y$ if X and Y satisfy the first axiom of countability.

13. Prove that if X is a k-space and Y is a locally compact, then $X \times Y$ is a k-space. [Use 4.2.3 to represent X as an identification space of a topological sum of compact spaces, and then apply 4.3.2.]

14. A function $f : X \to Y$ of spaces is called k-*continuous* if $f \mid C$ is continuous for each compact subspace C of X. Prove that the following are equivalent. (i) $f : X \to Y$ is k-continuous, (ii) $f : kX \to Y$ is continuous, (iii) $f : kX \to kY$ is continuous. Prove also that the identity $X \to X$ is k-continuous and that the composite of k-continuous functions is again k-continuous. Prove that if $f : X \to Y$, $g : X \to Z$ are k-continuous, then so also if $(f, g) : X \to Y \times Z$.

15. A function $f : X \to Y$ is called a k-*identification map* if f is k-continuous and for all Z and all functions $g : Y \to Z$, g is k-continuous if gf is k-continuous. Prove that $f : X \to Y$ is a k-identification map if and only if $f : kX \to kY$ is an identification map.

16. Let $f : X \to Y$ be a k-identification map and let Q satisfy: each compact subspace of Q is locally compact. Prove that $f \times 1 : X \times Q \to Y \times Q$ is a k-identification map.

17. Let $f : X \to Y$ be an identification map. Suppose also that X, Q and $X \times Q$ are k-spaces. Prove that $Y \times Q$ is a k-space if and only if $f \times 1 : X \times Q \to Y \times Q$ is an identification map. Hence give an example of k-spaces Y, Q such that $Y \times Q$ is not a k-space.

4.4 Cells and spheres

In this section, we shall be considering real n-space $\mathbb{R}^n$ with its Euclidean norm, written $|\ |$, and its Cartesian norm, written $\|\ \|$. The map

$$\begin{aligned} i : \mathbb{R}^n &\to \mathbb{R}^{n+1} \\ x &\mapsto (x, 0) \end{aligned}$$

is called the *natural inclusion*—it is a linear homeomorphism onto the closed subspace $\mathbb{R}^n \times \{0\}$ of $\mathbb{R}^{n+1}$. Also i is norm preserving, that is

$$\begin{aligned} |i(x)| &= |x|, \\ \|i(x)\| &= \|x\|, \end{aligned} \qquad x \in \mathbb{R}^n. \tag{4.4.1}$$

We have already defined the standard n-cell, n-ball, and $(n-1)$-sphere as

$$\begin{aligned}\mathbb{E}^n &= \{x \in \mathbb{R}^n : |x| \leqslant 1\},\\ \mathbb{B}^n &= \{x \in \mathbb{R}^n : |x| < 1\},\\ \mathbb{S}^{n-1} &= \{x \in \mathbb{R}^n : |x| = 1\}.\end{aligned}$$

For $n = 0$, this gives $\mathbb{E}^0 = \mathbb{B}^0 = \{0\}$, $\mathbb{S}^{-1} = \varnothing$. Notice that $\mathbb{S}^{n-1}, \mathbb{E}^n$ are closed, bounded subsets of $\mathbb{R}^n$, and so are compact.

It is clear from Equation (4.4.1) that

$$\begin{aligned}i[\mathbb{E}^n] &= \mathbb{E}^{n+1} \cap (\mathbb{R}^n \times \{0\}),\\ i[\mathbb{S}^{n-1}] &= \mathbb{S}^n \cap (\mathbb{R}^n \times \{0\}).\end{aligned}$$

We call $i[\mathbb{S}^{n-1}]$ the *equatorial* $(n-1)$*-sphere* of $\mathbb{S}^n$. This $(n-1)$-sphere divides $\mathbb{S}^n$ into two parts called the northern and southern hemispheres

$$\begin{aligned}E^n_+ &= \{(x, t) \in \mathbb{S}^n : x \in \mathbb{R}^n, t \geqslant 0\},\\ E^n_- &= \{(x, t) \in \mathbb{S}^n : x \in \mathbb{R}^n, t \leqslant 0\}.\end{aligned}$$

Thus $\mathbb{S}^n = E^n_+ \cup E^n_-$ and $E^n_+ \cap E^n_- = i[\mathbb{S}^{n-1}]$.

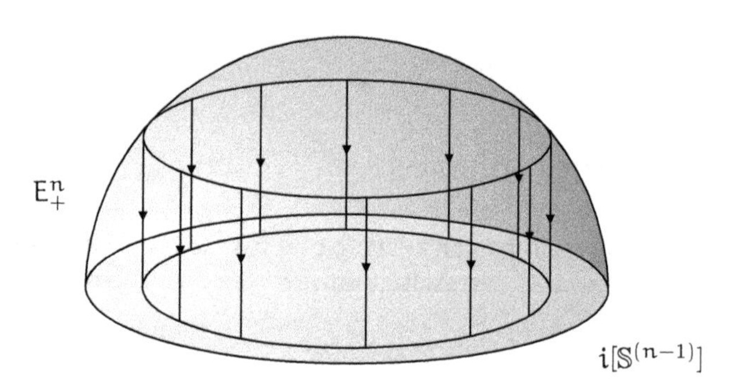

Fig. 4.11

We shall continue to regard $\mathbb{R}^{n+1}$ as $\mathbb{R}^n \times \mathbb{R}$, so that the projection $p : \mathbb{R}^{n+1} \to \mathbb{R}^n$ simply omits the last coordinate. If $x \in \mathbb{S}^n$, then $|x| = 1$ whence $|p(x)| \leqslant 1$, i.e., $p[\mathbb{S}^n] \subseteq \mathbb{E}^n$.

4.4.2 *The projection* $p : \mathbb{R}^{n+1} \to \mathbb{R}^n$ *maps both* E^n_+ *and* E^n_- *homeomorphically onto* $\mathbb{E}^n$.

Proof The function

$$\begin{aligned} \mathbb{E}^n &\to E^n_+ \\ x &\mapsto (x, \sqrt{1-|x|^2}) \end{aligned}$$

is well-defined and is a continuous inverse to $p \mid E^n_+, \mathbb{E}^n$ [Fig. 4.11]. This proves the result for E^n_+, and the proof for E^n_- is similar. □

The sets $\mathbb{B}^n, \mathbb{E}^n$ are convex and hence path-connected.

4.4.3 $\mathbb{S}^n$ *is path-connected for* $n \geqslant 1$.

Proof From 4.4.2, E^n_+, E^n_- are path-connected. But $E^n_+ \cap E^n_-$ is non-empty for $n \geqslant 1$, and so $\mathbb{S}^n$ is path-connected by 3.4.2. □

There is an important connection between cells, spheres, and the cone construction. We recall that for any topological space X, the cone on X is

$$CX = X \times \mathbb{I}/X \times \{0\}.$$

4.4.4 *There is a homeomorphism* $h : C\mathbb{S}^{n-1} \to \mathbb{E}^n$.

Proof The function

$$\begin{aligned} k : \mathbb{S}^{n-1} \times \mathbb{I} &\to \mathbb{E}^n \\ (x, t) &\mapsto tx \end{aligned}$$

is continuous and surjective. Since all the spaces concerned are compact and Hausdorff, k is an identification map. Also k shrinks $\mathbb{S}^{n-1} \times \{0\}$ to the point 0 of $\mathbb{E}^n$. It follows that k defines a homeomorphism $h : C\mathbb{S}^{n-1} \to \mathbb{E}^n$. □

For any space X, let the *suspension* of X be

$$SX = CX/X \times \{1\}.$$

Thus SX is obtained from $X \times \mathbb{I}$ by shrinking to a point first $X \times \{0\}$ and then $X \times \{1\}$. Let l be the composite of the identification maps $X \times \mathbb{I} \to CX \to SX$; by 4.2.5, l is an identification map.

The picture for SX is Fig. 4.12 in which each section $X \times \{t\}$, $t \neq 0, 1$ is homeomorphic to X under the projection $X \times \{t\} \to X$. The subspaces $X \times [\frac{1}{2}, 1]$, $X \times [0, \frac{1}{2}]$ are mapped by l to subspaces C^+X, C^-X. By an easy application of (d) of 4.3.1 (Corollary 1), l restricts to an identification map $X \times [0, \frac{1}{2}] \to C^-X$ and so we have an identification map

$$\begin{aligned} X \times \mathbb{I} &\to X \times [0, \tfrac{1}{2}] \to C^-X \\ (x, t) &\mapsto (x, t/2) \mapsto l(x, t/2) \end{aligned}$$

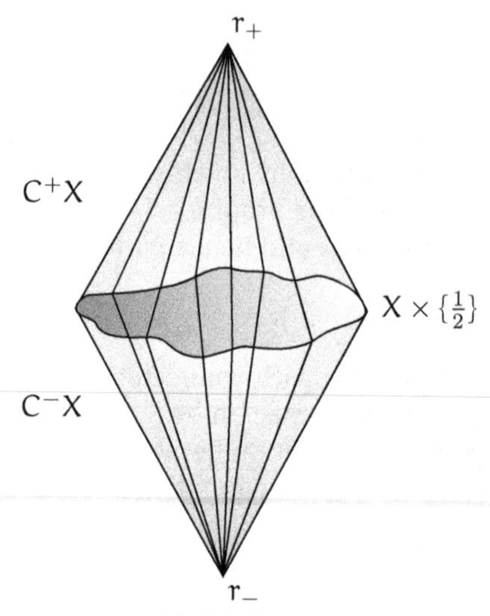

Fig. 4.12

which shrinks to a point $X \times \{0\}$. It follows that there is a homeomorphism $CX \to C^-X$. A similar argument (using the homeomorphism $\mathbb{I} \to [\frac{1}{2}, 1]$ given by $t \mapsto 1 - t/2$) shows that there is a homeomorphism $CX \to C^+X$; on $X \times \{1\}$ both of these homeomorphisms send $(x, 1) \mapsto (x, \frac{1}{2})$.

The argument of the last paragraph generalises the representation of $\mathbb{S}^n$ as the union of its Northern and Southern hemispheres. In fact, by the last paragraph, 4.4.2, and 4.4.4, we can construct the homeomorphisms

$$E^n_+ \to C^+\mathbb{S}^{n-1}, \quad E^n_- \to C^-\mathbb{S}^{n-1}$$

which on $i[\mathbb{S}^{n-1}]$ are given by $(x, 0) \mapsto (x, \frac{1}{2})$. By the gluing rule for homeomorphisms [2.7.3] these homeomorphisms define a homeomorphism

$$\mathbb{S}^n \to S\mathbb{S}^{n-1}.$$

The composite

$$\mathbb{E}^n \xrightarrow{h^{-1}} C\mathbb{S}^{n-1} \xrightarrow{p} S\mathbb{S}^{n-1}$$

(where p is the identification map) is injective on $\mathbb{B}^n$ and shrinks the boundary $\mathbb{S}^{n-1}$ of $\mathbb{E}^n$ to a point. It follows that $\mathbb{E}^n/\mathbb{S}^{n-1}$ is homeomorphic to $S\mathbb{S}^{n-1}$, and hence there is a homeomorphism

$$k : \mathbb{E}^n/\mathbb{S}^{n-1} \to \mathbb{S}^n.$$

This homeomorphism sends the shrunken $\mathbb{S}^{n-1}$ to the South Pole $P = (0, \ldots, 0, -1)$ of $\mathbb{S}^n$. Further, $\mathbb{B}^n$ is a subspace of $\mathbb{E}^n/\mathbb{S}^{n-1}$ and k maps $\mathbb{B}^n$ homeomorphically onto $\mathbb{S}^n \setminus \{P\}$. For this reason, we write

$$\mathbb{S}^n = e^0 \cup e^n$$

where e^r denotes an *open* r*-cell*, or r*-ball*, that is, a space homeomorphic to $\mathbb{B}^r$. In the case of $\mathbb{S}^n$, $e^0 = \{P\}$, $e^n = \mathbb{S}^n \setminus \{P\}$.

It might be expected that $\mathbb{S}^n \setminus \{x\}$ is homeomorphic to $\mathbb{B}^n$ for any point x of $\mathbb{S}^n$. This is in fact true, and is a consequence of the *homogeneity* of $\mathbb{S}^n$ by which is meant that for any points x, y of $\mathbb{S}^n$ there is a homeomorphism $\sigma : \mathbb{S}^n \to \mathbb{S}^n$ such that $\sigma(x) = y$ [cf. Exercise 5 of Section 5.4].

We conclude this section by proving a simple result which will have a more general analysis in chapter 5.

4.4.5 *There is a homeomorphism* $\mathbb{E}^m \times \mathbb{E}^n \to \mathbb{E}^{m+n}$.

Proof Let $\|\ \|$ denote the Cartesian norm on $\mathbb{R}^n$ and let

$$\mathbb{J}^n = \{x \in \mathbb{R}^n : \|x\| \leqslant 1\}.$$

Thus $\mathbb{J}^n$ is the n-fold product of the interval $[-1, 1]$ with itself. The associativity of the product shows that $\mathbb{J}^m \times \mathbb{J}^n$ is homeomorphic to $\mathbb{J}^{m+n}$. So our result is proved if we exhibit a homeomorphism from $\mathbb{E}^m$ to $\mathbb{J}^m$.

Consider the function

$$f : \mathbb{E}^m \to \mathbb{J}^m$$
$$x \mapsto \begin{cases} \|x\|^{-1}|x|x, & x \neq 0 \\ 0, & x = 0. \end{cases}$$

Then f is a bijection, and is continuous certainly at all $x \neq 0$. We prove f is continuous at 0.

The closed ball about 0 of radius r in the Euclidean norm is $r\mathbb{E}^m$, and in the Cartesian norm is $r\mathbb{J}^m$. Also

$$f^{-1}[r\mathbb{J}^m] = r\mathbb{E}^m$$

But the sets $r\mathbb{E}^m$, and also the sets $r\mathbb{J}^m$, taken for all $r > 0$ form bases for the neighbourhoods of 0. Therefore, f is continuous at 0.

A similar argument shows that f^{-1} is continuous. □

Now $\mathbb{E}^i$ is a subset of $\mathbb{R}^i$, $i = m, n, m+n$, so that $\mathbb{E}^m \times \mathbb{E}^n$ is a subset of $\mathbb{R}^{m+n}$. Let 'interior' mean the interior operator on these Euclidean spaces. Then the homeomorphism $h : \mathbb{E}^m \times \mathbb{E}^n \to \mathbb{E}^{m+n}$ constructed above maps the interior of $\mathbb{E}^m \times \mathbb{E}^n$ bijectively onto the interior of $\mathbb{E}^{m+n}$, that is, it maps $\mathbb{B}^m \times \mathbb{B}^n$ bijectively onto $\mathbb{B}^{m+n}$. Therefore, h maps the boundary of $\mathbb{E}^m \times \mathbb{E}^n$ bijectively onto the boundary of $\mathbb{E}^{m+n}$; hence h restricts to a homeomorphism

$$\mathbb{S}^{m-1} \times \mathbb{E}^n \cup \mathbb{E}^m \times \mathbb{S}^{n-1} \to \mathbb{S}^{m+n-1}. \tag{4.4.6}$$

EXERCISES

1. Let $x_0 \in \mathbb{S}^n$ and let Π be the hyperplane in $\mathbb{R}^{n+1}$ which is perpendicular to x_0 and which passes through the origin. Let $s : \mathbb{S}^n \setminus \{x_0\} \to \Pi$ be the *stereographic projection* defined as follows: if $x \in \mathbb{S}^n \setminus \{x_0\}$ then $s(x)$ is the unique point of Π such that $s(x)$, x, and x_0 are collinear. Let $p : \mathbb{R}^{n+1} \to \Pi$ be the perpendicular projection onto Π. Prove that for each $x \in \mathbb{S}^n \setminus \{x_0\}$

$$s(x) = p(x)/(1 - \cos\theta)$$

where θ is the angle between x and x_0. Prove also that s is a homeomorphism.
2. Prove that E^n_+ is a retract of $\mathbb{S}^n$. Prove also that each point x of $\mathbb{S}^n$ has a base $\mathcal{B}$ for the neighbourhoods (in $\mathbb{S}^n$) of x such that each element of $\mathcal{B}$ is a retract of $\mathbb{S}^n$.
3. Let the points of $\mathbb{R}^{p+q+2}$ be denoted by (x, y) where $x \in \mathbb{R}^{p+1}$, $y \in \mathbb{R}^{q+1}$. Define subsets A_+, A_- of $\mathbb{S}^{p+q+1}$ by

$$A_+ = \{(x, y) \in \mathbb{S}^{p+q+1} : |x| \geqslant |y|\},$$
$$A_- = \{(x, y) \in \mathbb{S}^{p+q+1} : |x| \leqslant |y|\}.$$

Prove that A_+ is homeomorphic to $\mathbb{S}^p \times \mathbb{E}^{q+1}$, A_- is homeomorphic to $\mathbb{E}^{p+1} \times \mathbb{S}^q$, and $A_+ \cap A_-$ is homeomorphic to $\mathbb{S}^p \times \mathbb{S}^q$.
4. Let X be a topological space and let $x_0 \in X$. Define the *reduced cone* and *reduced suspension* respectively by

$$\Gamma X = X \times \mathbb{I}/(X \times \{0\} \cup \{x_0\} \times \mathbb{I}),$$
$$\Sigma X = X \times \mathbb{I}/(X \times \{0, 1\} \cup \{x_0\} \times \mathbb{I}).$$

Prove that $\Gamma\mathbb{S}^{n-1}$ is homeomorphic to $\mathbb{E}^n$, and that $\Sigma\mathbb{S}^{n-1}$ is homeomorphic to $\mathbb{S}^n$.

4.5 Adjunction spaces

Arbitrary identification spaces are so general that it is difficult to say anything about them. We therefore concentrate attention on a special kind of identification space—the adjunction spaces—which occur commonly and for which we can prove useful theorems.

Suppose there is given a topological space X, a closed subspace A of X and a map $f : A \to B$. The *adjunction space*

$$B \,{}_f\!\sqcup X$$

is obtained by 'gluing' X to B by means of f. More precisely, we have a relation from X to B given by

$$x \sim b \Leftrightarrow x \in A \text{ and } fx = b.$$

The elements of $B \,{}_f\!\sqcup X$ are then the equivalence classes of $B \sqcup X$ under the equivalence relation generated by $\sim$.

Let $F : B \sqcup X \to B \,{}_f\!\sqcup X$ be the projection. We have a diagram

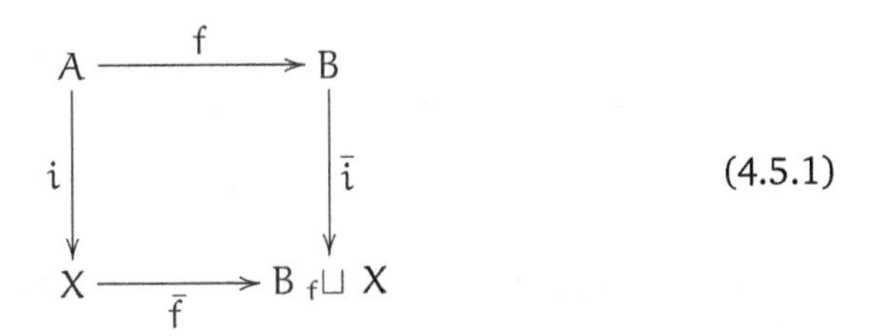

$$\begin{array}{ccc} A & \xrightarrow{f} & B \\ {\scriptstyle i}\downarrow & & \downarrow{\scriptstyle \bar{i}} \\ X & \xrightarrow[\bar{f}]{} & B \,{}_f\!\sqcup X \end{array} \qquad (4.5.1)$$

in which i is the inclusion and $\bar{i}, \bar{f}$ are composites

$$B \xrightarrow{i_1} B \sqcup X \xrightarrow{F} B \,{}_f\!\sqcup X, \qquad X \xrightarrow{i_2} B \sqcup X \xrightarrow{F} B \,{}_f\!\sqcup X.$$

Clearly, diagram (4.5.1) is commutative, that is, $\bar{f}i = \bar{i}f$, since if $a \in A$

$$i_2(a) \sim i_1 f(a) \text{ in } B \sqcup X.$$

An equivalence class of $B \sqcup X$ either contains a single element $i_2(x)$ for x in $X \setminus A$ or is $\{i_1(b)\} \cup i_2 f^{-1}[b]$ for b in B. It is usual, therefore, to identify x in $X \setminus A$ and b in B with their corresponding equivalence classes in $B \sqcup X$, and so to regard $B \,{}_f\!\sqcup X$ as the union of the sets B and $X \setminus A$. This convention causes confusion only when B meets $X \setminus A$.

The topology on $B \,{}_f\!\sqcup X$ is to be the final topology with respect to $\bar{i}, \bar{f}$ or, equivalently, the identification topology with respect to

$$F : B \sqcup X \to B \,{}_f\!\sqcup X.$$

If $g : B \,{}_f\!\sqcup X \to Y$ is any map, then we can construct a commutative

diagram of maps

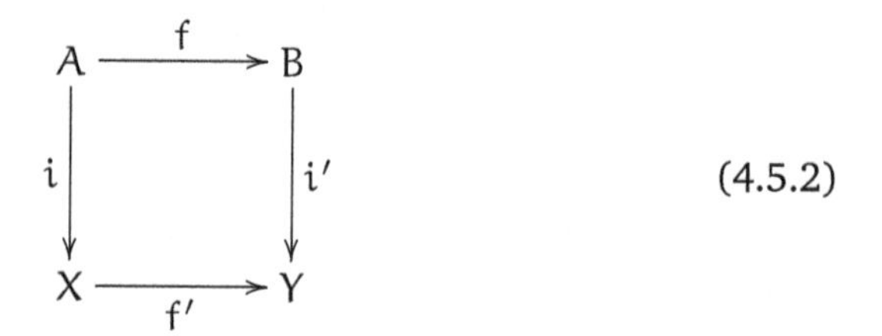

(4.5.2)

where $i' = g\bar{i}$, $f' = g\bar{f}$. The converse of this statement gives a method of constructing maps from $B\,{}_f\!\sqcup X$ to Y.

4.5.3 *If* $i' : B \to Y$, $f' : X \to Y$ *are maps such that* $f'i = i'f$, *then there is a unique map* $g : B\,{}_f\!\sqcup X \to Y$ *such that*

$$g\bar{i} = i', \qquad g\bar{f} = f'.$$

Proof Let $x \in X$, $b \in B$. The condition $f'i = i'f$ ensures that

$$x \sim b \Rightarrow f'(x) = i'(b).$$

The result is immediate from Example 1 of Section 4.2. □

In the usual way, the 'universal' property 4.5.3 characterises $\bar{i}, \bar{f}$ up to a homeomorphism—if the maps f', i' of diagram (4.5.2) are universal then the map $g : B\,{}_f\!\sqcup X \to Y$ of 4.5.3 is a homeomorphism. This type of universal property occurs in diverse situations and so deserves a name. Suppose we are given a commutative diagram of maps of topological spaces

$$\begin{array}{ccc} X_0 & \xrightarrow{\;i_1\;} & X_1 \\ {\scriptstyle i_2}\downarrow & & \downarrow{\scriptstyle u_1} \\ X_2 & \xrightarrow[\;u_2\;]{} & X \end{array} \qquad (4.5.4)$$

Then we say (u_1, u_2) is a *pushout* of (i_1, i_2), and also that the square (4.5.4) is a *pushout*, if the following property holds: *if* $u_1' : X_1 \to X'$, $u_2' : X_2 \to X'$ *are maps such that* $u_1'i_1 = u_2'i_2$, *then there is a unique map* $u : X \to X'$ *such that* $uu_1 = u_1'$, $uu_2 = u_2'$. As usual, this property characterises the pair (u_1, u_2) up to a homeomorphism of X. For this reason, it is common to make an abuse of language and refer to X as *the* pushout of (i_1, i_2). In such case, we write

$$X = X_1\,{}_{i_1}\!\sqcup_{i_2} X_2.$$

(In chapter 7 we shall need the notion of a *weak pushout*: the definition of this is the same as that of pushout except that the word *unique* is omitted. Of course, the property that (4.5.4) is a weak pushout does not characterise (u_1, u_2) up to a homeomorphism of X.)

We shall, as above, restrict the term adjunction space to a pushout of a pair (f, i) in which i is an inclusion of a closed subspace; and, of course, we are abbreviating $B\,{}_f\!\sqcup_i X$ to $B\,{}_f\!\sqcup X$. The map f is called the *attaching map* of the adjunction space.

One of the good features of adjunction spaces is shown in the following result.

4.5.5 (a) $\bar{i}$ *is a closed map.* (b) $\bar{f} \mid X \setminus A$ *is an open map.*

Proof (a) Let C be closed in B and let $C' = \bar{i}[C]$. Then

$$\bar{i}^{-1}[C'] = C, \quad \bar{f}^{-1}[C'] = f^{-1}[C].$$

The first set is closed in B (trivially) and the second set is closed in X because f is continuous and A is closed in X. Since $B\,{}_f\!\sqcup X$ has the final topology with respect to $\bar{i}, \bar{f}$, it follows from 4.2.2(b) that C' is closed in $B\,{}_f\!\sqcup X$.
(b) Let U be open in $X \setminus A$ and let $U' = \bar{f}[U]$. Then

$$\bar{i}^{-1}[U'] = \varnothing, \quad \bar{f}^{-1}[U'] = U.$$

These sets are open in B, X respectively, and so U' is open in $B\,{}_f\!\sqcup X$. □

It is immediate from 4.5.5 that $\bar{i}$ is a homeomorphism onto a closed subspace, and $\bar{f} \mid X \setminus A$ is a homeomorphism onto an open subspace of $B\,{}_f\!\sqcup X$. So if we identify B and $X \setminus A$ with the corresponding subsets of $B\,{}_f\!\sqcup X$, we obtain:

4.5.5 *(Corollary 1) B is a closed subspace, and* $X \setminus A$ *is an open subspace, of* $B\,{}_f\!\sqcup X$.

In some cases, adjunction spaces can be regarded as identification spaces.

4.5.6 *If* $f : A \to B$ *is an identification map, then so also is* $\bar{f} : X \to B\,{}_f\!\sqcup X$.

Proof Clearly $\bar{f}$ is surjective. We prove that $B\,{}_f\!\sqcup X$ has the final topology with respect to $\bar{f}$. Suppose $g : B\,{}_f\!\sqcup X \to Y$ is such that $g\bar{f} : X \to Y$ is continuous. Then $g\bar{f}i = g\bar{i}f$ is continuous. Since f is an identification map, the continuity of $g\bar{i}f$ implies the continuity of $g\bar{i}$. Finally, the continuity of $g\bar{f}$ and $g\bar{i}$ implies the continuity of g. □

EXAMPLES

1. If A is empty, then $B\ {}_f{\sqcup}\ X = B \sqcup X$. If $A = X$, then $B\ {}_f{\sqcup}\ X = B$. If $X = B$, then $B\ {}_1{\sqcup}\ X = X$.

2. Suppose $X = B \cup C$, $A = B \cap C$ where B, C are closed in X. Let $j : A \to B$ be the inclusion. Then

$$X = B\ {}_j{\sqcup}\ C.$$

The only problem here is one of topology; the fact that X has the final topology with respect to the inclusions $B \to X$, $C \to X$ is a consequence of the Gluing Rule [2.5.12].

3. Let B be the space consisting of a single point b, let A be non-empty, and let $f : A \to B$ be the unique map. Clearly, f is an identification map, and therefore so also is $\bar{f} : X \to B\ {}_f{\sqcup}\ X$. Since $\bar{f}$ simply shrinks A to a point it follows that $B\ {}_f{\sqcup}\ X$ is homeomorphic to X/A; the only difference between these spaces is that in $B\ {}_f{\sqcup}\ X$ the point A of X/A has been replaced by b.

4. Many common identification spaces can be regarded as adjunction spaces. For example, let A_1, A_2 be closed in X and let $\varphi : A_1 \to A_2$ be a homeomorphism which is the identity on $A_1 \cap A_2$. Let $f : A_1 \cup A_2 \to A_2$ be the identity on A_2 and φ on A_1—the given conditions ensure that f is well-defined and continuous. Let $Y = A_2\ {}_f{\sqcup}\ X$. It is easy to prove that f is an identification map, and it follows that $\bar{f} : X \to Y$ is an identification map. Thus Y is obtained from X by the identifications $a_1 \sim \varphi a_1$ for all $a_1 \in A_1$.

5. One of our principal examples of adjunction spaces is when $X = \mathbb{E}^n$, the standard n-cell, and $A = \mathbb{S}^{n-1}$, the standard $(n-1)$-sphere. Thus, let $n = 1$. Then $\mathbb{E}^1 = [-1, 1]$, $\mathbb{S}^0 = \{-1, +1\}$ and $B\ {}_f{\sqcup}\ \mathbb{E}^n$ can be pictured as one of the spaces in Fig. 4.13.

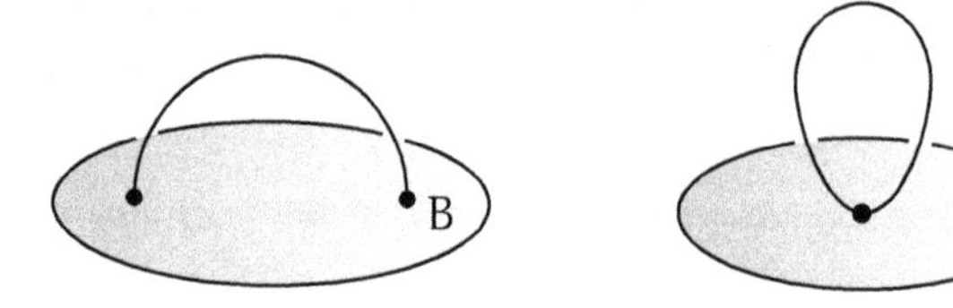

Fig. 4.13

For $n > 1$, and even for $n = 2$, it is usually impossible to visualise $B\ {}_f{\sqcup}\ \mathbb{E}^n$. But some special cases are of interest. First of all let B consist of a single point: $B\ {}_f{\sqcup}\ \mathbb{E}^n$ is homeomorphic to $\mathbb{E}^n/\mathbb{S}^{n-1}$ which, as we have seen in section 4.4, is homeomorphic to $\mathbb{S}^n$.

Second, consider the function

$$\begin{aligned} f : \mathbb{S}^1 &\to B \\ z &\mapsto \mathrm{Re}(z) \end{aligned}$$

where $B = [-1, 1]$. Then $B\ {}_f\!\sqcup\ \mathbb{E}^2$ is homeomorphic to $\mathbb{S}^1$. The picture for this is Fig. 4.14 (which may be thought of as illustrating the operation of zipping up a purse).

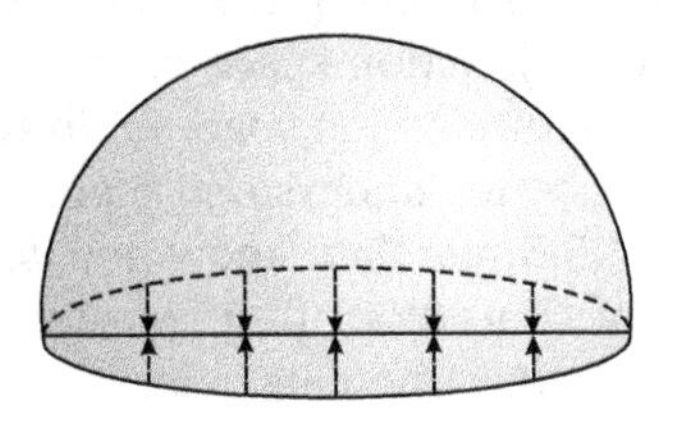

Fig. 4.14

Third, let $f : \mathbb{S}^1 \to \mathbb{S}^1$ be the function $z \mapsto z^2$. Then f identifies the points $\pm z$ for z in $\mathbb{S}^1$ and f is an identification map. Therefore $\mathbb{S}^1\ {}_f\!\sqcup\ \mathbb{E}^2$ is homeomorphic to $\mathbb{S}^2$ with antipodal points identified. This space is a model of the real projective plane $P^2(\mathbb{R})$ of chapter 5.

6. A *pointed space* is a topological space X and a point x_0 of X, called the *base point* such that $\{x_0\}$ is closed in X. Let X, Y be pointed spaces. The *wedge* of X and Y is the space $X \vee Y$ obtained from X and Y by identifying x_0 with y_0. Fig. 4.15 illustrates $\mathbb{S}^1 \vee \mathbb{S}^2$ (for some choice of base points) and $\mathbb{S}^2 \vee \mathbb{S}^2$ can be thought of as two tangential 2-spheres in $\mathbb{R}^3$.

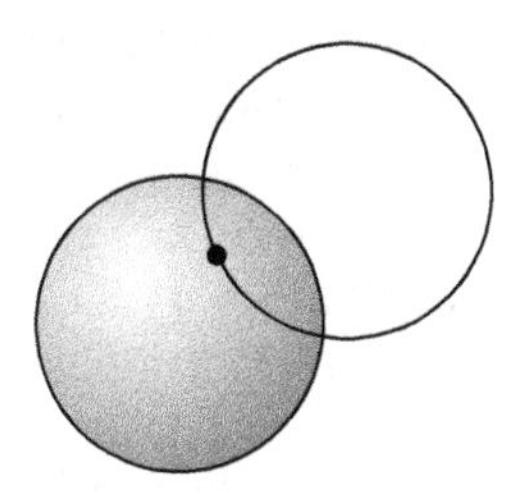

Fig. 4.15

If we take the boundary of a square and identify opposite sides according to the schemes of either Fig. 4.3 or Fig. 4.4 on p. 98, then the result is essentially $\mathbb{S}^1 \vee \mathbb{S}^1$.

7. Let T^2 be the torus and let $g : \mathbb{I}^2 \to T^2$ be the identification map by which opposite edges of $\mathbb{I}^2$ are identified. Let $h : \mathbb{E}^2 \to \mathbb{I}^2$ be a homeomorphism which maps $\mathbb{S}^1$ onto the boundary of $\mathbb{I}^2$ (in fact any homeomorphism $\mathbb{E}^2 \to \mathbb{I}^2$ does this), and let $\bar{f} = gh : \mathbb{E}^2 \to T^2$. Then $\bar{f}[\mathbb{S}^1]$ is essentially a subset $\mathbb{S}^1 \vee \mathbb{S}^1$ of T^2 (the the last example). Since $\bar{f}$ is an identification map we can say that T^2 is homeomorphic to $(\mathbb{S}^1 \vee \mathbb{S}^1)\ {}_f\!\sqcup\ \mathbb{E}^2$ where f is a map $\mathbb{S}^1 \to \mathbb{S}^1 \vee \mathbb{S}^1$.

We have shown how to construct spaces using adjunction spaces. On the other hand, it may help us to grasp the topology of a given space Q if we can represent Q as an adjunction space or, at least, a homeomorph of $B\,{}_f\!\sqcup X$. This can give a 'structure' to the space Q, in showing how it is made up from smaller and often better understood pieces put together in a well defined way. Further, by 4.5.3, we then know how to construct continuous functions on Q in terms of continuous functions on these smaller pieces B and X. We will see that constructing particular mathematical structures as pushouts occurs widely in mathematics.

Suppose there is given then a closed subspace B of Q, a closed subspace A of X and a map $f' : X \to Q$ such that $f'[A] \subseteq B$. Consider the diagram

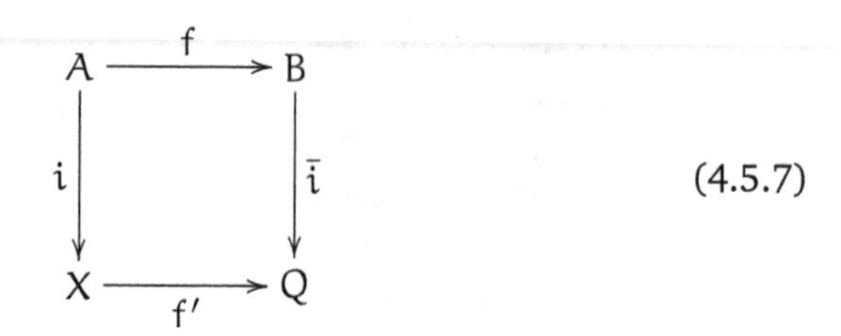

in which $i, \bar{i}$ are the given inclusions, and f is the restriction of f'. We then ask for conditions which ensure that (4.5.7) is a pushout.

Suppose (4.5.7) is a pushout of spaces and maps. Then there is a homeomorphism $k : Q \to B\,{}_f\!\sqcup X$ such that $kf' = \bar{f}$, $k\bar{i} = \bar{i}$. Because of the construction of $B\,{}_f\!\sqcup X$ it follows that $h = \bar{f} \mid X \setminus A, Q \setminus B$ is a bijection. Also, if this map h is a bijection, then (4.5.7) is a pushout of sets—if $f' : X \to Y$, $i' : B \to Y$ are functions such that $f'i = i'f$, then there is a unique function $g : Q \to Y$ such that $g\bar{i} = i'$, $g\bar{f} = f'$, where g is i' on B and $x \mapsto f'h^{-1}(x)$ on $Q \setminus B$.

Suppose then $h = \bar{f} \mid X \setminus A, Q \setminus B$ is a bijection and, as before, that B is closed in Q, A is closed in X.

4.5.8 *The square* (4.5.7) *is a pushout if* $\bar{f}[X]$ *is closed in* Q *and* $\bar{f} \mid X, \bar{f}[X]$ *is an identification map. These conditions hold if* X *is compact and* Q *is Hausdorff.*

Proof The last statement is clear. For the first statement, we have to prove that Q has the final topology with respect to $\bar{i}, \bar{f}$.

Let C be a subset of Q such that $\bar{i}^{-1}[C], \bar{f}^{-1}[C]$ are closed in B, X respectively. Then $B \cap C = \bar{i}^{-1}[C]$ is closed in B and hence is closed in Q. Also if

$$C' = C \cap \bar{f}[X] = \bar{f}\bar{f}^{-1}[C],$$

then C' is closed in $\bar{f}[X]$ (since $\bar{f} \mid X, \bar{f}[X]$ is an identification map) and therefore C' is closed in Q). Therefore

$$C = (B \cap C) \cup C'$$

is closed in Q. □

We shall need the notion of attaching a number of spaces: this is a simple extension of previous results, so we sketch the theory, leaving the details as exercises.

Let B be a topological space, let $(X_\lambda)_{\lambda \in \Lambda}$ be a family of spaces, and for each λ in Λ let A_λ be a closed subset of X_λ and let $f_\lambda : A_\lambda \to B$ be a map. The adjunction space

$$Q = B \,_{(f_\lambda)}\!\sqcup (X_\lambda)$$

is obtained by identifying each a_λ of A_λ with $f_\lambda(a_\lambda)$ in B. Thus Q is an identification space of the sum of B and all the X_λ. Providing the X_λ are disjoint and do not meet B, the underlying set of Q can be identified with the union of B and all the $X_\lambda \setminus A_\lambda$. There is an inclusion $\mathrm{i} : B \to Q$, and there are maps $\bar{f}_\lambda : X_\lambda \to Q$. The topology of Q is the final topology with respect to the family consisting of $\bar{\mathrm{i}}$ and all $\bar{f}_\lambda$. The maps $\bar{\mathrm{i}}$ and $\bar{f}_\lambda \mid X_\lambda \setminus A_\lambda$ are closed and open maps respectively. The adjunction space Q is characterised by the universal property: if $i_\lambda : A_\lambda \to X_\lambda$ is the inclusion, and $i' : B \to Y$, $g_\lambda : X_\lambda \to Y$ are maps such that $g_\lambda i_\lambda = i' f_\lambda$, $\lambda \in \Lambda$, then there is a unique map $g : Q \to Y$ such that $g\mathrm{i} = i'$, $g\bar{f}_\lambda = g_\lambda$, $\lambda \in \Lambda$. In the case Λ is finite, say $\Lambda = \{1, \ldots, n\}$, we write

$$Q = B \,_{f_1}\!\sqcup X_1 \cdots \,_{f_n}\!\sqcup X_n.$$

This notation is unambiguous since f_i is a map to B.

Exercises

1. Let Q be the subspace of $\mathbb{R}^2$ which is the union of $\{0\} \times [0, 2]$ and the set $\{(x, y) \in \mathbb{R}^2 : 0 < x \leqslant 1/\pi, \sin 1/x \leqslant y \leqslant 2\}$. Let B be the boundary of the subset Q of $\mathbb{R}^2$. Prove that B is closed in Q, that $Q \setminus B$ is homeomorphic to $\mathbb{B}^2$, but that Q is not homeomorphic to $B \,_f\!\sqcup \mathbb{E}^2$ for any $f : \mathbb{S}^1 \to B$.
2. Let B be a topological space and let $f_\lambda : A_\lambda \to B$ be maps, where A_λ is a closed subset of X_λ and $\lambda = 1, 2$. Let $\bar{\mathrm{i}}_1 : B \to B \,_{f_1}\!\sqcup X_1$ be the inclusion. Characterise the adjunction space

$$Q' = (B \,_{f_1}\!\sqcup X_1) \,_{\bar{\mathrm{i}} f_2}\!\sqcup X_2$$

by a universal property and prove that Q' is homeomorphic to $B \,_{f_1}\!\sqcup X_1 \,_{f_2}\!\sqcup X_2$. Generalise this result to arbitrary finite indexed families (f_λ), (X_λ).

3. Let B be a closed subspace of Q. For each $\lambda = 1, \ldots, n$ let $\bar{f}_\lambda : X_\lambda \to Q$ be a map, and let A_λ be a closed subspace of X_λ such that (i) $\bar{f}_\lambda[A_\lambda] \subseteq B$, (ii) $\bar{f}_\lambda \mid X_\lambda \setminus A_\lambda$ is injective, (ii) the sets $\bar{f}_\lambda[X_\lambda \setminus A_\lambda]$ are disjoint and cover $Q \setminus B$, (iv) $\bar{f}_\lambda \mid X_\lambda, \bar{f}_\lambda[X_\lambda]$ is an identification map. Prove that a function $g : Q \to Y$ is continuous if and only if $g \mid B,\ g\bar{f}_\lambda$ is continuous, $\lambda = 1, \ldots, n$. Prove also that there is a homeomorphism

$$Q \to B \ {}_{f_1}\sqcup X_1 \cdots {}_{f_n}\sqcup X_n$$

which is the identity on B.
4. Let $B \ {}_f\sqcup X$ be an adjunction space, where $f : A \to B$. Prove that B is a *retract* (see p. 127) of $B \ {}_f\sqcup X$ if and only if there is a map $g : X \to B$ such that $g \mid A = f$. Deduce that if A is a retract of X then B is a retract of $B \ {}_f\sqcup X$.

4.6 Properties of adjunction spaces

Throughout this section we suppose given an adjunction space $B \ {}_f\sqcup X$ and the pushout square

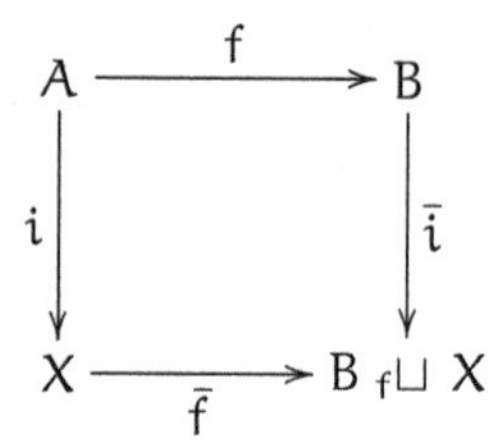

Let $F : B \sqcup X \to B \ {}_f\sqcup X$ be the identification map.

4.6.1 *Let* B', X' *be subspaces of* B, X *respectively such that* $X' \supseteq A$ *and* $B' \supseteq f[A]$. *Let* $g = f \mid A, B'$. *Then* $B' \ {}_g\sqcup X'$ *is a subspace of* $B \ {}_f\sqcup X$.

Proof We have to prove that the restriction of F

$$F' : B' \sqcup X' \to F[B' \sqcup X']$$

is an identification map, and for this we can use 4.3.1.

Notice that F is injective on $(B \setminus B') \sqcup (X \setminus X')$. Suppose that $U \subseteq B \sqcup X$ and $U' = U \cap (B' \sqcup X')$. Then U is F-saturated $\Leftrightarrow$ U' is F'-saturated (since all the identifications take place in $B' \sqcup X'$). So 4.3.1 applies, and F' is an identification map. □

We now consider the connectivity of $B \ {}_f\sqcup X$.

4.6.2 *If* A *is non-empty and* X *and* B *are connected then* $B \ {}_f\sqcup X$ *is connected.*

Proof The space $B \ {}_f\sqcup X$ is the union of the sets B and $\bar{f}[X]$ both of which are connected. But B meets $\bar{f}[X]$ (since A is non-empty); therefore $B \ {}_f\sqcup X$ is connected [3.2.5]. □

4.6.3 *If* A *is non-empty and* X *and* B *are path-connected then* $B\ {}_f\!\sqcup\ X$ *is path-connected.*

The proof of 4.6.3 is similar to that of 4.6.2.

4.6.4 *If* $B\ {}_f\!\sqcup\ X$ *and* A *are connected then* B *is connected.*

Proof Suppose that B is not connected and that $i' : B \to Y$ is a map onto the discrete space $Y = \{1, 2\}$. Since A is connected, $i'f : A \to Y$ is constant with value 1 say. Let $f' : X \to Y$ be the constant function with value 1. Then $i'f = f'i$.

By the pushout property 4.5.3, there is a map $g : B\ {}_f\!\sqcup\ X \to Y$ such that $g\bar{f} = f'$, $g\bar{i} = i'$. The last condition shows that g is surjective; hence $B\ {}_f\!\sqcup\ X$ is not connected. □

If B, X are compact then so also is the sum $B \sqcup X$ and hence $B\ {}_f\!\sqcup\ X$ is compact.

The question of whether $B\ {}_f\!\sqcup\ X$ is Hausdorff is important but rather delicate. It is true that if B and X are normal then so also is $B\ {}_f\!\sqcup\ X$ [cf. Exercise 6 and also [Hu64]]. We shall prove not this but instead a result involving the placing of A in X which is satisfied in many cases [cf. Section 7.3].

We recall that a *retraction* of X onto A is a map $r : X \to A$ such that $r \mid A$ is the identity. If such a retraction exists we say A is a *retract* of X. To prove that A is a retract of X it is of course sufficient to produce a retraction $X \to A$—it is rather harder to prove that A is not a retract of X. A retraction $X \to A$ is surjective: so A is not a retract of X if X is connected and A is not connected. For example, $\mathbb{S}^0$ is not a retract of $\mathbb{E}^1$. We shall later be able to prove that $\mathbb{S}^1$ is not a retract of $\mathbb{E}^2$; the proof that $\mathbb{S}^n$ is not a retract of $\mathbb{E}^{n+1}$ for any n is beyond the scope of this book.

We say A is a *neighbourhood retract* of X if A is a retract of some neighbourhood of A in X. This neighbourhood may always be taken to be taken to be open since, if $A \subseteq \operatorname{Int} N$ and $r : N \to A$ is a retraction, then so also is $r \mid \operatorname{Int} N$.

EXAMPLES

1. A space X is a retract of itself. If $x \in X$ then $\{x\}$ is a retract of X. The product of retractions is a retraction; hence for any Y, $\{x\} \times Y$ is a retract of $X \times Y$.
2. $\mathbb{S}^n$ is a neighbourhood retract of $\mathbb{E}^{n+1}$ since the map

$$\begin{aligned} \mathbb{E}^{n+1} \setminus \{0\} &\to \mathbb{S}^n \\ x &\mapsto x/|x| \end{aligned}$$

is a retraction.

3. $\mathbb{L}$ is not a neighbourhood retract of $\mathbb{I}$. For suppose N is a neighbourhood of $\mathbb{L}$ in $\mathbb{I}$ and $r : N \to \mathbb{L}$ is continuous. For some $\varepsilon > 0$, N contains $[0, \varepsilon]$, which is connected. Hence r is constant on $[0, \varepsilon]$ and so cannot be a retraction.

4.6.5 *The adjunction space $B\ {}_f\!\sqcup\ X$ is Hausdorff if the following conditions hold:*

(a) B *and* X *are Hausdorff,*
(b) *Each* $x \in X \setminus A$ *has a neighbourhood closed in* X *and not meeting* A,
(c) A *is a neighbourhood retract of* X.

Proof Let $Q = B\ {}_f\!\sqcup\ X$ and let $y_1, y_2 \in Q$. In order to find disjoint neighbourhoods W_1, W_2 of y_1, y_2, we distinguish three cases.

(i) Suppose $y_1, y_2 \in X \setminus A$. Since $X \setminus A$ is Hausdorff there are disjoint sets W_1, W_2 which are neighbourhoods in $X \setminus A$ of y_1, y_2. But $X \setminus A$ is open in Q [4.5.5 (Corollary 1)]. So W_1, W_2 are also neighbourhoods in Q of y_1, y_2.
(ii) Suppose $y_1 \in X \setminus A$, $y_2 \in B$. Let W_1 be a neighbourhood of y_1 closed in X and not meeting A. Then W_1 is also a neighbourhood in Q of y_1.

Let $W_2 = Q \setminus W_1$. Then

$$\bar{\imath}^{-1}\imath[W_2] = B, \quad \bar{f}^{-1}[W_2] = X \setminus W_1$$

which are open in B, X respectively. Hence W_2 is open in Q.
(iii) Suppose $y_1, y_2 \in B$. Since B is Hausdorff there are in B disjoint open neighbourhoods V_1, V_2 of y_1, y_2. Then $f^{-1}[V_1], f^{-1}[V_2]$ are open in A, but not necessarily open in X. We therefore use the neighbourhood retraction to enlarge them into sets open in X.

Let $r : N \to A$ be a retraction where N is open in X. Let

$$V_i' = r^{-1}f^{-1}[V_i], \quad i = 1, 2.$$

Then V_1', V_2' are disjoint open sets open in N and so open in X. Notice also that because r is a retraction

$$V_i' \setminus A = V_i' \setminus f^{-1}[V_i], \quad i = 1, 2.$$

Let W_i be the subset of Q

$$W_i = V_i \cup (V_i' \setminus A), \quad i = 1, 2.$$

Then $\bar{\imath}^{-1}[W_i] = V_i$, and

$$\begin{aligned} \bar{f}^{-1}[W_i] &= \bar{f}^{-1}[V_i] \cup (V_i' \setminus A) \\ &= V_i'. \end{aligned}$$

Therefore W_i is open in Q. Clearly, $y_1 \in W_1$, $y_2 \in W_2$. □

(The existence, for each A closed in X and each $x \in X \setminus A$, of a closed neighbourhood of x not meeting A is one of the standard separation axioms. Such a space X is called T_3 by [Bou66], who also calls X *regular* if it is T_3 and Hausdorff. The opposite terminology is used by [Kel55].)

EXAMPLES

4. If X is a metric space and A is closed in X then condition (b) of 4.6.5 is always satisfied. For suppose $x \in X \setminus A$; then $\mathrm{dist}(x, A) = \varepsilon > 0$ [2.9.2] and $E(x, \varepsilon/2)$ is a closed neighbourhood of x not meeting A. The condition is also satisfied if X is Hausdorff and A is compact [3.5.6 (Corollary 3)]. If follows (for either reason!) that, if $f : \mathbb{S}^{n-1} \to B$ and B is Hausdorff, then $B\ {}_f\!\sqcup\ \mathbb{E}^n$ is Hausdorff.

5. Let X, Y be topological spaces and let $f : X \to Y$ be a map. Let $f' : X \times \{0\} \to Y$ be the map $(x, 0) \mapsto f(x)$. Since $X \times \{0\}$ is a closed subspace of $X \times \mathbb{I}$ we can define the adjunction space [Fig. 4.16]

$$M(f) = Y\ {}_{f'}\!\sqcup\ (X \times \mathbb{I}).$$

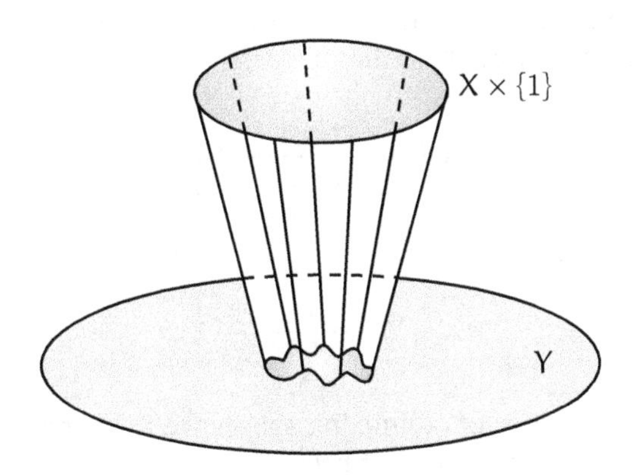

Fig. 4.16

This is the *mapping cylinder* of f. It is a construction of great use in chapter 7. If X is Hausdorff, then so also is $X \times \mathbb{I}$. It is easy to verify the conditions (b) and (c) of 4.6.5 with A, X replaced by $X \times \{0\}$, $X \times \mathbb{I}$. Hence, *if* X *and* Y *are Hausdorff, then so also is M(f).*

The *mapping cone* of f is

$$C(f) = M(f)/X \times \{1\}.$$

We can prove by a similar method that if X and Y are Hausdorff, then so also is $C(f)$.

Our next result will become vital when we come to consider homotopies of maps of adjunction spaces in chapters 7 and 9.

4.6.6 *Let Y be a locally compact space. In the following diagrams let the first square be a pushout; then the second square is a pushout.*

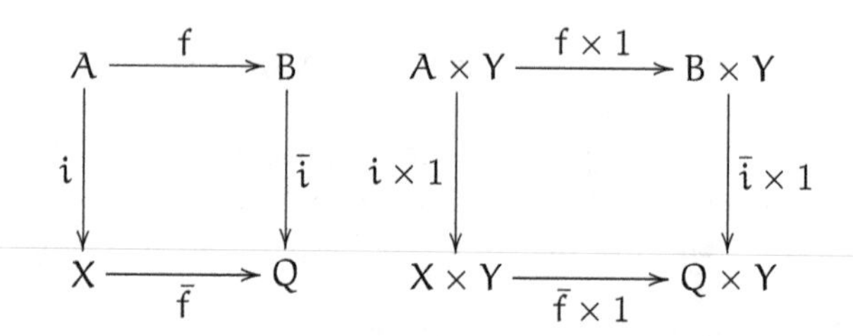

Proof As sets, $Q = B \sqcup (X \setminus A)$, $Q \times Y = (B \times Y) \sqcup (X \setminus A) \times Y$. So the only problem is the topology on $Q \times Y$. But since the map $F : B \sqcup X \to Q$ defined by $\bar{\imath}, \bar{f}$ is an identification map, so also is $F \times 1 : (B \sqcup X) \times Y \to Q \times Y$, by 4.3.2. This completes the proof. □

The main utility of this result is that to specify a map $Q \times Y \to Z$ it is sufficient to give a commutative diagram

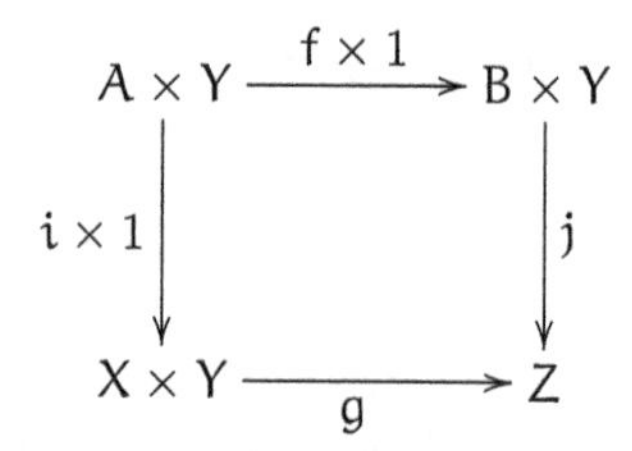

For further properties of adjunction spaces we refer the reader to §3 of chapter IV of [Hu64] and also to [Mic66].

Exercises

1. Prove that the adjunction space $B\ {}_f\!\sqcup\ X$ is a T_1-space if B and X are T_1-spaces.
2. Let X be a non-normal space. Prove that there are disjoint, closed subsets A, B of X such that $(X/A)/B$ is not Hausdorff.
3. Prove that $\mathbb{R}/\mathbb{Z}$ does not satisfy the first axiom of countability.
4. Prove the following generalisation of 4.6.5: an adjunction space $B\ {}_{(f_\lambda)}\!\sqcup\ (X_\lambda)$ is Hausdorff if for each λ in Λ (i) B and X_λ are Hausdorff, (ii) each x in $X_\lambda \setminus A_\lambda$ has a neighbourhood closed in X_λ and not meeting A_λ, (iii) A_λ is a neighbourhood retract of X_λ.

5. Let X, Y be spaces and let $x_0 \in X$, $y_0 \in Y$. Prove that if P is the subspace $X \times \{y_0\} \cup \{x_0\} \times Y$ of $X \times Y$ then the square of inclusions

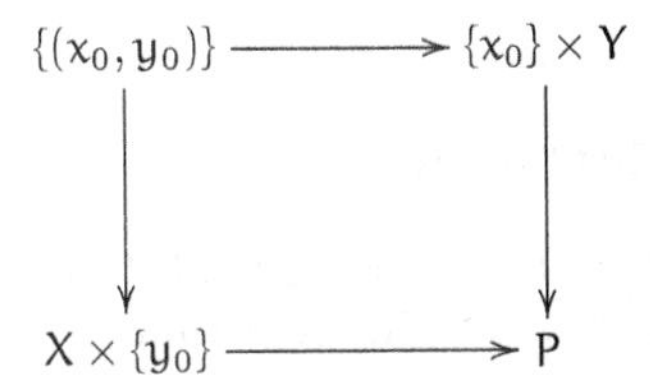

is a pushout square. Deduce that if x_0, y_0 are taken as base points of X, Y respectively, then P is homeomorphic to $X \vee Y$.
6. Prove that an adjunction space $B \,{}_f{\sqcup}\, X$ is normal if and only if B and X are normal. [Assume the Tietze Extension Theorem.]
7. Let the space X be the union of a family $(X_i)_{i \in \mathbb{N}}$ of subspaces such that (i) for each $i \in \mathbb{N}$, X_i is a closed subset of X_{i+1}, (ii) a set C is closed in X if and only if $C \cap X_i$ is closed in X_i for each i in $\mathbb{N}$. Prove that if each X_i is a normal space, then so also is X. [Assume the Tietze Extension Theorem.]
8. Let A be a closed subspace of X, and let $f : A \to B$, $g : B \to C$ be maps. Prove that the spaces

$$C \,{}_g{\sqcup}\, (B \,{}_f{\sqcup}\, X), \quad C \,{}_{gf}{\sqcup}\, X$$

are homeomorphic.
9. Let $i : A \to X$, $j : X \to Y$ be inclusions of closed subspaces, and let $f : A \to B$ be a map. Prove that the spaces

$$B \,{}_f{\sqcup}\, Y, \quad (B \,{}_f{\sqcup}\, X) \,{}_{\bar{f}}{\sqcup}\, Y$$

are homeomorphic.
10. Let A be a closed subspace of X and $f : A \to B$ a map. Prove that the subspace $L = (B \,{}_f{\sqcup}\, X) \times \{0\} \cup B \times \mathbb{I}$ of $M = (B \,{}_f{\sqcup}\, X) \times \mathbb{I}$ is homeomorphic to $(B \times \mathbb{I}) \,{}_{f \times 1}{\sqcup}\, (X \times \{0\} \cup A \times \mathbb{I})$, where $f \times 1 : A \times \mathbb{I} \to B \times \mathbb{I}$. Hence show that if $X \times \{0\} \cup A \times \mathbb{I}$ is a retract of $X \times \mathbb{I}$, then L is a retract of M.

4.7 Cell complexes

In order to emphasise the intuitive ideas, we shall restrict ourselves to finite cell complexes. Indeed the theory for infinite cell complexes involves, in the main, arranging the topologies so that theorems and proofs for the finite case carry over almost without change to the infinite case.

There are two useful ways of thinking about cell complexes: (a) constructive, (b) descriptive. In the first approach, we simply construct cell complexes by starting in dimension -1 with the empty set $\varnothing$, and then attach cells to $\varnothing$ in order of increasing dimension. In the second approach, we suppose there is given a topological space X, and seek to describe X in a useful way as the union of open cells. In both approaches, a vital role is played by the sequence of *skeletons*, the n-skeleton being the union of all cells of dimension $\leqslant n$.

The constructive approach

We construct a sequence of spaces K^n such that $K^n \subseteq K^{n+1}$, $n \geqslant -1$. The space K^{-1} is to be the empty set. Suppose that K^{n-1} has been constructed. We suppose given maps

$$f_1^n, \ldots, f_{r_n}^n : \mathbb{S}^{n-1} \to K^{n-1}.$$

Then K^n is to be the adjunction space

$$K^{n-1} {}_{f_1^n}\sqcup \mathbb{E}^n \cdots {}_{f_{r_n}^n}\sqcup \mathbb{E}^n.$$

For $n > 0$, we allow the case $r = 0$, when $K^n = K^{n-1}$. In fact, since we are dealing only with finite cell complexes, we shall assume that for some N, $K^n = K^{n-1}$ for $n > N$, and define $K = K^N$. Note that if K^0 is empty, so also is K.

Let us consider the intuitive picture, in low dimensions, of a non-empty cell complex. The 0-skeleton K^0 is a non-empty, finite, discrete space. The 1-skeleton K^1 is formed by attaching to K^0 a finite number (possibly 0) of 1-cells by means of maps $f_1^1, \ldots, f_{r_1}^1 : \mathbb{S}^0 \to K^0$. Now $\mathbb{S}^0$ is the discrete space $\{-1, +1\}$. So for each $i = 1, \ldots, r_1$ the map f_i^1 is either constant or maps to two distinct points of K^0. Thus a representative picture for K^1 is Fig. 4.17, where the dots denote elements of K^0.

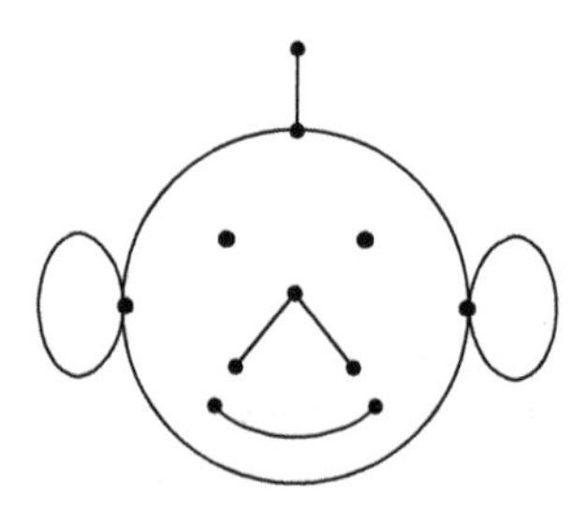

Fig. 4.17

Next we form K^2 by attaching a finite number of 2-cells to K^1 by means of maps $f_1^2, \ldots, f_{r_2}^2 : \mathbb{S}^1 \to K^1$. It can be shown that if K^1 has 1-cells (i.e., if $K^1 \neq K^0$) then there is an uncountable number of maps $\mathbb{S}^1 \to K^1$. It is even possible to construct an uncountable set of spaces of the form $\mathbb{I}\ {}_f\sqcup\ \mathbb{E}^2$, no two of which are homeomorphic.

The pictures in Fig. 4.18 are of two simple cell complexes of the form $\mathbb{E}^1\ {}_f\sqcup\ \mathbb{E}^2$, where $K^0 = \{-1, +1\}$, $K^1 = \mathbb{E}^1$. In (a), $f[\mathbb{S}^1] = \{+1\}$; in (b), $f[\mathbb{S}^1] = \{0\}$.

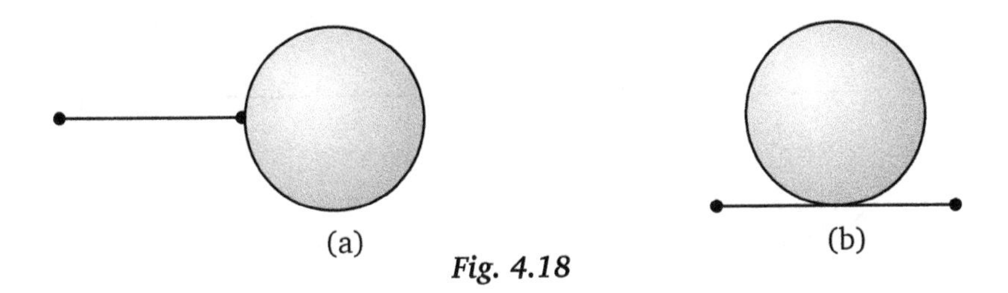

Fig. 4.18

It may be felt that, because of the wide latitude in the attaching maps, cell complexes are bizarre spaces, and that the attaching maps should be restricted in some way, for example to be the embeddings. However, many important spaces, for example the projective spaces, have natural cell structures in which the attaching maps are not embeddings. Also we shall be considering homotopies of attaching maps in chapter 7 and any restriction on these maps would be inconvenient.

The burden of this section is that cell complexes are always 'good' spaces, although additional restrictions on the attaching maps might make them 'better'. In any case, by results of the last section, a cell complex is always a compact, Hausdorff space.

The descriptive approach

We recall that an open n-cell (also called an n-ball) is a space e homeomorphic to the standard n-ball $\mathbb{B}^n$. The *dimension* of e is the natural number $\dim e = n$. However the proof that this dimension is well-defined is non-trivial, and depends on the Invariance of Dimension: if $f : \mathbb{R}^m \to \mathbb{R}^n$ is a homeomorphism, then $m = n$. Since $\mathbb{B}^n$ is homeomorphic to $\mathbb{R}^n$, it follows that if e is homeomorphic to $\mathbb{B}^m$ and to $\mathbb{B}^n$ then $m = n$; hence $\dim e$ is well-defined.

An idea which occurred early in topology is that of a decomposition of a space Q by open cells. In this, Q is given as the union $\bigcup_{\lambda \in \Lambda} e_\lambda$, where e_λ is an open n_λ-cell and the sets e_λ are disjoint.

The difficulty has always been what to do with such a decomposition, since it gives relatively little information on the space Q. For example, the 'bad' space Q of Exercise 1 of Section 4.5 has a decomposition with two 0-cells, two 1-cells and one 2-cell. The two spaces of Fig. 4.19 have decompositions with one 0-cell, and two 1-cells. Yet by simple local cut point arguments they are not homeomorphic.

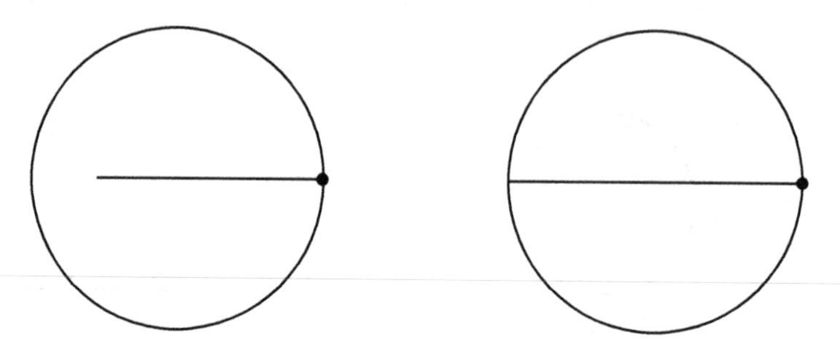

Fig. 4.19

Let us consider what we are given with a decomposition by open cells, $Q = \bigcup_{\lambda\in\Lambda} e_\lambda$. For each λ in Λ, there is a homeomorphism $k_\lambda : \mathbb{B}^{n_\lambda} \to e_\lambda$. Since $\dim e_\lambda = n_\lambda$ is uniquely defined, for each natural number n the n-*skeleton* of Q is well-defined by

$$Q^n = \bigcup_{n_\lambda \leqslant n} e_\lambda.$$

J. H. C. Whitehead realised that a satisfactory theory could be built by assuming the extra condition that the homeomorphism $k_\lambda : \mathbb{B}^{n_\lambda} \to e_\lambda$ extends to a map $h_\lambda : \mathbb{E}^{n_\lambda} \to Q$ such that $h_\lambda[\mathbb{S}^{n_\lambda - 1}] \subseteq Q^{n_\lambda - 1}$. This ensures that e_λ is part of the adjunction of a closed n_λ-cell to $Q^{n_\lambda-1}$; it takes us to the situation considered before and, from the technical point of view, allows proofs by induction on the skeletons.

The last sentence is not quite correct. The reason for this is that if the number of cells is infinite then the cells do not uniquely determine the topology of Q. For example, any Q has a decomposition in which each point of Q is a 0-cell. Therefore we restrict ourselves to the finite case, which retains the geometric ideas without additional topological difficulties.

It is convenient in defining a cell complex to keep the above maps h_λ as part of the structure. It is also necessary to assume that Q is Hausdorff.

Let Q be a non-empty Hausdorff topological space. A *cell structure* on Q is a finite set of maps

$$h_\lambda : \mathbb{E}^{n_\lambda} \to Q, \quad \lambda \in \Lambda,$$

called the *characteristic maps*. Let

$$e_\lambda = h_\lambda[\mathbb{B}^{n_\lambda}].$$

The n-*skeleton* of the cell structure is

$$Q^n = \bigcup_{n_\lambda \leqslant n} e_\lambda.$$

Then we require that the characteristic maps satisfy:
CS1 $h_\lambda \mid \mathbb{B}^{n_\lambda}, e_\lambda$ *is a bijection.*
CS2 *the* e_λ *are disjoint and* $Q = \bigcup_\lambda e_\lambda$,
CS3 $h_\lambda[\mathbb{S}^{n_\lambda - 1}] \subseteq Q^{n_\lambda - 1}$.
A *cell complex* is a space Q with a cell structure on it. It is usual to denote the cell complex simply by Q.

The space $\mathbb{E}^n$ is compact and Q^{n-1} is Hausdorff. Therefore Exercise 3 of Section 4.5 implies that Q^n is homeomorphic to an adjunction space,

$$Q^{n-1} {}_{f_1}\sqcup \mathbb{E}^n \cdots {}_{f_n}\sqcup \mathbb{E}^n$$

where f_i is a map $\mathbb{S}^{n-1} \to Q^{n-1}$. This shows how the present definition links with the constructive approach considered first.

Another easy consequence of the definition is that

$$\overline{e_\lambda} = h_\lambda[\mathbb{E}^{n_\lambda}] :$$

for the continuity of h_λ implies that

$$h_\lambda[\mathbb{E}^{n_\lambda}] = h_\lambda\overline{[\mathbb{B}^{n_\lambda}]} \subseteq \overline{h_\lambda[\mathbb{B}^{n_\lambda}]};$$

on the other hand $h[\mathbb{E}^{n_\lambda}]$ is a compact, and hence closed, set containing e_λ.

4.7.1 *Let* Q *be a cell complex and* $g : Q \to X$ *a function to a topological space* X. *The following conditions are equivalent.*

(a) g *is continuous.*
(b) gh_λ *is continuous for each characteristic map* h_λ.
(c) $g \mid \overline{e_\lambda}$ *is continuous for each cell* e_λ.

Proof Obviously (a) $\Rightarrow$ (b) and (a) $\Rightarrow$ (c). Also (c) $\Rightarrow$ (b) since $h_\lambda[\mathbb{E}^{n_\lambda}] = \overline{e_\lambda}$. Since $\mathbb{E}^{n_\lambda}$ is compact and Q is Hausdorff, the map $h_\lambda \mid \mathbb{E}^{n_\lambda}, \overline{e_\lambda}$ is an identification map.

To complete the proof we show that (b) $\Rightarrow$ (a). The proof is by induction on the skeletons.

Since Q^0 is discrete, $g \mid Q^0$ is certainly continuous. But if $g \mid Q^{n-1}$ is continuous, then by Exercise 3 of Section 4.5 so also is $g \mid Q^n$. Since $Q = Q^n$ for some n, it follows that g is continuous. □

In order to extend the present theory to infinite cell complexes, it is necessary to take the last result as one of the properties characterising the topology of Q.

The smallest integer n such that $Q = Q^n$ is called the *dimension* of the cell complex Q. The methods which are used to prove the invariance of dimension can also prove that the dimension of a cell complex is a topological invariant.

Our next result shows one of the ways in which cell complexes are 'nice' spaces.

4.7.2 *If Q is a cell complex, then the following conditions are equivalent.*

(a) Q^1 *is connected,*
(b) Q *is path-connected,*
(c) Q *is connected.*

Proof Clearly, (b) $\Rightarrow$ (c). We prove also (c) $\Rightarrow$ (a) $\Rightarrow$ (b).

(c) $\Rightarrow$ (a) The n-sphere $\mathbb{S}^n$ is connected for $n \geqslant 1$. Further Q^{n+1} is homeomorphic to a space obtained by attaching $(n+1)$-cells to Q^n. Therefore by 4.6.4, if Q^{n+1} is connected and $n \geqslant 1$, then Q^n is connected. But $Q^{n+1} = Q$ for some n. So the result follows by downward induction.

(a) $\Rightarrow$ (b) By 4.6.3 and induction on the skeletons, if Q^1 is path-connected then so also is Q. So it is sufficient to prove the more general result that each path-component of Q^1 is both open and closed in Q^1.

Let P be a path-component of Q^1. If P contains no 1-cells of Q, the P consists of an isolated point, which is certainly both open and closed in Q^1. We suppose then that P contains more than one point.

A point x of P must belong to $\bar{e}^1$ for some open cell e^1, for otherwise x would be an isolated point. On the other hand, each $\bar{e}^1$ is path-connected (since it is the continuous image of $\mathbb{E}^1$) and so $\bar{e}^1$ is either contained in or disjoint from P. Therefore P is the union of a finite number of $\bar{e}^1$, and so P is closed.

We now prove P is open. Let $x \in P$. If $x \in Q^0$, then the set N, consisting of x and all open 1-cells e^1 such that $x \in \bar{e}^1$, is a path-connected neighbourhood of x, and so $N \subseteq P$. If $x \in Q^1 \setminus Q^0$, then x belongs to some open 1-cell e^1; clearly e^1 is contained in P, and therefore P is a neighbourhood of x. □

Let Q be a cell complex with characteristic maps h_λ, $\lambda \in \Lambda$. Let P be a non-empty subset of Q and let M be the set of λ in Λ such that the image of h_λ is contained in P. For each λ in M, let $g_\lambda : \mathbb{E}^{n_\lambda} \to P$ be the restriction of $h_\lambda : \mathbb{E}^{n_\lambda} \to Q$. We say P is a *subcomplex* of Q if the characteristic maps g_λ, $\lambda \in M$, form a cell structure on P. In this case P is covered by the open cells e_λ for λ in M and $Q \setminus P$ is covered by the open cells e_λ for λ in $M \setminus \Lambda$.

EXAMPLES
1. The n-skeleton Q^n of Q is a subcomplex of Q.
2. The intersection and the union of a family of subcomplexes of Q are again subcomplexes of Q.
3. Let X be any subset of Q. Then the intersection of all subcomplexes of Q containing X is the smallest subcomplex of Q containing X.

Let P, R be cell complexes and let $f : P \to R$ be a continuous function. We say f is a *cellular map* if

$$f[P^n] \subseteq R^n, \quad n \geqslant 0.$$

4.7.3 *Let* Q, R *be cell-complexes, let* P *be a subcomplex of* Q *and let* $f : P \to R$ *be a cellular map. Then the adjunction space* $R \,{}_f\!\sqcup\, Q$ *can be given the structure of a cell-complex.*

Proof The open cells of $R \,{}_f\!\sqcup\, Q$ are the open cells of R and of $Q \setminus P$. In order to describe the characteristic maps, let $\bar{f} : Q \to R \,{}_f\!\sqcup\, Q$, $\bar{i} : R \to R \,{}_f\!\sqcup\, Q$ be the usual maps. Let h_λ, $\lambda \in \Lambda$, be the characteristic maps of Q; g_λ, $\lambda \in M$, those of P; and $k_\nu, \nu \in N$, those of R. If e_ν is an open cell of R, then its characteristic map in $R \,{}_f\!\sqcup\, Q$ is to be $\bar{i}k_\nu$. If e_λ is an open cell of $Q \setminus P$, then its characteristic map in $R \,{}_f\!\sqcup\, Q$ is to be $\bar{f}h_\lambda$.

The conditions CS1 and CS2 are obviously satisfied, as is CS3 for the maps $\bar{i}k_\nu$, $\nu \in N$. Since f is cellular, so also is $\bar{f}$. Hence CS3 is satisfied for the maps $\bar{f}h_\lambda$, $\lambda \in \Lambda \setminus M$. □

We now show that the product of cell complexes is a cell complex. Let P, Q be cell complexes with characteristic maps f_λ, $\lambda \in \Lambda$, g_μ, $\mu \in M$ respectively. Since P is the union of disjoint open cells e_λ, and Q is the union of disjoint open cells e_μ, $P \times Q$ is the union of the disjoint sets $e_\lambda \times e_\mu$. But $e_\lambda \times e_\mu$ is an open $(n_\lambda + n_\mu)$-cell. So the n-skeleton $(P \times Q)^n$ is well defined and

$$P^m \times Q^n \subseteq (P \times Q)^{m+n}.$$

Suppose that e_λ is an open m-cell of P, and e_μ is an open n-cell of Q. Let $h_{mn} : \mathbb{E}^{m+n} \to \mathbb{E}^m \times \mathbb{E}^n$ be the homeomorphism constructed in 4.4.5—we recall that

$$h_{mn}[\mathbb{S}^{m+n-1}] = \mathbb{S}^{m-1} \times \mathbb{E}^n \cup \mathbb{E}^m \times \mathbb{S}^{n-1}.$$

Let $k_{\lambda\mu}$ be the composite

$$\mathbb{E}^{m+n} \xrightarrow{h_{mn}} \mathbb{E}^m \times \mathbb{E}^n \xrightarrow{f_\lambda \times g_\mu} P \times Q.$$

Then $k_{\lambda\mu}$ maps $\mathbb{B}^{m+n}$ bijection onto $e_\lambda \times e_\mu$ and $k_{\lambda\mu}[\mathbb{S}^{m+n-1}]$ is contained in $P^{m-1} \times Q^n \cup P^m \times Q^{n-1}$ which is itself contained in $(P \times Q)^{m+n-1}$. This shows that the $k_{\lambda\mu}$ are characteristic maps for a cell structure on $P \times Q$.

When exhibiting a space as a cell complex it is common practice to write it simply as the union of open cells, and say afterwards what are the characteristic maps. For example, the m-sphere has a cell structure with one m-cell and one 0-cell, and we therefore write

$$\mathbb{S}^m = e^0 \cup e^m.$$

Also, if e^m is an open m-cell of a cell complex Q with characteristic map $h : \mathbb{E}^m \to Q$, we make an abuse of language and call $h \mid \mathbb{S}^{m-1}, Q^{m-1}$ the attaching map of e^m. In particular, we say that the cell e^m of $\mathbb{S}^m$ is attached by the constant map.

Consider now $\mathbb{S}^m \times \mathbb{S}^n$. If we take $\mathbb{S}^m = e^0 \cup e^m$, $\mathbb{S}^n = e^0 \cup e^n$, then we can write

$$\mathbb{S}^m \times \mathbb{S}^n = e^0 \cup e^m \cup e^n \cup e^{m+n}.$$

Here both e^m and e^n are attached by constant maps, so that $e^0 \cup e^m \cup e^n$ is homeomorphic to $\mathbb{S}^m \vee \mathbb{S}^n$. Thus $\mathbb{S}^m \times \mathbb{S}^n$ is, up to homeomorphism, $(\mathbb{S}^m \vee \mathbb{S}^n) \cup e^{m+n}$. The attaching maps of the $(m+n)$-cell is a map $w : \mathbb{S}^{m+n-1} \to \mathbb{S}^m \vee \mathbb{S}^n$ called the *Whitehead product map*. [cf. Exercise 2 of Section 5.7 for a generalisation of this map].

EXERCISES

1. Prove that the composite, and product, of cellular maps is cellular. Is the diagonal map $Q \to Q \times Q$ cellular?
2. Prove that if Q_1, Q_2 are cell complexes, then so also is $Q_1 \sqcup Q_2$.
3. Let Q, R be cell complexes, let $i : P \to Q$ be the inclusion of the subcomplex P of Q, and let $f : P \to R$ be cellular. Prove that if K is a cell complex and $i' : R \to K$, $f' : Q \to K$ are cellular maps such that $i'f = f'i$, then the unique map $g : R \sqcup_f Q \to K$ such that $g\bar{f} = f'$, $g\bar{i} = i'$ is cellular.
4. Read an account of the classification of surfaces [for example in [Cai61]] and give cell structures for the normal forms of surfaces.

The following exercises outline a part of the theory of infinite cell complexes.

5. Let Q be a non-empty, not necessarily Hausdorff, space, and suppose given on Q a cell structure $\{h_\lambda\}_{\lambda \in \Lambda}$ in the sense of the definition on p. 131 except that Λ is not supposed finite. We say Q is a *CW-complex* if the following axioms hold:

CW1 A set $C \subseteq Q^n$ is closed in Q^n if and only if $C \cap Q^{n-1}$ is closed in Q^{n-1} and $h_\lambda^{-1}[C]$ is closed in $\mathbb{E}^n$ for each λ such that $n_\lambda = n$.

CW2 A set $C \subseteq Q$ is closed in Q if and only if $C \cap Q^n$ is closed in Q^n for each n.

Let $\Lambda_n = \{\lambda \in \Lambda : n_\lambda = n\}$ and let Λ_n have the discrete topology. Define $q : Q^{n-1} \sqcup (\Lambda_n \times \mathbb{E}^n) \to Q^n$ to be $x \mapsto x$ on Q^{n-1} and $(\lambda, e) \mapsto h_\lambda e$ on $\Lambda_n \times \mathbb{E}^n$.

Prove that CW1 is equivalent to: Q^n has the identification topology with respect to q.

In the following exercises, Q denotes a CW-complex with cell structure $\{h_\lambda\}_{\lambda \in \Lambda}$.

6. Prove that a function $f : Q^n \to Y$ is continuous if and only if $f \mid Q^{n-1}$ is continuous and $fh_\lambda : \mathbb{E}^n \to Y$ is continuous for each λ in Λ with $n_\lambda = n$. Prove that a function $f : Q \to Y$ is continuous if and only if $f \mid Q^n$ is continuous for each n. Deduce that $f : Q \to Y$ is continuous if and only if fh_λ is continuous for each λ in Λ.

7. Let C be a closed subset of Q and $g : C \to \mathbb{I}$ any map. Prove that g extends to a map $f : Q \to \mathbb{I}$ and thus prove that Q is normal. Prove that each point of Q is closed, and hence show that Q is Hausdorff. [You may assume the Tietze extension theorem.]

8. Prove that any compact subset of Q is contained in a finite union of cells e_λ. [If $C \subset Q$ meets an infinite number of cells, choose points c_i of C, $i = 1, 2, \ldots$, which lie in distinct cells e_{λ_i}. Define $gc_i = 1/i$ and extend g over Q by Exercise 7.]

9. A CW-complex Q is said to be *locally finite* if each point x of Q has a neighbourhood N such that N is contained in a finite subcomplex of Q. Prove that Q is locally finite if and only if it is, as a topological space, locally compact.

10. Prove that a CW-complex is a k-space. Prove that if P, Q are CW-complexes then $P \times Q$, with the cell structure given in the present section, is a CW-complex if and only if $P \times Q$ is a k-space. Hence show that $P \times Q$ is a CW-complex if P or Q is locally finite. [Prove first that the weak product space $P \times_W Q$ of Exercise 12 of Section 4.3 is always a CW-complex.]

NOTES

The idea of adjunction space is due to J. H. C. Whitehead, for whom the concept evolved over a period of about ten years. I heard it said that he took a year to prove that the product of CW-complexes one of which is locally finite is also a CW-complex. The result 4.6.6 on the product of an identification map is essentially due to Whitehead.

There are various books which deal with the case of infinite cell complexes, that is, CW-complexes. For applications of these complexes in algebraic topology, see [Spa62] and [Whi78], as well as the basic paper on the topic, [Whi49].

The fact that 4.6.6 is false without the locally compact assumption lead R. Brown in his thesis ([Bro61]) to suggest Hausdorff k-spaces as an appropriate tool for topology. See also [Bro63], [Bro64] and section 5.9. Thus 4.6.6 gives one indication that the notion of topological space should not be regarded as the final setting for the concept of continuity.

Chapter 5

Projective and other spaces

5.1 Quaternions

In this section, we construct the algebra $\mathbb{H}$ of quaternions, and we show that $\mathbb{H}$ is a field. The word field is here used in a slightly more general sense than is usual, since the multiplication of $\mathbb{H}$ is non-commutative. In fact, the term skew field is often used in this context.

As a set, and in fact as a vector space over $\mathbb{R}$, $\mathbb{H}$ is just $\mathbb{R} \times \mathbb{R}^3$. Thus, if $q \in \mathbb{H}$ then $q = (\lambda, x)$ where λ is real and x is in $\mathbb{R}^3$; we call λ the *real part* of q, x the *vector part* of q and write

$$\operatorname{Re}(q) = \lambda, \quad \operatorname{Ve}(q) = x.$$

It is convenient to identify $(\lambda, 0)$ with λ and $(0, x)$ with x, and so to write

$$q = \lambda + x.$$

We shall, as far as possible, keep this convention of using Greek letters for the real part, and Roman letters for the vector part, of a quaternion. Thus the addition and scalar multiplication of the vector space structure of $\mathbb{H}$ is given by

$$(\lambda + x) + (\lambda' + x') = \lambda + \lambda' + x + x'.$$
$$\lambda'(\lambda + x) = \lambda'\lambda + \lambda' x.$$

We now define a distributive multiplication on $\mathbb{H}$. To do this it suffices to define the product xy of vectors, since the product of quaternions $q = \lambda + x$, $r = \mu + y$ must then, by distributivity, be defined by

$$qr = \lambda\mu + \lambda y + \mu x + xy \tag{5.1.1}$$

However, xy will not be a vector—in terms of the usual scalar and vector product in $\mathbb{R}^3$ we set

$$xy = -x \cdot y + x \times y \tag{5.1.2}$$

The scalar and vector product of vectors are bilinear. It follows from this and (5.1.1), (5.1.2) that multiplication of quaternions is bilinear—that is, for any quaternions q, r, s and real number λ

$$\begin{aligned} q(r + \lambda s) &= qr + \lambda(qs), \\ (r + \lambda s)q &= rq + \lambda(sq). \end{aligned} \tag{5.1.3}$$

One of the points of definition (5.1.2) is that it replaces the non-associative vector product by an associative product. By well-known rules

$$\begin{aligned} x(yz) &= x(-y \cdot z + y \times z) \\ &= -(y \cdot z)x - x \cdot y \times z + x \times (y \times z) \\ &= -x \cdot y \times z - (y \cdot z)x + (x \cdot z)y - (x \cdot y)z \end{aligned}$$

and a similar computation gives the same value for $(xy)z$. Hence $x(yz) = (xy)z$. It follows from this and (5.1.1) that for any quaternions q, r, s

$$q(rs) = (qr)s;$$

that is, multiplication of quaternions is associative.

A vector a of $\mathbb{R}^3$ is a unit vector if $a \cdot a = 1$; this is clearly equivalent to $a^2 = -1$. Two vectors a, b are orthogonal, that is $a \cdot b = 0$, if and only if ab is a vector. In such case

$$ab = a \times b = -b \times a = -ba.$$

Conversely, if $ab = -ba$, then by (5.1.2) $a \cdot b = 0$ and so a and b are orthogonal.

The ordered set a, b, c of vectors is said to form a *right-handed orthonormal system* if a, b, c are of unit length, are mutually orthogonal, and $a \times b = c$.

5.1.4 *The set of vectors a, b, c is a right-handed orthonormal system if and only if*

(a) $a^2 = b^2 = c^2 = -1$,
(b) $abc = -1$.

Proof Condition (a) holds if and only if a, b, c are of unit length. If further a, b are orthogonal and $a \times b = c$, then

$$-1 = c^2 = (a \times b)c = abc.$$

Conversely, given (a) and (b), then

$$ab = -abc^2 = c.$$

Thus ab is a vector and so $a \cdot b = 0$, $a \times b = c$. Clearly c is orthogonal to a and b. □

Suppose that a, b, c form a right-handed orthonormal system. Then clearly

$$ab = c = -ba, \quad bc = a = -cb, \quad ca = b = -ac. \tag{5.1.5}$$

Now any vector x of $\mathbb{R}^3$ can be written uniquely as

$$x = x_1 a + x_2 b + x_3 c, \quad x_1, x_2, x_3 \in \mathbb{R}.$$

Therefore, the product xy of vectors is determined uniquely by 5.1.4(a), (5.1.5) and the condition of bilinearity—in fact, these rules imply the expression (5.1.2) for xy. Now any quaternion q can be written uniquely as

$$q = q_0 + q_1 a + q_2 b + q_3 c, \quad q_i \in \mathbb{R}.$$

The rules given allow us to work out the product of q and $q' = q_0' + q_1' a + q_2' b + q_3' c$—we write out the complete formula for the one and only time:

$$\begin{aligned} qq' = {} & q_0 q_0' - q_1 q_1' - q_2 q_2' - q_3 q_3' \\ & +(q_0 q_1' + q_1 q_0' + q_2 q_3' - q_3 q_2')a \\ & +(q_0 q_2' + q_2 q_0' + q_3 q_1' - q_1 q_3')b \\ & +(q_0 q_3' + q_3 q_0' + q_1 q_2' - q_2 q_1')c. \end{aligned} \tag{5.1.6}$$

The real number 1 acts on the quaternions as identity: $1q = q1 = q$ for any quaternion q. We prove that for any non-zero quaternion q there is a quaternion q^{-1} such that $qq^{-1} = q^{-1}q = 1$.

Let $q = \lambda + x$. The *conjugate* of q is defined by

$$\bar{q} = \lambda - x.$$

If x, y are vectors then

$$\begin{aligned} \overline{xy} &= x \cdot y - x \times y \\ &= y \cdot x + y \times x \\ &= yx \\ &= \bar{y}\bar{x}. \end{aligned}$$

It follows from this and (5.1.1) that for any quaternions q, r,

$$\overline{qr} = \bar{r}\bar{q}. \tag{5.1.7}$$

Notice also that $\overline{q+r} = \bar{q} + \bar{r}$ and that if λ is real then $\overline{\lambda q} = \lambda \bar{q}$.

Let $|x|$ denote, as usual, the square root of $x \cdot x$. Then, if $q = \lambda + x$, we have

$$q\bar{q} = \lambda^2 + x \cdot x = \bar{q}q. \tag{5.1.8}$$

So we may define the *modulus* of q to be

$$|q| = (q\bar{q})^{\frac{1}{2}}.$$

Thus $|q|$ is the Euclidean norm of q when q is considered as an element of $\mathbb{R}^4$ (under the identification $\mathbb{R} \times \mathbb{R}^3 = \mathbb{R}^4$).

Let $q, r \in \mathbb{H}$. Then

$$\begin{aligned} |qr|^2 &= (qr)(\overline{qr}) = qr\bar{r}\bar{q} \\ &= q|r|^2\bar{q} = q\bar{q}|r|^2 \\ &= |q|^2|r|^2. \end{aligned}$$

This proves the important rule

$$|qr| = |q||r|.$$

Clearly, $q = 0$ if and only if $|q| = 0$. Therefore, if $q \neq 0$

$$q(\bar{q}/|q|^2) = (\bar{q}/|q|^2)q = 1;$$

so we write $q^{-1} = \bar{q}/|q|^2$, and call this quaternion the inverse of q.

The quaternions satisfy all the usual axioms for a field, except that the multiplication is not commutative. We also regard $\mathbb{H}$ as carrying the structures of vector space over $\mathbb{R}$, the conjugation function $q \mapsto \bar{q}$, the modulus $q \mapsto |q|$ and also the topology induced by this modulus (this topology is, of course, the usual topology for $\mathbb{R} \times \mathbb{R}^3$). Thus $\mathbb{H}$, like $\mathbb{R}$ and $\mathbb{C}$, has a rich structure and this is its advantage and interest.

For any unit vector x the set of quaternions $\lambda + \mu x$ for $\lambda, \mu \in \mathbb{R}$ is a subfield of $\mathbb{H}$ isomorphic to the complex numbers under the function $\lambda + \mu x \mapsto \lambda + \mu i$. In particular, if we let i be the vector $(1, 0, 0)$ of $\mathbb{H}$ and identify i with the complex number i, then we can regard $\mathbb{C}$ as a subfield of $\mathbb{H}$. We emphasise, however, that the elements of $\mathbb{C}$ do not commute with the elements of $\mathbb{H}$ since, if j is the vector $(0, 1, 0)$ of $\mathbb{H}$, then $ij = -ji$.

There are two generalisations of the quaternions. The *Cayley numbers*, or *octonions*, $\mathbb{O}$, are the elements of $\mathbb{R}^8$ with a distributive multiplication with identity which also satisfies $|xy| = |x||y|$, $x, y \in \mathbb{R}^8$. Also, for any $x \neq 0$ in $\mathbb{R}^8$, there is an element x^{-1} such that $xx^{-1} = x^{-1}x = 1$. However, this multiplication is non-associative. An account of the Cayley numbers, and also a proof that the *only multiplications on* $\mathbb{R}^n$ *which are bilinear and satisfy* $|xy| = |x||y|$, *are for* $n = 1, 2, 4, 8$, *and in these cases the resulting algebras are isomorphic to* $\mathbb{R}, \mathbb{C}, \mathbb{H}$, *and* $\mathbb{O}$, is given in [Alb63] (cf. also [Kur63]).

A multiplication on $\mathbb{R}^n$ is said to have *divisors of zero* if there are non-zero elements x, y of $\mathbb{R}^n$ such that $xy = 0$. It is true that $\mathbb{R}^n$ *has a bilinear multiplication with no divisors of zero only for* $n = 1, 2, 4,$ *or* 8. [cf. [Mil58].]

A different type of generalisation of the quaternions is the sequence of Clifford algebras C_n, $n \geqslant 1$. These are associative, but have divisors of zero for $n > 2$. They are closely related to orthogonal transformations of $\mathbb{R}^m$. For an account of octonions and Clifford algebras, see [Por69].

EXERCISES

1. Let q be a quaternion and let y be a vector orthogonal to $\mathrm{Ve}(q)$. Prove that $qy = y\bar{q}$.
2. Let q, r be quaternions such that $|q| = |r|$, $\mathrm{Re}\, q = \mathrm{Re}\, r$. Prove that there is a unit quaternion s such that $qs = sr$.
3. Let y be a unit vector. Prove that the mapping $\mathbb{R}^3 \to \mathbb{R}^3$ given by $x \mapsto yxy$ is reflection in the plane through the origin and perpendicular to y.
4. For any quaternion q, let $L_q : \mathbb{H} \to \mathbb{H}$ be the function $x \mapsto qx$. Prove that L_q is $\mathbb{R}$-linear and that

$$L_q L_{\bar{q}} = |q|^2 L_1.$$

Let L_q denote also the 4×4 real matrix of L_q with respect to a basis $1, a, b, c$ of $\mathbb{H}$ where a, b, c is a right-handed orthonormal set. Prove that $L_{\bar{q}}$ is the transpose of L_q and deduce that $\det(L_q) = |q|^4$.

5. Since $\mathbb{C}$ is a subfield of $\mathbb{H}$, we can regard $\mathbb{H}$ as a right vector space over $\mathbb{C}$. Prove that $\mathbb{H}$ is of dimension 2 over $\mathbb{C}$ with basis $1, j$ (where $j = (0, 1, 0)$). Prove that the function $L_q : \mathbb{H} \to \mathbb{H}$ of the previous exercise is $\mathbb{C}$-linear and that if $q = z + jw$ $(z, w \in \mathbb{C})$ then L_q has matrix

$$M_q = \begin{bmatrix} z & -\bar{w} \\ w & \bar{z} \end{bmatrix}$$

Prove that $M_{q'} M_q = M_{qq'}$. Hence such that $\mathbb{H}$ is isomorphic to the vector space $\mathbb{C}^2$ over $\mathbb{C}$ with multiplication given by

$$(z', w')(z, w) = (z'z - \bar{w}'w, w'z + \bar{z}'w).$$

6. An integer n is said to be a 4-square if there are integers n_1, n_2, n_3, n_4 such that $n = n_1^2 + n_2^2 + n_3^2 + n_4^2$. Prove that the product of 4-square integers is 4-square. [It may be proved also that any prime number is 4-square.]
7. Let a, b, c be quaternions. Prove that the equation

$$aq + qb = c$$

has a unique solution for q if $2a\,\mathrm{Re}\,(b) + a^2 = |b|^2 \neq 0$.

5.2 Normed vector spaces again

Let $\mathbb{K}$ denote one of the fields $\mathbb{R}, \mathbb{C}$, or $\mathbb{H}$. We write $d = d(\mathbb{K})$ for the dimension of $\mathbb{K}$ as a vector space over $\mathbb{R}$, so that

$$d(\mathbb{R}) = 1, \quad d(\mathbb{C}) = 2, \quad d(\mathbb{H}) = 4. \tag{5.2.1}$$

Let V be a vector space over $\mathbb{K}$. The theory of linear dependence, bases and dimension of V is usually given for the case $\mathbb{K}$ is commutative and V is a left vector space over $\mathbb{K}$. However, the change from left to right vector spaces is purely formal, and the usual proofs of the basis theorems do not use the commutativity of the field. We therefore assume this theory as known. The reason for using right vector spaces will be clear later when discussing the matrix of a linear transformation.

An example of a finite dimensional vector space over $\mathbb{K}$ is of course $\mathbb{K}^n$; the *standard ordered basis* of $\mathbb{K}^n$ consists of the elements $e^1, \ldots, e^n$ where the j-th coordinate of e^i is δ_{ij}. Any n-dimensional vector space over $\mathbb{K}$ is isomorphic to $\mathbb{K}^n$.

Now $\mathbb{R}$ is a subfield of $\mathbb{K}$ and so $\mathbb{K}^n$ can be considered as a vector space over $\mathbb{R}$. The dimension of this vector space is nd: as vector spaces over $\mathbb{R}$, $\mathbb{K}$ is isomorphic to $\mathbb{R}^d$, whence, by associativity, $\mathbb{K}^n$ is isomorphic to $\mathbb{R}^{nd}$.

Let V be a vector subspace (over $\mathbb{K}$) of $\mathbb{K}^n$. If we consider $\mathbb{K}^n$ as a vector space over $\mathbb{R}$, then V is also a vector subspace over $\mathbb{R}$. We distinguish the two notions of a subspace as $\mathbb{K}$-*subspace* and $\mathbb{R}$-*subspace* respectively. This distinction is necessary since when $\mathbb{K} \neq \mathbb{R}$ not all $\mathbb{R}$-subspaces are $\mathbb{K}$-subspaces.

Thus $\mathbb{C}^2$ is a vector space of $\mathbb{R}$-dimension 4. Any $\mathbb{C}$-subspace of $\mathbb{C}^2$ is of $\mathbb{R}$-dimension $0, 2$, or 4, but there are $\mathbb{R}$-subspaces of $\mathbb{C}^2$ of $\mathbb{R}$-dimension 1 or 3. It is not even true that every 2-dimensional $\mathbb{R}$-subspace is a $\mathbb{C}$-subspace. For example, the $\mathbb{R}$-subspace U spanned by $(1,0)$ and $(0,i)$ contains $(1,0)$, but not $(i,0)$. Therefore, U is not a $\mathbb{C}$-subspace of $\mathbb{C}^2$.

For the rest of this section let V be a finite dimensional normed vector space over $\mathbb{K}$. Then V is also a finite dimensional normed vector space over $\mathbb{R}$; therefore, any two norms on V are equivalent [Exercise 12 of Section 3.5], and any linear function from V to a normed vector space W is continuous. (We are here assuming a theorem not proved in the text—the reader who does not wish to use this theorem may instead assume that V has, for example, the Euclidean norm with respect to some basis.)

If V is n-dimensional over $\mathbb{K}$, then V is nd-dimensional over $\mathbb{R}$, and so [cf. Exercise 11 of Section 2.8] there are homeomorphisms

$$E(V) \approx \mathbb{E}^{nd}, \quad B(V) \approx \mathbb{B}^{nd}, \quad S(V) \approx \mathbb{S}^{nd-1}.$$

The intersection of an m-dimensional $\mathbb{K}$-subspace U with $S(V)$ is the sphere $S(U)$, which we call a *great* $\mathbb{K}$*-sphere* in $S(V)$; here $S(U)$ is homeomorphic to $\mathbb{S}^{md-1}$.

Let U be a $\mathbb{K}$-subspace of V. We can, by choosing a basis for U and extending it over V, find a linear function $p : V \to V$ such that $pp = p$ and $\operatorname{Im} p = U$. If $q = 1 - p$, then q is linear and $U = q^{-1}[0]$. Therefore U is closed in V. Also, if $U^c = p^{-1}[0]$, then V is the direct sum $U \oplus U^c$—we call U^c a *complementary subspace* of U.

We end this section with a remark on the special case $V = \mathbb{K}^n$ with the Euclidean norm. An element $\varphi \in \mathbb{K}$ can be written as a d-tuple $(\varphi_1, \dots, \varphi_d)$ of elements of $\mathbb{R}$, and an element of $\mathbb{K}^n$ is an n-tuple $(\varphi^1, \dots, \varphi^n)$ of elements of $\mathbb{K}$. The forementioned isomorphism $\Phi : \mathbb{K}^n \to \mathbb{R}^{nd}$ of vector spaces over $\mathbb{R}$ is simply

$$(\varphi^1, \dots, \varphi^n) \mapsto (\varphi^1_1, \varphi^1_2, \dots, \varphi^n_d).$$

Since $|\varphi|^2 = |\varphi_1|^2 + \cdots + |\varphi_d|^2$, this isomorphism is norm preserving, that is, $|\Phi(x)| = |x|$. Therefore, in this case, it is reasonable to *identify* $E(\mathbb{K}^n)$ with $\mathbb{E}^{nd}$, $B(\mathbb{K}^n)$ with $\mathbb{B}^{nd}$, and $S(\mathbb{K}^n)$ with $\mathbb{S}^{nd-1}$.

5.3 Projective spaces

Let V be an $(n+1)$-dimensional normed vector space over $\mathbb{K}$. We write V_* for $V \setminus \{0\}$.

By a *line* in V_* is meant the set $U_* = U \setminus \{0\}$ for any $\mathbb{K}$-subspace U of V of dimension 1. More precisely, this is a line in V through the origin 0 but with 0 excluded; however we shall have no need of other lines. The exclusion of 0 is a technical convenience for describing the topology on the set of lines.

If $x \in V_*$, we write px for the line in V_* spanned by x, that is,

$$px = \{x\varphi : \varphi \in \mathbb{K}_*\}.$$

If l is any line in V_* and $x \in l$ then $l = px$.

The *projective space* of V is $P(V)$, the set of all lines in V_*. In particular, if $V = \mathbb{K}^{n+1}$ (with the Euclidean norm) then $P(V)$ is written $P^n(\mathbb{K})$ and is called n*-dimensional projective space over* $\mathbb{K}$.

We relate this definition of $P^n(\mathbb{K})$ with the common definition in terms of homogeneous coordinates. It is often said that a point in $P^n(\mathbb{K})$ is given by homogeneous coordinates $[x_1, \dots, x_{n+1}]$ where $x_i \in \mathbb{K}$, not all of the x_i are 0, and the convention is made that for any $\varphi \in \mathbb{K}_*$

$$[x_1\varphi, \dots, x_{n+1}\varphi] = [x_1, \dots, x_{n+1}].$$

This is the same as saying that the homogeneous coordinates of a point is the *set* of $(n+1)$-tuples $(x_1\varphi, \ldots, x_{n+1}\varphi)$ for all $\varphi \in \mathbb{K}_*$. This set is simply px where $x = (x_1, \ldots x_{n+1})$. Thus the two definitions of projective space agree.

The *fundamental map* is the function

$$\begin{aligned} p : V_* &\to P(V) \\ x &\mapsto px. \end{aligned}$$

Clearly, p is a surjection. We give $P(V)$ the identification topology with respect to p.

5.3.1 *The fundamental map is an open map.*

Proof Let A be open in V_*. We prove that $p^{-1}p[A]$ is open in V_*. Now $p^{-1}p[A]$ is the union of the sets $A\varphi = \{a\varphi : a \in A\}$ for all $\varphi \in \mathbb{K}_*$. So we have only to show that each such $A\varphi$ is open.

Let $x\varphi \in A\varphi$ so that $x \in A$. Then $B(x, r) \subset A$ for some $r > 0$. This implies that $B(x\varphi, r|\varphi|) \subseteq A\varphi$—hence $A\varphi$ is open. □

5.3.2 *If* U *is any* $\mathbb{K}$*-subspace of* V*, then* $P(U)$ *is a closed subspace of* $P(V)$.

Proof Any line in U_* is also a line in V_*—therefore $P(U)$ is a subset of $P(V)$. Further, the fundamental map $p_V : V_* \to P(V)$ sends U_* onto $P(U)$. Now U_* is a closed, saturated subset of V_*. Therefore by 4.3.1 (Corollary 1) the restriction $p_V \mid U_*, P(U)$ is an identification map. □

5.3.3 *Let* U, W *be* K*-subspaces of* V *such that* $V = U \oplus W$ *and* W *is of* $\mathbb{K}$*-dimension* 1. *Then* $U^\dagger = P(V) \setminus P(U)$ *is homeomorphic to* U.

Proof Let w be a non-zero element of W. Each element of V can be written uniquely in the form $x + w\varphi$ where $x \in U$ and $\varphi \in \mathbb{K}$; and $x + w\varphi$ belongs to U if and only if $\varphi = 0$.

The inverse image of $U^\dagger$ under the fundamental map p is the open set $V_* \setminus U_*$. Let $r : V_* \setminus U_* \to U$ be defined by $r(x+w\varphi) = x\varphi^{-1}$, $x \in U, \varphi \in \mathbb{K}$. Then r is continuous and $r(v) = r(v\psi)$ for any $v \in V_* \setminus U_*$, $\psi \in \mathbb{K}_*$. Therefore r defines a map $r' : U^\dagger \to U$ such that $r'p = r$, since $V_* \setminus U_* \to U^\dagger$ is an identification map, by 4.3.1.

Let $q : U \to U^\dagger$ be defined by $x \mapsto p(x + w)$. Then q is continuous and $r'q = 1$, $qr' = 1$ [cf. Fig. 5.1.] □

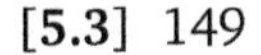

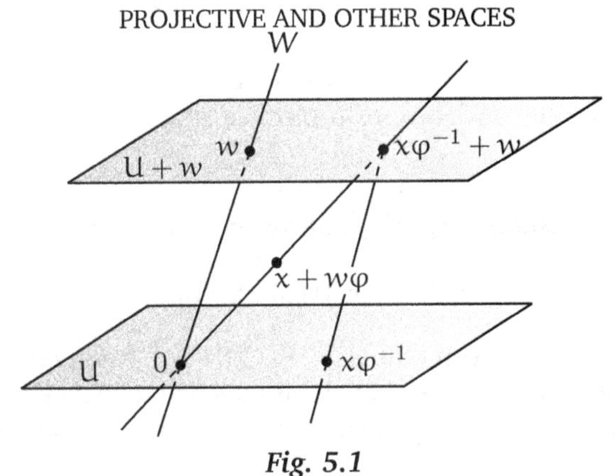

Fig. 5.1

Remark As a special case of the last result, let $V = \mathbb{K}^{n+1}$, $U = \mathbb{K}^n \times \{0\}$, and let us identify $P(U)$ with $P^{n+1}(\mathbb{K})$. We can then state 5.3.3 as: $P^n(\mathbb{K})$ is obtained from the n-dimensional space U by adding 'at infinity' a projective $(n-1)$-space $P^{n-1}(\mathbb{K})$. The reason for the words 'at infinity' here are that if φ tends to 0, then $p(x + w\varphi)$ tends to the point px of $P(U)$, while the corresponding point $x\varphi^{-1}$ of U 'tends to infinity'.

5.3.4 $P(V)$ *is a Hausdorff space.*

Proof By 5.3.3, it is sufficient to prove that any two (distinct) points of $P(V)$ belong to a set $U^\dagger$ for U a $\mathbb{K}$-subspace of V of dimension n.

Let px, py be distinct points of $P(V)$. Let $x^1, \ldots, x^{n+1}$ be a basis for V such that $x^1 = x$, $x^2 = x + y$, and let U be the $\mathbb{K}$-subspace spanned by $x^2, x^3, \ldots, x^{n+1}$. Clearly U is of dimension n, and $x, y \in V_* \setminus U_*$, so that $px, py \in U^\dagger$. □

Remark The preceding results extend on the whole to the case that V is of infinite dimension. The main difficulty that occurs is that a subspace U of V need not be closed in V, and that even if U is closed there may not be a continuous projection map $p : V \to V$ such that $pp = p$ and $\operatorname{Im} p = U$. However, it is a consequence of the Hahn-Banach theorem [cf. [Sim63]] that, if U is a finite dimensional subspace of V, then such a projection map does exist, and this is the essential fact for the previous proofs.

5.3.5 $P(V)$ *is a path-connected, compact space.*

Proof If $\dim V = 1$, then $P(V)$ has only one point and so is path-connected. If $\dim V > 1$, then V_* is path-connected, and hence so also is $P(V)$.

The sphere $S(V)$ is homeomorphic to $\mathbb{S}^{d(n+1)-1}$ and so is compact. If px is a point of $P(V)$ then $px = p(x|x|^{-1})$, and $x|x|^{-1} \in S(V)$. Therefore p maps $S(V)$ onto $P(V)$, whence $P(V)$ is compact. □

The map $h_V = p \mid S(V) : S(V) \to P(V)$ is called the *Hopf map of* V. Its role in describing the structure of projective spaces is shown by the next result.

5.3.6 *Let* U *be a* $\mathbb{K}$*-subspace of* V *of dimension* n. *There is a homeomorphism*

$$P(V) \to P(U)\, {}_h\sqcup\, E(U).$$

Proof We have already seen that $U^\dagger = P(V) \setminus P(U)$ is homeomorphic to U; but it is difficult from this to see how $U^\dagger$ is attached to $P(U)$ since there is no more of U to describe the attaching. However there is a homeomorphism

$$\begin{aligned} g : B(U) &\to U, \\ x &\mapsto x(1 - \|x\|)^{-1}. \end{aligned}$$

Let $q : U \to U^\dagger$ be the map $x \mapsto p(x + w)$ of 5.3.3, where w is a point of $V \setminus U$. Then if $x \in B(U)$

$$\begin{aligned} qg(x) &= p(x(1 - \|x\|)^{-1} + w) \\ &= p(x + w(1 - \|x\|)), \end{aligned}$$

and this suggests considering the map

$$\begin{aligned} f : E(U) &\to P(V) \\ x &\mapsto p(x + w(1 - \|x\|)). \end{aligned}$$

Clearly, $f \mid S(U), P(U)$ is the Hopf map $x \mapsto p(x)$, while

$$f \mid B(U), P(V) \setminus P(U) = qg$$

which is a homeomorphism. Since $E(U)$ is compact, and $P(V)$ is Hausdorff, the theorem follows from 4.5.8. □

As an important corollary we have:

5.3.7 *Projective* n*-space* $P^n(\mathbb{K})$ *has a cell structure in which*

$$P^n(\mathbb{K}) = e^0 \cup e^d \cup e^{2d} \cup \cdots \cup e^{nd}.$$

Proof The proof is by induction. The result is clearly true for $n = 0$, since $P^0(\mathbb{K})$ consists of a single point. Suppose the result is true for $n - 1$. Let $U = \mathbb{K}^n \times \{0\}$, let $k : \mathbb{E}^{dn} \to E(U)$ be a homeomorphism, and let $f : E(U) \to P^n(\mathbb{K})$ be the map constructed in the proof of 5.3.6. Then we can take $fk : \mathbb{E}^{dn} \to P^n(\mathbb{K})$ as characteristic map for the dn-cell of $P^n(\mathbb{K})$. □

The case $n = 1$ of 5.3.7 is particularly simple. A cell complex $e^0 \cup e^d$ is homeomorphic to $\mathbb{S}^d$; hence there is a homeomorphism

$$P^1(\mathbb{K}) \to \mathbb{S}^d.$$

The attaching map of the n-cell in 5.3.7 is the Hopf map $h : \mathbb{S}^{nd-1} \to P^{n-1}(\mathbb{K})$ (provided we identify $P^{n-1}(\mathbb{K})$ and $P(\mathbb{K}^n \times \{0\})$). The inverse images of h of points of $P^{n-1}(\mathbb{K})$ are exactly the intersection of $S(\mathbb{K}^n) = \mathbb{S}^{nd-1}$ with lines in $\mathbb{K}^n$; each such sphere is a great $\mathbb{K}$-sphere, and *a fortiori* a great $\mathbb{R}$-sphere, homeomorphic to $\mathbb{S}^{d-1}$; these spheres are disjoint and cover $\mathbb{S}^{nd-1}$. This is represented symbolically by the diagram

$$\mathbb{S}^{d-1} \to \mathbb{S}^{nd-1} \xrightarrow{h} P^{n-1}(\mathbb{K}).$$

In particular, let $n = 2$. We have a covering of $\mathbb{S}^{2d-1}$ by disjoint great $\mathbb{K}$-spheres homeomorphic to $\mathbb{S}^{d-1}$. The identification space determined by this covering is homeomorphic to $P^1(\mathbb{K})$ and so to $\mathbb{S}^d$. Therefore we have a diagram

$$\mathbb{S}^{d-1} \to \mathbb{S}^{2d-1} \to \mathbb{S}^d$$

in which the map $\mathbb{S}^{2d-1} \to \mathbb{S}^d$ is also called a Hopf map, and is written h.

The previous constructions can be followed through in detail to give a formula for $h : \mathbb{S}^{2d-1} \to \mathbb{S}^d$, but it is a rather complicated one. It is better to define maps a, b

$$\begin{array}{ccccc} \mathbb{S}^d & \xleftarrow{b} & \mathbb{E}^d & \xrightarrow{a} & P^1(\mathbb{K}) \\ (s\varphi, 1-2t^2) & \leftarrow\!\shortmid & t\varphi & \mapsto & p(\varphi, \sqrt{(1-t^2)}) \end{array}$$

where $s = 2t\sqrt{(1-t^2)}$, $0 \leqslant t \leqslant 1$, $\varphi \in \mathbb{K}$ and $|\varphi| = 1$. Then $ax = ay$ if and only if $bx = by$. Therefore a and b define a homeomorphism $P^1(\mathbb{K}) \to \mathbb{S}^d$. We leave the reader to check that the composite $\mathbb{S}^{2d-1} \to P^1(\mathbb{K}) \to \mathbb{S}^d$ is

$$\begin{array}{ccc} \mathbb{S}^{2d-1} & \to & \mathbb{S}^d \\ (\varphi, \psi) & \mapsto & (2\varphi\overline{\psi}, |\psi|^2 - |\varphi|^2) \end{array} \tag{5.3.8}$$

where φ, ψ are elements of $\mathbb{K}$ such that $|\varphi|^2 + |\psi|^2 = 1$.

In particular, let $\mathbb{K} = \mathbb{R}$, and identify $\mathbb{R}^2$ with $\mathbb{C}$. Then the map (5.3.8) is identical with $\mathbb{S}^1 \to \mathbb{S}^1$, $z \mapsto -iz^2$.

EXERCISES

1. Let V be a normed vector space over $\mathbb{K}$ and let x, y in V be of modulus 1. Prove that if $r = \text{dist}(x, py)$, and $s = r/(r+2)$, then the two sets $B(x,s), B(y,s)$ are mapped into disjoint open neighbourhoods of px, py by $p : V_* \to P(V)$.

2. If U is a $\mathbb{K}$-subspace of V of dimension $m+1$, then $P(U)$ is called an m-dimensional projective subspace of $P(V)$. Prove that if P_1, P_2 are projective subspaces of $P(V)$, then, under the ordering of subspaces by inclusion, P_1, P_2 have a least upper bound $P_1 \vee P_2$ and a greatest lower bound $P_1 \wedge P_2$, and that

$$\dim P_1 + \dim P_2 = \dim P_1 \wedge P_2 + \dim P_1 \vee P_2.$$

3. Prove that if $V = \mathbb{R}^n$, then V_* is homeomorphic to $\mathbb{R} \times \mathbb{S}^{n-1}$.
4. Let $H_{n,p,q}$ denote the 'quadric' in $\mathbb{R}^n$ defined by the equation

$$x_1^2 + x_2^2 + \cdots + x_p^2 - x_{p+1}^2 - \cdots - x_{p+q}^2 = 1 \quad (p+q \leqslant n).$$

Prove that $H_{n,p,q}$ is homeomorphic to $\mathbb{S}^{p-1} \times \mathbb{R}^{n-p}$.
5. Let $H'_{n,p}$ denote the 'quadric' in $P^n(\mathbb{R})$ defined by the equation in homogeneous coordinates

$$x_0^2 + x_1^2 + \cdots + x_{p-1}^2 - x_p^2 - \cdots - x_n^2 = 0 \quad (1 \leqslant p \leqslant n).$$

Prove that $H'_{n,1}$ and $H'_{n,n}$ are homeomorphic to $\mathbb{S}^{n-1}$; for $2 \leqslant p \leqslant n-1$, $H'_{n,p}$ is homeomorphic to the subspace obtained by identifying each point (y, z) of $\mathbb{S}^{p-1} \times \mathbb{S}^{n-p}$ with its opposite $(-y, -z)$. Prove that every point $H'_{n,p}$ has a neighbourhood homeomorphic to $\mathbb{R}^{n-1}$. Prove also that $H'_{3,2}$ is homeomorphic to $\mathbb{S}^1 \times \mathbb{S}^1$.
6. A *topological group* consists of a topological space G and a group structure on G such that the function $G \times G$, $(x, y) \mapsto xy^{-1}$, is continuous. Prove that $\mathbb{K}_*$ and $S(\mathbb{K})$ are topological groups.
7. An *action* (or *operation*) of the topological group G on the right of a space X is a function $X \times G \to X$, written $(g, g) \mapsto xg$, such that if e is the identity element of G then (i) $xe = x$ for all x in X, (ii) if $x \in X$, $g, h \in G$, then $x(gh) = (xg)h$. Given such an action, the *orbit space* of X is the space X/G whose elements are the equivalence classes under the relation $x \sim y \Leftrightarrow$ there is a g in G such that $xg = y$; the topology of X/G is the identification topology with respect to the projection $p : X \to X/G$. Prove that p is an open map. Show that 5.3.1 is a consequence (when $G = \mathbb{K}_*$, $X = V_*$).
8. Let $U, U^\dagger$ be as in 5.3.3. Prove that there is a commutative diagram

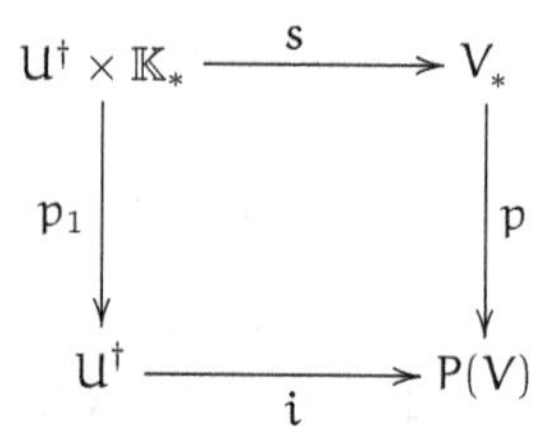

in which s is an embedding, i is the inclusion and p_1 is the projection of the product.

5.4 Isometries of inner product spaces

The object of this section is to give a brief description of some spaces of isometries—these spaces are among the central objects of mathematics.

Since there are very good treatments of inner product spaces available (e.g., [Hal60], [Por69], [Sim63]) we shall state without proofs the results we need. But first we want to record a remark about matrices over a non-commutative field.

Let V be a (right) vector space over $\mathbb{K}$, let $f : V \to V$ be a linear function and let $v^1, \dots, v^n$ be a basis for V. Then for $i = 1, \dots, n$ we can write

$$fv^j = \sum_{i=1}^{n} v^i f_{ij}$$

where the elements f_{ij} belong to $\mathbb{K}$. The function $(i, j) \mapsto f_{ij}$ is an $n \times n$ matrix $\tilde{f}$ over $\mathbb{K}$. Suppose further $g : V \to V$ is linear with matrix $\tilde{g}$. Then it is easy to check that

$$gfv^j = \sum_{h,i} v^h g_h f_{ij}$$

and it follows that $\widetilde{gf}$, the matrix of gf, is the product $\tilde{g}\tilde{f}$ of the matrices of g and f. This result is false for left vector spaces over a non-commutative field.

Let V be a (right) vector space over $\mathbb{K}$. An *inner product* on V is a function

$$\begin{aligned} V \times V &\to \mathbb{K} \\ (x, y) &\mapsto (x \mid y) \end{aligned}$$

satisfying the following axioms: (all $x, y, z \in V$, $\varphi \in \mathbb{K}$).

IPS 1 $(x \mid y + z) = (x \mid y) + (x \mid z)$,
IPS 2 $(x \mid y\varphi) = (x \mid y)\varphi$,
IPS 3 $\overline{(x \mid y)} = (y \mid x)$,
IPS 4 $x \neq 0 \Rightarrow (x \mid x) > 0$.

These are the usual axioms for an inner product with due allowance made for the fact that we have right instead of left vector spaces. It is easy to prove from the axioms that $|\,|$ defined by $|x| = \sqrt{(x \mid x)}$ is a norm on V. The *standard inner product* on $\mathbb{K}^n$ is defined by

$$(x \mid y) = \sum_{i=1}^{n} \overline{\varphi_i}\psi_i$$

for $x = (\varphi_1, \dots, \varphi_n)$, $y = (\psi_1, \dots, \psi_n)$.

Let $x, y \in V$. We say x, y are *orthogonal* if $(x \mid y) = 0$. A subset X of V is a *orthogonal set* if any two distinct elements of X are orthogonal; and

X is an *orthonormal set* if it is orthogonal and each x in X satisfies $|x| = 1$. In particular, an *orthonormal basis* for a subspace U of V is an orthonormal set which is also a basis for U. If X is a subset of V, then $X^\perp$ is the subspace of V of all y such that $(y \mid x) = 0$ for all x in X.

We now state without proof the basic results on inner product spaces.

5.4.1 If U is a vector subspace of V with a finite orthonormal basis, then V is the direct sum of U and $U^\perp$.

5.4.2 If V is finite dimensional then V has an orthonormal basis.

Let W be another inner product space. A function $f : V \to W$ is called an *isometry* if (a) f is a (linear) isomorphism of vector spaces, (b) for all x, y in V, $(fx \mid fy) = (x \mid y)$.

5.4.3 Let $f : V \to W$ be a linear isomorphism. The f is an isometry $\Leftrightarrow |fx| = |x|$ for all $x \in V$.

5.4.4 Let $v^1, \dots, v^n$ be a an orthonormal basis for V. (a) If $f : V \to W$ is linear then $fv^1, \dots, fv^n$ is an orthonormal basis for W if and only if f is an isometry. (b) If $w^1, \dots, w^n$ is an orthonormal basis for W, then there is a unique isometry $f : V \to W$ such that $fv^i = w^i$, $i = 1, \dots, n$.

5.4.5 If V is n-dimensional, then there is an isometry $f : V \to \mathbb{K}^n$.

5.4.6 Let $f : V \to V$ be linear, let $v^1, \dots, v^n$ be an orthonormal basis for V and let A be the matrix of f with respect to this basis. Then f is an isometry if and only if $\overline{A}^t A = I$ (where I is the unit matrix).

The set of all isometries $V \to V$ is written $G(V)$—it is clear that $G(V)$ is a group under composition.

Let $M_n(\mathbb{K})$ denote the (right) vector space of all $n \times n$ matrices over $\mathbb{K}$. Then $M_n(\mathbb{K})$ is isomorphic to $\mathbb{K}^{n^2}$ and this isomorphism determines an inner product structure on $M_n(\mathbb{K})$—thus $M_n(\mathbb{K})$ becomes a topological space. With this topology, the product function $M_n(\mathbb{K}) \times M_n(\mathbb{K}) \to M_n(\mathbb{K})$ which sends $(A, B) \mapsto AB$ is continuous since it is $\mathbb{R}$-bilinear [Example 13 of Section 2.8].

The group $G(\mathbb{K}^n)$ will be identified with the topological subspace of $M_n(\mathbb{K})$ of matrices A such that $\overline{A}^t A = I$. The group $G(\mathbb{K})$ is particularly simple. If $[\varphi]$ is a 1×1 matrix, then

$$[\varphi] \in G(\mathbb{K}) \Leftrightarrow \overline{\varphi}\varphi = 1 \Leftrightarrow |\varphi| = 1.$$

Therefore $G(\mathbb{K})$ is homeomorphic to the sphere $S(\mathbb{K}) = \mathbb{S}^{d-1}$.

If $\mathbb{K}$ is commutative, that is, if $\mathbb{K} = \mathbb{R}$ or $\mathbb{C}$, then the determinant function $\det : M_n(\mathbb{K}) \to \mathbb{K}$ is defined. If $A \in G(\mathbb{K}^n)$, then $\overline{A}^t A = I$ and it follows easily that $\varphi = \det A$ satisfies $\overline{\varphi}\varphi = 1$. Hence det defines a morphism of groups

$$\det : G(\mathbb{K}^n) \to S(\mathbb{K}).$$

The kernel of this morphism, that is the set of A in $G(\mathbb{K}^n)$ with determinant $+1$, is called the *special* group of isometries and is written $SG(\mathbb{K}^n)$.

The groups $G(\mathbb{K}^n)$ are given particular names in the three cases $\mathbb{K} = \mathbb{R}$, $\mathbb{C}$, or $\mathbb{H}$.

(a) $G(\mathbb{R}^n)$ is called the *orthogonal group* and is written $O(n)$. In this case det is a morphism $O(n) \to \{-1, +1\}$, and the special orthogonal group $SO(n)$ is also called the group of rotations of $\mathbb{R}^n$.

(b) $G(\mathbb{C}^n)$ is called the *unitary group* and is written $U(n)$. In this case, det is a morphism $U(n) \to \mathbb{S}^1$ whose kernel is $SU(n)$.

(c) $G(\mathbb{H}^n)$ is called the *symplectic group* and is written $Sp(n)$. (There is another family of groups, the group of *spinors* and *special spinors*—these groups are written $Pin(n)$ and $Spin(n)$ and are closely related to $O(n)$ and $SO(n)$ respectively; cf. [Por69].)

These three families of groups are related. In fact, $\mathbb{C}^n$ is isomorphic as *normed vector space over* $\mathbb{R}$ to $\mathbb{R}^{2n}$; hence, by 5.4.3, any isometry $f : \mathbb{C}^n \to \mathbb{C}^n$ defines an isometry $\lambda_n(f) : \mathbb{R}^{2n} \to \mathbb{R}^{2n}$. So λ_n defines an injection morphism of groups

$$\lambda_n : U(n) \to O(2n).$$

5.4.7 *There is an isomorphism of groups*

$$\mu : \mathbb{S}^1 \to SO(2).$$

Proof Let μ be the composite of λ_1 with the isomorphism $\mathbb{S}^1 \to U(1)$ defined previously. Let $\mathbb{R}^2$ have $\mathbb{R}$-basis the complex numbers $1, i$. If $z \in \mathbb{S}^1$, then $\mu(z)$ is the function $\mathbb{R}^2 \to \mathbb{R}^2$, $x \mapsto zx$. So if $z = \cos\alpha + i\sin\alpha$, we find that $\mu(z)$ has matrix

$$\begin{bmatrix} \cos\alpha & -\sin\alpha \\ \sin\alpha & \cos\alpha \end{bmatrix}.$$

On the other hand, it is easily checked (using the equations $A^{-1} = A^t$, $\det A = 1$) that any element A of $SO(2)$ must be of this form. □

A similar argument works for $\mathbb{H}$ and $\mathbb{C}$: since $\mathbb{H}^n$ is isomorphic as normed vector space over $\mathbb{C}$ to $\mathbb{C}^{2n}$, there is an injective morphism of groups $Sp(n) \to U(2n)$.

5.4.8 *There is an isomorphism of groups*

$$\mathbb{S}^3 \to SU(2).$$

The proof is similar to that of 5.4.7 and is left to the reader.

Next we show how to determine $SO(3)$ in terms of $\mathbb{S}^3$. We first need to show that every element of $SO(3)$ is a rotation about an axis.

5.4.9 *Let* $f \in SO(3)$. *Then there is an orthonormal basis* a, b, c *for* $\mathbb{R}^3$ *such that with respect to this basis* f *has matrix*

$$\begin{bmatrix} 1 & 0 & 0 \\ 0 & \cos\alpha & -\sin\alpha \\ 0 & \sin\alpha & \cos\alpha \end{bmatrix}.$$

Proof Let A be the matrix of f with respect to some orthonormal basis. Let $A^tA = I$ and $\det A = 1$. Hence $A^t(A - I) = I - A^t$ and so

$$\begin{aligned} \det(A - I) &= \det A^t(A - I) \\ &= \det(I - A^t) \\ &= -\det(A - I). \end{aligned}$$

Therefore $\det(A - I) = 0$. Hence there is a non-zero element a of $\mathbb{R}^3$, such that $(A - I)a = 0$, that is $fa = a$. Replacing a by $a/|a|$ if necessary, we may suppose $|a| = 1$.

Let b, c be an orthonormal basis of the plane U through the origin and orthogonal to a. Since f is an isometry and $fa = a$, we have $f[U] \subseteq U$. Therefore $g = f \mid U, U$ is defined and is an isometry. The equation $fa = a$ implies that $\det g = \det f = 1$. Therefore g has matrix with respect to b, c

$$\begin{bmatrix} \cos\alpha & -\sin\alpha \\ \sin\alpha & \cos\alpha \end{bmatrix}$$

and the matrix for f follows. □

The form of the matrix of 5.4.9 shows that f is a rotation through angle α about the axis a.

5.4.10 *There is a surjective morphism*

$$\nu : \mathbb{S}^3 \to SO(3)$$

with kernel the quaternions $+1, -1$.

Proof Let q be a unit quaternion, and let $x \in \mathbb{R}^3$. Define

$$r = qx\bar{q}.$$

Then

$$\begin{aligned} 2\,\mathrm{Re}(r) &= r + \bar{r} \\ &= qx\bar{q} + q\bar{x}\bar{q} \\ &= q(x + \bar{x})\bar{q} \\ &= 0 \ \text{ since } \mathrm{Re}(x) = 0. \end{aligned}$$

Therefore r is a vector. It follows that $x \mapsto qx\bar{q}$ defines a function $\nu_q : \mathbb{R}^3 \to \mathbb{R}^3$—clearly ν_q is linear. It is also an isomorphism since it has inverse $x \mapsto \bar{q}xq$.

Let $x \in \mathbb{R}^3$. Then

$$|\nu_q x| = |qx\bar{q}| = |q||x||\bar{q}| = |x|.$$

Therefore ν_q is an isometry.

In fact, we can find a formula for ν_q.

Suppose $q = \cos\alpha + a\sin\alpha$ where a is a unit vector. Let a, b, c be a right-handed orthonormal system in $\mathbb{R}^3$. Then an easy check shows that

$$\begin{aligned} \nu_q a &= a \\ \nu_q b &= b\cos 2\alpha + c\sin 2\alpha \\ \nu_q c &= -b\sin 2\alpha + c\cos 2\alpha. \end{aligned} \qquad (*)$$

Therefore ν_q is a rotation about the axis a through an angle 2α. In particular, ν_q belongs to $SO(3)$ and any element of $SO(3)$ is of the form ν_q for some q in $\mathbb{S}^3$.

The function ν is a homomorphism, since if $q, r \in \mathbb{S}^3$, then

$$\begin{aligned} \nu_{qr}(x) &= qrx\,\overline{qr} \\ &= qrx\,\bar{r}\bar{q} \\ &= \nu_q\nu_r(x). \end{aligned}$$

It follows that $\nu_q = \nu_r \Leftrightarrow \nu_{qr^{-1}} = 1$. If $q = \cos\alpha + a\sin\alpha$, then it follows from (*) that $\nu_q = 1$ if and only if $\alpha = n\pi$, $n \in \mathbb{Z}$. Therefore, $\nu_q = 0 \Leftrightarrow q = \pm 1$, and so $\nu_q = \nu_r \Leftrightarrow q = \pm r$. □

5.4.10 *(Corollary 1) There is a homeomorphism*

$$SO(3) \to P^3(\mathbb{R}).$$

Proof The space $SO(3)$ is Hausdorff since it is a subspace of the normed vector space $M_3(\mathbb{R})$. Since $\mathbb{S}^3$ is compact, and ν is continuous, the space $SO(3)$ is obtained, like $P^3(\mathbb{R})$, by identifying antipodal points of $\mathbb{S}^3$. □

This result gives, of course, a cell structure for $SO(3)$. In fact, cell structures have been given for all the coset spaces $G(\mathbb{K}^n)/G(\mathbb{K}^m)$, $0 \leqslant m < n$; for an excellent account of this, see [SE62].

EXERCISES

1. Prove the assertions 5.4.1—5.4.6.
2. Prove that $G(\mathbb{K}^n)$ is closed, bounded subset of $M_n(\mathbb{K})$. Deduce that $G(\mathbb{K}^n)$ is compact. [Use the equation $\overline{A}^t A = I$.]
3. Let a_n be the point $(0, \ldots, 0, 1)$ of K^n. Define

$$\rho_n : G(\mathbb{K}^n) \to S(\mathbb{K}^n), \quad f \mapsto f a_n.$$

Prove that ρ_n is continuous and surjective, and that $\rho_n^{-1}[a_n]$ is a subgroup of $G(\mathbb{K}^n)$ isomorphic to $G(\mathbb{K}^{n-1})$. Prove also that if $\mathbb{K}$ is commutative then ρ_n restricts to a continuous surjection $q_n : SG(\mathbb{K}^n) \to S(\mathbb{K}^n)$ such that $q_n^{-1}[a_n]$ is a subgroup of $SG(\mathbb{K}^n)$ isomorphic to $SG(\mathbb{K}^{n-1})$.
4. Prove that the groups $SO(n)$, $SU(n)$, $U(n)$, $Sp(n)$ are connected. [Use the previous result and induction.]
5. Let U, V be two vector subspaces of $\mathbb{K}^n$ of the same dimension. Prove that there is an isometry $\sigma : \mathbb{K}^n \to \mathbb{K}^n$ such that $\sigma[U] = V$. Prove also that if $x, y \in \mathbb{S}^{n-1}$, then there is a rotation σ of $\mathbb{R}^n$ such that $\sigma x = y$.
6. Let U be a subspace of $\mathbb{R}^n$ of dimension $(n-1)$, so that $\mathbb{R}^n = U \oplus U^\perp$. The function $x + y \mapsto x - y$ ($x \in U$, $y \in U^\perp$) is called *reflection* in the hyperplane U. Prove that such a reflection is an isometry of $\mathbb{R}^n$, and also that any isometry of $\mathbb{R}^n$ with U as its set of fixed points is a reflection in U.
7. Prove that all elements of $O(n)$ are products of reflections. [Use induction.]

5.5 Simplicial complexes

Although cell complexes form a highly useful class of spaces, they are not restrictive enough for many purposes. In this section we discuss the simplicial complexes—these can be regarded as cell complexes but with a particular kind of attaching map of the cells. For applications of simplicial complexes we refer the reader to [HW60], [Lef49], [Cai61], or [Spa66]. The purpose of this section is mainly to link our results with the theories described in these books.

A subset A of $\mathbb{R}^k$ is called an n-*simplex* if there are points $a_0, \ldots, a_n$ in A such that A consists of all points which can be written *uniquely* in the form

$$a = t_0 a_0 + \cdots + t_n a_n$$

where

$$0 \leqslant t_i, t_0 + \cdots + t_n = 1.$$

In such case A is said to be *spanned* by $a_0, \ldots, a_n$ and the numbers $t_0, \ldots, t_n$ are called the *barycentric coordinates* of the point $a = t_0 a_0 + \cdots + t_n a_n$. (The reason for the latter name is that a is the centre of gravity, or barycentre, of the particle system with weight t_i at a_i, $i = 0, \ldots, n$.) The points $a_0, \ldots, a_n$ are called the *vertices* of A—we will see below that the vertices are determined by the set A.

5.5.1 *Let A be an n-simplex with vertices $a_0, \ldots, a_n$. Let $b_0, \ldots, b_m$ be points of A and let $s_0, \ldots, s_m$ be positive real numbers whose sum is 1. Then the point $b = s_0 b_0 + \cdots + s_m b_m$ belongs to A, and $b = a_{i_0}$ if and only if $b_0 = \cdots = b_m = a_{i_0}$.*

Proof Since $b_j \in A$, we can write

$$b_j = \sum_{i=0}^{n} t_{ji} a_i, \qquad t_{ji} \geqslant 0, \qquad \sum_j t_{ji} = 1.$$

It follows that

$$b = \sum_{i,j} s_j t_{ji} a_i.$$

But $s_j t_{ji} \geqslant 0$ and

$$\sum_{i,j} s_j t_{ji} = \sum_j s_j \sum_i t_{ji} = \sum_j s_j = 1.$$

Therefore $b \in A$.

Suppose that $b = a_{i_0}$. Then for $i \neq i_0$

$$\sum_j s_j t_{ji} = 0.$$

Since the s_j are positive, it follows that $t_{ji} = 0$, $i \neq i_0$, and so

$$b_0 = \cdots = b_m = a_{i_0}.$$

Conversely, this last condition clearly implies $b = a_{i_0}$. □

5.5.1 *(Corollary 1) If A is a simplex spanned by both $a_0, \ldots, a_n$ and $b_0, \ldots, b_m$, then $m = n$ and the b_j are a rearrangement of the a_i.*

Proof Since $b_0, \ldots, b_m$ are vertices of A, we can write uniquely

$$a_i = \sum_j s_j b_j, \quad s_i \geqslant 0, \quad s_0 + \cdots + s_m = 1.$$

By 5.5.1, if $s_{j_i} \neq 0$ then $b_j = a_i$. Hence there is a unique j_i such that $b_{j_i} = a_i$. □

If A is an n-simplex then n is determined by A—we call n the *dimension* of A.

Another obvious consequence of 5.5.1 is that an n-simplex A is convex, and is in fact the smallest convex subset of $\mathbb{R}^k$ containing the vertices of A.

Let A be the simplex with vertices $a_0, \dots, a_n$. A *face* of A is a simplex spanned by any subset of $\{a_0, \dots, a_n\}$. For example, the vertices $a_0, \dots, a_{i-1}, a_{i+1}, \dots a_n$ span an $(n-1)$-simplex which is often written $\partial_i A$—it is the face opposite the i-th vertex.

Suppose that $a = t_0 a_0 + \cdots + t_n a_n$, where the t_i are barycentric coordinates, and suppose $t_0 \neq 1$. Let

$$a' = (1 - t_0)^{-1}(t_1 a_1 + \cdots + t_n a_n).$$

Then a' belongs to the face $\partial_0 A$ of A opposite a_0 and

$$a = t_0 a_0 + (1 - t_0) a'.$$

Thus the points of A are the points of the line segments joining a_0 to the points of the $(n-1)$-simplex $\partial_0 A$. This gives an inductive description of an n-simplex and justifies the following pictures of n-simplexes for $n \leqslant 3$.

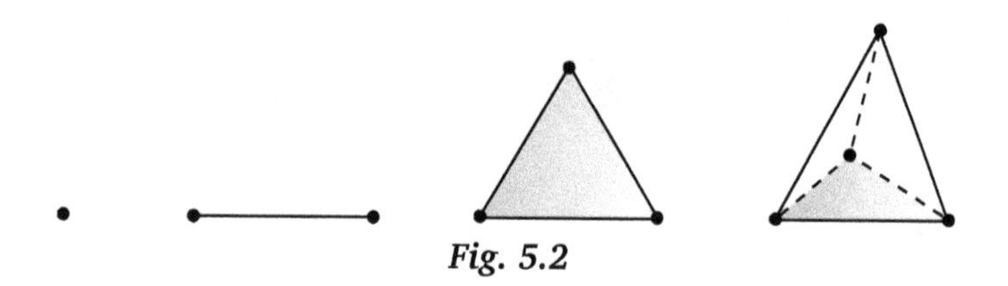

Fig. 5.2

If A is an n-simplex, then $\dot{A}$, the *boundary* of A, is the union of all faces of A of dimension $< n$ (this is not necessary the topological boundary Bd A of the set A).

5.5.2 *If A is an n-simplex, then there is a homeomorphism $A \to \mathbb{E}^n$ which maps $\dot{A}$ homeomorphically onto $\mathbb{S}^{n-1}$.*

Proof We first construct a homeomorphism from A to an n-simplex B in $\mathbb{R}^n$. Let B have vertices $b_0, \dots, b_n$ where $b_0 = n^{-\frac{1}{2}}(-1, \dots, -1)$ and $b_1, \dots, b_n$ is the standard basis of $\mathbb{R}^n$ [cf. Fig. 5.3]. If A has vertices $a_0, \dots, a_n$ define

$$f : A \to B, \qquad t_0 a_0 + \cdots + t_n a_n \mapsto t_0 b_0 + \cdots + t_n b_n;$$

clearly f is a homeomorphism which maps A onto B.

A homeomorphism $g : B \to \mathbb{E}^n$ is now constructed by radial projection from the origin. □

It follows from 5.5.2 that instead of using $(\mathbb{E}^n, \mathbb{S}^{n-1})$ as the standard model from which to construct cell complexes, we could use instead $(A, \dot{A})$ for any n-simplex A—this is the method we shall use when showing that a simplicial complex can be given a cell structure.

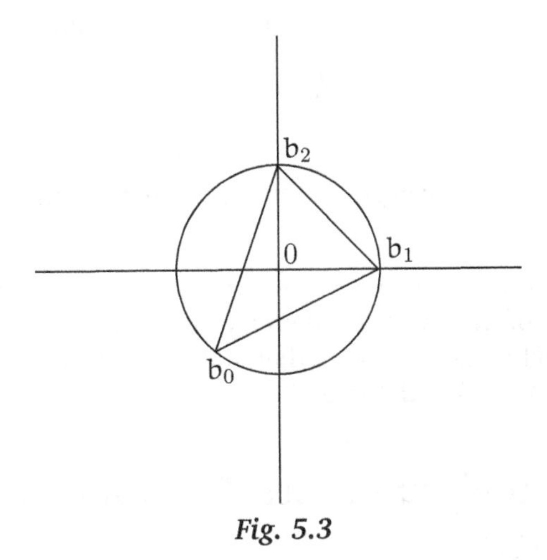

Fig. 5.3

One further point will explain the utility of simplicial methods. A map $f : A \to B$ from an n-simplex A to an m-simplex B is called *linear* if the barycentric coordinates of fa are linear functions of the barycentric coordinates of a in A. It is obvious that an m-simplex is *linearly* homeomorphic to an n-simplex if and only if $m = n$. The corresponding statement without the word linearly is true but is much more difficult to prove—it constitutes in fact the Invariance of Dimension.

A *simplicial complex* K is a finite set of simplices of $\mathbb{R}^k$ (for some k) with the following properties:

SIM 1 If $A \in K$ then any face of A belongs to K.

SIM 2 Any two simplices of K meet in a common face (possibly empty).

The space of K is $|K|$, the union of all the simplices of K, with the topology as a subspace of $\mathbb{R}^k$.

In Fig. 5.4, (a) and (b) are pictures of the simplicial complexes but (c) is not; however (c) can be made into a simplicial complex by adding extra simplices.

If K is a simplicial complex, then K^n is the subset of K of all simplices of dimensions $\leqslant n$.

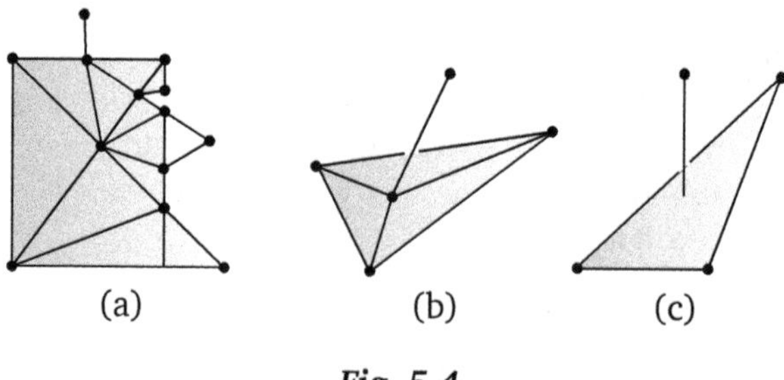

Fig. 5.4

5.5.3 *If K is a simplicial complex, then $|K|$ has a cell structure whose open cells are $A \setminus \dot{A}$ for each simplex A of K.*

Proof It is clear from 5.5.2 that $A \setminus \dot{A}$ is an open cell, for each simplex A of K. Also if A is of dimension n, then we can take A itself as a model of the n-cell; the inclusion $A \to |K|$ is then a characteristic map for A which maps $\dot{A}$ into the $(n-1)$-skeleton of $|K|$. □

5.6 Bases and sub-bases for open sets; initial topologies

The aim of this section is the construction of the initial topology on X with respect to a family of functions from X; this important construction is a kind of 'dual' to that of final topologies in chapter 4. Before proceeding with the definitions we need some of the theory of bases and sub-bases.

If $\mathcal{B}$ is a base for the neighbourhoods of a topological space X then $\mathcal{B}$ defines the topology on X completely (since a subset N of X is a neighbourhood of a point x of X if and only if N contains a set B of $\mathcal{B}(x)$). We now consider the converse problem: let X be a set and $\mathcal{B} : x \mapsto \mathcal{B}(x)$ a function assigning to each x in X a non-empty set of subsets of X; under what conditions is $\mathcal{B}$ a base for the neighbourhoods of a topology on X?

5.6.1 *For a function $\mathcal{B}$ as above, $\mathcal{B}$ is a base for the neighbourhoods of a topology on X if and only if the following conditions hold:*

(a) *If $x \in X$ and $B \in \mathcal{B}(x)$ then $x \in B$.*
(b) *If $B, B' \in \mathcal{B}(x)$ then $B \cap B'$ contains a set of $\mathcal{B}(x)$.*
(c) *If $B \in \mathcal{B}(x)$ then B contains a set M such that $x \in M$ and also if $y \in M$ then M contains a set B' of $\mathcal{B}(y)$.*

Proof The topology defined by $\mathcal{B}$, if it exists, is such that N is a neighbourhood of x if and only if N contains a set of $\mathcal{B}(x)$. Then (a), (b), and (c) are simply restatements of the Axioms N1, N3, and N4 respectively for neighbourhoods. (Notice, by the way, that (a) is a consequence of (c).) □

In the case when each element of $\mathcal{B}(x)$ is an open neighbourhood of x in the topology defined by $\mathcal{B}$, then we call $\mathcal{B}$ an *open base* for the neighbourhoods of X: the conditions for $\mathcal{B}$ to be an open base for the neighbourhoods of a topology on X are clearly 5.6.1 (a), (b), and also (c'): *if* $B \in \mathcal{B}(x)$ *then* $B \in \mathcal{B}(y)$ *for each* y *in* B.

We now consider similar questions for open sets.

5.6.2 *Let* X *be a topological space and* $\mathcal{U}$ *an open cover of* X. *The following conditions are equivalent:*

(a) *each open set of* X *is a union of elements of* $\mathcal{U}$,
(b) *for each* x *in* X, *the set* $\mathcal{U}(x)$ *of elements of* $\mathcal{U}$ *which contain* x *is a base for the neighbourhoods of* x.

Proof (a) $\Rightarrow$ (b) If N is a neighbourhood of x, then N contains an open set U such that $x \in U$. Since U is a union of elements of $\mathcal{U}$ there is a set U_x in $\mathcal{U}$ such that $x \in U_x$. Hence $x \in U_x \subseteq N$.

(b) $\Rightarrow$ (a) If U is an open set of X, then for each x in U there is an element U_x of $\mathcal{U}(x)$ such that $x \in U_x \subseteq U$. Hence $U = \bigcup_{x \in U} U_x$. □

If $\mathcal{U}$ is an open cover of X satisfying either of the conditions (a), (b) of 5.6.2, then we say $\mathcal{U}$ is a *base for the open sets* of the topology of X, or, simply, a base for the topology of X. Conversely, given a set $\mathcal{U}$ of subsets of X, we wish to know under what circumstances $\mathcal{U}$ is a base for a topology on X—if this topology exists it will clearly be unique.

5.6.3 *A cover* $\mathcal{U}$ *of* X *by subsets of* X *is a base for a topology on* X *if and only if for each* U, V *in* $\mathcal{U}$ *and* x *in* $U \cap V$ *there is a* W *in* $\mathcal{U}$ *such that*

$$x \in W \subseteq U \cap V.$$

Proof The forward implication is easy, by 5.6.1(b).

For the converse implication, let $\mathcal{U}(x)$ be for each x in X the set of elements of $\mathcal{U}$ which contain x. Then it is immediate that $x \mapsto \mathcal{U}(x)$ is an open base for the neighbourhoods of a topology on X. □

EXAMPLES

1. Let X, Y be topological spaces. The sets $U \times V$ for U open in X, V open in Y form a base for the product topology on $X \times Y$.
2. Let X be a metric space. The open balls $B(x, r)$ for all x in X and $r > 0$ form a base for the metric topology on X.

3. The intervals $]a, b[$ for a, b in $\mathbb{Q}$ form a base for the usual topology of $\mathbb{R}$.

A generalisation of the notion of base for a topology is that of *sub-base*: this is a set $\mathcal{V}$ of subset of a topological space X such that the set of finite intersections of elements of $\mathcal{V}$ is a base for the topology of X.

5.6.4 *If X is a set and $\mathcal{V}$ any set of subsets of X, then $\mathcal{V}$ is a sub-base for a unique topology on X.*

Proof Let $\mathcal{U}$ be the set of finite intersections of elements of $\mathcal{V}$. Then $X \in \mathcal{U}$ (since X is the intersection of the empty set of elements of $\mathcal{V}$!) and so $\mathcal{U}$ covers X. Also, the intersection of two elements of $\mathcal{U}$ again belongs to $\mathcal{U}$—so $\mathcal{U}$ is a base for a topology $\mathcal{T}$ on X, by 5.6.3.

Any topology on X which has $\mathcal{V}$ as a sub-base has $\mathcal{U}$ as a base, and therefore coincides with $\mathcal{T}$. This proves uniqueness of the topology. □

We shall next characterise continuity of functions in terms of bases and sub-bases. However, for the applications of our results that we have in mind, it is helpful to have a more general kind of function than that considered before.

Let X, Y be sets. By a *partial function from* X *to* Y, written

$$f : X \rightarrowtail Y,$$

we shall mean a triple consisting of X, Y and a subset F of $X \times Y$ with the property that if $(x, y), (x, y') \in F$, then $y = y'$. If $(x, y) \in F$, we write $y = fx$. The *domain* $\mathcal{D}f$ of f is the set of x in X such that fx is defined—thus we have extended the definition of function given in the Appendix by allowing the domain of f to be any subset of X.

If $f : X \rightarrowtail Y$, $g : Y \rightarrowtail Z$ are partial functions as above, then the composite $gf : X \rightarrowtail Z$ has domain the set of x in X such that $fx \in \mathcal{D}g$, and gf sends $x \mapsto gfx$. The definitions of $f[A]$ and $f^{-1}[A]$ for a set A and function $f : X \rightarrowtail Y$ apply without change.

Let X, Y be topological spaces and $f : X \rightarrowtail Y$ a partial function. For our purposes it is convenient to say that f is *continuous* if $f^{-1}[U]$ is open in X for each open set U of Y. Since $\mathcal{D}f = f^{-1}[Y]$, this implies that $\mathcal{D}f$ is open in X. It is easy to prove that the composite of continuous partial functions is again continuous.

5.6.5 *Let $f : X \rightarrowtail Y$ be a partial function where X, Y are topological spaces. The following conditions are equivalent:*

(a) f *is continuous,*

(b) *if $\mathcal{U}$ is a base for the topology of Y, then $f^{-1}[U]$ is open in X for each U in $\mathcal{U}$,*

(c) *if* $\mathcal{V}$ *is a sub-base for the open sets of* Y, *then* $f^{-1}[V]$ *is open in* X *for each* V *in* $\mathcal{V}$.

Proof The implications (a) ⇒ (b) ⇒ (c) follow from the definition of continuity.

The implication (b) ⇒ (a) follows from the (easily verified) fact that the inverse image of a union of sets is the union of their inverse images. The implication (c) ⇒ (b) follows from the fact that the inverse image of an intersection of sets is the intersection of their inverse images. □

Suppose now that we are given a set X, a family $(X_\lambda)_{\lambda\in\Lambda}$ of topological spaces and for each λ in Λ a partial function

$$f_\lambda : X \rightarrowtail X_\lambda.$$

A topology $\mathcal{I}$ on X is *initial with respect to* (f_λ) if it has the following property: for any topological space Y a function $k : Y \rightarrowtail X_{\mathcal{I}}$ is continuous if and only if the composite $f_\lambda k : Y \rightarrowtail X_\lambda$ is continuous for each λ in Λ.

5.6.6 *If* $\mathcal{I}$ *is an initial topology on* X *with respect to* (f_λ) *then* $\mathcal{I}$ *is the coarsest of the topologies* $\mathcal{J}$ *on* X *such that each* $f_\lambda : X_{\mathcal{J}} \rightarrowtail X_\lambda$ *is continuous.*

Proof Since the identity function $1 : X_{\mathcal{I}} \to X_{\mathcal{I}}$ is continuous, it follows that $f_\lambda = f_\lambda 1$ is continuous. Suppose $\mathcal{T}$ is any topology on X such that each $f_\lambda : X_{\mathcal{T}} \rightarrowtail X_\lambda$ is continuous. Let $k : X_{\mathcal{T}} \to X_{\mathcal{I}}$ be the identity function. Then $f_\lambda k = f_\lambda : X_{\mathcal{T}} \rightarrowtail X_\lambda$, and so k is continuous. □

It follows from 5.6.6 that there is at most one initial topology on X with respect to (f_λ).

5.6.7 *The initial topology on* X *with respect to* (f_λ) *exists and is the topology which has as a sub-base the sets*

$$f_\lambda^{-1}[U], \text{ all } U \text{ open in } X_\lambda, \text{ all } \lambda \text{ in } \Lambda.$$

Proof Let $\mathcal{I}$ be the topology on X with the above sub-base. We prove that $\mathcal{I}$ is initial.

Let $k : Y \rightarrowtail X_{\mathcal{I}}$ be a function where Y is a topological space, and suppose first that k is continuous. By 5.6.5 each $f_\lambda : X_{\mathcal{I}} \rightarrowtail X_\lambda$ is continuous and hence $f_\lambda k : Y \rightarrowtail X_\lambda$ is continuous.

Suppose, conversely, that each $f_\lambda k : Y \rightarrowtail X_\lambda$ is continuous. Then again, because the sets $f_\lambda^{-1}[U]$ for U open in X_λ form a sub-base for $\mathcal{I}$, the function k is continuous. □

In the first two of the following examples the initial topologies are taken with respect to functions whose domains are X itself.

EXAMPLES

4. Let X be a subset of the topological space X_1 and let $i : X \to X_1$ be the inclusion. The initial topology on X with respect to i has a sub-base the sets $i^{-1}[U]$ for U open in X_1. But $i^{-1}[U] = U \cap X$; so the initial topology with respect to i is simply the relative topology on X.

More generally, if X_1 is a topological space and $i : X \to X_1$ is an injection, then the initial topology with respect to i is that which makes i an embedding.

5. Let $(X_\lambda)_{\lambda \in \Lambda}$ be a family of topological spaces, and let X be the product of their underlying sets. The *product topology* on X is the initial topology with respect to the family of projections $p_\lambda : X \to X_\lambda$.

Suppose that $U \subseteq X_\lambda$. Then $p_\lambda^{-1}[U]$ consists of all points x in X such that $x_\lambda \in U$ (the other coordinates of x being unrestricted). That is, $p_\lambda^{-1}[U]$ is the product

$$\prod_{\mu \in \Lambda} U'_\mu \quad \text{where} \quad U'_\mu = \begin{cases} X_\mu, & \mu \neq \lambda \\ U, & \mu = \lambda. \end{cases}$$

A finite intersection of such sets, say $p_{\lambda_1}^{-1}[U_1] \cap \cdots \cap p_{\lambda_n}^{-1}[U_n]$, is the product

$$\prod_{\mu \in \Lambda} U''_\mu \quad \text{where} \quad U''_\mu = \begin{cases} X_\mu, & \mu \neq \lambda_1, \ldots, \lambda_n \\ U_i, & \mu = \lambda_i, \ldots, \lambda_n. \end{cases}$$

Thus if Λ is finite, say $\Lambda = \{1, \ldots, n\}$, then a base for the open sets of X consists of all products $U_1 \times \cdots \times U_n$ for U_i open in X_i, and the product topology is that defined in section 2.3. However, if $\Lambda = \{1, 2, \ldots\}$, then a base for the open sets of X consists of all products $U_1 \times U_2 \times \cdots$ in which U_i is open in X_i and $U_i = X_i$ for all but a finite number of i.

6. Let M be a topological space, and let X be a set. An M-*chart* on X is an injective partial function $f : X \rightarrowtail M$ whose image is open in M. An M-*atlas* for X is a family $\mathcal{A} = \{f_\alpha\}_{\alpha \in A}$ of M-charts for X such that if $f_\alpha, f_\beta : X \rightarrowtail M$ are charts in $\mathcal{A}$, then $f_\alpha, f_\beta^{-1} : M \rightarrowtail M$ is continuous. Given such an M-atlas $\mathcal{A}$, let X have the initial topology with respect to all M-charts in $\mathcal{A}$. Then $f_\alpha^{-1} : M \rightarrowtail X$ is continuous, since $f_\beta f_\alpha^{-1}$ is continuous for all $\beta \in A$. Hence f_α maps its domain homeomorphically to its image. Topologies constructed from atlases in this way are a basic part of the theory of *manifolds*, in which the space M, the *model space* for the manifold, is usually taken to be a Euclidean space $\mathbb{R}^n$ or a suitable Banach space. Such a space is called a *manifold modelled on* V—these spaces are important in many branches of topology, geometry, and analysis. In fact, the notion of a topology arose from the need to describe such spaces as Riemann surfaces, where the notion of neighbourhood is clear locally, as in the definition of manifold.

There is a useful 'transitive law' for initial topologies.

5.6.8 *Suppose there is given a set* X*; a family*

$$(f_\lambda : X \rightarrowtail X_\lambda)_{\lambda \in \Lambda}$$

of partial functions from X*; and for each* λ *in* Λ *a family*

$$(g_{\lambda\mu} : X_\lambda \to X_{\lambda\mu})_{\mu \in M_\lambda}$$

of partial functions from X_λ. *If* X_λ *has the initial topology with respect to* $(g_{\lambda\mu})_{\mu \in M_\lambda}$*, then the initial topologies on* X *with respect to* (f_λ) *and* $(g_{\lambda\mu} f_\lambda)$ *coincide.*

Proof Let $k : Y \rightarrowtail X$ be a function where Y is a topological space. Since X_λ has the initial topology the functions $g_{\lambda\mu} f_\lambda k$, $\mu \in M_\lambda$, are continuous if and only if $f_\lambda k$ is continuous; the result follows easily. □

5.6.8 *(Corollary 1) Let* $i : X \to X_1$ *be an inclusion function and let* $f_\lambda : X_1 \rightarrowtail X_\lambda$, $\lambda \in \Lambda$ *be a family of functions where* X_λ *is a topological space. If* X_1 *has the initial topology with respect to* (f_λ)*, then the initial topology on* X *with respect to* $(f_\lambda i)$ *is the subspace topology.*

EXERCISES

1. A space X is *second countable*, or *satisfies the second axiom of countability*, if there is a countable base for the open sets of X. Prove that if a space is second countable then it is separable and first countable. Prove also that a separable metric space is second countable.
2. Prove that any open cover of a second countable space has a countable sub-cover.
3. Let $\mathcal{U}$ be an open cover of X. For each U in $\mathcal{U}$ let $i_U : X \rightarrowtail X$ be the inclusion $x \mapsto x$, $x \in U$. Prove that X has the initial topology with respect to the family $(i_U)_{U \in \mathcal{U}}$.
4. Let X be a set, $(X_\lambda)_{\lambda \in \Lambda}$ a family of topological spaces, and $(f_\lambda : X \rightarrowtail X_\lambda)_{\lambda \in \Lambda}$ a family of functions. Let Z be the topological product of the family $(X_\lambda)_{\lambda \in \Lambda}$ and let $f : X \rightarrowtail Z$ be the unique function such that $p_\lambda f = f_\lambda$, $\lambda \in \Lambda$. Prove that the initial topology with respect to f is the same as the initial topology with respect to (f_λ). Given this topology on X, is f an open map?
5. Let $\mathcal{C}$ be a set of functions $Y \to X$ where Y, X are spaces. If $C \subseteq Y$, $U \subseteq X$ let $M_{\mathcal{C}}(C, U)$ denote the set of functions f in $\mathcal{C}$ such that $f[C] \subseteq U$ (when $\mathcal{C}$ can be understood from the context we abbreviate this to $M(C, U)$). The *compact-open* topology on $\mathcal{C}$ is that which has as a sub-basis for the open sets the sets $M_{\mathcal{C}}(C, U)$ for all compact subsets C of Y and open subsets U of X. In the following, $\mathcal{C}$ will have the compact-open topology. Prove that (i) if $\mathcal{D}$ is a subset of $\mathcal{C}$ and $\mathcal{D}$ has the compact-open topology, then $\mathcal{D}$ is a subspace of $\mathcal{C}$; (ii) if X is Hausdorff, then $\mathcal{C}$ is

Hausdorff; (iii) if $\mathcal{C}$ contains all the constant functions $Y \to X$, and $\mathcal{C}$ is Hausdorff, then X is Hausdorff; (iv) if Y is discrete, and $\mathcal{C}$ consists of all functions $Y \to X$, then $\mathcal{C}$ is homeomorphic to the topological product $\prod_{y \in Y} X$ of the family $y \mapsto X$, $y \in Y$.
6. Let X be the topological product of the family of spaces X_λ, $\lambda \in \Lambda$, and let A be the topological product of the subspace A_λ of X_λ for $\lambda \in \Lambda$. Prove that A is a subspace of X. Which of the following are true? (i) If each A_λ is open in X_λ then A is open in X. (ii) If each A_λ is closed in X_λ then A is closed in X. (iii) The closure of A is the product of the closures of the A_λ. (iv) The interior of A is the product of the interiors of the A_λ. (v) The boundary of A is the product of the boundaries of the A_λ.
7. Let X be the product of uncountably many copies of the real line $\mathbb{R}$. Let $\mathbf{0}$ be the element of X with all components 0. Prove Prove that $\{\mathbf{0}\}$ is not a G_δ-set in X.
8. Read a proof of the Tychonoff theorem: *any topological product of compact spaces is compact.* In particular, examine proofs which use ultrafilters, and investigate further the uses of ultrafilters in constructing 'non-standard' models of the natural numbers. [See many books on general topology.]

5.7 Joins

The join $X * Y$ of two topological spaces X, Y arises in a natural way when X, Y are disjoint subspaces of some normed vector space V. Consider the line segments $[x, y]$ for x in X, y in Y and suppose X, Y are so placed in V that two such line segments $[x, y]$, $[x', y']$ meet, if they meet at all, only in end points.

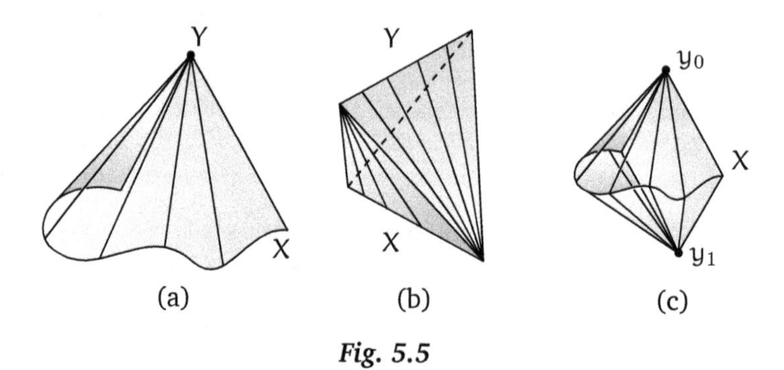

Fig. 5.5

The union of the line segments $[x, y]$, $x \in X, y \in Y$ with its topology as a subspace of V, is then called the *join* of X and Y and is written $X * Y$—the points of $X * Y$ are thus the sums

$$rx + sy, \quad x \in X,\ y \in Y, \quad r, s \geqslant 0, \quad r + s = 1.$$

This construction has several awkward features. First of all it is only defined for subsets of a normed space. Second, it is not even defined for all such subsets—for example, to define $\mathbb{S}^1 * \mathbb{S}^1$ it is necessary to embed two copies of $\mathbb{S}^1$ in some space in such a way that the above curious condition on line segments holds. Thirdly, it is not obvious that the resulting space is independent of the embeddings.

We shall generalise the above construction in two ways. First, we give a canonical definition for topological spaces. Second, we give the definition for the join of n spaces $X_1, \dots, X_n$.

Let $X_1, \dots, X_n$ be topological spaces. The *join*

$$X = X_1 * \cdots * X_n$$

shall as a set consist of all $2m$-tuples, $1 \leqslant m \leqslant n$,

$$x = (r_{i_1}, x_{i_1}, \dots, r_{i_m}, x_{i_m})$$

where $1 \leqslant i_1 < i_2 < \cdots < i_m \leqslant n$, $r_{i_j} > 0$, $r_{i_1} + \cdots + r_{i_j} = 1$, and $x_{i_j} \in X_{i_j}$. This looks rather formidable—we therefore write x in the form

$$x = r_1 x_1 + \cdots + r_n x_n$$

where $r_i \geqslant 0$, $r_1 + \cdots + r_n = 1$, $x_i \in X_i$ and it is agreed that if $r_i = 0$ then the term $r_i x_i$ is to be ignored. This convention agrees well with our earlier description of $X * Y$ for subspaces of a normed vector space. Notice also that the conditions on the r_i say that the point $(r_1, \dots, r_n)$ belongs to the $(n-1)$-simplex Δ^{n-1} in $\mathbb{R}^n$ spanned by the standard basis vectors of $\mathbb{R}^n$.

We have also to define a topology on X—we do this by means of initial topologies as defined in the previous section. Consider the functions

$$\xi_i : X \to \mathbb{I} \qquad\qquad \eta_i : X \rightarrowtail X_i$$

$$r_1 x_1 + \cdots + r_n x_n \mapsto r_i \qquad\qquad r_1 x_1 + \cdots + r_n x_n \mapsto x_i$$

Here η_i has domain $\xi_i^{-1}]0,1]$. If $x \in X$, we call the points $\xi_i x$ and $\eta_i x$ (when defined) the *coordinates* of x; the functions ξ_i, η_i are called the *coordinate functions*. If $f : Z \to X$ is a function, then the functions $\xi_i f, \eta_i f$ are called the *coordinates* of f.

The *join topology* on $X = X_1 * \cdots * X_n$ is the initial topology with respect to the coordinate functions. Thus a function $f : Z \to X$ is continuous if and only if its coordinates are continuous; so with this topology we are well placed for deciding continuity of a function *into* the join. However the only functions out of the join of whose continuity we can be assured are the coordinate functions ξ_i, η_i. The difficulties this leads to will be mentioned later.

As an application of this topology we prove:

5.7.1 *Let* $1 \leqslant i \leqslant n$. *There is a canonical homeomorphism*

$$a : (X_1 * \cdots * X_i) * (X_{i+1} * \cdots * X_n) \to X_1 * \cdots * X_n.$$

Proof Let a be the function which sends

$$\begin{aligned} x &= r(r_1x_1 + \cdots + r_ix_i) + s(r_{i+1}x_{i+1} + \cdots + r_nx_n) \\ &\mapsto rr_1x_1 + \cdots + rr_ix_i + sr_{i+1}x_{i+1} + \cdots + sr_nx_n \end{aligned}$$

where of course $x_i \in X_i$, $(r, s) \in \Delta^1$, $(r_1, \ldots, r_i) \in \Delta^{i-1}$, $(r_{i+1}, \ldots, r_n) \in \Delta^{n-i-1}$. The coordinates of a (considered as a function into $X_1 * \cdots * X_n$) are given by

$$\xi_j : x \mapsto \begin{cases} rr_j, & j \leqslant i \\ sr_j, & j > i \end{cases} \qquad \eta_j : x \mapsto x_j, \quad j = 1, \ldots, n$$

where x is as above; notice that if $r = 0$, then the term $r(r_1x_1 + \cdots + r_ix_i)$ does not occur, but we interpret rr_j as 0.

Suppose that $j \leqslant i$. Then η_j is the composite of the coordinate functions

$$x \mapsto r_1x_1 + \cdots + r_ix_i \mapsto x_j$$

and so η_j is continuous. The points x for which $r \neq 0$ form an open set and on this set ξ_j is the product of the coordinate function $x \mapsto r$ and the composite $x \mapsto r_1x_1 + \cdots + r_ix_i \mapsto r_j$; hence ξ_i is continuous on this set. If $r = 0$, then ξ_j is given by $x \mapsto 0$, and the continuity of ξ_j at such points is proved in a similar way to that of the sandwich rule [Exercise 2 of Section 1.2]; we leave details to the reader. The proof for $j > i$ is similar.

The inverse of a sends the point

$$y = r_1y_1 + \cdots + r_ny_n \quad (y_i \in X_i,\ (r_1, \ldots, r_n) \in \Delta^{n-1})$$

to the point

$$r(s_1y_1 + \cdots + s_iy_i) + s(s_{i+1}y_{i+1} + \cdots + s_ny_n),$$

where $r = r_1 + \cdots + r_i$, $s = r_{i+1} + \cdots + r_n$,

$$s_j = \begin{cases} r^{-1}r_j & \text{if } r \neq 0 \\ 0 & \text{if } r = 0 \end{cases}, \ 1 \leqslant j \leqslant i, \quad s_k = \begin{cases} s^{-1}r_k & \text{if } s \neq 0 \\ 0 & \text{if } s = 0 \end{cases}, \ i+1 \leqslant k \leqslant n.$$

The coordinates of a^{-1}, as a function into the two-fold join, are thus $y \mapsto r$, $y \mapsto s$, which are both clearly continuous, and also $y \mapsto s_1y_1 + \cdots + s_iy_i$, $y \mapsto s_{i+1}y_{i+1} + \cdots + s_ny_n$. The checking of the continuity of these last two functions into the i-fold and $(n-i)$-fold joins respectively is left to the reader. □

In the future we shall leave details proofs of continuity which are similar to the above as exercises for the reader.

5.7.1 *(Corollary 1) There is a homeomorphism*

$$X_1 * (X_2 * X_3) \to (X_1 * X_2) * X_3.$$

Proof This is immediate from 5.7.1 since both spaces are homeomorphic to $X_1 * X_2 * X_3$. □

(We leave the reader to work out the formula for this homeomorphism.)

5.7.2 *If $X_1, \ldots, X_n$ are Hausdorff, then so also is $X_1 * \cdots * X_n$.*

Proof By 5.7.1 and induction it is sufficient to prove the result for the case $n = 2$. We say sets A, A' *separate* x, x' if A, A' are disjoint neighbourhoods of x, x' respectively. Let

$$x = rx_1 + sx_2, \quad x' = r'x_1' + s'x_2'$$

be points of $X_1 * X_2$. If $r \neq r'$, then the set ${\xi_1}^{-1}[U], {\xi_1}^{-1}[U']$ for any sets U, U' which separate r, r' in $\mathbb{I}$, separate x, x'. Suppose $r = r' \neq 0$, $x_1 \neq x_1'$. Let V_1, V_1' separate x_1, x_1'. Then ${\eta_1}^{-1}[V_1], {\eta_1}^{-1}[V_1']$ separate x, x'. A similar argument applies for the remaining case

$$s = s' \neq 0, \ x_2 \neq x_2'.$$

This completes the proof. □

We define the function

$$\begin{aligned} J : X_1 \times \cdots \times X_n \times \Delta^{n-1} &\to X_1 * \cdots * X_n \\ (x_1, \ldots, x_n, r_1, \ldots r_n) &\mapsto r_1x_1 + \cdots + r_nx_n. \end{aligned}$$

5.7.2 *(Corollary 1) If $X_1, \ldots, X_n$ are compact and Hausdorff, then J is an identification map.*

Proof Clearly J is always surjective; it is also continuous because its components are continuous. The $(n-1)$-simplex Δ^{n-1} is a closed, bounded subset of $\mathbb{R}^n$ and so is compact. Therefore J is a continuous surjection from a compact space to a Hausdorff space, and so J is an identification map. □

Without the condition of compactness 5.7.2 (Corollary 1) is false [cf. Exercise 3], and so we have two topologies for the join, one convenient for maps *into* the join, the other convenient for maps *from* the join, and in general these topologies are distinct. Not only that, the join, when given the identification topology, need not be associative. This awkwardness can be resolved by working not in topological spaces and continuous maps but in k-spaces and continuous maps, with a revised notion of product (see section 5.9). So we have an interesting example of the way in which a technical awkwardness can suggest an entirely new approach. However a discussion of any of these topics would take us too far afield, and so we shall often assume when dealing with joins that the spaces concerned are compact and Hausdorff.

We now give some particular examples. Let e denote the unique point of $\mathbb{E}^0$. The following function

$$\begin{aligned} X \times \mathbb{I} &\to X \times \mathbb{E}^0 \\ (x, r) &\mapsto rx + (1-r)e \end{aligned}$$

is continuous because its coordinates are continuous; also it shrinks $X \times \{0\}$ to a single point e of $X \times \mathbb{E}^0$. So this map induces a continuous bijection

$$CX \to X * \mathbb{E}^0.$$

Hence when X is compact and Hausdorff, CX is homeomorphic to $X * \mathbb{E}^0$ [cf. Fig. 5.5(a)].

Suppose now $Y = \mathbb{S}^0$ [cf. Fig. 5.5(c)]. For convenience, let us denote the points $1, -1$ of $\mathbb{S}^0$ by e_+, e_- respectively. The function

$$X \times \mathbb{I} \to X * \mathbb{S}^0$$

$$(x, r) \mapsto \begin{cases} (2-2r)x + (2r-1)e_+, & r \geqslant \frac{1}{2} \\ 2rx + (1-2r)e_-, & r \leqslant \frac{1}{2} \end{cases}$$

is continuous because its coordinates are continuous; also it shrinks $X \times \{1\}$ to the point e_+, and $X \times \{0\}$ to the point e_- of $X \times \mathbb{S}^0$. Hence this map induces a continuous bijection

$$SX \to X * \mathbb{S}^0$$

which is a homeomorphism if X is compact and Hausdorff.

An alternative way of dealing with CX and SX is to use the bijections $CX \to X * \mathbb{E}^0$, $SX \to X * \mathbb{S}^0$, together with the join topologies, to define topologies on CX and SX. These topologies we call the *coarse* topologies. As pointed out before, the coarse topologies are convenient for maps into, rather than from, these spaces.

The join is also convenient when dealing with spheres. From the previous paragraphs and 5.7.1 we see that the n-fold join of $\mathbb{S}^0$ with itself is homeomorphic to the $(n-1)$-th suspension of $\mathbb{S}^0$; by section 4.4 this n-fold join is homeomorphic to $\mathbb{S}^{n-1}$.

There is another relationship between joins and spheres. The sphere $\mathbb{S}^{p+q+1}$ may be taken as consisting of all pair (x, y) such that

$$x \in \mathbb{R}^{p+1}, \quad y \in \mathbb{R}^{q+1} \quad \text{and} \quad |x|^2 + |y|^2 = 1.$$

With this coordinatisation, we define

$$\begin{aligned} k : \mathbb{S}^{p+q+1} &\to \mathbb{S}^p * \mathbb{S}^q \\ (x, y) &\mapsto rx' + sy' \end{aligned}$$

where $x' = x/|x|$, $y' = y/|y|$, $r\pi/2 = \sin^{-1}|x|$, $s\pi/2 = \sin^{-1}|y|$. Notice that these definitions imply that $r + s = 1$, that x' is defined only for $x \neq 0$ (and so for $r \neq 0$), and that y' is defined for $y \neq 0$ (and so for only $s \neq 0$). It follows that k is well-defined; k is continuous because its coordinates are continuous. Also k is a bijection because it has inverse

$$rx' + sy' \mapsto (x' \sin r\pi/2, y' \sin s\pi/2)$$

where, as usual, $x' \sin r\pi/2$, $y' \sin s\pi/2$ are taken as 0 when, respectively, r, s are 0. Since all the spaces concerned are compact and Hausdorff, it follows that k is a homeomorphism. We shall see that this result is related to the existence of the homeomorphism of 4.4.5.

5.7.3 *If* A, B *are subspaces of* X, Y *respectively, then* $A * B$ *is a subspace of* $X * Y$.

Proof This is an easy consequence of the transitive law for initial topologies. □

5.7.3 *(Corollary 1) The maps*

$$\begin{aligned} X &\to X * Y \qquad & Y &\to X * Y \\ x &\mapsto 1.x \qquad & y &\mapsto 1.y \end{aligned}$$

are embeddings.

Proof This follows from 5.7.3 since, for example, the image of X under the given map is the subspace $X * \varnothing$ of $X * Y$. □

We shall use the maps of 5.7.3 (Corollary 1) to identify X, Y with the corresponding subspaces of $X * Y$.

Let us consider again the picture of $X * Y$ [Fig. 5.6]. By the *top half* of $X*Y$ we mean the subspace of points $rx+sy$ such that $s \geqslant \frac{1}{2}$. By the *bottom half* of $X * Y$ we mean the subspace of points $rx + sy$ such that $r \geqslant \frac{1}{2}$. It is intuitively clear from the picture that the top half of $X * Y$ is bijective with $CX \times Y$, and the bottom half of $X * Y$ is bijective with $X \times CY$. Actually we prove the following result.

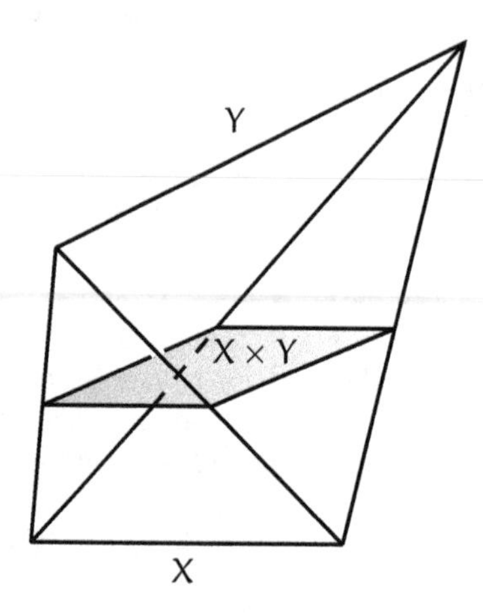

Fig. 5.6

5.7.4 *There is a homeomorphism*

$$\nu : X * Y * \mathbb{E}^0 \to (X * \mathbb{E}^0) \times (Y * \mathbb{E}^0)$$

which restricts to a homeomorphism

$$X * Y \to (X * \mathbb{E}^0) \times Y \cup X \times (Y * \mathbb{E}^0).$$

Proof The unique point of $\mathbb{E}^0$ is written e.

Consider first the case $X = \{x\}$, $Y = \{y\}$. We have then to produce, in essence, a homeomorphism

$$\Delta^1 * \mathbb{E}^0 \to \Delta^1 \times \Delta^1$$

that is, a homeomorphism from a triangle to a square.

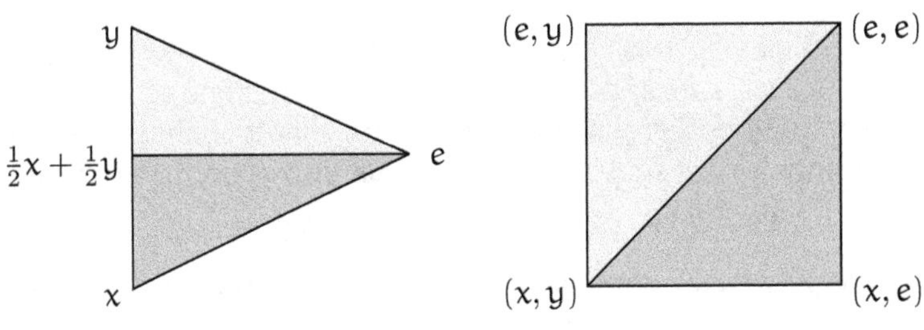

Fig. 5.7

This is done by splitting the triangle into two and mapping each half linearly onto one of the triangles into which the square is divided by a diagonal [Fig. 5.7].

We now consider the general case. Points of $X * Y * \mathbb{E}^0$ are $rx + sy + te$ where $(r, s, t) \in \Delta^2$, $x \in X$, $y \in Y$; points of $(X * \mathbb{E}^0) \times (Y * \mathbb{E}^0)$ are $(ux + ve, u'y + v'e)$ where $(u, v), (u', v') \in \Delta^1$, $x \in X$, $y \in Y$. We require a function which sends

$$x \mapsto (x, e), \qquad y \mapsto (e, y),$$
$$e \mapsto (e, e), \qquad \tfrac{1}{2}x + \tfrac{1}{2}y \mapsto (x, y).$$

Hence for $r \leqslant s$ (and so for $r \leqslant \frac{1}{2}$) we have

$$\begin{aligned} rx + sy + te &= 2r(\tfrac{1}{2}x + \tfrac{1}{2}y) + (s - r)y + te \\ &\mapsto 2r(x, y) + (s - r)(e, y) + t(e, e) \\ &= (2rx + (s - r + t)e, (r + s)y + te) \end{aligned}$$

while for $s \leqslant r$ (and so for $s \leqslant \frac{1}{2}$) we have

$$rx + sy + te \mapsto ((r + s)x + te, 2sy + (r - s + t)e).$$

These formulae agree for $r = s$ and so define ν. The detailed proof of continuity of ν (as a function into the product) is left to the reader.

The inverse of ν is defined as follows. If $u' \leqslant u$ then

$$\begin{aligned} (ux + ve, u'y + v'e) &= u'(x, y) + v(e, e) + (u - u')(x, e) \\ &\mapsto u'(\tfrac{1}{2}x + \tfrac{1}{2}y) + ve + (u - u')x \\ &= (u - \tfrac{1}{2}u')x + \tfrac{1}{2}u'y + ve \end{aligned}$$

while, if $u \leqslant u'$, then

$$(ux + ve, u'y + v'e) \mapsto \tfrac{1}{2}ux + (u' - \tfrac{1}{2}u)y + v'e.$$

Again the proof of continuity is left to the reader.

Notice also that $\nu(rx + sy + te) \in (X * \mathbb{E}^0) \times Y \cup X \times (Y * \mathbb{E}^0)$ if and only if t is either 0, or $r - s$ for $r \leqslant s$, or $s - r$ for $s \leqslant r$, and one of the last two conditions can hold only if $r = s$ (since $t \geqslant 0$), that is, only if $t = 0$; but $t = 0$ if and only if $rx + sy + te \in X * Y$. This proves the last part of the result. □

5.7.4 *(Corollary 1) If X, Y are compact Hausdorff, there is a homeomorphism*

$$\mu : C(X * Y) \to CX \times CY$$

which restricts to a homeomorphism

$$X * Y \to CX \times Y \cup X \times CY.$$

Proof This is immediate from 5.7.4, the associativity of the join and the fact that $Z * \mathbb{E}^0$ is homeomorphic to CZ if Z is compact and Hausdorff. □

Since $\mathbb{S}^{p+q+1}$ is homeomorphic to $\mathbb{S}^p * \mathbb{S}^q$, and $C\mathbb{S}^n$ is homeomorphic to $\mathbb{E}^{n+1}$, there is also a homeomorphism

$$\mathbb{S}^{p+q+1} \to \mathbb{E}^{p+1} \times \mathbb{S}^q \cup \mathbb{S}^p \times \mathbb{E}^{q+1},$$

a fact we have prove by a different method in section 4.4.

EXERCISES

1. Prove that if K, L are cell complexes, then the topological space $K * L$ can be given the structure of a cell complex.
2. Let X, Y be compact, Hausdorff spaces. Prove that if SX, SY have base point the 'top' vertex, then there is a homeomorphism

$$SX \times SY \to (SX \vee SY) \,{}_w\!\sqcup\, C(X * Y)$$

for some map $w : X * Y \to SX \vee SY$. (This map w is called the *Whitehead product* map.) [Let $p : CX \to SX = CX/X$ be the identification map. Use 4.5.8 on the composite $C(X * Y) \xrightarrow{\mu} CX \times CY \xrightarrow{p \times p} SX \times SY$].
3. Show that the map $J : X_1 \times X_2 \times \Delta^1 \to X_1 * X_2$ of 5.7.2 (Corollary 1) is not in general an identification map. [Use 5.7.3.]
4. Let X, Y be compact and Hausdorff. Let $p : X \times \{1\} \times Y \to X$ be the projection. Prove that $X * Y$ is homeomorphic to

$$X \,{}_p\!\sqcup\, (CX \times Y).$$

5. Let Z_n denote the space obtained from

$$\mathbb{E}^n \times \mathbb{S}^{n-1} \cup \mathbb{S}^{n-1} \times \mathbb{E}^n$$

by the identifications $(x, y) = (y, x)$ $(x \in \mathbb{E}^n, y \in \mathbb{S}^{n-1})$. Prove that Z_n is homeomorphic to $\mathbb{S}^{n-1} * P^{n-1}(\mathbb{R})$. [Define $f : \mathbb{E}^n \times P^{n-1}(\mathbb{R}) \to Z_n$ as follows. Let $x \in \mathbb{E}^n$, $y \in P^{n-1}(\mathbb{R})$. Draw a line through x parallel to y. This will meet $\mathbb{S}^{n-1}$ in z, z' say, where z, z' are named so that z is at least as near to x as z' is. Define $f(x, y)$ to be the equivalence class of $(2x - z, z)$. Now use Exercise 4 and 4.5.8.]
6. The *symmetric square* of a space X is obtained from $X \times X$ by identifying (x, y) with (y, x) for all x, y in X. Prove that if A_n denotes the symmetric square of $\mathbb{E}^n$, then A_n is homeomorphic to CZ_n, where Z_n is as in the previous exercise. Deduce that the symmetric square of $\mathbb{S}^n$ is homeomorphic to a mapping cone $C(f)$ for a map $f : \mathbb{S}^{n-1} * P^{n-1}(\mathbb{R}) \to \mathbb{S}^n$. [cf. [JTTW63]]

5.8 The smash product

We recall that a pointed space is a topological space X and point x_0 of X, called the base point, such that $\{x_0\}$ is closed in X. We shall find it convenient to denote both the base point x_0 of a pointed space X and the set $\{x_0\}$ by $\cdot$. Note also that we use the same symbol for a pointed space and the underlying topological space.

The *wedge* of pointed spaces X, Y is the subspace of the product $X \times Y$ defined by

$$X \vee Y = X \times \cdot \cup \cdot \times Y.$$

(It is easy to prove that $X \vee Y$ is homeomorphic to the space obtained from $X \sqcup Y$ by identifying the two base points to a single point.) Thus the wedge consists of the 'axes' of the product $X \times Y$.

The *smash product* of pointed spaces X, Y is the identification space

$$X \mathbin{⁂} Y = X \times Y / X \vee Y$$

with the set $X \vee Y$ taken as base point of $X ⁂ Y$. (This space has also been called the *reduced join, smashed product, collapsed product*; other notations used are $X \wedge Y$, $X \bowtie Y$.)

5.8.1 *If* X, Y *are Hausdorff spaces then so also is* $X ⁂ Y$.

Proof Clearly $X \vee Y$ is closed in $X ⁂ Y$ and $(X \times Y) \setminus (X \vee Y)$ is an open subset of $X ⁂ Y$. Hence any two points of this subset are separated in $X ⁂ Y$. To complete the proof we have only to show that the base point of $X ⁂ Y$ is separated from any other point (x, y).

Let $U_\cdot, U$ be disjoint open neighbourhoods of $\cdot, x$ respectively, and let $V_\cdot, V$ be disjoint open neighbourhoods of $\cdot, y$ respectively. Then $U_\cdot \times Y \cup X \times V_\cdot$, $U \times V$ are disjoint open neighbourhoods of $X \vee Y$, (x, y) respectively. Also these neighbourhoods are saturated with respect to the identification map $X \times Y \to X ⁂ Y$. Therefore their images separate $\cdot$ and (x, y). □

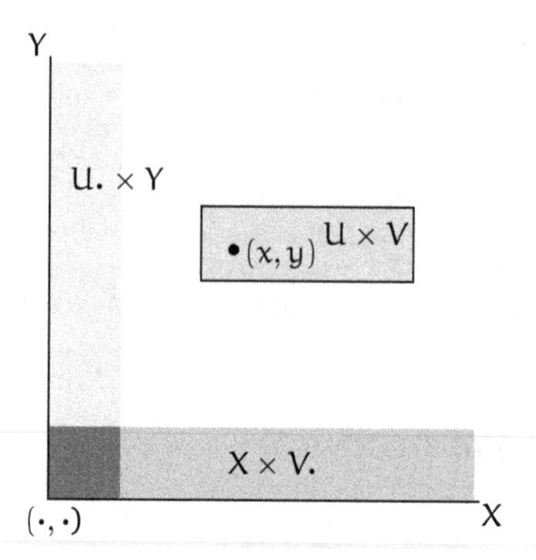

Fig. 5.8

5.8.2 *For any pointed spaces* X, Y, Z *there is a bijection*

$$a : X \divideontimes (Y \divideontimes Z) \to (X \divideontimes Y) \divideontimes Z$$

which is continuous if X *is locally compact, or* Y *and* Z *are compact and Hausdorff. Further* a^{-1} *is continuous if* Z *is locally compact or* X *and* Y *are compact and Hausdorff. Hence* a *is a homeomorphism if* X *and* Z *are locally compact, or if two of* X, Y, Z *are compact and Hausdorff.*

Proof Each of the composites

$$X \times Y \times Z \xrightarrow{1 \times p} X \times (Y \divideontimes Z) \xrightarrow{p} X \divideontimes (Y \divideontimes Z),$$

$$X \times Y \times Z \xrightarrow{p \times 1} (X \divideontimes Y) \times Z \xrightarrow{p} (X \divideontimes Y) \divideontimes Z$$

shrinks to a point of the subspace

$$\cdot \times Y \times Z \cup X \times \cdot \times Z \cup X \times Y \times \cdot$$

of $X \times Y \times Z$. The existence of the bijection a is immediate. By 4.6.6, $1 \times p$ is an identification map if X is locally compact, and $p \times 1$ is an identification map if Z is locally compact. By 3.6.3 (Corollary 1), $1 \times p$ is closed if Y and Z are compact and Hausdorff, and hence, in this case, $1 \times p$ is an identification map. A similar result holds if X and Y are compact and Hausdorff. The continuity of a and a^{-1} in the various cases follows (since also a compact Hausdorff space is locally compact). □

For any pointed space X, the *reduced suspension* of X is

$$\Sigma X = X \ast \mathbb{S}^1.$$

The reason for this name is that if $p' : \mathbb{I} \to \mathbb{S}^1$ is the map $t \mapsto e^{2\pi i t}$, then the composite

$$X \times \mathbb{I} \xrightarrow{1 \times p'} X \times \mathbb{S}^1 \xrightarrow{p} X \ast \mathbb{S}^1$$

is an identification map ($1 \times p'$ is an identification map by 3.6.3 (Corollary 1)) which shrinks to a point the subspace $X \times \dot{\mathbb{I}} \cup \cdot \times \mathbb{I}$ of $X \times \mathbb{I}$. But the suspension SX is obtained from $X \times \mathbb{I}$ by shrinking each of $X \times \{0\}$, $X \times \{1\}$ to a point. It follows that ΣX is homeomorphic to

$$SX / \cdot \times \mathbb{I}.$$

We shall prove below that $\Sigma \mathbb{S}^{n-1}$ is homeomorphic to $\mathbb{S}^n$ where $\mathbb{S}^{n-1}$ has base point $e_{n-1} = (1, 0, \ldots, 0)$. In the case $n = 2$, the following picture shows how $\mathbb{S}^2$ is represented as $\mathbb{S}^1 \ast \mathbb{S}^1$—each point of $\mathbb{S}^2$ is described by coordinates (x, y) for $x, y \in \mathbb{S}^1$, with $(x, y) = e_2$ the base point of $\mathbb{S}^2$ if x or y is e_1.

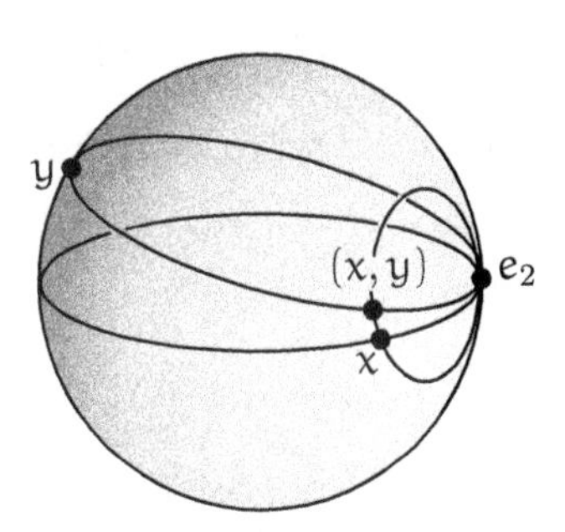

Fig. 5.9

More generally we prove

5.8.3 *There is a homeomorphism*

$$\mathbb{S}^m \ast \mathbb{S}^n \to \mathbb{S}^{m+n}.$$

Proof By the last part of section 4.7, the product $\mathbb{S}^m \times \mathbb{S}^n$ has a cell structure

$$e^0 \cup e^m \cup e^n \cup e^{m+n}$$

in which the subcomplex $e^0 \cup e^m \cup e^n$ is $\mathbb{S}^m \vee \mathbb{S}^n$. It follows that $\mathbb{S}^m \divideontimes \mathbb{S}^n$ has a cell structure $e^0 \cup e^{m+n}$. The attaching map of the $(m+n)$-cell is constant, and so $e^0 \cup e^{m+n}$ is homeomorphic to $\mathbb{S}^{m+n}$. □

In the case where X, Y are compact and Hausdorff there is a simple relation between the join $X * Y$ and $\Sigma(X \divideontimes Y)$. Let R be the subspace of $X * Y$ of points $rx + sy$ where x, or y, or both x and y, are at the base point.

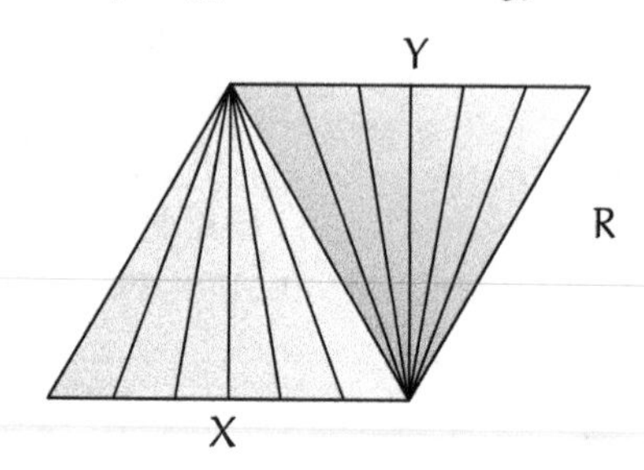

Fig. 5.10

5.8.4 *There is a bijection*

$$k : \Sigma(X \divideontimes Y) \to (X * Y)/R$$

which is continuous and which is a homeomorphism if X, Y *are compact and Hausdorff.*

Proof The identification map

$$X \times Y \times \Delta^1 \to X \times Y \times \mathbb{S}^1 \to (X \divideontimes Y) \times \mathbb{S}^1 \to \Sigma(X \divideontimes Y)$$

has exactly the effect of shrinking to a point the subspace

$$\cdot \times Y \times \Delta^1 \cup X \times \cdot \times \Delta^1 \cup X \times Y \times \dot{\Delta}^1.$$

The composite

$$X \times Y \times \Delta^1 \to X * Y \to (X * Y)/R$$

is continuous and has exactly the same effect. However it can be shown using Exercise 5 of Section 4.3 that $(X * Y)/R$ is Hausdorff if X and Y are Hausdorff. So if X and Y are compact Hausdorff, then the latter composite is an identification map. This proves the result. □

The conditions imposed for this result show the desirability of some fresh approach, for example that outlined in the next section.

EXERCISES

1. Prove that any space with base point X is homeomorphic to $X \divideontimes \mathbb{S}^0$.
2. Prove that for any spaces with base point X, Y, the spaces $X \divideontimes Y$ and $Y \divideontimes X$ are homeomorphic.
3. Prove that if K, L are cell complexes with base points vertices of K, L respectively, then $K \divideontimes L$ can be given the structure of a cell complex.

5.9 Spaces of functions, and the compact-open topology

The problem of finding a suitable topology for spaces of continuous functions $Y \to Z$ has occupied a large amount of literature on general topology. It is a problem crucial in many applications. For example we like to say that the output of a radio is a continuous function of the position of the volume knob. But the output of a radio is a signal, which itself can be described as a function of time. Again, we would like to know if the solution of a differential equation which itself has parameters depends continuously on these parameters. The solution is some kind of function. Thus in both cases we need to know what it means for a family of functions to be a *continuous* family.

One criterion for topologising sets of functions arises as follows. Suppose X, Y, Z are topological spaces and $f : X \times Y \to Z$ is a continuous function. If $x \in X$, then the function $f(x, -) : Y \to Z$ given by $y \mapsto f(x, y)$ is continuous. Let Z^Y denote the set of all continuous functions $Y \to Z$. Then $x \mapsto f(x, -)$ defines a function

$$\hat{f} : X \to Z^Y.$$

The problem is whether there is a topology on all such sets Z^Y such that the correspondence $f \mapsto \hat{f}$ defines a bijection

$$e : Z^{X \times Y} \to (Z^Y)^X.$$

If this can be done, then the continuous functions $f : X \times Y \to Z$ are precisely those which define continuous functions $\hat{f} : X \to Z^Y$.

For example, if $f : \mathbb{R} \times \mathbb{R} \to \mathbb{R}$ is given by $(x, y) \mapsto x^2 + y^2$, then in order to picture f we draw the contours $f^{-1}[c]$, for $c \in \mathbb{R}$, as in Fig. 5.11(i). However for each $x \in \mathbb{R}$, $\hat{f}(x) : \mathbb{R} \to \mathbb{R}$ is a function and the graphs of the functions $y \mapsto x^2 + y^2$ of this family are given in Fig. 5.11(ii). The aim is to be able to say in mathematical terms what is clear from the picture, that this family of parabolas $\hat{f}(x)$ '*varies continuously with* x'.

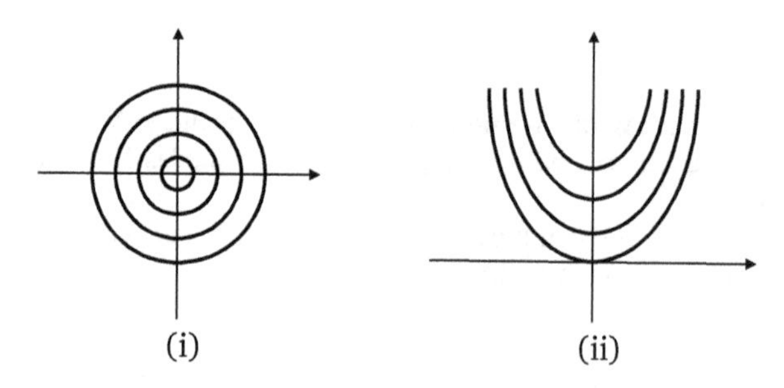

Fig. 5.11

Strangely, the general case of this problem has proved not entirely tractable within the traditional outlook of general topology, that is, using topological spaces and continuous functions. In this section we describe one method of dealing with this difficulty.

A space X is called a k-space if X has the final topology with respect to all maps $C \to X$ for all compact Hausdorff spaces C. In such case, a function $f : X \to Y$ is continuous if and only if $ft : C \to Y$ is continuous for all compact Hausdorff spaces C and maps $t : C \to X$. (The use of the letter k is traditional. It was introduced because the German for compact is *kompakte*.)

At first sight this definition seems ridiculous, since a property of a space X is described by reference to a large class of spaces. However, as should be clear from earlier chapters, this procedure is in the modern spirit and can be convenient precisely because of this global reference. The following result shows however that in order to test for a k-space we need look only at a *set* of test spaces C.

5.9.1 *Let X be a space. Then the following are equivalent:*
(a) *X is a k-space;*
(b) *there is a set $\mathcal{C}_X$ of maps $t : C_t \to X$ for compact Hausdorff spaces C_t such that a set A is closed in X if and only if $t^{-1}(A)$ is closed in C_t for all $t \in \mathcal{C}_X$;*
(c) *X is an identification space of a space which is a sum of compact Hausdorff spaces.*

Proof (a) $\Rightarrow$ (b) The set $\mathcal{C}_X$ is constructed as follows. Since X is a k-space, for each non-closed subset B of X there is a compact Hausdorff space C_B and map $t : C_B \to X$ such that $t^{-1}[B]$ is not closed in C_B. Choose one such C_B and one such t for each non-closed B, and let $\mathcal{C}_X$ be the set of all these t. That this set has the required property is clear.

(b) $\Rightarrow$ (c) Suppose $\mathcal{C}_X$ given as in (b). Let K be the sum of the spaces C_t for all $t \in \mathcal{C}_X$, and let $i_t : C_t \to K$ be the inclusion. Let $p : K \to X$ be the unique map such that $pi_t = t$ for all $t \in \mathcal{C}_X$. Then property (b), and the defining property of the sum implies that a function $f : X \to Y$ to a space Y is continuous if and only if fp is continuous. Hence p is an identification map.

(c) $\Rightarrow$ (a) Suppose $p : K \to X$ is an identification map where K is a sum $\bigsqcup_{\alpha \in A} C_\alpha$ of compact Hausdorff spaces C_α. Let $i_\alpha : C_\alpha \to K$ be the inclusion. Let $f : X \to Y$ be a function to a topological space Y such that $ft : C \to X$ is continuous for all test maps $t : C \to X$. Then in particular, $fi_\alpha : C_\alpha \to Y$ is continuous for all $a \in A$. Hence $fp : K \to Y$ is continuous. Since p is an identification map, f is continuous. Hence X is a k-space. □

5.9.1 *(Corollary 1) Let X be a k-space and let Y be locally compact and Hausdorff. Then $X \times Y$ is a k-space.*

Proof Let $p : K \to X$, $q : M \to Y$ be identification maps where K is a sum of compact Hausdorff spaces C_α, $\alpha \in A$, and M is a sum of compact Hausdorff spaces D_β, $\beta \in B$. Since K is locally compact, it follows from 4.3.2 that $r = 1 \times q : K \times M \to K \times Y$ is an identification map. Similarly, $s = p \times 1 : K \times Y \to X \times Y$ is an identification map. Hence $sr = p \times q : K \times M \to X \times Y$ is an identification map. But $K \times M$ is a sum of compact Hausdorff spaces $C_\alpha \times D_\beta$. By 5.9.1(c), $X \times Y$ is a k-space. □

5.9.1 *(Corollary 2) Any identification space of a k-space is a k-space.*

Proof Let X be a k-space and let $q : X \to Y$ be an identification map. By 5.9.1 there is an identification map $p : K \to X$ such that K is a sum of compact Hausdorff spaces. Then $qp : K \to Y$ is an identification map. Hence Y is a k-space, by 5.9.1. □

Examples are known of pairs of k-spaces whose product is not a k-space [Exercise 10].

It is clear that a compact Hausdorff space is a k-space. The following result gives a useful family of other examples of k-spaces. We say that a space is *first countable* if it satisfies the first axiom of countability, that is, if each point of the space has a countable base of neighbourhoods.

5.9.2 *The following are k-spaces:* (a) *any locally compact, Hausdorff space;* (b) *any first countable space.*

Proof (a) Let X be a locally compact, Hausdorff space. Then each point of X has a basis of compact, Hausdorff neighbourhoods. It follows that X has the final topology with respect to all inclusions of compact Hausdorff subspaces of X. Therefore X is a k-space.

(b) Let X be a first countable space. This means that each point x of X

has a countable base $\mathcal{B}(x)$ of neighbourhoods of x. Let these neighbourhoods be written $B_n(x)$, $n \in \mathbb{N}$. Then it is easy to define a new base $B'_n(x)$, $n \in \mathbb{N}$, of neighbourhoods of x such that $B'_n(x) \supseteq B'_{n+1}(x)$ for all n. All one has to do is set $B'_0(x) = B_0(x)$ and $B'_n(x) = B'_{n-1}(x) \cap B_n(x)$, for all $n \in \mathbb{N}$. Now let $\mathbb{N}^+$ be $\mathbb{N} \cup \{\omega\}$ with the topology in which if $n \in \mathbb{N}$ then any subset of $\mathbb{N}^+$ containing n is a neighbourhood of n, while a subset N of $\mathbb{N}$ is a neighbourhood of ω if and only if $\omega \in N$ and N contains all but a finite number of elements of $\mathbb{N}$. Essentially, $\mathbb{N}^+$ is the Alexandroff one-point compactification of the discrete space $\mathbb{N}$ (cf. Exercise 6 of Section 3.6). Let $\mathcal{A}$ be the set of all continuous functions $\mathbb{N}^+ \to X$. We claim X has the final topology with respect to this set $\mathcal{A}$ of functions.

For the proof, let $g : X \to Y$ be any function and suppose that $gf : \mathbb{N}^+ \to Y$ is continuous for all f in $\mathcal{A}$. Suppose g is not continuous. Then there is a point x in X and neighbourhood W of $g(x)$ such that for no neighbourhood M of x is it true that $g[M] \subseteq W$. In particular, for each $n \in \mathbb{N}$ there is a point x_n in $B'_n(x)$ such that $g(x) \notin W$. Define $f : \mathbb{N}^+ \to X$ by $f(n) = x_n$, $f(\omega) = x$. By the construction of the sets $B'_n(x)$, the function f is continuous. However, the function gf is clearly not continuous, and so we have a contradiction. □

Since all metric spaces satisfy the first axiom of countability, we have a lot of useful examples of k-spaces. The following are some known examples of spaces which are not k-spaces: (a) an uncountable product of copies of the real line $\mathbb{R}$ (the proof is not easy, but may be found as part of a stronger result in [Bro61]); (b) consider the identification space, often called the wedge,

$$W_A = \left(\bigsqcup_{\alpha \in A} \mathbb{I} \times \{\alpha\} \right) \Big/ \left(\bigsqcup_{\alpha \in A} \{0\} \times \{\alpha\} \right)$$

where $\mathbb{I}$ is the unit interval and A is a set. Then W_A is a k-space since it is an identification of such a space. However if A is countable and B is uncountable then the product $W_A \times W_B$ is not a k-space. A proof is in essence given in [Dow52], which says that this space is not a CW-complex; (c) the space X of Exercise 8 of Section 2.10. The last example is the one of which it is easiest to verify that it is not a k-space, but we do not give the proof here. It is because the examples of spaces which are not k-spaces are somewhat weird that the restriction to k-spaces is not going to be unreasonable. A further justification for the method is the idea that compact Hausdorff spaces are the most easily comprehended, and so it is reasonable to use these as test spaces for more general kinds of spaces.

There is a method of constructing for any space X an associated k-space kX. We simply let kX have the same underlying set as X, but the topology on kX is the final topology with respect to all continuous maps $C \to X$

for all compact Hausdorff spaces C, and where X has its given topology. By the basic properties of final topologies as in section 4.2, the identity map $i : kX \to X$ is continuous and, for all compact Hausdorff spaces C, i determines by composition a bijection between the set of continuous maps $C \to kX$ and the set of continuous maps $C \to X$. It follows that kX itself is a k-space. In other words, $kkX = kX$.

For any spaces Y, Z let $\mathcal{K}(Y, Z)$ denote the set of functions $Y \to Z$ such that $ft : C \to Z$ is continuous for all compact Hausdorff spaces C and continuous functions $t : C \to Y$. In other words,

$$\mathcal{K}(Y, Z) = Z^{kY}.$$

The elements of $\mathcal{K}(Y, Z)$ will be called k-maps $Y \to Z$. For example, the identity function $Y \to kY$ is a k-map but in general is not a map. If $f : Y \to Z$ is a k-map, then so also is $f : Y \to kZ$, while $f : kY \to kZ$ is a map, i.e. is continuous. It is convenient to topologise in the first instance the set of k-maps.

By a *test map* (on Y) we mean a map $t : C \to Y$ for some compact Hausdorff space C. Given such a test map and an open set U of Z, we define

$$\mathcal{W}(t, U) = \{f \in \mathcal{K}(Y, Z) : ft[C] \subseteq U\}.$$

These sets are to form a sub-basis for the *test-open topology* on $\mathcal{K}(Y, Z)$.

Note that if Y is Hausdorff, then the image $t[C]$ of a test map $t : C \to Y$ is also compact Hausdorff. Thus in this case the set of $\mathcal{W}(t, U)$ defines the same topology as does the sub-base consisting of the set of $\mathcal{W}(C, U) = \{f \in \mathcal{K}(Y, Z) : f[C] \subseteq U\}$ for all compact subsets C of Y and open subsets U of Z; this topology on the function space $\mathcal{K}(Y, Z)$, or the corresponding relative topology on a subspace, is known as the *compact-open topology*.

However, it *is* necessary to deal with non-Hausdorff spaces. For example, they arise commonly as identification spaces. It may not be easy or even possible to prove that a given space is, or is not, Hausdorff. The problems of dealing with spaces which are not Hausdorff force one to the test-open topology, and this often allows the technical question of whether or not a given space is Hausdorff simply to be avoided. Most texts on topology deal with the compact-open topology, but the more general case causes only a slight extra difficulty in the proofs.

The intuitive idea for the compact-open topology is illustrated for the case $Y = Z = \mathbb{R}$ as follows. Let $C = [a, b]$ be a closed interval in $\mathbb{R}$ and let $U = \,]c, d[$ be an open interval. Then $\mathcal{W}(C, U)$ consists of those maps $f : \mathbb{R} \to \mathbb{R}$ such that $f[C] \subseteq U$. The graph of such a map is shown in the following picture.

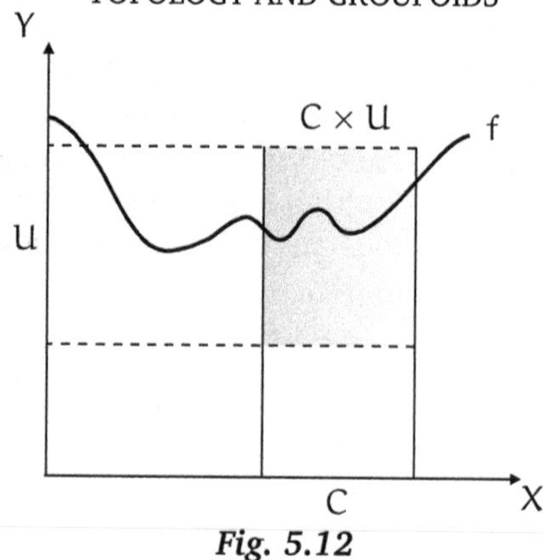

Fig. 5.12

Now let us go to our main problem. We suppose given spaces X, Y, Z. Our aim is to construct a bijection

$$e : \mathcal{K}(X \times Y, Z) \to \mathcal{K}(X, \mathcal{K}(Y, Z)), \tag{1}$$

where all the function spaces have the test-open topology. The proof is given in a series of partial results.

5.9.3 *Let* $x \in X$, $f \in \mathcal{K}(X \times Y, Z)$. *Then the function* $f(x, -) : y \mapsto f(x, y)$ *is an element of* $\mathcal{K}(Y, Z)$.

Proof Let $t : C \to Y$ be a test map. Then $i \times t : \{x\} \times C \to X \times Y$, $(x, y) \mapsto (x, ty)$, is also a test map, and hence $f(i \times t)$ is continuous. It follows easily that $f(x, -)t : C \to Z$ is continuous. □

5.9.4 *Let* $f \in \mathcal{K}(X \times Y, Z)$, *and let the function* $\hat{f} : X \to \mathcal{K}(Y, Z)$ *be given by* $x \mapsto f(x, -)$. *Then* $\hat{f} \in \mathcal{K}(X, \mathcal{K}(Y, Z))$.

Proof Let $s : B \to X$ be a test map. We have to prove that $\hat{f}s : B \to \mathcal{K}(Y, Z)$ is continuous. Let $b \in B$, and suppose $\mathcal{W}(t, U)$ is a sub-basic neighbourhood of $\hat{f}s(b) = f(sb, -)$, where U is open in Z and $t : C \to Y$ is a test map. Then $s \times t : B \times C \to X \times Y$ is also a test map. Hence $g = f(s \times t) : B \times C \to Z$ is continuous. Further, $g[\{b\} \times C] \subseteq U$. Hence $\{b\} \times C$ is contained in the open set $g^{-1}[U]$. By 3.5.6 (Corollary 2), b has an open neighbourhood V such that $V \times C \subseteq g^{-1}[U]$. It follows that for each $v \in V$, $f(sv, -)$ maps $t[C]$ into U. Hence $\hat{f}s[V] \subseteq \mathcal{W}(t, U)$. That is, we have proved that $\hat{f}s$ is continuous. □

Up to now, we have not used the Hausdorff property of the test spaces C. This is used at the next and most subtle point, namely in proving that the function e of (1) is surjective. For this we introduce the *evaluation map*

$$\begin{aligned} \varepsilon : \mathcal{K}(Y, Z) \times Y &\to Z \\ (f, y) &\mapsto fy. \end{aligned}$$

Note here an important aspect of function spaces. In the early days of functions, people always thought of $f(x)$ as meaning that f was fixed and x was a variable. Now we think of both f and x as varying, and may even on occasion want to keep x fixed and let f vary.

5.9.5 *Let* $t : C \to Y$ *be a test map. Then the composite*

$$\varepsilon(1 \times t) : \mathcal{K}(Y, Z) \times C \to Z$$

is continuous.

Proof Let $\delta = \varepsilon(1 \times t)$, let $(f, c) \in \mathcal{K}(Y, Z) \times C$, and let U be an open neighbourhood of $z = \delta(f, c)$. Then $ft(c) = z \in U$. Since ft is continuous, there is an open neighbourhood V of c such that $ft[V] \subseteq U$. Since C is compact Hausdorff, it is locally compact, and so there is a compact neighbourhood B of c contained in V. Then $ft[B] \subseteq U$. Let $s = t \mid B : B \to Y$. Then B is compact Hausdorff and so s is a test map. By definition of $\mathcal{W}(s, U)$,

$$\delta[\mathcal{W}(s, U) \times B] \subseteq U.$$

Since $\mathcal{W}(s, U)$ is a neighbourhood of f and B is a neighbourhood of c, it follows that δ is continuous. □

5.9.5 *(Corollary) If* Y *is locally compact and Hausdorff, then*

$$\varepsilon : \mathcal{K}(Y, Z) \times Y \to Z$$

is continuous.

Proof If $y \in Y$ then y has a compact Hausdorff neighbourhood N_y, since Y is locally compact and Hausdorff. By 5.9.5, $\varepsilon \mid \mathcal{K}(Y, Z) \times N_y$ is continuous. Hence ε is continuous. □

5.9.6 *A function* $f : X \times Y \to Z$ *is in* $\mathcal{K}(X \times Y, Z)$ *if and only if* $f(s \times t) : B \times C \to Z$ *is continuous for all test maps* $s : B \to X$, $t : C \to Y$.

Proof $\Rightarrow$ If s, t are test maps, then so also is $s \times t : B \times C \to X \times Y$, and hence $f(s \times t)$ is continuous.

$\Leftarrow$ Let $u : D \to X \times Y$ be a test map. Then $s = p_1 u : D \to X$, $t = p_2 u : D \to Y$ are also test maps. By assumption,

$$f(s \times t) : D \times D \to Z$$

is continuous. But $fu = f(s \times t)\Delta$, where $\Delta : D \to D \times D$ is the diagonal. Hence fu is continuous. □

We give a corollary of 5.9.6 which will be used later, although it is not necessary for the immediate problem.

5.9.6 *(Corollary) If X and Y are spaces, then*

$$k(kX \times kY) = k(X \times Y).$$

Proof Since the identity functions $kX \to X$, $kY \to Y$ are continuous, so also is the identity function

$$kX \times kY \to X \times Y,$$

and hence the identity $k(kX \times kY) \to k(X \times Y)$ is continuous. To prove the identity $f : k(X \times Y) \to k(kX \times kY)$ continuous we use 5.9.6. Let $s : B \to X$, $t : C \to Y$ be test maps. Then

$$s \times t : B \times C \to kX \times kY$$

is a test map, as is $s \times t : B \times C \to k(kX \times kY)$. By 5.9.6, f is continuous. □

5.9.7 *Let $g \in \mathcal{K}(X, \mathcal{K}(Y, Z))$. Let $\bar{g} : X \times Y \to Z$ be defined by $(x, y) \mapsto (gx)y$. Then $\bar{g} \in \mathcal{K}(X \times Y, Z)$.*

Proof Using 5.9.4, it is sufficient to suppose given test maps $s : B \to X$, $t : C \to Y$, and prove that $\bar{g}(s \times t)$ is continuous. Consider the commutative diagram

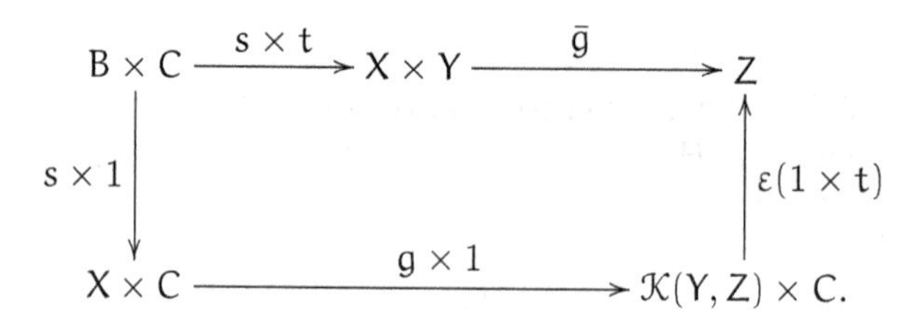

We are given that gs is continuous, and by 5.9.3 $\varepsilon(1 \times t)$ is continuous. Hence $\bar{g}(s \times t)$ is continuous. □

It is now simple to prove our main result.

5.9.8 (Exponential law for k-maps) *The exponential map*

$$\begin{aligned} e : \mathcal{K}(X \times Y, Z) &\to \mathcal{K}(X, \mathcal{K}(Y, Z)) \\ f &\mapsto (x \to (y \mapsto f(x, y)) \end{aligned}$$

is a well defined bijection.

Proof It is proved in 5.9.2 that $e : f \mapsto \hat{f}$ is well defined. The previous result, 5.9.5, shows that $e' : g \mapsto \bar{g}$ is also well defined. Clearly e and e' are mutual inverses. Therefore e is a bijection. □

Useful applications for this theorem come from specialising to cases where the set of k-maps coincides with the set of maps. First note that if Y and Z are spaces then the set Z^Y of continuous maps $Y \to Z$ is a subset of $\mathcal{K}(Y, Z)$. We therefore give Z^Y the relative topology.

5.9.9 *If* X, Y *are* k*-spaces such that* $X \times Y$ *is a* k*-space, then the 'topological exponential map'*

$$e_T : Z^{X \times Y} \to (Z^Y)^X$$

is defined and is a bijection. In particular, e_T *is a bijection if*
(a) X *and* Y *are both first countable spaces, or*
(b) X *is a* k*-space and* Y *is locally compact, or*
(c) X *is locally compact and* Y *is a* k*-space.*

Proof The conditions given first ensure that $Z^{X \times Y} = \mathcal{K}(X \times Y, Z)$, that $Z^Y = \mathcal{K}(Y, Z)$, and that $(Z^Y)^X = \mathcal{K}(X, \mathcal{K}(Y, Z))$. This implies that e_T is a bijection. By 5.9.2, conditions (a), (b) or (c) ensure that X and Y are k-spaces. It remains to show that they ensure that $X \times Y$ is a k-space.

If X and Y are first countable spaces, so also is $X \times Y$. Hence in this case $X \times Y$ is a k-space. In cases (b) and (c), $X \times Y$ is a k-space, by 5.9.1 (Corollary 1). □

One can do a little better than 5.9.9(b) and show that e_T is bijective assuming only that Y is locally compact and Hausdorff.

It turns out that instead of looking for conditions which ensure that the topological exponential map e_T is a bijection, it is more useful to sidestep these technical problems and work in a *convenient class of spaces*. This is a class in which all the properties one would reasonably like to use do in fact hold.

This raises the general question: *what is the purpose of the theory of topological spaces?* A rough and ready answer is that the theory is intended to give a *useful, convenient and adequate* setting for our notions of continuity. These vague words can be defined only in terms of the way the theory works in practice, and the sort of applications that arise in other areas of mathematics. There is a danger that, when a theory has been around a good while, then it acquires a certain sanctity—in spite of increasing complications in the use of the theory, it can still be difficult to see how a theory can be found which works better. At such a stage, it is important to consider the basic requirements for, and purposes of, the theory. That is, why was it worked out in the first place? The study of this kind of question is an area which tends to be neglected in the teaching of mathematics, where it can be forgotten that often the important question is not: *What is the answer?*, but instead is: *What is the question?* In research, evaluation of work done, and problem formulation, can be more important than problem solution.

The advance from metric spaces to topological spaces was important for several reasons. One was that the proofs often were improved and clearer, because the essential features of the problems were revealed by working with a topology rather than a metric. The other, and the clincher, was that many spaces arose in a natural way with a topology rather than a metric, and even when a metric did exist, it was not always possible to find one which was related to the geometric situation.

Since Hausdorff's famous book *Mengenlehre* was published, topological spaces have come to play a central role in mathematics. The problem is that now the theory is becoming a bit ragged round the edges, as shown by the difficulty of finding an adequate and convenient theory of function spaces within the context of topological spaces and continuous maps.

Attempting to find a function space topology satisfying the exponential correspondence leads to a messy theory for topological spaces and continuous maps, with all sorts of technical conditions which seem inappropriate and which divert attention from what is really going on. Thus we have an *anomaly*, and it is right that an anomaly should be a starting point for a new direction rather than an irritant.

One appropriate method in this case turns out to be to replace all spaces by k-spaces, by applying k to everything in sight. This idea is best expressed using the categorical language of the next chapter. But we can say enough here to be able to see how the method works out.

Let Y, Z be k-spaces. We define $K(Y, Z)$ to be the space obtained from the space $\mathcal{K}(Y, Z)$ by applying k. So the underlying sets of $\mathcal{K}(Y, Z)$ and of $K(Y, Z)$ are the same but the second space has the larger topology. The important point is that $K(Y, Z)$ is now a k-space.

Further we have to change the product topology, because there is no guarantee that the ordinary product of k-spaces is a k-space. So we define for any spaces

$$X \times_{\mathsf{k}} Y = \mathsf{k}(X \times Y).$$

This we call the k-product of X and Y. See Exercise 1 for its key property.

5.9.10 *Let* X, Y, *and* Z *be* k-*spaces. Then the exponential correspondence*

$$e_K : K(X \times_k Y, Z) \to K(X, K(Y, Z))$$

is defined and is a homeomorphism.

Proof Since X and Y are k-spaces, we have equalities of sets

$$K(X \times_k Y, Z) = \mathcal{K}(X \times_k Y, Z), \quad K(X, K(Y, Z)) = \mathcal{K}(X, K(Y, Z)).$$

It follows from 5.9.8 that e_K is a well defined bijection. It remains to prove that e_K is a homeomorphism. Here we use the fact that e_K is defined and is a bijection for all k-spaces X, Y, Z, and in particular we can replace X, Y and Z by function spaces of the form $K(Y, Z)$.

We wish to prove that e_K is continuous. But e_K corresponds to a map

$$e' : K(X \times_k Y, Z) \times_k X \to K(Y, Z)$$

which itself corresponds to the evaluation map

$$e'' : K(X \times_k Y, Z) \times_k X \times_k Y \to Z.$$

Since e'' is continuous, it follows that e' is continuous and hence, by the same argument, that e_K is continuous.

Let d be the inverse bijection to e_K, so that d is a function

$$K(X, K(Y, Z)) \to K(X \times_k Y, Z).$$

To prove that d is continuous it is sufficient to prove that the corresponding function

$$d' : K(X, K(Y, Z)) \times_k X \times_k Y \to Z$$

is continuous. Now for $x \in X$, $y \in Y$, we have

$$(dg)(x, y) = (gx)(y).$$

It follows that

$$d'(g, x, y) = (gx)(y).$$

This says that d' is the composite

$$K(X, K(Y, Z)) \times_k X \times_k Y \xrightarrow{\varepsilon_X \times_k 1_Y} K(Y, Z) \times_k Y \xrightarrow{\varepsilon_Y} Z$$

where ε_X and ε_Y are the evaluation maps. But evaluation maps are continuous on these spaces. So d' is continuous. Hence d is continuous. This proves that e_K is a homeomorphism. □

We now make a deduction from the fact that e_K is always a bijection.

5.9.10 *(Corollary) Let* X *and* Y *be* k*-spaces and let* $f : Y \to Z$ *be an identification map. Then* $1 \times_k f : X \times_k Y \to X \times_k Z$ *and* $f \times_k 1 : Y \times_k X \to Z \times_k X$ *are identification maps.*

Proof It is sufficient to prove that $f \times_k 1$ is an identification map. By 5.9.1 (Corollary 2) we know that Z is a k-space. Let $g : Z \times_k X \to W$ be any function to a k-space W such that $g(f \times_k 1)$ is continuous. We have to prove that g is continuous.

Consider the diagrams

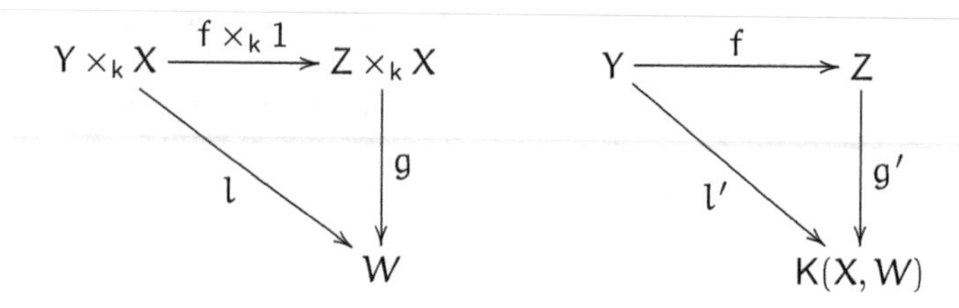

By assumption, $l = g(f \times_k 1)$ is continuous. Hence the corresponding map $l' : Y \to K(X, W)$ is continuous. Let $g' : Z \to K(X, W)$ be the function corresponding to g. Then $g'f = l'$, and so $g'f$ is continuous. Since f is an identification map, g' is continuous. Hence g is continuous. This proves that $f \times_k 1$ is an identification map. □

We shall not use greatly here the fact that e_K is a homeomorphism rather than just a bijection. However the proof of the previous results does illustrate nicely how the fact that 5.9.8 is valid for all spaces can be used. We now give some further illustrations of the use of the powerful result 5.9.10.

For the remainder of this section, we assume that all spaces concerned are k-spaces.

We first examine the composition function.

5.9.11 *The composition function*

$$\begin{aligned} c : K(Z, W) \times_k K(Y, Z) &\to K(Y, W) \\ (g, f) &\mapsto gf \end{aligned}$$

is continuous.

Proof Consider the function

$$\begin{aligned} c' : K(Z, W) \times_k K(Y, Z) \times_k Y &\to W \\ (g, f, y) &\mapsto gf(y). \end{aligned}$$

Then c' is the composition

$$K(Z, W) \times_k K(Y, Z) \times_k Y \xrightarrow{1 \times_k \varepsilon_Y} K(Z, W) \times_k Z \xrightarrow{\varepsilon_Z} W.$$

It follows that c' is continuous. But under the exponential correspondence, c' corresponds to c. Hence c is continuous. □

Now we can show how $K(Y, Z)$ behaves with regard to maps on Y and on Z. Let $g : Z \to W$, $f : Y \to Z$ be maps of k-spaces. Then g and f induce by composition functions

$$g_* : K(Y, Z) \to K(Y, W), \qquad f^* : K(Z, W) \to K(Y, W)$$
$$h \mapsto gh, \qquad k \mapsto kf.$$

5.9.12 *The induced functions g_* and f^* are continuous. Further,*
(a) *if g is a homeomorphism into, so also is g_*;*
(b) *if f is an identification map, then f^* is a homeomorphism into.*

Proof These induced functions are the composites of the composition function c of 5.9.11 with the maps

$$K(Y, Z) \to K(Z, W) \times_k K(Y, Z), \qquad K(Z, W) \to K(Z, W) \times_k K(Y, Z)$$
$$h \mapsto (g, h) \qquad k \mapsto (k, f)$$

respectively.

We now prove (a). Let $\bar{g} : g[Z] \to Z$ be the inverse of the restriction of g. Then $\bar{g}$ is continuous, by the assumption on g. Let M be the image of g_*. Since g is injective, so also is g_* and hence g_* restricts to a continuous bijection $K(Y, Z) \to M$. Let $m : M \to K(Y, Z)$ be the inverse of this restriction. We wish to prove m is continuous. For this it is sufficient to prove that the corresponding map $m' : M \times_k Y \to Z$ is continuous. But $m(gh)(y) = \bar{g}h(y)$, so that m' is the composition

$$M \times_k Y \xrightarrow{\varepsilon} g[Z] \xrightarrow{\bar{g}} Z$$

which is continuous. It follows that m is continuous.

We now prove (b). Let N be the image of f^*. Since f is surjective, f^* is injective, and so restricts to a continuous bijection $K(Z, W) \to N$. Let $n : N \to K(Z, W)$ be the inverse of this restriction. We wish to prove n is continuous. For this it is sufficient to prove that the corresponding map $n' : N \times_k Z \to W$ is continuous. Now n' is given by $(kf, z) \mapsto k(z)$ so that

n' is defined by the commutative diagram

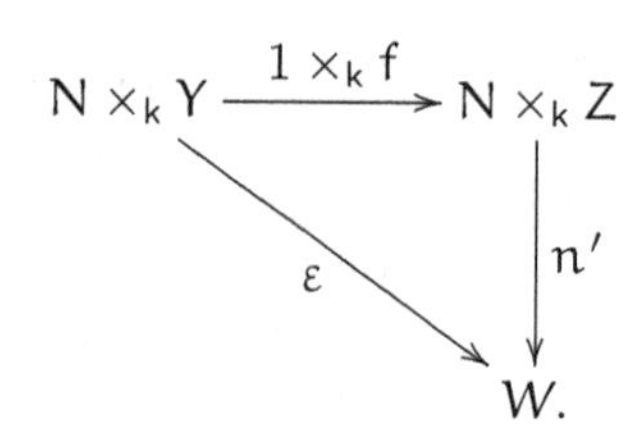

But ε is continuous and $1 \times_k f$ is an identification map, by 5.9.10 (Corollary). It follows that n' is continuous. Hence n is continuous. □

One application of the last result is to make the results of the previous two sections more convenient. By working entirely with k-spaces, and using the identification topology on joins and smash products, one can obtain associativity of the join and also relations between the join and smash product without making the compact Hausdorff assumptions used earlier. Indeed we do not have to bother to verify whether or not spaces might be Hausdorff, since the methods work in any case. This makes the results applicable in a wide variety of situations. However we shall not pursue this topic in this text.

We shall not use the results on function spaces elsewhere, except to hint at them in chapter 7. But the basic ideas are so important, and so widely applied, yet difficult to find in textbooks, that it seemed essential to include an introduction to these topics. As explained above, they are also interesting as showing some limitations of a traditional approach. It is hoped that this will encourage the reader in a critical and questioning attitude.

Another reason for including an account of function spaces is that the notion of an 'object of maps' is proving to be increasingly important in mathematics. For applications in mechanics see [Law66], and in computer science, see [PAPR86], where the exponential map is referred to as *currying*.

In many situations where one has a structure of a specific kind, then for two objects X and Y with this structure, one looks for an object $S(X, Y)$ with the same kind of structure and whose underlying set is the set of structure preserving maps $X \to Y$. For example, if X and Y are abelian groups, then the set of group homomorphisms $X \to Y$ has the structure of an abelian group, by addition of values.

There are however many situations when such an object $S(X, Y)$ does not seem to exist. For example, if X and Y are groups, then the set of homomorphisms $X \to Y$ is not even closed under the product defined by multiplication of values. In such cases, there are the options of accepting the situation, or of seeking a more convenient kind of structure in which to work. In the case of groups, it is possible to enlarge the point of view to that of groupoids, and recover a 'groupoid of morphisms'.

There is one extension of the exponential law which is becoming increasingly important, and which deals with the problem of topologising spaces of partial maps. Let X and Y be topological spaces. The set of continuous partial functions $X \rightarrowtail Y$ is written $\mathcal{P}(X, Y)$. We would like to know what it may mean for a family of partial maps to be continuous, that is, we would like a topology on $\mathcal{P}(X, Y)$ satisfying a kind of exponential law.

This is obtained by a device which first came into extensive use in the the theory of toposes ([Joh02], [BW85]). Let Y be a set. By Y^+ we mean $Y \cup \{\omega\}$ where ω is a point not belonging to Y. A partial function $f : X \rightarrowtail Y$ determines uniquely a function $f^+ : X \to Y^+$ defined by

$$f^+(x) = \begin{cases} f(x) & \text{if } x \in \mathcal{D}_f \\ \omega & \text{otherwise.} \end{cases}$$

This gives a bijection between the functions $X \to Y^+$ and the partial functions $X \rightarrowtail Y$.

This device enables us to topologise certain spaces of partial maps by topologising Y^+ and using standard topologies on spaces of maps to Y^+. Two standard topologies on Y^+ have been used in this situation, both given in [BB78b].

Let Y be a topological space. The space $Y^\sim$ is the set Y^+ with the topology on which C is closed in $Y^\sim$ if and only if $C = Y^\sim$ or C is closed in Y . It is easy to verify that this does indeed define a topology on $Y^\sim$. Note that apart from the empty set, all open sets of $Y^\sim$ include the point ω. So $Y^\sim$ is a Hausdorff space if and only if Y is empty. If X is a topological space, then the maps $X \to Y^\sim$ are bijective with the partial maps $X \rightarrowtail Y$ with closed domain. This gives a procedure for topologising spaces of maps with closed domain; applications of this method are given in [BB78b]. Note that we obtain an exponential law of a different form: let $\mathcal{K}_{\mathcal{PC}}(X, Y)$ be the set of partial k-maps with closed domain. Then for all spaces X, Y, Z

$$\mathcal{K}_{\mathcal{PC}}(X \times Y, Z) \cong \mathcal{K}(X \times Y, Z^\sim) \cong \mathcal{K}(X, \mathcal{K}(Y, Z^\sim)) \cong \mathcal{K}(X, \mathcal{K}_{\mathcal{PC}}(Y, Z)).$$

Another topology on Y^+ gives a space $Y^\wedge$, say, in which a set U is open in $Y^\wedge$ if and only if $U = Y^\wedge$ or U is open in Y. Then the maps $X \to Y^\wedge$ are bijective with the partial maps $X \rightarrowtail Y$ with open domain. This method is elaborated in [AB80], and it gives the same topology on the space of such partial maps as was first introduced in [Ehr80]. In this setup it is not so clear how best to handle partial k-maps with open domain. The difficulty is that $t : C \to X$ is a test map and V is open in X, then $t^{-1}[V]$ need not be compact. Of course it will be locally compact, and this possibly gives the method of dealing with the situation. (cf. [Vog71]). However, just to give an idea of the theory we stick to the case of maps, so that the exponential law is valid only under restrictive circumstances.

Suppose then that X and Y are Hausdorff. Then the test-open topology on $Z^{X\times Y}$ coincides with the compact-open topology. Let $\mathcal{PO}(X,Z)$ be the set of partial maps $X \to Y$ with open domain. Topologise $\mathcal{PO}(X,Z)$ so that the natural map $\mathcal{PO}(X,Z) \to (Z^{\wedge})^X$ is a homeomorphism. Then the topology on $\mathcal{PO}(X,Z)$ has as a sub-base the sets $\mathcal{W}_{\mathcal{O}}(C,U)$ of partial maps $f : X \rightarrowtail Z$ such that C is a subset of the domain of f and $f[C] \subseteq U$, for all compact subsets C of X and open subsets U of Z. The exponential correspondence is of the form

$$\mathcal{PO}(X\times Y,Z) \cong (Z^{\wedge})^{X\times Y} \cong ((Z^{\wedge})^Y)^X \cong (\mathcal{PO}(Y,Z))^X.$$

This is valid if for example X and Y satisfy the conditions of 5.9.9. This result has its uses even when Z is a singleton $\{1\}$. Then $\mathcal{PO}(Y,Z)$ is bijective with the set $\mathcal{O}(Y)$ of open sets of Y, since a partial map $Y \rightarrowtail \{1\}$ is entirely determined by its domain. This gives a topology on $\mathcal{O}(Y)$ with sub-base the set of

$$\mathcal{W}(C) = \{U \in \mathcal{O}(Y) : C \subseteq U\}$$

for all compact subsets C of Y . Hence if X and Y are Hausdorff and locally compact we get a homeomorphism

$$\mathcal{O}(X\times Y) \cong (\mathcal{O}(Y))^X.$$

This says that an open set U in $X\times Y$ can in these circumstances be regarded as a continuously X-indexed family $x \mapsto U_x$ of open sets of Y, where $U_x = \{y \in Y : (x,y) \in U\}$.

The case for introducing spaces of partial maps is strong. Such maps occur in elementary analysis. For example the functions

$$\sin^{-1} x, \ \sqrt{x}, \ \log x, \ (x-1)^{-1} + (x+3)^{-1}$$

are all sensibly considered as partial maps $\mathbb{R} \rightarrowtail \mathbb{R}$. Thus it is surprising that the algebra and topology of such partial maps has been relatively little considered. For example, it is difficult to find a book on functional analysis in which partial functions rate a mention. One of the reasons may be the difficulty of such a study, which is illustrated by the fact that it is not known how to topologise in a sensible way the set of all partial maps $X \rightarrowtail Y$.

If Y is a singleton, then the partial maps $X \rightarrowtail Y$ are bijective with the subsets of X . So the problem of topologising spaces of partial maps includes the problem of topologising the set $\mathcal{P}(X)$ of subsets of a space X . There is a large literature on the topologies for spaces of closed subsets (such spaces are called *hyperspaces*), but not so much on the general problem. Thus it is not so clear what it might mean for a family of closed to have limit point an open set, say.

An even more worrying question is the following. Let A_λ be the subset of $X = \mathbb{S}^1 \times \mathbb{S}^1$ of points $(e^{2\pi i t}, e^{2\pi i \lambda t})$ for all $t \in \mathbb{R}$. It is known that A_λ is closed in X if λ is rational, and is dense in X if λ is irrational. What is not so clear is whether or not the function $\lambda \mapsto A_\lambda$ should be expected to be continuous. On the one hand, the set A_λ seems to jump about from closed to dense, and so seems unlikely to be conveniently regarded as a continuous function of λ. On the other hand, A_λ is given by a formula involving standard and well-behaved functions (e.g. the exponential function e^z) and so ought, as a function of λ, to be not only continuous, but even analytic!

One attempt to begin a resolution of these questions is the thesis [Har86], which is based on work in [Joh83]. Again, it seems that the problem is not resolvable within the context of topological spaces. But the 'correct' solution is not clear.

EXERCISES

1. Let X, Y, Z be k-spaces. Prove that the projections $p_1 : X \times_k Y \to X$ and $p_2 : X \times_k Y \to Y$ are continuous, and that if $f_1 : Z \to X$ and $f_2 : Z \to Y$ are maps, then there is a unique map $f : Z \to X \times_k Y$ such that $p_1 f = f_1$, $p_2 f = f_2$.
2. Let $\mathcal{U}$ be a sub-base for the topology of Y. Prove that the sets $\mathcal{W}(t, U)$ for all test maps $t : C \to X$ and all $U \in \mathcal{U}$ form a sub-base for the test-open topology on $\mathcal{K}(X, Y)$. [Take the proof of a similar result from [Dug68] or [Jam84] and modify it for the new situation.]
3. Use Exercises 1 and 2 to prove that the map

$$\mathcal{K}(X, Y \times Z) \to \mathcal{K}(X, Y) \times \mathcal{K}(X, Z)$$

given by $f \mapsto (p_1 f, p_2 f)$ is a homeomorphism.
4. Prove that for all spaces X, Y, Z, the test-open topology on $\mathcal{K}(X \times Y, Z)$ has as a sub-base the sets $W(s \times t, U)$ for all test maps $s : C \to X$, $t : D \to Y$. [The hint here is the same as for Exercise 2. I do not know if there is a result combining the results of Exercise 2 and Exercise 3.]
5. Use Exercise 4 to prove that the function

$$\mathcal{K}(X, Z) \times \mathcal{K}(Y, T) \to \mathcal{K}(X \times Y, Z \times T)$$

which sends $(f, g) \mapsto f \times g$ is a homeomorphism into.
6. Use Exercises 2 and 4 to prove that the exponential map e of 5.9.8 is a homeomorphism.

7. Let X and Y be topological spaces and let $\mathcal{C}$ be any set of functions $X \to Y$. For any test map $t : C \to X$, and open subset U of Y, let $\mathcal{W}_{\mathcal{C}}(t, U)$ be the set of elements f of $\mathcal{C}$ such that $ft[C] \subseteq U$. The test-open topology on $\mathcal{C}$ is defined like the test-open topology on $\mathcal{K}(X, Y)$ but using $\mathcal{W}_{\mathcal{C}}$ in place of $\mathcal{W}$. Prove that if $\mathcal{D}$ is a subset of $\mathcal{C}$ and $\mathcal{D}$ also has the test-open topology, then $\mathcal{D}$ is a subspace of $\mathcal{C}$. Prove also that: (i) if Y is Hausdorff, then $\mathcal{C}$ is Hausdorff; (ii) if $\mathcal{C}$ contains all constant functions, and $\mathcal{C}$ is Hausdorff, then Y is Hausdorff; (iii) if X is discrete, then $\mathcal{K}(X, Y)$ is homeomorphic to the topological product $\prod_{x \in X} Y$ of the constant family $x \mapsto Y$.
8. Let $f : X \to W$ be a k-map and let $g : Z \to Y$ be continuous. Prove that f and g induce by composition continuous maps $\mathcal{K}(X, g) : \mathcal{K}(X, Z) \to \mathcal{K}(X, Y)$, $\mathcal{K}(f, Y) : \mathcal{K}(W, Y) \to \mathcal{K}(X, Y)$. Prove that if g is a homeomorphism into, then so also is $\mathcal{K}(X, g)$; and that $\mathcal{K}(f, Y)$ is a homeomorphism into if f is surjective and for each test map $s : B \to W$ there is a test map $t : C \to X$ and a surjective map $r : C \to D$ such that $sr = ft$.
9. Prove that the composition map $c : \mathcal{K}(Z, W) \times \mathcal{K}(Y, Z) \to \mathcal{K}(Y, W)$ is a k-map. [Use the methods of 5.9.11.]
10. Let $X \times_S Y$ be the product set $X \times Y$ with the final topology with respect to all inclusions $\{x\} \times Y \to X \times Y$ for all $x \in X$ and all maps $1 \times t : X \times C \to X \times Y$ for all test maps $t : C \to Y$. Prove that $k(X \times_S Y) = k(X \times Y)$, and that $Z^{X \times_S Y}$ is a subspace of $\mathcal{K}(X \times Y, Z)$. Hence show that the exponential map

$$e_S : Z^{X \times_S Y} \to (Z^Y)^X$$

is a homeomorphism. [This is the exponential law for topological spaces proved in the Hausdorff case in [Bro64] and in the general case in [BT80].]
11. Sketch the family of curves $f_t : x \mapsto \log(x + t)$ for all $t \in \mathbb{R}$, and show how the exponential law for partial maps with open domain makes $t \mapsto f_t$ a continuous family of such maps.
12. Let $\mathbb{Q}$ be the space of rational numbers and let $\mathbb{Q}/\mathbb{Z}$ be the space $\mathbb{Q}$ with the subspace of integers shrunk to a point. Prove that $\mathbb{Q}$ and $\mathbb{Q}/\mathbb{Z}$ are k-spaces such that the usual cartesian product $\mathbb{Q} \times \mathbb{Q}/\mathbb{Z}$ is not a k-space.
13. Show how to obtain general versions of the results on joins and smash products by using the k-product and identification topologies on k-spaces and joins.

NOTES

The combination of algebra, topology and geometry given in sections 5.1 to 5.5 is further developed in the text *Topological geometry* ([Por69]). The study of groups of isometries leads naturally to the study of Lie groups ([Che46]), topological groups ([Pon46]; [Bou66, chapter 4]; [Hig63]; for example). For a survey of some of their uses, see [Her66]. For examples of the uses of the groups SU(1) and SU(3) in the physics of fundamental particles, see the story of the 'eight-fold way' in, for example, [Pic86].

The idea of a group, because of its utility for expressing in mathematical terms our intuitive ideas of symmetry, is a fundamental concept in mathematics and its applications. But, as explained in the preface to the second edition of this book, there are strong indications that the idea of a group should be properly subsumed in the idea of 'groupoid' (see the next few chapters, the survey article [Bro82], and the books [Hig66] and [Mac05]).

The ideas of section 5.9 have a curious history. It was J. H. C. Whitehead who first recognised the importance of dealing with products of identification maps, and so formulated essentially 4.3.2. The idea of a Hausdorff k-space began to crop up in the late 1950s, the weak product $X \times_k Y$ began to be used, and [Kel55] uses functions continuous on compact subsets in dealing with function spaces.

The author in his thesis ([Bro61], which was widely circulated) proved results analogous to those of 5.9.10 and its corollaries, but in the Hausdorff case, as a by-product of studies of the homotopy type of function spaces (for the meaning of the words *homotopy type*, see section 6.5). The paper [Bro63] writes "It may be that the category of Hausdorff k-spaces is adequate and convenient for all purposes of topology", while [Bro64] states "These results show that we cannot obtain a convenient category of spaces simply by changing the product topology." (For the notion of *category*, see the next chapter.) The paper [Bro64] points out that "the category of Hausdorff spaces and continuous maps does not have all the formal properties one would like. Specifically, there is no product and function space topology such that

$$X^{Z\times Y} \cong (X^Y)^Z, \quad (X\times Y)^Z \cong X^Z \times Y^Z$$

for all X, Y, Z. (Another difficulty, discussed in [Bro63] is that the product of identification maps is not in general an identification map.)" The paper goes on to show that "these difficulties disappear in the category of Hausdorff spaces and functions continuous on compact subspaces", and that an alternative strategy is "replacing the usual subspace, product and function-space topologies by the corresponding weak topologies".

The term 'convenient category of spaces' was adopted in the paper [Ste67]. This paper lists the desirable properties of a convenient category of spaces to be that: $(Y \times Z)^X = Y^X \times Z^X$; $Z^{X\times Y} = (Z^Y)^X$; the product of identification maps is an identification map; a product of a sum is a sum of products; a sum of identification maps is an identification map (in fact a new terminology is used). The papers [Bro63], [Bro64] are cited in [Ste67].

The method of dealing with the non-Hausdorff case was found by a number of writers—see for example [Cla68], [Day68, Day72], [Vog71], [BT80] and the references in [Her72, Her71]. See [Whi78] for a consistent use of (Hausdorff) k-spaces in homotopy theory.

The study in the large of categories relevant to the considerations of topology is part of the subject of *categorical topology*. One part of this subject deals with categories with a suitable 'function space object'. These

Chapter 6

The fundamental groupoid

The topological invariants discussed in chapter 3 were not very subtle. One reason for this lies in the nature of the invariants; a set of numbers is not a subtle mathematical structure, and in order to probe further we need better structures to model the geometric properties of spaces.

The invariants we want come from a study of paths in a space. The paths on X with their addition form a *category*—this is an algebraic object, defined by [EM45], which is basic to the study of much recent mathematics. From the category PX of paths on X we obtain another category πX which, because of its extra properties, is called a *groupoid*. This *fundamental groupoid* πX, and its various subgroupoids, give a useful algebraic model of X.

We have now use the word 'model' twice—a precise expression of the idea is given by the notion of a functor which is a sort of homomorphism between categories. Any functor of topological spaces, such as the fundamental groupoid, gives rise to a topological invariant of any space X. Algebraic topology may be defined as the study of functors from the category of topological spaces to a category of algebraic objects (e.g., groups, rings, groupoids, etc.).

The importance of the study of categories and functors is that it gives a mathematics of mathematical structures. The algebraic structures which have been found useful to systematise our mathematical techniques have also been found to lead to new mathematical structures which are interesting and useful in their own right. This multiple level of mathematical discourse is one of the fascinations of the subject.

Categories and functors are a good tool for making analogies across areas of mathematics, such as between topology and algebra. As for most mathematics, such analogies are not between the structures, the objects, themselves, but between the relations among the objects of each kind. That is, we make analogies between structures of relations. In particular, the notion of pushout which we have used previously for topological spaces will now be developed for groupoids.

The high point of this chapter is in section 6.7, where we use methods built up in this and previous chapters to show that the fundamental group $\pi(\mathbb{S}^1, 1)$ of the circle $\mathbb{S}^1$ is an infinite cyclic group, i.e. is isomorphic to the additive group of integers. The next two diagrams show an analogy between topology and algebra:

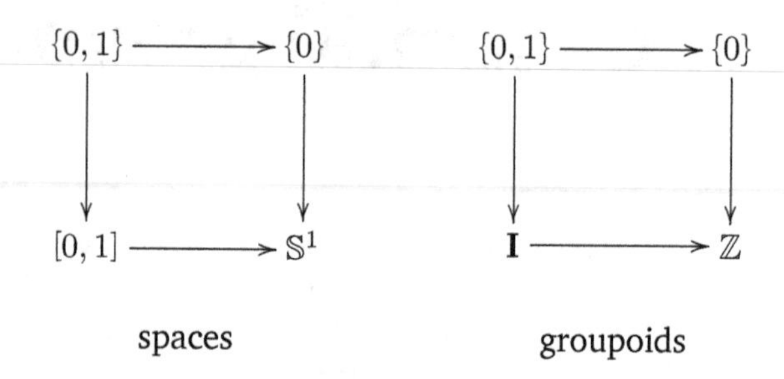

The left hand diagram shows the circle as obtained from the unit interval $[0, 1]$ by identifying, in the category of spaces, the two end points $0, 1$. The right hand diagram shows the infinite group of integers as obtained from the finite groupoid **I** again by identifying $0, 1$, but this time in the category of groupoids. So we have one verification that groupoids can give a useful modelling of the topology.

It was finding this modelling that led to the commitment of this book to the theory of groupoids, and which in turn led to new higher dimensional developments treated elsewhere.

Of course the above result on the circle is just a start. The modelling of the geometry of pushouts of spaces by pushouts of groupoids is continued in chapters 8 and 9.

6.1 Categories

A category $\mathcal{C}$ consists of
(a) a class[‡] $\mathrm{Ob}(\mathcal{C})$, called the class of *objects* of $\mathcal{C}$,
(b) for each x, y in $\mathrm{Ob}(\mathcal{C})$ a set $\mathcal{C}(x, y)$ called the set of *morphisms in* $\mathcal{C}$ *from* x *to* y,

---

[‡]For the distinction between *sets* and *classes*, see the Glossary of terms from set theory.

(c) a function, called *composition*, which to each g in $\mathcal{C}(y,z)$ and to each f in $\mathcal{C}(x,y)$ assigns an element gf in $\mathcal{C}(x,z)$; that is, composition is a function

$$\mathcal{C}(y,z) \times \mathcal{C}(x,y) \to \mathcal{C}(x,z).$$

These terms must satisfy the axioms:

CAT 1 (Associativity) If $h \in \mathcal{C}(z,w)$, $g \in \mathcal{C}(y,z)$, $f \in \mathcal{C}(x,y)$ then

$$h(gf) = (hg)f.$$

CAT 2 (Existence of identities) For each x in $\mathrm{Ob}(\mathcal{C})$ there is an element 1_x in $\mathcal{C}(x,x)$ such that if $g \in \mathcal{C}(w,x)$, $f \in \mathcal{C}(x,y)$ then

$$1_x g = g, \quad f1_x = f.$$

We shall always assume that the various sets $\mathcal{C}(x,y)$ are disjoint and shall write $\mathcal{C}$ also for the union of these sets: thus 'f is an element of $\mathcal{C}$' and '$f \in \mathcal{C}$' both mean '$f \in \mathcal{C}(x,y)$ for some objects x, y of $\mathcal{C}$'. Then the structure of a category can be roughly stated as: a category is a set $\mathcal{C}$ with a multiplication which is associative and has two-sided identities, but such that the multiplication is partial, i.e., is not everywhere defined.

If $f \in \mathcal{C}(x,y)$ then we also write $f : x \to y$, or $x \xrightarrow{f} y$; the notation $x \to y$ simply denotes some element of $\mathcal{C}(x,y)$. For each x in $\mathrm{Ob}(\mathcal{C})$ the identity in $\mathcal{C}(x,x)$ is unique, since if $1_x, 1'_x$ are both identities in $\mathcal{C}(x,x)$ then

$$1_x = 1_x 1'_x = 1'_x.$$

It is usually convenient to abbreviate 1_x to 1 (this means that from equations such as $fg = 1$, $gf = 1$ we cannot deduce $fg = gf$ since 1 may denote different identities in each equation).

The definition of a category, and the notation, is suggested by Examples 2-5 below. In the first example, we show that the paths in a topological space form a category in which composition of paths is written as addition.

EXAMPLES

1. Let X be a topological space. The category PX of paths on X has the set X as its set of objects, and for any x, y in X the set $PX(x,y)$ is the set of paths in X from x to y. Composition of paths b, a is written, as before, $b + a$. The identity in $PX(x,x)$ is the zero path 0_x. Finally, addition of paths is associative since if c, b, a are paths of lengths r, q, p respectively, then $c + (b + a)$ is defined if and only if $(c + b) + a$ is defined, and both paths

are given by

$$t \mapsto \begin{cases} at, & \text{if } 0 \leqslant t \leqslant p \\ b(t-p), & \text{if } p \leqslant t \leqslant p+q \\ c(t-p-q), & \text{if } p+q \leqslant t \leqslant p+q+r. \end{cases}$$

2. Let Set be the category whose objects are all sets and whose morphisms $X \to Y$ are simply the functions from X to Y, and whose composition is the usual composition of functions. The axioms for a category are clearly satisfied, the identity in $\mathsf{Set}(X, X)$ being the identity function 1_X.
3. Let Top be the category whose objects are all topological spaces and whose morphisms $X \to Y$ are the maps, that is, the continuous functions, $X \to Y$. Again the axioms are obviously satisfied.

Here we already see the double use of the idea of category. (a) General statements about topological spaces and continuous functions can in many cases be regarded as statements of an algebraic character about the category Top, and this is often convenient, particularly when it brings out analogies between constructions for topological spaces and constructions for other mathematical objects. (b) The category PX of paths on X is regarded as an algebraic object in its own right, as much worthy of study as an example of a category as are examples of groups, rings or fields.
4. The category of all groups and all morphisms (i.e., homomorphisms) of groups is written Grp.
5. On the other hand, if G is group, then G is also a category with one object—namely, the identity 1 of G—with morphisms $1 \to 1$ the elements of G, and with composition the multiplication of G. Actually, for this construction one needs only that G is a *monoid*, that is, the multiplication is associative and has a two-sided identity.

Let $\mathcal{C}, \mathcal{D}$ be categories. We say $\mathcal{D}$ is a *subcategory* of $\mathcal{C}$ if
(a) each object of $\mathcal{D}$ is an object of $\mathcal{C}$, i.e., $\mathrm{Ob}(\mathcal{D}) \subseteq \mathrm{Ob}(\mathcal{C})$.
(b) for each x, y in $\mathrm{Ob}(\mathcal{D})$, we have $\mathcal{D}(x, y) \subseteq \mathcal{C}(x, y)$.
(c) composition of morphisms in $\mathcal{D}$ is the same as that for $\mathcal{C}$, and
(d) for each x in $\mathrm{Ob}(\mathcal{D})$ the identity in $\mathcal{D}(x, x)$ is the identity in $\mathcal{C}(x, x)$.

The subcategory $\mathcal{D}$ of $\mathcal{C}$ is called *full* if

$$\mathcal{D}(x, y) = \mathcal{C}(x, y)$$

for all objects x, y of $\mathcal{D}$; and $\mathcal{D}$ is a *wide* subcategory if $\mathrm{Ob}(\mathcal{D}) = \mathrm{Ob}(\mathcal{C})$. For example, we can obtain full subcategories of any category $\mathcal{C}$ by taking $\mathrm{Ob}(\mathcal{D})$ to be any class of objects of $\mathcal{C}$, and then defining $\mathcal{D}(x, y) = \mathcal{C}(x, y)$ for all x, y in $\mathrm{Ob}(\mathcal{D})$. In this way, we obtain the full subcategories of Top whose objects are all Hausdorff spaces, all metrisable spaces, or all compact spaces. On the other hand, we obtain a wide subcategory of Top by suitably restricting the maps, for example to be open, or closed, or identification maps. In fact, any property $\mathcal{P}$ of continuous functions defines a wide subcategory of Top (in which the morphisms $X \to Y$ are the continuous functions $X \to Y$ which have property $\mathcal{P}$) provided only that any identity map has property $\mathcal{P}$, and that the composite of two maps with property $\mathcal{P}$ also has property $\mathcal{P}$.

Let $\mathcal{C}$ be a category, and suppose f, g are morphisms in $\mathcal{C}$ such that $gf = 1$, an identity morphism. Then we call g a *left-inverse* of f and f a *right-inverse* of g; we also say that g is a *retraction*, and f a *co-retraction*.

6.1.1 *Let $f : x \to y$, $g_1, g_2 : y \to x$ be morphisms in $\mathcal{C}$ such that*

$$g_1 f = 1_x, \quad f g_2 = 1_y.$$

Then $g_1 = g_2$. If, further, $gf = 1_x$, then $g = g_1$.

Proof $g_1 = g_1 1_y = g_1(fg_2) = (g_1 f)g_2 = 1_x g_2 = g_2$. Similarly, $g = g_2$ and so $g = g_1$. □

This result can be stated: if f has a left and a right inverse, then f has an unique two-sided inverse. Such a morphism f is called *invertible*, or an *isomorphism*, and the unique inverse of f is written f^{-1} or, when using additive notation, $-f$. If there is an isomorphism $x \to y$, then we say x and y are *isomorphic*. It is easy to prove that any identity is an isomorphism; the inverse of an isomorphism is an isomorphism; and the composite gf of two isomorphisms is an isomorphism. Note that in the last case $(gf)^{-1} = f^{-1}g^{-1}$. It follows from these remarks that the relation 'x is isomorphic to y' is an equivalence relation on the objects of $\mathcal{C}$.

A category whose objects form a set and in which every morphism is an isomorphism is called a *groupoid*. For example, a group, regarded as a category with one object, is also a groupoid.

The category PX of paths on X is not a groupoid since if a is a path in X of positive length then there is no path b such that $b + a$ is a zero path. This is an awkward feature of PX. Another awkward feature is that even for simple spaces (e.g., $X = \mathbb{I}$) $PX(x, y)$ can be uncountable. In the next section we shall show how to avoid both of these difficulties by constructing from PX the *fundamental groupoid* πX of X.

Exercises

1. Prove that (i) the composite of retractions is a retraction, (ii) the composite of co-retractions is a co-retraction, (iii) the composite of isomorphisms is an isomorphism.

2. Let a, b, c be morphisms such that ba, cb are defined and are isomorphisms. Prove a, b, c are isomorphisms.
3. A *graph* Γ is a set $\mathrm{Ob}(\Gamma)$ and for each x, y in $\mathrm{Ob}(\Gamma)$ a set $\Gamma(x, y)$ called the set of *edges* from x to y—the sets $\Gamma(x, y)$ are supposed disjoint. A *path* in Γ from x to y consists of either the empty sequence $\varnothing \to \Gamma(x, x)$ if $x = y$ or a sequence $(a_n, \dots, a_1)$, such that (i) $a_i \in \Gamma(x_i, x_{i+1})$, (ii) $x_1 = x$, $x_{n+1} = y$; the set of paths from x to y is written $P\Gamma(x, y)$. These paths are multiplied by the rule that if

$$a = (a_n, \dots, a_1), \quad b = (b_m, \dots, b_1) \quad (a \in P\Gamma(x, y),\ b \in P\Gamma(y, z))$$

then $ba = (b_m, \dots, b_1, a_n, \dots, a_1)$; also the empty sequence in $P\Gamma(x, x)$ is to act as identity. Prove that $P\Gamma$ is a category. (This is the category *freely generated* by Γ.)
4. Let $\mathcal{C}$ be a category. Prove that a category $\mathcal{C}^{op}$, the opposite or *dual* of $\mathcal{C}$, is defined as follows. (i) $\mathrm{Ob}(\mathcal{C}^{op}) = \mathrm{Ob}(\mathcal{C})$, (ii) if $x, y \in \mathrm{Ob}(\mathcal{C}^{op})$ then $\mathcal{C}^{op}(x, y) = \mathcal{C}(y, x)$ (however, the elements of $\mathcal{C}^{op}(x, y)$ are written f^* for each f in $\mathcal{C}(y, x)$), (iii) the composition in $\mathcal{C}^{op}$ is defined by $g^* f^* = (fg)^*$.
5. Let $\mathcal{C}$ be a category. A morphism $f : C \to D$ in $\mathcal{C}$ is called *monic* (and a *mono*) if for all A in $\mathrm{Ob}(\mathcal{C})$ and $g, h : A \to C$ in $\mathcal{C}$, the relation $fg = fh$ implies $g = h$; f is called *epic* (and an *epi*) if for all B in $\mathrm{Ob}(\mathcal{C})$ and all $g, h : D \to B$, the relation $gf = hf$ implies $g = h$ [cf. Exercises 2, 3 of Section A.1]. Prove that (i) a coretraction is monic and a retraction is epic, (ii) an isomorphism is both epic and monic, (iii) the composition of monos is monic, the composition of epics is epic. Give an example of a category in which some morphism is epic and monic but not an isomorphism.
6. Prove that f in $\mathcal{C}$ is monic $\Leftrightarrow$ f^* in $\mathcal{C}^{op}$ is epic.
7. An object P of a category $\mathcal{C}$ is called *initial* in $\mathcal{C}$ if $\mathcal{C}(P, X)$ has exactly one element for all objects X of $\mathcal{C}$; and P is *final* if $\mathcal{C}(X, P)$ has exactly one element for all objects X of $\mathcal{C}$. Prove that all initial objects in $\mathcal{C}$ are isomorphic, as are all final objects. If P is both initial and final, then P is called a *zero object*. Prove that (i) the categories Set and Top have initial and final objects, but no zero, (ii) the category of groups, and the category of vector spaces over a given field, both have a zero object.
8. Prove that in the category Grp of groups, a morphism $f : G \to H$ is monic if and only if it is injective; less trivially, f is epic if and only if f is surjective. [Suppose f is not surjective and let $K = \mathrm{Im}\, f$. If the set of cosets H/K has two elements, then K is normal in H and it is easy to prove f is not epic. Otherwise there is a permutation γ of H/K whose only fixed point is K. Let $\pi : H \to H/K$ be the projection and choose a function $\theta : H/K \to H$ such that $\pi\theta = 1$. Let $\tau : H \to K$ be such that $x = (\tau x)(\theta\pi x)$ for all x in H and define $\lambda : H \to H$ by $x \mapsto (\tau x)(\theta\gamma\pi x)$. The morphisms α, β of H into the group P of all permutations of H, defined by $\alpha(h)(x) = hx$, $\beta(h) = \lambda^{-1}\alpha(h)\lambda$ satisfy $\alpha h = \beta h$ if and only if $h \in K$. Hence $\alpha f = \beta f$].
9. Prove that in the category of Hausdorff spaces and continuous functions, a map $f : X \to Y$ is epic if and only if $\mathrm{Im}\, f$ is a dense subset of Y.
10. For sets X, Y define a *relation from* X *to* Y to be a triple (X, Y, R) where R is a subset of $X \times Y$. If R is a subset of $X \times Y$, S is a subset of $Y \times Z$, let SR be the subset of $X \times Z$ of pairs (x, z) such that for some y in Y, $(x, y) \in R$ and $(y, z) \in S$. Using this product, define the composite of a relation from X to Y and a relation from Y

to Z, and prove that sets and relations between sets form a category containing Set as a wide subcategory. [An obvious question seems to be: what do the conditions epic, monic, iso imply about a relation in this category?]
11. Let $\mathcal{C}$ be a category and let $f : C \to D$ be a morphism in $\mathcal{C}$. Prove that f induces functions for each X in $\mathrm{Ob}(\mathcal{C})$

$$f_X : \mathcal{C}(X, C) \to \mathcal{C}(X, D) \qquad f^X : \mathcal{C}(D, X) \to \mathcal{C}(C, X)$$
$$g \mapsto fg \qquad\qquad g \mapsto gf.$$

Prove that the following conditions are equivalent: (i) f is an isomorphism, (ii) f_X is a bijection for each X in $\mathrm{Ob}(\mathcal{C})$, (iii) f^X is a bijection for each X in $\mathrm{Ob}(\mathcal{C})$. Prove also that f is monic if f_X is injective for all X, and f is epic if and only if f^X is injective for all X. Under what conditions is f^X surjective for all X, f_X surjective for all X?

6.2 Construction of the fundamental groupoid

The fundamental groupoid πX will be a groupoid such that the set $\pi X(x, y)$ is a set of equivalence classes of $PX(x, y)$. In order to define the equivalence relation, we consider first two paths a, b in $PX(x, y)$ of the same length r. A *homotopy rel end points of length* q *from* a *to* b is defined to be a map

$$F : [0, r] \times [0, q] \to X$$

such that

$$\begin{array}{lll} F(s, 0) = a(s), & F(s, q) = b(s), & s \in [0, r] \\ F(0, t) = x, & F(r, t) = y, & t \in [0, q]. \end{array} \tag{6.2.1}$$

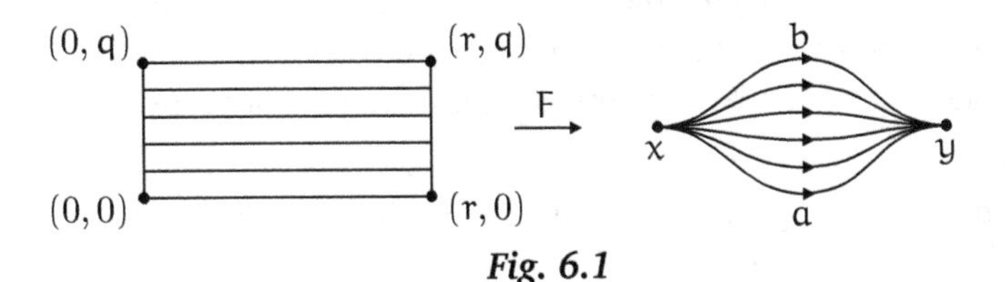

Fig. 6.1

Notice that for each t in $[0, q]$ the path $F_t : s \mapsto F(s, t)$ is a path in $PX(x, y)$; the family (F_t) can be thought of as a 'continuous family of paths' between $F_0 = a$ and $F_1 = b$. Alternatively, we can think of F as a 'deformation' of a into b.

We use the notation $F : a \sim b$ to mean that F is a homotopy rel end points from a to b (of some length). There is a unique homotopy of length 0 from a to a. If $F : a \sim b$ is a homotopy of length q, then $-F$, defined by

$(s,t) \mapsto F(s, q-t)$, is a homotopy $b \sim a$. If $F : a \sim b$, $G : b \sim c$ are of length q, q' respectively where a, b, c are of length r, then the *sum* of F and G

$$G + F : [0, r] \times [0, q + q'] \to X$$
$$(s,t) \mapsto \begin{cases} F(s,t), & \text{if } 0 \leqslant t \leqslant q \\ G(s, t-q), & \text{if } q \leqslant t \leqslant q + q' \end{cases}$$

is continuous by the gluing rule [2.5.12] and is a homotopy $a \sim c$.

Two paths a, b of the same length are called *homotopic* rel end points, written $a \sim b$, if there is a homotopy $F : a \sim b$. We abbreviate homotopic rel end points to homotopic since in the case of paths we have no need of other homotopies. Then it is clear from the previous paragraph that the relation $a \sim b$ is an equivalence relation.

Let $F : [0, r] \times [0, q] \to X$ be a homotopy $a \sim b$. Then there is a homotopy $F' : a \sim b$ of length 1, namely

$$F' : [0, r] \times \mathbb{I} \to X$$
$$(s,t) \mapsto F(s, qt).$$

So for the rest of this chapter we restrict our attention to homotopies of length 1.

It is convenient to have a flexible notation for homotopies. We think of a homotopy F (of length 1) as a function $t \mapsto F_t$ where F_t is a path, and then abbreviate $t \mapsto F_t$ to F_t. This enables us to say, for example, that if F_t is a homotopy $a \sim b$, then F_{1-t} is a homotopy $b \sim a$.

For any real number $r \geqslant 0$ and x in X, let r_x denote the constant path at x of length r. When no confusion can be caused, we abbreviate r_x to r. In particular, for any path a and $r \geqslant 0$, the paths $a + r$, $r + a$ are well defined.

We can now state the basic lemmas on homotopies of paths.

6.2.2 *Let* $a, b \in PX(x, y)$, $c, d \in PX(y.z)$ *where* $|a| = |b|$, $|c| = |d|$.
(a) *If* $a \sim b$, *then* $-a \sim -b$.
(b) *If* $a \sim b$, *and* $c \sim d$, *then* $c + a \sim d + b$.
(c) *For any* $r \geqslant 0$, $a + r \sim r + a$.

Proof (a) Let F be a homotopy $a \sim b$. Then

$$(s,t) \mapsto F(|a| - s, t)$$

is a homotopy $-a \sim -b$.

(b) Let $F : a \sim b$, $G : c \sim d$. Then

$$H : [0, |c| + |a|] \times \mathbb{I} \to X$$
$$(s,t) \mapsto \begin{cases} F(s,t), & \text{if } s \leqslant |a| \\ G(s-|a|,t), & \text{if } |a| \leqslant s \end{cases}$$

is a homotopy $c + a \sim d + b$ [cf. Fig. 6.2].

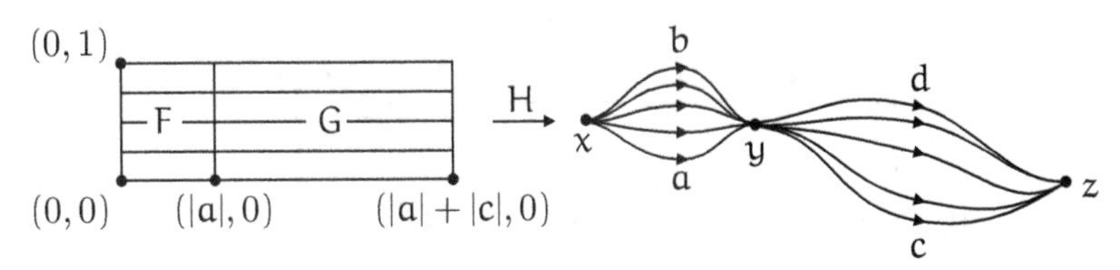

Fig. 6.2

(c) Let $|a| = r'$. We define [cf. Fig. 6.3]

$$F : [0, r + r'] \times \mathbb{I} \to X$$
$$(s,t) \mapsto \begin{cases} x, & \text{if } 0 \leqslant s \leqslant tr \\ a(s-tr), & \text{if } tr \leqslant s \leqslant tr + r' \\ y, & \text{if } tr + r' \leqslant s \leqslant r + r'. \end{cases}$$ □

It should be noticed that in the homotopy H of 6.2.2(b) the point y is fixed (that is, $H(|a|, t) = y$ for all t in $\mathbb{I}$). However, there are homotopies $c + a \sim d + b$ which do not have this property. This fact is exploited in the next result.

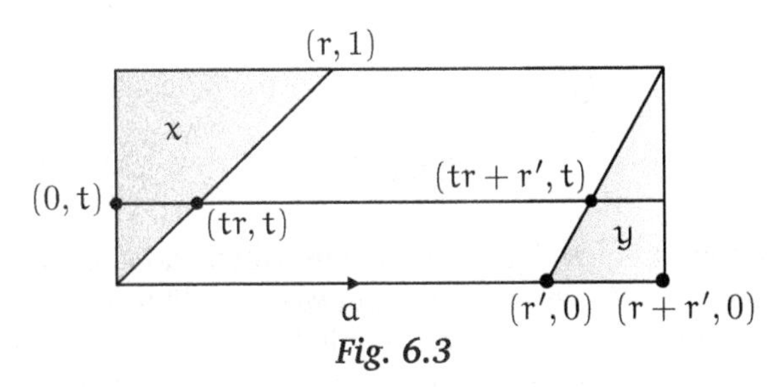

Fig. 6.3

6.2.3 *If* $a \in PX(x, y)$ *and* $|a| = r$, *then*

$$-a + a \sim 2r_x, \qquad a - a \sim 2r_y.$$

Proof It is a little simpler to define a homotopy $2r_x \sim -a + a$. We define [cf. Fig. 6.4]

$$F : [0, 2r] \times \mathbb{I} \to X$$
$$(s, t) \mapsto \begin{cases} a(s), & \text{if } 0 \leqslant s \leqslant rt \\ a(2rt - s), & \text{if } rt \leqslant s \leqslant 2rt \\ x, & \text{if } 2rt \leqslant s \leqslant 2r. \end{cases}$$

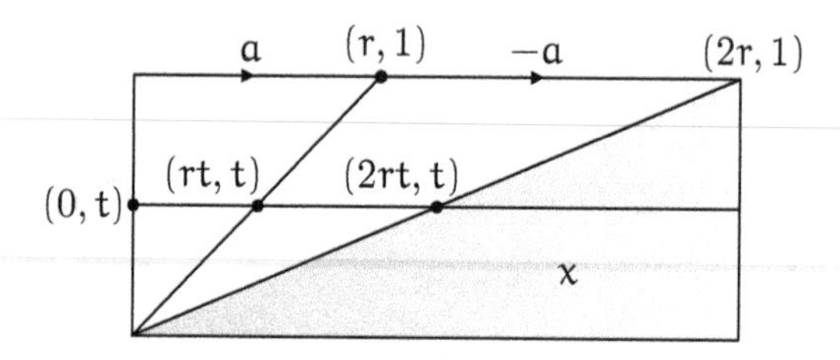

Fig. 6.4

Clearly F is well-defined, continuous, and a homotopy $2r_x \sim -a + a$. It follows that $-a + a \sim 2r_x$. On replacing a by $-a$ we find $a - a \sim 2r_y$. □

The path F_t of the proof of 6.2.3 is depicted for various t in Fig. 6.5. Of course, a, and $-a$ should really be superimposed.

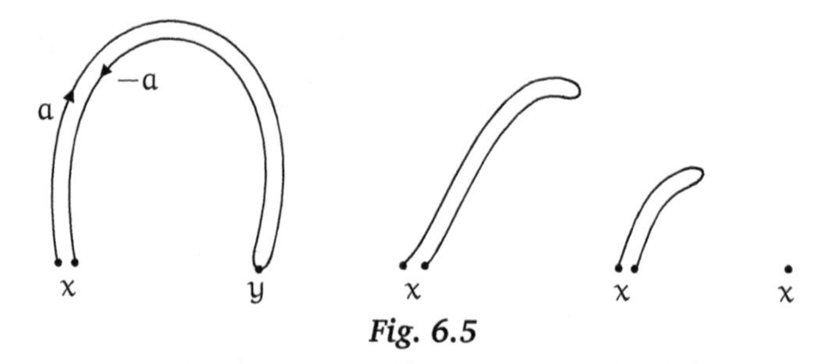

Fig. 6.5

We now define an equivalence relation between paths of various lengths. Let $a, b \in PX(x, y)$. We say a, b are *equivalent* if there are real numbers, $r, s \geqslant 0$, such that $r + a$, $s + b$ are homotopic (in which case, r, s must satisfy $|a| + r = |b| + s$). This relation is obviously reflexive and symmetric (since homotopy is reflexive and symmetric). It is also transitive; for given homotopies $r + a \sim s + b$, $s' + b \sim t + c$ (where a, b, c are paths and $r, s, s', t \geqslant 0$) then there are homotopies

$$s' + r + a \sim s' + s + b = s + s' + b \sim s + t + c.$$

The definition of equivalence of paths is non-canonical (that is, it involves choices) and this is perhaps unaesthetic. An alternative definition yielding the same equivalence classes is suggested in Exercise 5.

EXAMPLES

1. Let a in $P(x, y)$ be of length $r \geqslant 0$. Let $r' > 0$. Then a is equivalent to a path of length r', namely, the path $b : s \mapsto a(sr/r')$. A specific homotopy $r' + a \sim r + b$ is given by

$$F : (s,t) \mapsto \begin{cases} a(rs/\lambda_t), & \text{if } 0 \leqslant s \leqslant \lambda_t \\ y, & \text{if } \lambda_t \leqslant s \leqslant r + r' \end{cases}$$

where $\lambda_t = r(1-t) + r't$ [cf. Fig. 6.6].

This argument does not show that any path a in $PX(x, x)$ is equivalent to a path of length 0 since, if we take $r' = 0$ in the formula for $F(s, t)$, then F is continuous if and only if a is constant. However, any constant path r is equivalent to a zero path since $r = r + 0$.

2. Equivalent paths of the same length are in fact homotopic. This can be proved by constructing for any $r, s \geqslant 0$ a homeomorphism $G : [0, r] \times \mathbb{I} \to [0, r+s] \times \mathbb{I}$ which is the identity on $\{0\} \times \mathbb{I} \cup [0, r] \times \mathbb{I}$ and which maps $\{r\} \times \mathbb{I}$ homeomorphically onto $[r, r+s] \times \dot{\mathbb{I}} \cup \{r+s\} \times \mathbb{I}$. So if $F : [0, r+s] \times \mathbb{I} \to X$ is a homotopy $s + a \sim s + b$ where a, b have length r, then the composite FG is a homotopy $a \sim b$.

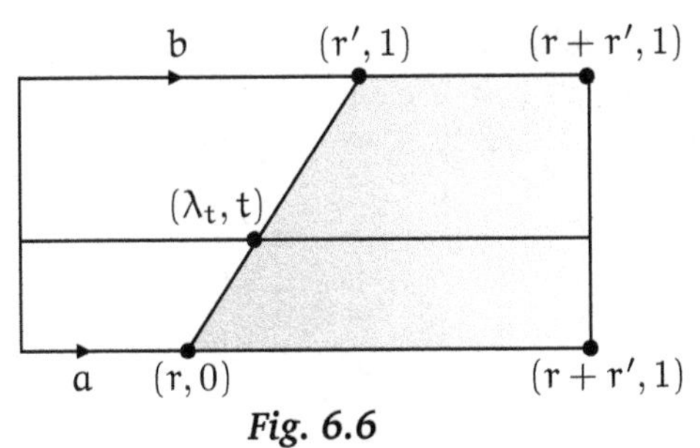

Fig. 6.6

However this result, although it is interesting and will be used in chapter 7, is not essential to this chapter. So we only state that in Fig. 6.7 G will be the identity on the shaded areas, and on the remaining part will map each line segment $[z, w]$ linearly onto $[z, w']$. Further details are left to the reader.

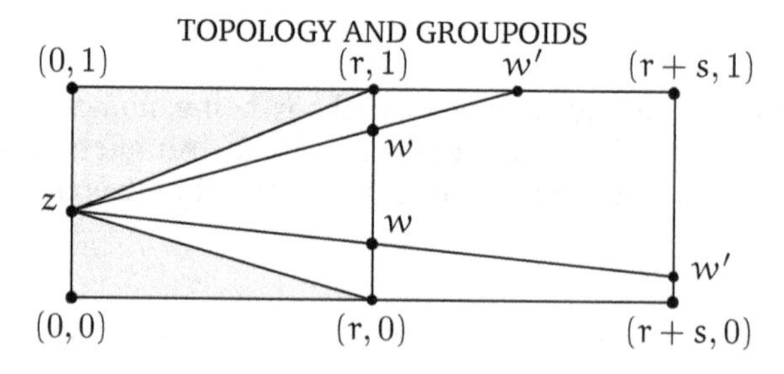

Fig. 6.7

If two paths a, b are equivalent we write $a \sim b$—the last two examples show that this will cause no confusion. The equivalence classes of paths from x to y are called *path classes* from x to y, and the set of these path classes is written $\pi X(x, y)$. The class of the zero path at x is written 0, or to avoid ambiguity, 0_x. By Example 1, the zero path class 0_x includes all constant paths at x.

6.2.4 *A negative and sum of path classes is defined by*

$$\begin{aligned} -\operatorname{cls} a &= \operatorname{cls}(-a), \qquad a \in PX(x, y), \quad b \in PX(y.z) \\ \operatorname{cls} b + \operatorname{cls} a &= \operatorname{cls}(b + a). \end{aligned}$$

Proof Suppose a, a' are equivalent paths in $PX(x, y)$; then there are constant paths r, r' such that $r + a$, $r' + a'$ are homotopic. Hence

$$r - a \sim -a + r = -(r + a) \sim -(r' + a') \sim r' - a'.$$

Therefore $-\operatorname{cls} a$ is well-defined.

Suppose further that b, b' are equivalent paths in $PX(y.z)$ and that $s + b$, $s' + b'$ are homotopic, where $s, s' \geqslant 0$. Then

$$\begin{aligned} r + s + b + a &\sim r + s' + b' + a \\ &\sim s' + b' + r + a \\ &\sim s' + b' + r' + a' \\ &\sim s' + r' + b' + a'. \end{aligned}$$

□

6.2.5 *Addition of path classes is associative. Further if $\alpha \in \pi X(x, y)$ then*

$$\begin{gathered} \alpha + 0_x = 0_y + \alpha = \alpha, \\ -\alpha + \alpha = 0_x, \quad \alpha - \alpha = 0_y. \end{gathered}$$

Proof The first statement is obvious, since addition of paths is associative. The equations $\alpha + 0_x = 0_y + \alpha = \alpha$ are immediate from the relations $a + 0_x = 0_y + a = a$ for paths a in $P(x, y)$.

The last equations are immediate from 6.2.3 and the fact that the zero path class at x contains all constant paths at x. □

We have now shown that πX is a groupoid whose objects are the points of X and whose morphisms $x \to y$ are the path classes from x to y—this groupoid is called the *fundamental groupoid* of X.

EXAMPLES

3. If X consists of a single point, then πX has one object x say and $\pi X(x, x)$ consists only of the zero path class. More generally, if the path-components of X consist of single points, then

$$\pi X(x, y) = \begin{cases} \varnothing, & \text{if } x \neq y \\ \{0_x\}, & \text{if } x = y. \end{cases}$$

A groupoid with this property is called *discrete*.

4. Let X be a convex subset of a normed vector space, and let a, b be two paths in X from x to y of the same length r. Then a and b are homotopic, since

$$\begin{aligned} F : [0, r] \times \mathbb{I} &\to X \\ (s, t) &\mapsto (1 - t)a(s) + b(s) \end{aligned}$$

is a homotopy $a \sim b$. It follows easily that any two paths from x to y are equivalent; so $\pi X(x, y)$ has exactly one element for all x, y in X. A groupoid with this property is called 1-*connected*, and also a *tree* groupoid; and if πX is a tree groupoid we say X is 1-*connected* (this, of course, implies X is path connected). For example, the unit interval $\mathbb{I}$ is 1-connected, since it is a convex subset of $\mathbb{R}$.

5. If each path-component of X is 1-connected then, for each x, y in X, $\pi X(x, y)$ contains not more than one element; we then say X, and also πX, is *simply-connected*. Thus X is simply-connected means that any two paths in X with the same end points are equivalent. A groupoid G is called *simply-connected* if $G(x, y)$ has not more than one element for all objects x, y of G.

6. We have at this stage no techniques for showing that $\pi X(x, y)$ can ever contain more than one element. That is, we cannot yet show that $PX(x, y)$ can contain non-equivalent paths. However, anyone who has tied elastic round sticks will find it reasonable to suppose that the two paths in $\mathbb{R}^2 \setminus \mathbb{E}^2$ shown in Fig. 6.8 are not equivalent. A proof of this fact will appear later when we have techniques for computing the fundamental groupoid.

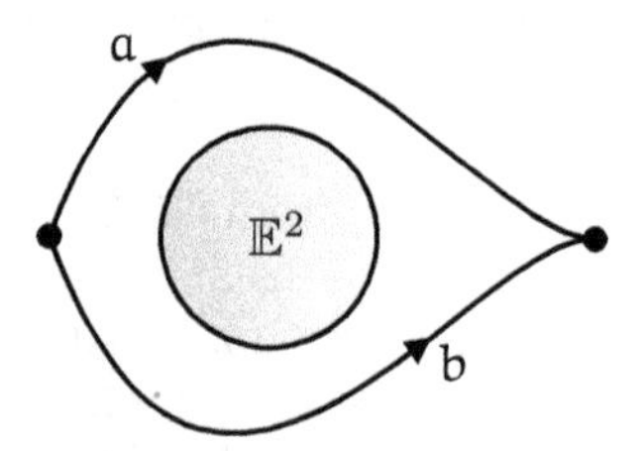

Fig. 6.8

EXERCISES

In these exercises, X is assumed to be a topological space.

1. Let $x, y, z \in X$, and let a in $PX(x, y)$, b in $PX(y.z)$ be paths of length 1. Define the path $b.a$ of length 1 in $PX(x, z)$ by

$$(b.a)(t) = \begin{cases} a(2t), & \text{if } 0 \leqslant t \leqslant \frac{1}{2} \\ b(2t-1), & \text{if } \frac{1}{2} \leqslant t \leqslant 1. \end{cases}$$

Prove that $b.a$ is well-defined, and that $b.a \sim b+a$. Let 1 denote as usual a constant path of length 1. Prove that $a.1 \sim a \sim 1.a$, that $a.(-a) \sim 1$, $(-a).a \sim 1$, and that if also $c \in P(z, w)$ then

$$c.(b.a) \sim (c.b).a.$$

2. If $\alpha : \mathbb{I} \to \mathbb{R}^{\geqslant 0}$ is continuous, let

$$J(\alpha) = \{(s, t) \in \mathbb{R}^2 : 0 \leqslant t \leqslant 1,\ 0 \leqslant s \leqslant \alpha(t)\}.$$

Prove that $J(\alpha)$ is compact.

Let a, b be paths from x to y in X. A *wavy homotopy* $F : a \rightsquigarrow b$ is a map $F : J(\alpha) \to X$ for some map $\alpha : \mathbb{I} \to \mathbb{R}^{\geqslant 0}$ such that $\alpha(0) = |a|, \alpha(1) = |b|$ and for $0 \leqslant t \leqslant 1$

$$\begin{aligned} F(0,t) &= x, & F(s,0) &= a(s), & 0 &\leqslant s \leqslant |a| \\ F(\alpha(t),t) &= y, & F(s,1) &= b(s), & 0 &\leqslant s \leqslant |b|. \end{aligned}$$

Prove that $\rightsquigarrow$ is an equivalence relation, and that $a \rightsquigarrow b$ if and only if a and b are equivalent.

3. Let a, b be paths from x to y of the same length r. A *weak homotopy* $F : a \sim_W b$ is a function $F : [0, r] \times \mathbb{I} \to X$ satisfying the usual conditions for a homotopy $a \sim b$ except that F is only *separately* continuous, that is, for each s the function $t \mapsto F(s, t)$ is continuous, and for each t the function $s \mapsto F(s, t)$ is continuous.

Let $a, b : \mathbb{I} \to \mathbb{S}^1$ be paths $s \mapsto e^{2\pi i s}$, $s \mapsto 1$, respectively. Prove that the map $(s, t) \mapsto \exp(2\pi i s^t)$, $(s \neq 0)$, $(0, t) \mapsto 1$ is a weak homotopy $b \sim_W a$. [This example shows that weak homotopy gives an uninteresting theory.]

4. Let a, b be paths of length r in X, Y respectively and let c be the path $X \times Y$, $s \mapsto (a(s), b(s))$. Prove that c is equivalent to $b' + a'$ where a' is the path $s \mapsto (a(s), b(0))$ and b' is the path $s \mapsto (a(r), b(s))$.

5. If $a : [0, r] \to X$ is a path in X, let $\bar{a}$ be the map

$$\mathbb{R}^{\geqslant 0} \to X$$
$$s \mapsto \begin{cases} as, & \text{if } 0 \leqslant s \leqslant r \\ ar, & \text{if } r \leqslant s. \end{cases}$$

Prove that the two paths a, b in $PX(x, y)$ are equivalent if and only if there is a map $F : \mathbb{R}^{\geqslant 0} \times \mathbb{I} \to X$ such that there is an $r_0 \geqslant 0$ such that for all $(s, t) \in \mathbb{R}^{\geqslant 0} \times \mathbb{I}$

$$F(s, 0) = \bar{a}s; \qquad F(s, 1) = \bar{b}s;$$
$$F(0, t) = x;$$
$$F(s, t) = y \text{ for all } s \geqslant r_0.$$

6. Let Y be a subspace of X and $i : Y \to X$ the inclusion. Assume that if a, b are paths in Y with the same end points, then the paths ia, ib are equivalent in X. Let $P(X, Y)$ be the set of paths in X whose end points lie in Y. For $a, b \in P(X, Y)$ write $a \approx b$ if there are real numbers $r, s \geqslant 0$ and a map $F : [0, u] \times \mathbb{I} \to X$ such that

$$F_0 = r + a, \qquad F_1 = s + b,$$

$$F_t(0),\ F_t(u) \in Y.$$

Prove that $\approx$ is an equivalence relation on $P(X, Y)$ and that the set of equivalence classes is a groupoid.

7. Define an equivalence relation $\equiv$ among paths in PX as follows. (i) If a is a path and $r \geqslant 0$, then $a + r \equiv r + a \equiv a$. (ii) If $a \in PX(x, y)$ then $a - a \equiv 0_y$, $-a + a \equiv 0_x$. (iii) If $a, b \in PX(x, y)$, then $a \equiv b$ if we can write $a = a_n + \cdots + a_1$, $b = b_n + \cdots + b_1$ and $a_i = b_i$, $i = 1, \ldots n$ by rules (i) and (ii). Prove that addition of paths induces an addition of equivalence classes by which these equivalence classes form a groupoid. [This should perhaps be called the *semi-fundamental groupoid of* X.]

6.3 Properties of groupoids

In this section, we prove some basic properties of groupoids and apply them to the fundamental groupoid of a space. We shall use additive notation and, in particular, $-a$ will denote the inverse (or negative) of an element a of a groupoid G. This enables us to write $a - b$ for $a + (-b)$ (when defined); and to write $a - b - c$ for both $(a - b) - c$ and $a - (c + b) = a + (-b - c)$. Of course we shall also in this chapter write the operation in a group as addition, but this does *not* mean that the groups which arise will all be commutative—we do not suppose $a + b = b + a$.

Let G be a groupoid. A *subgroupoid* of G is a subcategory H of G such that $a \in H \Rightarrow -a \in H$; that is, H is a subcategory which is also a groupoid. We say H is *full* (*wide*) if H is a full (wide) subcategory. We can construct full subgroupoids H of G by taking H to be the full subcategory of G on any subset of $\mathrm{Ob}(G)$. Thus we allow the empty subgroupoid of G. Further, the full subgroupoid of G on one object x of G is written $G(x)$—this groupoid has one zero and $a + b$ is defined for all a, b in $G(x)$. A groupoid with only one object is called a *group* [cf. Example 5 of Section 6.1] and, in particular, $G(x)$ is called the *object group* (or *vertex group*) at x.

If X is a topological space and $x \in X$, then $\pi X(x)$ is a group called the *fundamental group of* X *at* x (this group will often be written $\pi(X, x)$). More generally, if A is any set then the full subgroupoid of πX on the set $A \cap X$ is written πXA. The elements of πXA are all path classes, that is all equivalence classes of paths in X, joining points of $A \cap X$. The use of πXA for sets A not contained in X will prove convenient later.

A groupoid G is called *connected* if $G(x, y)$ is non-empty for all objects x, y of G. In particular, πX is connected if and only if X is path-connected.

Let x_0 be an object of G, and let Cx_0 be the full subgroupoid of G on all objects y of G such that $G(x, y)$ is not empty. If x, y are objects of Cx_0, then $G(x, y)$ is non-empty since it contains elements $b + a$ for b in $G(x_0, y)$, a in $G(x, x_0)$. It follows that Cx_0 is connected. Clearly, Cx_0 is the maximal, connected subgroupoid of G with x_0 as an object, and so we call Cx_0 the *component* of G containing x_0.

6.3.1 *Let* x, y, x', y' *be objects of the connected groupoid* G. *There is a bijection*

$$\varphi : G(x, y) \to G(x', y')$$

which if $x = y$, $x' = y'$ *can be chosen to be an isomorphism of groups.*

Proof Since G is connected we can choose $a : x \to x'$, $b : y \to y'$ in G. We define

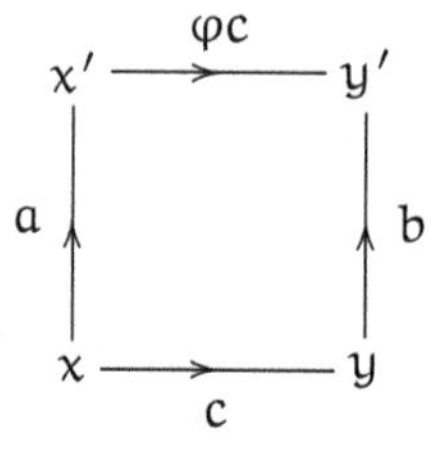

$$\begin{array}{ccc} x' & \xrightarrow{\varphi c} & y' \\ {\scriptstyle a}\uparrow & & \uparrow{\scriptstyle b} \\ x & \xrightarrow[c]{} & y \end{array} \qquad \begin{aligned} \varphi : G(x, y) &\to G(x', y') \\ c &\mapsto b + c - a \\ \psi : G(x', y') &\to G(x, y) \\ d &\mapsto -b + d + a. \end{aligned}$$

Clearly $\varphi\psi = 1$, $\psi\varphi = 1$ and so φ is a bijection.

If $x = y$, $x' = y'$ then let $a = b$ so that φ sends $c \mapsto a + c - a$. If $c, c' \in G(x, x)$ (which is $G(x)$) then

$$\begin{aligned} \varphi c + \varphi c' &= a + c - a + a + c' - a \\ &= a + c + c' - a \\ &= \varphi(c + c'). \end{aligned}$$

Therefore φ is an isomorphism. □

Thus the object groups of a connected groupoid are all isomorphic. For this reason we shall sometimes speak loosely of *the* object group of a connected groupoid. The isomorphism $G(x) \to G(x')$ which sends $c \mapsto a+c-a$ is written $a_{\mathbf{x}}$. If $x = x'$, then $a_{\mathbf{x}}$ is simply an inner automorphism of $G(x)$.

6.3.2 *Let x, x' belong to the same component of G. Then $a_{\mathbf{x}} = b_{\mathbf{x}} : G(x) \to G(x')$ for all $a, b : x \to x'$ if and only if $G(x)$ is abelian.*

Proof We first note that

$$(-b + a)_{\mathbf{x}} = (-b)_{\mathbf{x}} a_{\mathbf{x}} = (b_{\mathbf{x}})^{-1} a_{\mathbf{x}}.$$

Thus $b_{\mathbf{x}}^{-1} a_{\mathbf{x}}$ is an inner automorphism of $G(x)$. Also if $c : x \to x$, then

$$(b_{\mathbf{x}})^{-1}(b + c)_{\mathbf{x}} = c_{\mathbf{x}}$$

and every inner automorphism of $G(x)$ is of the form $b_{\mathbf{x}}^{-1} a_{\mathbf{x}}$. But the inner automorphisms of a group are trivial if and only if the group is abelian. □

These definitions and results apply immediately to the fundamental groupoid πX. The components of πX are the groupoids πX_0 for X_0 a path-component of X. If α is a path class in πX from x to x' then α determines an isomorphism $\alpha_{\mathbf{x}} : \pi(X, x) \to \pi(X, x')$ of fundamental groups, and $\alpha_{\mathbf{x}}$ is independent of the choice of α if and only if $\pi(X, x)$ is abelian.

We now give some examples of groupoids.

EXAMPLES

1. Let X be any set. The *discrete groupoid on* X is also written X; it has X as its set of objects, one zero for each element of X, and no other elements. Notice that if X is given the discrete topology, then this groupoid is πX.
2. Let G be a tree groupoid, so that $G(x, y)$ has exactly one element say a_{yx} for all objects x, y of G. Then $a_{zy} + a_{yx}$ is the unique element of $G(x, z)$ and so we have the addition rule

$$a_{zy} + a_{yx} = a_{zx}. \tag{*}$$

Conversely, given any set X we can form an essentially unique tree groupoid G such that $\mathrm{Ob}(G) = X$ by choosing distinct elements a_{yx} for each x, y in X and then taking $G(x, y)$ to consist solely of a_{yx}, with the addition rule (*). Notice that a tree groupoid with n objects as n^2 elements of which n are zeros. Tree groupoids with two objects $0, 1$ will be important, and will be denote ambiguously by $\mathbf{I}$, the unique element of $\mathbf{I}(0, 1)$ being written ι.

If X is 1-connected and $A \subseteq X$, then πXA is a tree groupoid. In particular, the fundamental groupoid of $\mathbb{I}$ on the set $\{0, 1\}$ is exactly $\mathbf{I}$—in symbols

$$\pi\mathbb{I}\{0, 1\} = \mathbf{I}.$$

3. Let G be a connected groupoid and T a tree groupoid which is a wide subgroupoid of G (we recall that wide means that T, G have the same objects). Let x_0 be an object of G and for each object x of G let τ_x be the unique element of $T(x_0, x)$. If $a \in G(x, y)$, then there is a unique element a' of $G(x)$ such that

$$a = \tau_y + a' - \tau_x.$$

If, further, $b \in G(y, z)$ then

$$\begin{aligned} b + a &= \tau_z + b' - \tau_y + \tau_y + a' - \tau_x \\ &= \tau_z + b' + a' - \tau_x. \end{aligned}$$

Therefore

$$(b + a)' = b' + a'.$$

This shows that G can be recovered from T and the group $G(x_0)$.

4. A groupoid G is *totally disconnected* if

$$G(x, y) = \varnothing \quad \text{for } x \neq y.$$

Such a groupoid is determined entirely by the family $(G(x))$, $x \in \mathrm{Ob}(G)$, of groups. If X is a space, and A consists of exactly one point in each path-component of X, then πXA is totally disconnected. A totally disconnected groupoid is sometimes called a *bundle of groups*.

EXERCISES

1. Let E be a subset of $X \times X$. Let $\mathcal{C}$ be defined by $\mathrm{Ob}(\mathcal{C}) = X$ and for each x, y in X, let

$$\mathcal{C}(x, y) = \begin{cases} \varnothing, & (x, y) \notin E \\ \{(y, x)\}, & (x, y) \in E. \end{cases}$$

Prove that if E is a transitive relation on X, then an associative, partial multiplication on $\mathcal{C}$ is defined by the rule

$$(z, y)(y, x) = (z, x) \quad \text{whenever } (x, y), (y, z) \in E.$$

Prove that if, further, E is reflexive then $\mathcal{C}$ is a category and that if E is an equivalence relation then $\mathcal{C}$ is a groupoid.

2. Find conditions on a groupoid G for G to have initial or final objects.

3. To what extent can 6.3.1, 6.3.2 be generalised to categories?

*4. Let $\mathcal{A}$ be the category of complete, Archimedean, ordered fields in which the morphisms $K \to L$ are the Archimedean functions $f : K \to L$ such that $f(1) = 1$, $f(x + y) = fx + fy$, $f(xy) = fxfy$, for all $x, y \in K$. Prove that $\mathcal{A}$ is a 1-connected groupoid.

6.4 Functors and morphisms of groupoids

It is usual when defining an algebraic object to define the mappings or morphisms of that object—that is, it is usual to define a category of the given objects. Now we agree to regard categories themselves as particular algebraic objects. Therefore we must define the mappings or morphisms of categories—these morphisms are known as functors.

Groupoids are special cases of categories and functors of groupoids will be called *groupoid morphisms*. Actually, the main line of our applications is to the case of groupoids. However, we will have applications of functors of categories; further, many of our results are of general interest for categories and have applications outside the topics of this book. Some of these applications will be indicated in the exercises, and in this way we hope to show how these ideas run into the main stream of mathematics.

Let $\mathcal{C}, \mathcal{D}$ be categories. A *functor* $\Gamma : \mathcal{C} \to \mathcal{D}$ assigns to each object x of $\mathcal{C}$ an object Γx of $\mathcal{D}$ and to each morphism $f : x \to y$ in $\mathcal{C}$ a morphism $\Gamma f : \Gamma x \to \Gamma y$ in $\mathcal{D}$; Γf is often called the morphism *induced* by f. These must satisfy the axioms.

FUN 1 If $1 : x \to x$ is an identity in $\mathcal{C}$ then $\Gamma 1 : \Gamma x \to \Gamma x$ is the identity in $\mathcal{D}$; that is, $\Gamma 1_x = 1_{\Gamma x}$.

FUN 2 If $f : x \to y$, $g : y \to z$ are morphisms in $\mathcal{C}$, then

$$\Gamma(gf) = \Gamma g \Gamma f.$$

Clearly we have an *identity functor* $1 : \mathcal{C} \to \mathcal{C}$ and if $\Gamma : \mathcal{C} \to \mathcal{D}$, $\Delta : \mathcal{D} \to \mathcal{E}$ are functors, then we can form the composite functor $\Delta\Gamma : \mathcal{C} \to \mathcal{E}$. Thus we can form the category Cat of all categories and functors. (There is a logical difficulty here, since it seems that Cat being a category must be one of its own objects, and from this one can obtain a contradiction by considering the category of all categories which do not have themselves as objects. For a brief mention of ways round this difficulty, see the Glossary, under *class*.)

Before giving examples of functors we prove one elementary result.

6.4.1 *Let* $\Gamma : \mathcal{C} \to \mathcal{D}$ *be a functor. Then* Γf *is a retraction, a co-retraction or isomorphism if* f *is respectively a retraction, co-retraction or isomorphism.*

Proof A relation $gf = 1$ implies $\Gamma g \Gamma f = 1$. □

EXAMPLES

1. If X is a space, then PX is a category. Suppose $f : X \to Y$ is a map of spaces. If a is a path in X from x to y then the composite fa is a path in Y

from fx to fy. If a is a zero path, then so also is fa. If $b + a$ is defined in X then $fb + fa$ is defined in Y and

$$fb + fa = f(b + a).$$

Therefore f determines a functor $Pf : PX \to PY$.
2. Let us proceed further with the last example. If f is the identity $X \to X$, then so also is $Pf : PX \to PX$. Further, it is easy to check that if $f : X \to Y$, $g : Y \to Z$ are maps then $P(gf) = PgPf$. Thus P is a functor $\mathsf{Top} \to \mathsf{Cat}$. From 6.4.1 we deduce that if X is homeomorphic to Y, then PX is isomorphic to PY. Thus PX is a topological invariant of X (though not a very tractable one).
3. For any space X, there is a functor $p : PX \to \pi X$ which is the identity on objects and sends each path in X to its equivalence class.
4. Let G, H be groupoids. A functor $G \to H$ will also be called a *morphism* of groupoids. So we obtain the category Grpd of all groupoids and morphisms of groupoids.

6.4.2 *The fundamental groupoid is a functor*

$$\pi : \mathsf{Top} \to \mathsf{Grpd}\,.$$

Proof Let $f : X \to Y$ be a map of spaces and $Pf : PX \to PY$ the corresponding functor of path categories. Suppose first of all that a, b are two homotopic paths in X of length r from x to x'. Then there is a map $F : [0, r] \times \mathbb{I} \to X$ such that $F_0 = a$, $F_1 = b$ and $F(0, t) = x$, $F(r, t) = x'$ for all $t \in \mathbb{I}$. It is easily checked that the composite $fF : [0, r] \times \mathbb{I} \to Y$ is a homotopy $fa \sim fb$.

If a, b are equivalent paths in X from x to x', then there are constant paths r, s such that $r + a$, $s + b$ are homotopic, whence

$$r + fa = f(r + a) \sim f(s + b) = s + fb$$

and so fa is equivalent to fb. Thus we have a well defined function

$$\begin{aligned} \pi f : \pi X &\to \pi Y \\ \mathrm{cls}\, a &\mapsto \mathrm{cls}\, fa \end{aligned}$$

and it is clear that πf is a morphism of groupoids. The verification of the functorial relations $\pi 1 = 1$, $\pi(gf) = \pi g \pi f$, is left to the reader. □

6.4.2 (*Corollary 1*) *If X is homeomorphic to Y, then πX is isomorphic to πY.*

Proof This is immediate from 6.4.1 and 6.4.2. □

Of course, before 6.4.2 (Corollary 1) can be used it is necessary to be able to compute πX—the techniques for this are given in chapter 9.

EXAMPLE

5. The fundamental group $\pi(X, x)$ is only defined for a space X and point x of X, so that the fundamental group is in no sense a functor Top $\to$ Grp. In order to obtain a functor, one introduces the category $\mathsf{Top}_\bullet$ of *pointed spaces* (or *spaces with base point*). A *pointed space* is a pair (X, x) where $x \in X$ and X is a topological space. A *pointed map* $(X, x) \to (Y, y)$ is determined by the two pointed spaces and a map $f : X \to Y$ such that $fx = y$ (but such a pointed map is usually written f). To any pointed space (X, x) we can assign the fundamental group $\pi(X, x)$ and to any pointed map $f : (X, x) \to (Y, y)$ we can assign $f_* : \pi(X, x) \to \pi(Y, y)$, the restriction of $\pi f : \pi X \to \pi Y$ to the appropriate object groups—this defines the fundamental group functor $\mathsf{Top}_\bullet \to \mathsf{Grp}$. Since it is easier to specify a group than a groupoid, the fundamental group is often the useful topological invariant (of a pointed space). For example, if X, Y are path-connected spaces, and X, Y have fundamental groups $\mathbb{Z}, \mathbb{Z}_2$ respectively, then we know immediately that X is not homeomorphic to Y.

6.4.3 *Let* $f : \mathcal{C} \to \mathcal{D}$ *be a functor. Then* f *is an isomorphism* $\Leftrightarrow$ *the functions*

$$\mathrm{Ob}(\mathcal{C}) \to \mathrm{Ob}(\mathcal{D}), \qquad \mathcal{C}(x, y) \to \mathcal{D}(fx, fy), \qquad x, y \in \mathrm{Ob}(\mathcal{C})$$

induced by f *are all bijections.*

Proof The implication $\Rightarrow$ is easy since an inverse g of f induces an inverse to each of the functions induced by f.

For the converse, define $g : \mathcal{D} \to \mathcal{C}$ to be the given inverse of f on $\mathrm{Ob}(\mathcal{D})$ and on each $\mathcal{D}(x', y')$. Then it is easy to check that g is a functor, and is an inverse to the functor f. □

Consider now the case of groupoids. If $f : G \to H$ is a morphism, then f induces a morphism of object groups $G(x) \to H(fx)$ which is written f_x or, simply, f. If f is an isomorphism then so also is each f_x; but it is not true that if each f_x is an isomorphism and f is bijective on objects then f is an isomorphism—for example, G could be totally disconnected and H connected.

The groupoid **I** gives rise to some simple and useful morphisms. Let G be any groupoid and let $a \in G(x, y)$. Then a morphism $\hat{a} : \mathbf{I} \to G$ is defined on objects by $\hat{a}0 = x$, $\hat{a}1 = y$, and on non-zero elements by

$$\hat{a}\iota = a, \qquad \hat{a}(-\iota) = -a$$

where, as defined earlier, ι is the unique element of $\mathbf{I}(0,1)$. The check that $\hat{a}$ is a morphism is easy since $\mathbf{I}$ has only four elements, two of them zeros. Notice that $\hat{a}$ is the only morphism $\mathbf{I} \to G$ which sends ι to a.

Products of categories

Let $\mathcal{C}_1, \mathcal{C}_2$ be categories. The *product* $\mathcal{C}_1 \times \mathcal{C}_2$ is defined to have as objects all pairs (x_1, x_2) for x_1 in $\mathrm{Ob}(\mathcal{C}_1)$, x_2 in $\mathrm{Ob}(\mathcal{C}_2)$ and to have as elements the pairs (a_1, a_2) for a_1 in $\mathcal{C}_1$, a_2 in $\mathcal{C}_2$—thus the set $\mathcal{C}_1 \times \mathcal{C}_2$ is just the cartesian product of the two sets. Also, if $a_1 : x_1 \to y_1$ in $\mathcal{C}_1$, $a_2 : x_2 \to y_2$ in $\mathcal{C}_2$, then we take in $\mathcal{C}_1 \times \mathcal{C}_2$

$$(a_1, a_2) : (x_1, x_2) \to (y_1, y_2).$$

The composition is defined as one would expect by

$$(b_1, b_2)(a_1, a_2) = (b_1 a_1, b_2 a_2)$$

whenever both $b_1 a_1, b_2 a_2$ are defined. It is very easy to show that $\mathcal{C}_1 \times \mathcal{C}_2$ is a category.

Notice also that if a_1, a_2 have inverses a_1^{-1}, a_2^{-1} then (a_1, a_2) has inverse (a_1^{-1}, a_2^{-1}). It follows that if $\mathcal{C}_1, \mathcal{C}_2$ are both groupoids then so also is $\mathcal{C}_1 \times \mathcal{C}_2$.

Let $p_1 : \mathcal{C}_1 \times \mathcal{C}_2 \to \mathcal{C}_1$, $p_2 : \mathcal{C}_1 \times \mathcal{C}_2 \to \mathcal{C}_2$ be the obvious projection functors. Then we have the universal property: *if* $f_1 : \mathcal{D} \to \mathcal{C}_1$, $f_2 : \mathcal{D} \to \mathcal{C}_2$ *are functors then there is a unique functor* $f : \mathcal{D} \to \mathcal{C}_1 \times \mathcal{C}_2$ *such that* $p_1 f = f_1$, $p_2 f = f_2$. The proof is easy and is left to the reader. As usual, this property characterises the product up to isomorphism.

The functor $\pi : \mathsf{Top} \to \mathsf{Grpd}$ preserves products in the following sense.

6.4.4 *If* $X = X_1 \times X_2$ *then* πX *is isomorphic to* $\pi X_1 \times \pi X_2$.

Proof The projections $p_r : X \to X_r$ $(r = 1, 2)$ induce morphisms

$$\pi p_r : \pi X \to \pi X_r$$

which, by the universal property determine

$$f : \pi X \to \pi X_1 \times \pi X_2.$$

In fact f is the identity on objects and is defined on elements by

$$f(\mathrm{cls}\, a) = (\mathrm{cls}\, p_1 a, \mathrm{cls}\, p_2 a)$$

for any path a in X.

Let $x = (x_1, x_2)$, $y = (y_1, y_2)$ belong to X. We prove that f induces a bijection

$$\pi X(x,y) \to \pi X_1(x_1,y_1) \times \pi X_2(x_2,y_2).$$

Let a, b be paths in X from x to y and suppose first that there are homotopies $F^1 : p_1 a \sim p_1 b$, $F^2 : p_2 a \sim p_2 b$. Then it is easy to check that

$$\begin{aligned} F : [0,r] \times \mathbb{I} &\to X_1 \times X_2 \\ (s,t) &\mapsto (F^1(s,t), F^2(s,t)) \end{aligned}$$

is a homotopy $F : a \sim b$. Suppose next only that

$$f(\operatorname{cls} a) = f(\operatorname{cls} b).$$

Then we know that $p_1 a$ is equivalent to $p_1 b$, $p_2 a$ is equivalent to $p_2 b$. Because $p_1 a, p_2 a$ have the same length, as do $p_1 b, p_2 b$, we can find real numbers $r, s \geqslant 0$ large enough so that both $r + p_1 a$ is homotopic to $s + p_1 b$ and $r + p_2 a$ is homotopic to $s + p_2 b$. It follows that $r + a$ is homotopic to $s + b$ and so $\operatorname{cls} a = \operatorname{cls} b$. Thus f is injective.

In order to prove that f is surjective, let $\operatorname{cls} a_1 \in \pi X_1$, $\operatorname{cls} a_2 \in \pi X_2$; we may suppose a_1, a_2 are both of length 1. The path $a = (a_1, a_2)$ then satisfies $p_1 a = a_1$, $p_2 a = a_2$, and so

$$f(\operatorname{cls} a) = (\operatorname{cls} a_1, \operatorname{cls} a_2).$$

Therefore f is surjective. □

The proof of 6.4.4 shows a little more than stated—in fact the morphisms

$$\pi X_1 \xleftarrow{\pi p_1} \pi(X_1 \times X_2) \xrightarrow{\pi p_2} \pi X_2$$

are a product of groupoids in the sense of the universal property.

The *coproduct* $G = G_1 \sqcup G_2$ of groupoids G_1, G_2 is a simple construction. For simplicity, let us suppose that G_1, G_2 have no common elements or objects. Then we define

$$\operatorname{Ob}(G) = \operatorname{Ob}(G_1) \cup \operatorname{Ob}(G_2)$$

and define

$$G(x,y) = \begin{cases} G_1(x,y) & \text{if } x, y \in \operatorname{Ob}(G_1) \\ G_2(x,y) & \text{if } x, y \in \operatorname{Ob}(G_2) \\ \varnothing & \text{otherwise.} \end{cases}$$

(The modification of this construction when G_1, G_2 are not disjoint is left to the reader.)

It is very easy to prove

6.4.5 *There is an isomorphism of groupoids*

$$\pi(X_1 \sqcup X_2) \to \pi X_1 \sqcup \pi X_2.$$

The proof is left to the reader.

EXERCISES

1. Let $\mathcal{C}$ be a category and let $X \in \mathrm{Ob}(\mathcal{C})$. Define $\mathcal{C}_X : \mathcal{C} \to \mathsf{Set}$ by $\mathcal{C}_X C = \mathcal{C}(X, C)$ for $C \in \mathrm{Ob}(\mathcal{C})$, and if $f : C \to D$ is a morphism in $\mathcal{C}$, let

$$\mathcal{C}_X f : \mathcal{C}(X, C) \to \mathcal{C}(X, D)$$
$$g \mapsto fg.$$

Prove that $\mathcal{C}_X$ is a functor.
2. Let $\mathcal{C}, \mathcal{D}$ be categories. A *contravariant functor* $\Gamma : \mathcal{C} \to \mathcal{D}$ assigns to each object C of $\mathcal{C}$ an object ΓC of $\mathcal{D}$ and to each morphism $f : C \to D$ in $\mathcal{C}$ a morphism $\Gamma f : \Gamma D \to \Gamma C$ in $\mathcal{D}$ subject to the axioms (i) $\Gamma 1_C = 1_{\Gamma C}$, (ii) if $f : C \to D$, $g : D \to E$ in $\mathcal{C}$, then $\Gamma(gf) = \Gamma f \Gamma g$. Prove that the contravariant functors $\mathcal{C} \to \mathcal{D}$ are determined by the functors $\mathcal{C}^{op} \to \mathcal{D}$ (where $\mathcal{C}^{op}$ is the dual category of $\mathcal{C}$). [A functor as defined in the text is often called *covariant*.]
3. Prove that if $X \in \mathrm{Ob}(\mathcal{C})$, then there is a contravariant functor $\mathcal{C}^X : \mathcal{C} \to \mathsf{Set}$ such that $\mathcal{C}^X C = \mathcal{C}(C, X)$.
4. Let $\Gamma : \mathcal{C} \to \mathsf{Set}$ be a functor and let $C \in \mathrm{Ob}(\mathcal{C})$. An element u of ΓC is called *universal* (for Γ) if the function

$$\mathcal{C}(C, X) \to \Gamma X, \qquad f \mapsto (\Gamma f)u$$

is bijective for all X in $\mathrm{Ob}(\mathcal{C})$. If such a u exists, we say Γ is *representable* and that (C, u)—or simply C—*represents* Γ. Prove that if $(C, u), (C^1, u^1)$ represent Γ, then there is a unique morphism $f : C \to C^1$ such that $(\Gamma f)u = u^1$, and this f is an isomorphism.
5. The definition of the previous exercise is applied to a contravariant functor $\mathcal{C} \to \mathsf{Set}$ by considering the corresponding (covariant) functor $\mathcal{C}^{op} \to \mathsf{Set}$. Write out the definition in detail.
6. Let $\mathcal{C}$ be a category and let $X \in \mathrm{Ob}(\mathcal{C})$. Which objects of $\mathcal{C}$ represent $\mathcal{C}_X, \mathcal{C}^X$?
7. Let R be an equivalence relation in a topological space C and for each space X let ΓX be the set of maps $f : C \to X$ such that $cRc^1 \Rightarrow fc = fc^1$. If $g : X \to Y$ is a map, let $\Gamma g : \Gamma X \to \Gamma Y$ be defined by composition. Prove that Γ is a functor represented by (C, p) where $p : C \to C/R$ is the identification map.
8. For which functors are the product, sum of sets universal?
9. Let $\Gamma : \mathsf{Grp} \to \mathsf{Set}$ be the functor which assigns to each group G its underlying set, and to each morphism the corresponding function. Prove that Γ is representable. (This functor is called the *forgetful functor* from Grp to Set.)
10. Let $\Gamma : \mathsf{Top} \to \mathsf{Set}$ be the functor which assigns to each topological space its underlying set and to each map the corresponding function. Prove that Γ is representable. (This functor is called the *forgetful functor* from Top to Set.)

11. Let $\mathcal{V}$ be the category of vector spaces over a given field and let Λ be a set. If V is a vector space, let ΓV be the set of all functions $\Lambda \to V$. Prove that this defines a representable functor $\Gamma : \mathcal{V} \to \mathsf{Set}$.
12. Let $\mathcal{P}^\bullet : \mathsf{Set} \to \mathsf{Set}$ be the contravariant functor assigning to each set X the set $\mathcal{P}(X)$ and to each functor $f : X \to Y$ the function $\mathcal{P}(Y) \to \mathcal{P}(X)$, $A \mapsto f^{-1}[A]$. Prove that $\mathcal{P}^\bullet$ is representable. Is the similar (covariant) functor $\mathcal{P}_\bullet$, in which $\mathcal{P}_\bullet(f) : \mathcal{P}(X) \to \mathcal{P}(Y)$ sends $A \mapsto f[A]$, representable?
13. Let R be an integral domain (i.e., a commutative ring with identity and no divisors of zero). Let $\mathcal{K}$ be the category of fields and morphisms of fields (i.e., functions such that $f(1) = 1$, $f(x+y) = f(x) + f(y)$, $f(xy) = fxfy$). Let $\Gamma : \mathcal{K} \to \mathsf{Set}$ assign to each field K the set of morphisms $R \to K$. Prove that Γ is representable.
14. The *product* of pointed spaces $(X, x), (Y, y)$ is the pointed space

$$(X \times Y, (x, y)).$$

Prove that the fundamental group of the product of pointed spaces is isomorphic to the product of their fundamental groups.
15. Define and construct coproducts of categories.
16. Let X, Y be topological spaces. Are the following statements true: $P(X \times Y)$ is isomorphic to $PX \times PY$? $P(X \sqcup Y)$ is isomorphic to $PX \sqcup PY$?

6.5 Homotopies

In defining the fundamental groupoid we have used homotopies of paths. In describing the invariance properties of the fundamental groupoid we must use homotopies of maps.

Let X, Y be topological spaces. A map $F : X \times [0, q] \to Y$ will be called a *homotopy of length* q; for such F, the *initial map* and the *final map* of F are respectively the functions

$$\begin{aligned} f : X &\to Y \\ x &\mapsto F(x, 0) \end{aligned} \qquad \begin{aligned} g : X &\to Y \\ x &\mapsto F(x, q). \end{aligned}$$

We say F is a homotopy from f to g, and we write

$$F : f \simeq g.$$

If a homotopy (of some length) $F : f \simeq g$ exists then we say f, g are *homotopic* and write $f \simeq g$.

6.5.1 *The relation* $f \simeq g$ *is an equivalence relation on maps* $X \to Y$.

The proof is a simple generalisation of the argument for homotopies of paths, and is left to the reader. It should be emphasised that the homotopies of section 6.2 were more restricted since the end points of the paths had to be fixed during the homotopy (this kind of homotopy is subsumed under the notion of homotopy rel A which we use in chapter 7).

There is a continuous surjection $\lambda : \mathbb{I} \to [0, q]$; so if $F : X \times [0, q] \to Y$ is a homotopy of length q, then $G = F(1 \times \lambda) : X \times \mathbb{I} \to Y$ is a homotopy of length 1. Also F and G have the same initial and the same final maps. So in discussing homotopies of maps it is sufficient to restrict ourselves to homotopies of length 1, and this we shall do for the rest of this chapter. We also denote a homotopy of length 1 by $F_t : X \to Y$ (where $t \in \mathbb{I}$).

6.5.2 *Let $f : W \to X$, $g_0, g_1 : X \to Y$, $h : Y \to Z$ be maps. If $g_0 \simeq g_1$, then*

$$hg_0f \simeq hg_1f : W \to Z.$$

Proof Let $g_t : g_0 \simeq g_1$ be a homotopy. Then hg_tf is a homotopy $hg_0f \simeq hg_1f$. □

6.5.3 *Let $f_0 \simeq f_1 : X \to Y$, $g_0 \simeq g_1 : Z \to W$. Then $f_0 \times g_0 \simeq f_1 \times g_1$.*

Proof If $f_t : f_0 \simeq f_1$, $g_t : g_0 \simeq g_1$, then the required homotopy is $f_t \times g_t$. The detailed proof of continuity of $f_t \times g_t$ (as a function $X \times Z \times \mathbb{I} \to Y \times W$) is left to the reader [cf. Example 2, p. 36]. □

Let $f : X \to Y$, $g : Y \to X$ be maps. If

$$fg \simeq 1_Y$$

then we say g is a *right homotopy inverse* of f, that f is a *left homotopy inverse* of g, and that X *dominates* Y. If g is both a left and a right homotopy inverse of f, then g is called simply a *homotopy inverse* of f; further f is called a *homotopy equivalence* and we write $f : X \simeq Y$. If a homotopy equivalence $X \to Y$ exists, then we say X, Y are homotopy equivalent (or of the same *homotopy type*) and we write $X \simeq Y$. (In much of the literature this relation is written $\equiv$). This relation is easily seen to be an equivalence relation on topological spaces.

EXAMPLES

1. Let Y be a convex subset of $\mathbb{R}^n$ and let $f_0, f_1 : X \to Y$ be maps. Then $f_t = (1 - t)f_0 + tf_1$ is a homotopy $f_0 \simeq f_1$.
2. A map is *inessential* if it is homotopic to a constant map—otherwise it is *essential*. For example, any map to a convex subset of $\mathbb{R}^n$ is inessential, by Example 1.

3. A space X is *contractible* if it is of the homotopy type of a space with only one point. In fact, X is *contractible if and only if the identity map* $1_X : X \to X$ *is inessential.*

Proof $\Rightarrow$ Suppose $f : X \simeq Y$ has homotopy inverse g, where Y is a single point space. Then $gf \simeq 1_X$ and gf is a constant map. Therefore 1_X is inessential.

$\Leftarrow$ Let $1_X \simeq f$ where $f : X \to X$ is a constant map with value x say. Let $Y = \{x\}$, let $i : Y \to X$ be the inclusion and let $f' : X \to Y$ be the unique map. Then $f'i = 1_Y$ and $if' = f \simeq 1_X$. □

It follows from these examples that any convex subset of $\mathbb{R}^n$ is contractible.

4. Let $f_t : X \to Y$ be a homotopy. Then for each x in X the map $t \mapsto f_t x$ is a path in Y from $f_0 x$ to $f_1 x$ and so $f_0 x, f_1 x$ lie in the same path-component of Y. It follows easily that, if X has more than one path-component, then $1 : X \to X$ is an essential map.

5. Let $y, z \in \mathbb{S}^1$ and let $g, h : \mathbb{S}^1 \to \mathbb{S}^1 \times \mathbb{S}^1$ be the maps $x \mapsto (x, y)$, $x \mapsto (x, z)$ respectively [Fig. 6.9]. Then it is easy to prove that $g \simeq h$—in fact if a is any path of length 1 in $\mathbb{S}^1$ from y to z, then $f_t : x \mapsto (x, at)$ is a homotopy $g \simeq h$. However, if k is the map $x \mapsto (y, x)$, then it is true that g is *not* homotopic to k, but the proof needs more theory than we have yet developed—in particular, we need to know that $\mathbb{S}^1$ is not simply-connected.

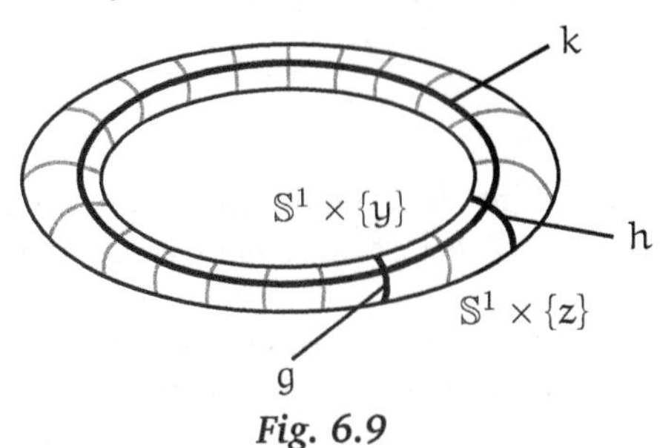

Fig. 6.9

We must now investigate the concept in the algebra of groupoids corresponding to homotopy of maps of spaces. For this it is convenient first to say something about functors on a product of categories.

Let $F : \mathcal{C} \times \mathcal{D} \to \mathcal{E}$ be a functor, where $\mathcal{C}, \mathcal{D}, \mathcal{E}$ are categories. If 1_x is the identity at x in $\mathcal{C}$, then let us write $F(x, b)$ for $F(1_x, b)$ where b is any morphism in $\mathcal{D}$. Similarly, let us write $F(a, y)$ for $F(a, 1_y)$ for any object y of $\mathcal{D}$ and a morphism a of $\mathcal{C}$. Then, as is easily verified, $F(x, \ )$ is a functor $\mathcal{D} \to \mathcal{E}$ (called the *x-section* of F) and $F(\ , y)$ is a functor $\mathcal{C} \to \mathcal{E}$ (called the *y-section* of F). These two functors determine F. If $a : x \to x'$, $b : y \to y'$ are morphisms in $\mathcal{C}, \mathcal{D}$ respectively then we have a commutative diagram

$$\begin{array}{ccc} F(x,y) & \xrightarrow{F(a,y)} & F(x',y) \\ {\scriptstyle F(x,b)}\downarrow & \searrow^{F(a,b)} & \downarrow{\scriptstyle F(x',b)} \\ F(x,y') & \xrightarrow[F(a,y')]{} & F(x',y') \end{array} \tag{6.5.4}$$

since $F(1_{x'}a, b1_y) = F(a,b) = F(a1_x, 1_{y'}b)$.

6.5.5 *Suppose for each* x *in* $\mathrm{Ob}(\mathcal{C})$ *and* y *in* $\mathrm{Ob}(\mathcal{D})$ *we are given functors*

$$F(x,\ ) : \mathcal{D} \to \mathcal{E},\ F(\ ,y) : \mathcal{C} \to \mathcal{E}$$

such that $F(x,y)$ *is a unique object of* $\mathcal{E}$. *Suppose for each* $a : x \to x'$ *in* $\mathcal{C}$ *and* $b : y \to y'$ *in* $\mathcal{D}$ *the outer square of* (6.5.4) *commutes. Then the diagonal composite* $F(x,b)$ *makes* F *a functor* $\mathcal{C} \times \mathcal{D} \to \mathcal{E}$. *All functors* $\mathcal{C} \times \mathcal{D} \to \mathcal{E}$ *arise in this way.*

Proof The verification of FUN 1 for F is easy since

$$\begin{aligned} F(1_x, 1_y) &= F(1_x, y)F(x, 1_y) \\ &= 1_{F(x,y)}1_{F(x,y)} \\ &= 1_{F(x,y)}. \end{aligned}$$

The verification of FUN 2 involves a diagram of four commutative squares, and is left to the reader. The last statement is clear from the discussion preceding 6.5.5. □

In order to model the notion of homotopy we need a model for categories of the unit interval. This is provided by the tree groupoid **I**.

Let $\mathcal{C}$ and $\mathcal{E}$ be categories. A *homotopy* (or *natural equivalence*) of functors from $\mathcal{C}$ to $\mathcal{E}$ is a functor

$$F : \mathcal{C} \times \mathbf{I} \to \mathcal{E}.$$

The *initial functor* of F is then $f = F(\ ,0)$ and the *finial functor* of F is $g = F(\ ,1)$; we say f is a homotopy from f to g and write $F : f \simeq g$. If such a homotopy from f to g exists then we say f, g are *homotopic* and write $f \simeq g$.‡

‡This definition of homotopy was pointed out to me by P. J. Higgins. I am grateful to W. F. Newns for suggesting that the emphasis be placed on this definition (rather than that by the function θ as below) and for other helpful comments on this section and section 6.7.

According to 6.5.5, in order to specify a homotopy F it is sufficient to give the initial and final functors f and g of F, and also for each object x of $\mathcal{C}$ a functor $F(x, \ ) : \mathbf{I} \to \mathcal{E}$ in such a way that the outside of (6.5.4) commutes. However, the functors $F(x, \ ) : \mathbf{I} \to \mathcal{E}$ are entirely specified by invertible elements θx of $\mathcal{E}$ where $\theta x = F(x, \iota)$. In these terms, (6.5.4) becomes (with $b = \iota$)

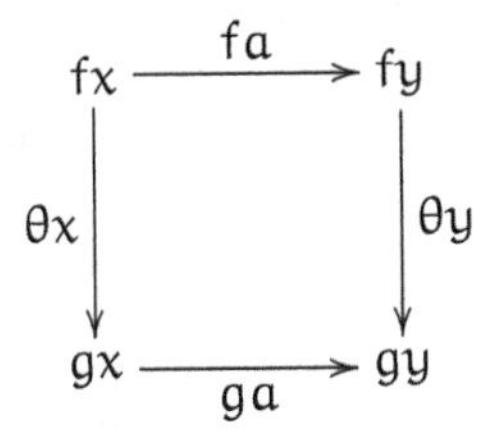

the commutativity of which asserts

$$(ga)(\theta x) = (\theta y)(fa). \tag{6.5.6}$$

Since θx is invertible this shows that g is determined by f and θ. Thus, given any functor $f : \mathcal{C} \to \mathcal{E}$ and for each object x of $\mathcal{C}$ an invertible element θx of $\mathcal{E}$ with initial point fx, then there is a homotopy $f \simeq g$ where g is defined by (6.5.6). We call θ a *homotopy function* from f to g and write also $\theta : f \simeq g$.

The function θ also gives rise to the useful diagram

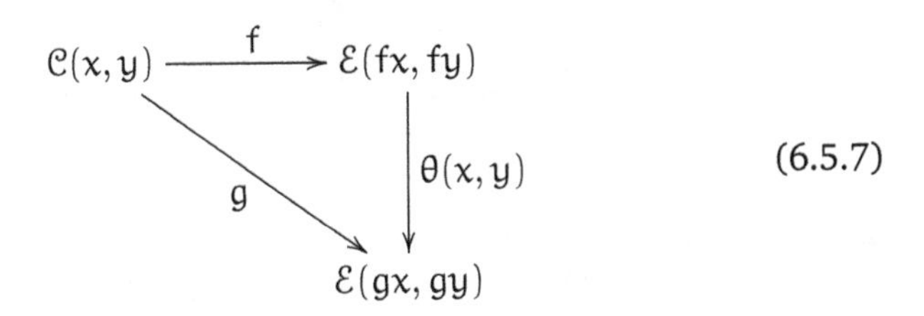

(6.5.7)

in which $\theta(x, y)$ is the bijection $a' \mapsto (\theta y)a'(\theta x)^{-1}$. The commutativity of (6.5.7) is immediate from (6.5.6) with $a' = fa$.

6.5.8 *Homotopy of functors is an equivalence relation.*

Proof (a) The function $x \mapsto 1_{fx}$ is a homotopy function $f \simeq f$.
(b) If θ is a homotopy function $f \simeq g$, then $x \mapsto (\theta x)^{-1}$ is a homotopy function $g \simeq f$.
(c) If θ, φ are homotopy functions $f \simeq g$, $g \simeq h$ respectively then $x \mapsto \varphi x \theta x$ is a homotopy function $f \simeq h$. □

The proof of the following proposition is left as an exercise.

6.5.9 *Let* $f : \mathcal{C} \to \mathcal{D}$, $g : \mathcal{D} \to \mathcal{E}$, $h : \mathcal{E} \to \mathcal{F}$ *be functors and suppose* $g \simeq g'$. *Then* $hgf \simeq hg'f$.

Exactly as for topological spaces we have the notions of *homotopy inverse* of a functor, and homotopy equivalence of categories. Further, all these notions apply *a fortiori* to groupoids and morphisms of groupoids.

The utility for topology of these notions is shown by:

6.5.10 *If* $f, g : X \to Y$ *are homotopic maps of spaces, then the induced morphisms* $\pi f, \pi g : \pi X \to \pi Y$ *are homotopic.*

Proof Let $F : X \times \mathbb{I} \to Y$ be a homotopy from f to g. Consider the composite

$$\pi X \times \mathbf{I} \to \pi X \times \pi\mathbb{I} \xrightarrow{\varphi} \pi(X \times \mathbb{I}) \xrightarrow{\pi F} \pi Y$$

in which the first morphism is inclusion (since $\mathbf{I} \subseteq \pi\mathbb{I}$) and the second morphism is the isomorphism constructed in 6.4.4. We prove that the composite G is a homotopy $\pi f \simeq \pi g$.

Let $\alpha \in \pi X(x, y)$, let $a \in \alpha$ be of length 1, and let c_ε be a constant path at ε of length 1 for $\varepsilon = 0, 1$. Then

$$\begin{aligned} G(\alpha, \varepsilon) &= (\pi F)\varphi(\alpha, \varepsilon) \\ &= (\pi F)(\mathrm{cls}(a, c_\varepsilon)) \\ &= \mathrm{cls}\, F(a, c_\varepsilon) \\ &= \begin{cases} \mathrm{cls}\, fa, & \text{if } \varepsilon = 0 \\ \mathrm{cls}\, ga, & \text{if } \varepsilon = 1. \end{cases} \end{aligned}$$

□

6.5.10 *(Corollary 1) If* $f : X \to Y$ *is a homotopy equivalence of spaces, then* $\pi f : \pi X \to \pi Y$ *is a homotopy equivalence of groupoids.*

6.5.11 *Let* $f : \mathcal{C} \to \mathcal{C}$ *be a functor such that* $f \simeq 1$. *Then for all objects* x, y *of* $\mathcal{C}$

$$f : \mathcal{C}(x, y) \to \mathcal{C}(fx, fy)$$

is a bijection.

Proof This is immediate from the commutativity of (6.5.7) and the fact that $\theta(x, y)$ is a bijection. □

6.5.12 *Let* $f : \mathcal{C} \to \mathcal{E}$ *be a homotopy equivalence of categories. Then for all objects* x, y *of* $\mathcal{C}$, $f : \mathcal{C}(x, y) \to \mathcal{E}(fx, fy)$ *is a bijection. Hence, if further* f *is bijective on objects then* f *is an isomorphism.*

Proof Let $g : \mathcal{E} \to \mathcal{C}$ be a homotopy inverse of f, so that $gf \simeq 1$, $fg \simeq 1$. Consider the functions,

$$\mathcal{C}(x, y) \xrightarrow{f} \mathcal{E}(fx, fy) \xrightarrow{g} \mathcal{C}(gfx, gfy) \xrightarrow{f} \mathcal{E}(fgfx, fgfy).$$

By 6.5.11, the composites of the first two, and of the last two, functions are bijections. It follows easily that each function is a bijection. The last part follows from this and 6.4.3. □

6.5.12 *(Corollary 1) If $f : X \to Y$ is a homotopy equivalence of spaces, then for each x in X, $\pi f : \pi(X, x) \to \pi(Y, fx)$ is an isomorphism of fundamental groups.*

The application of this result to spaces must wait until we can compute more readily the fundamental groupoid. However 6.5.10 (Corollary 1) focuses attention on the homotopy type of groupoids and so, more generally, of categories. The rest of this section is devoted to a simple result which enables us to replace a given category by a simpler but homotopically equivalent category. This process is especially useful for computations of the fundamental groupoid—in fact πX has an embarrassingly large number of objects, so any simplification of πX is welcome.

Let $F : \mathcal{C} \times \mathbf{I} \to \mathcal{E}$ be a homotopy, where $\mathcal{C}$ and $\mathcal{E}$ are categories. The homotopy is called *constant* if

$$F(a, 0) = F(a, t) = F(a, 1)$$

for all elements a of $\mathcal{C}$. Let θ be the homotopy function defined by F (so that $\theta x = F(x, t)$, $x \in \mathrm{Ob}(\mathcal{C})$). Then clearly F is constant if and only if θx is an identity of $\mathcal{E}$ for all x.

Suppose that $\mathcal{D}$ is a subcategory of $\mathcal{C}$. Then we say F is a *homotopy* rel $\mathcal{D}$ if $F \mid \mathcal{D} \times \mathbf{I}$ is a constant homotopy, or, equivalently, if $\theta x = 1$ for all x in $\mathrm{Ob}(\mathcal{D})$. In such case, the initial and final functors f, g of F are said to be *homotopic* rel $\mathcal{D}$, written $f \simeq g \operatorname{rel} \mathcal{D}$; these functors must of course agree on $\mathcal{D}$, but this alone, or even with the existence of a homotopy $f \simeq g$, is not enough to ensure $f \simeq g \operatorname{rel} \mathcal{D}$.

A subcategory $\mathcal{D}$ of $\mathcal{C}$ is a *deformation retract* of $\mathcal{C}$ if there is a functor $r : \mathcal{C} \to \mathcal{D}$ such that $ir \simeq 1_{\mathcal{C}} \operatorname{rel} \mathcal{D}$, where $i : \mathcal{D} \to \mathcal{C}$ is the inclusion. Such a functor r is called a *deformation retraction*. It is a retraction, since $ir \mid \mathcal{D} = 1_{\mathcal{C}} \mid \mathcal{D} = i$ and so $ri = 1_{\mathcal{D}}$. Further, if F is the homotopy $ir \simeq 1_{\mathcal{C}}$ and $a : x \to x'$ is an element of $\mathcal{C}$, then by (6.5.4) with $b = \iota : 0 \to 1$

$$F(x', \iota) ir(a) = aF(x, \iota).$$

In particular, let $a = F(x', \iota) : rx' \to x'$; then $F(x, \iota) = 1$ since F is rel $\mathcal{D}$. So we obtain

$$F(x', \iota) r(a) = a$$

whence $rF(x', \imath) = 1$. This shows that rF is the constant homotopy $r \simeq r$.

We note also that if $\mathcal{D}$ is a deformation retract of $\mathcal{C}$ then $i : \mathcal{D} \to \mathcal{C}$ is a homotopy equivalence.

To characterise deformation retracts we need a further definition. A subcategory $\mathcal{D}$ of $\mathcal{C}$ is *representative* in $\mathcal{C}$ if each object of $\mathcal{C}$ is isomorphic to an object of $\mathcal{D}$.

6.5.13 *A subcategory $\mathcal{D}$ of $\mathcal{C}$ is a deformation retract of $\mathcal{C}$ if and only if $\mathcal{D}$ is a full, representative subcategory of $\mathcal{C}$. In fact, if $\mathcal{D}$ is a full representative subcategory of $\mathcal{C}$ and if we define $\theta x = 1_x$ for each object x of $\mathcal{D}$, and choose θy for any other object y of $\mathcal{C}$ to be an isomorphism of some x in $\mathrm{Ob}(\mathcal{D})$ with y, then θ determines a deformation retraction $r : \mathcal{C} \to \mathcal{D}$ and a homotopy $ir \simeq 1 \operatorname{rel} \mathcal{D}$, where $i : \mathcal{D} \to \mathcal{C}$ is the inclusion.*

Proof We leave the proof of 'only if' as an exercise. That θ can be chosen follows from the fact that $\mathcal{D}$ is representative in $\mathcal{C}$. By the remarks following (6.5.6), θ^{-1} determines a homotopy $1_{\mathcal{C}} \simeq g$ for some functor g, and this homotopy is $\operatorname{rel} \mathcal{D}$ be construction. Now gy is an object of $\mathcal{D}$ for each object y of $\mathcal{C}$; since $\mathcal{D}$ is full it follows that $ga \in \mathcal{D}$ for each element a of $\mathcal{C}$. Therefore, we can write $g = ir$ for some functor $r : \mathcal{C} \to \mathcal{D}$. □

6.5.13 *(Corollary 1) Any groupoid is of the homotopy type of a totally disconnected groupoid.*

Proof Let G be a groupoid and H a full subgroupoid of G consisting of one object group in each component of G. By 6.5.13, the inclusion $H \to G$ is a homotopy equivalence. □

This shows that the interesting invariant of homotopy type of a space X is a groupoid consisting of one object group in each path-component of X. It would thus seem likely that groupoids which are not totally disconnected are of little interest.

We shall see later that this view is not valid. In fact, groupoids and groupoid morphisms carry information (essentially of a graph theoretic character) which it is more difficult to describe in terms of groups alone.

The situation is analogous to that of vector spaces. Vector spaces are of course characterised up to isomorphism by a single cardinal number, the dimension of the vector space. This does not imply that vector spaces can reasonably be excised from the mathematical literature. One of the reasons is that morphisms of vector spaces, and vector spaces with additional structure, are of considerable importance in mathematics.

EXERCISES

1. Prove that a category HTop is defined as follows: the objects of HTop are the topological spaces; the maps $X \to Y$ are the homotopy classes of maps $X \to Y$; composition is defined by

$$(\mathrm{cls}\, g)(\mathrm{cls}\, f) = \mathrm{cls}(gf).$$

Define HGrpd similarly and prove that π : Top $\to$ Grpd determines a functor HTop $\to$ HGrpd.
2. For any space X, let $\pi_0 X$ be the set of path components of X. Prove that π_0 determines a functor HTop $\to$ Set. Deduce that if $X \simeq Y$, then $\divideontimes(\pi_0 X) = \divideontimes(\pi_0 Y)$.
3. Prove that if X, X', Y, Y' are spaces and $X \simeq X'$, $Y \simeq Y'$ then $X \times Y \simeq X' \times Y'$, $X \sqcup Y \simeq X' \sqcup Y'$.
4. Prove that if Y is contractible then any maps $X \to Y$ are homotopic. Prove that if X and Y are contractible then any map $X \to Y$ is a homotopy equivalence. Prove also that a retract of a contractible space is contractible. Prove results for groupoids similar to those of this and the preceding two exercises.
5. Let $\mathcal{V}_K$ be the category of finite dimensional vector spaces over a given field K. Let K be the full subcategory of $\mathcal{V}_K$ on the vector spaces K^n, $n \geqslant 0$. Prove that K is a deformation retract of $\mathcal{V}_K$.
6. Let $\mathbf{I}^{\to}$ be the category with two objects $0, 1$ and only one non-identity element $\iota : 0 \to 1$. If $F : \mathcal{C} \times \mathbf{I}^{\to} \to \mathcal{E}$ is a functor, then $f = F(\ ,0)$ and $g = F(\ ,1)$ are called the *initial* and *final* functors of F, and F is called a *morphism*, or *natural transformation*, from f to g. Prove that there is a category $\mathsf{Fun}(\mathcal{C}, \mathcal{E})$, whose objects are the functors $\mathcal{C} \to \mathcal{E}$, whose morphisms are the morphisms of functors and such that the invertible elements of this category are the homotopies.
7. Let $\mathcal{V}$ be the category of vector spaces over a (commutative) field K. For each object V of $\mathcal{V}$ let $V^* = \mathcal{V}(V, K)$, considered again as a vector space over K, and let $gV = V^{**}$; let $\theta V : V \to V^{**}$ be the function $\nu \mapsto (\lambda \mapsto \lambda\nu)$. Prove that g is a functor $\mathcal{V} \to \mathcal{V}$ and that θ determines a morphism $1_{\mathcal{V}} \to g$ such that θV is invertible if V is of finite dimension. [This example was a crucial one in the development of category theory. For a discussion, see the fundamental paper [EM45] of Eilenberg and Mac Lane.]
8. Let $\Gamma : \mathcal{C} \to$ Set be a functor represented by (C, u) and let $\Delta : \mathcal{C} \to$ Set be any functor. Prove that the function

$$\begin{aligned} \mathsf{Fun}(\Gamma, \Delta) &\to \Delta C, \\ F &\mapsto F(C, \iota)(u) \end{aligned}$$

is a bijection. (This says that the natural transformations of a representable functor are entirely determined by their values on the universal element.)
9. Prove that, if C, D are objects of $\mathcal{C}$, then there is a bijection

$$\mathsf{Fun}(\mathcal{C}_C, \mathcal{C}_D \to \mathcal{C}(C, D).$$

10. Prove that a functor $\Gamma : \mathcal{C} \to$ Set is representable by C if and only if there is a homotopy $\Gamma \simeq \mathcal{C}_C$.

11. Let $\mathcal{D}$ be a category. Prove that if $\widehat{\mathcal{D}}$ assigns to each pair of x, y of $\mathcal{D}$ the set $\mathcal{D}(x, y)$ and to each pair of morphisms $a : x \to x'$, $b : y \to y'$ the morphism

$$\begin{aligned} \mathcal{D}(a, b) : \mathcal{D}(x', y) &\to \mathcal{D}(x, y') \\ c &\mapsto bca, \end{aligned}$$

then $\widehat{\mathcal{D}}$ is a functor $\mathcal{D}^{op} \times \mathcal{D} \to \mathsf{Set}$.
12. Let $\mathcal{C}, \mathcal{D}$ be categories and $\Gamma : \mathcal{C}^{op} \times \mathcal{D} \to \mathsf{Set}$ a functor such that for each object C of $\mathcal{C}$ the functor $\Gamma(C, \ )$ is representable by $(\Delta C, uC)$. Prove that the function $C \mapsto \Delta C$ of objects extends uniquely to a functor $\Delta : \mathcal{C} \to \mathcal{D}$ such that the bijection

$$\begin{aligned} \mathcal{D}(\Delta C, D) &\to \Gamma(C, D) \\ f &\mapsto \Gamma(C, f)(uC) \end{aligned}$$

is a homotopy $\widehat{\mathcal{D}}(\Delta C \times 1) \simeq \Gamma$ (where $\widehat{\mathcal{D}}$ is as in Exercise 11).
*13. Let $\mathcal{C}, \mathcal{D}, \mathcal{E}$ be categories. Prove that there is an isomorphism of categories

$$\mathsf{Fun}(\mathcal{C}, \mathsf{Fun}(\mathcal{D}, \mathcal{E})) \to \mathsf{Fun}(\mathcal{C} \times \mathcal{D}, \mathcal{E}).$$

14. A functor $f : \mathcal{C} \to \mathcal{D}$ induces for all objects x, y of $\mathcal{C}$ a function $f : \mathcal{C}(x, y) \to \mathcal{D}(fx, fy)$. If this function is injection for all x, y then f is called *faithful*; if it is surjective for all x, y then f is called *full*; finally if each object z of $\mathcal{D}$ is isomorphic to some object fx, then f is called *representative*. Prove that f is a homotopy equivalence of categories if and only if f is full, faithful and representative. Prove also that if g is a homotopy inverse of f, and if θ is a homotopy function $fg \simeq 1_{\mathcal{D}}$, then we can choose a homotopy function $\varphi : gf \simeq 1_{\mathcal{C}}$ such that $f\varphi = \theta f$, $g\theta = \varphi g$.
15. Let both the path functor P and the fundamental groupoid functor π be considered as functors $\mathsf{Top} \to \mathsf{Cat}$. Prove that the assignment to each topological space X of the projection $p_\bullet : PX \to \pi X$ defines a natural transformation $P \to \pi$.

6.6 Coproducts and pushouts

We have already used in several categories the idea of a coproduct. It seems reasonable to formulate now the general definition. We shall also define pushouts in a general category—there is a close relation between coproducts and pushouts which is presented briefly in the Exercises.

Let $\mathcal{C}$ be a category. A *coproduct* (or *sum*) of two objects C_1, C_2 of $\mathcal{C}$ is a diagram

$$C_1 \xrightarrow{i_1} C \xleftarrow{i_2} C_2 \qquad (6.6.1)$$

of morphisms of $\mathcal{C}$ with the following φ-universal property: for any diagram

$$C_1 \xrightarrow{\nu_1} C' \xleftarrow{\nu_2} C_2$$

of morphisms of $\mathcal{C}$, there is a unique morphism $v : C \to C'$ such that

$$vi_1 = v_1, \qquad vi_2 = v_2.$$

The usual universal argument shows that this property characterises coproducts up to isomorphism. If (6.6.1) is a coproduct in $\mathcal{C}$ it is usual to write

$$C = C_1 \sqcup C_2$$

and by an abuse of language, to refer to C (rather than i_1, i_2) as the coproduct of C_1 and C_2.

This definition is of course a simple extension to arbitrary categories of definitions encountered already in the categories Set, Top, and Grpd—in each of these categories, any two objects have a coproduct. This is also true in the category Grp, but the proof, which will be given in chapter 8, is non-trivial.

We now discuss pushouts. A diagram

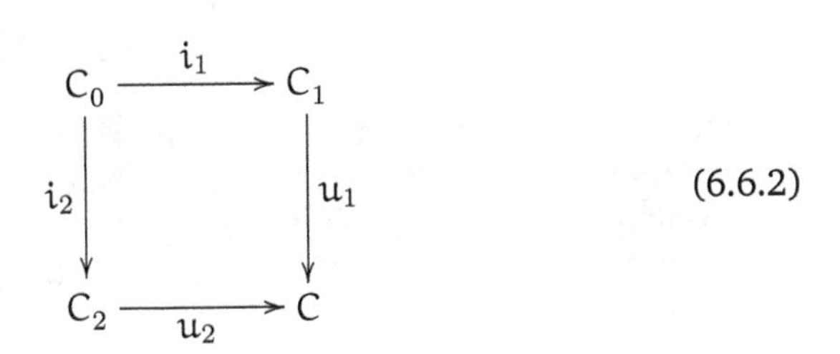

(6.6.2)

of morphisms of $\mathcal{C}$ is called a *pushout* (of i_1, i_2) if

(a) The diagram is commutative: that is $u_1 i_1 = u_2 i_2$.

(b) u_1, u_2 are φ-universal for property (a); that is, if the diagram

$$\begin{array}{ccc} C_0 & \xrightarrow{i_1} & C_1 \\ {\scriptstyle i_2}\downarrow & & \downarrow{\scriptstyle v_1} \\ C_2 & \xrightarrow[v_2]{} & C' \end{array} \qquad (6.6.3)$$

of morphisms of $\mathcal{C}$ is commutative, then there is a unique morphism $v : C \to C'$ such that $vu_1 = v_1$, $vu_2 = v_2$.

The usual universal argument shows that if (6.6.2) is a pushout, then (6.6.3) is a pushout if and only if there is an isomorphism $v : C \to C'$ such that $vu_\alpha = v_\alpha$, $\alpha = 1, 2$. Thus a pushout is determined up to isomorphism by i_1, i_2.

If (6.6.2) is a pushout it is usual to write

$$C = C_1 \,{}_{i_1}\!\sqcup_{i_2} C_2$$

and, by an abuse of language, to refer to C itself as the pushout of i_1, i_2.

It is important to note that we have not asserted that pushouts exist for any $\mathcal{C}$ and i_1, i_2. The existence of arbitrary pushouts is to be regarded as a good property of the category $\mathcal{C}$. In the exercises we give a condition for pushouts to exist—here we are more concerned with giving useful lemmas for proving that a particular diagram is a pushout.

Suppose given in $\mathcal{C}$ a commutative diagram

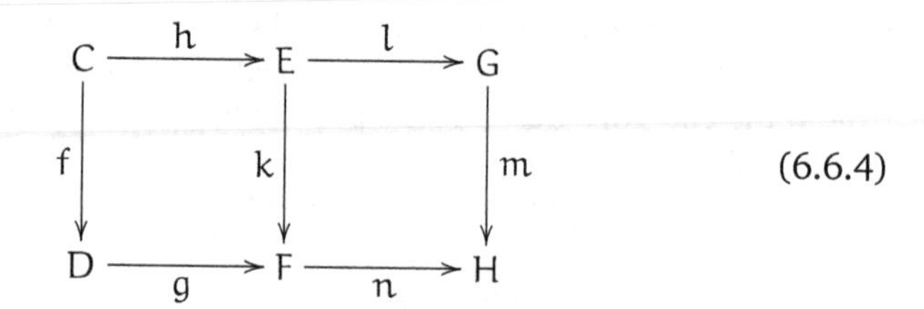

Then the outer part of this diagram is also a commutative square which we call the *composite* of the two individual squares. The abstract algebra of such compositions is a 'double category', since there are possible compositions both horizontally and vertically [cf. [Ehr65]]. But we shall need only the horizontal composition.

6.6.5 *In any category $\mathcal{C}$, the composite of two pushouts is a pushout.*

Proof We use the notation of (6.6.4). Suppose given a commutative diagram

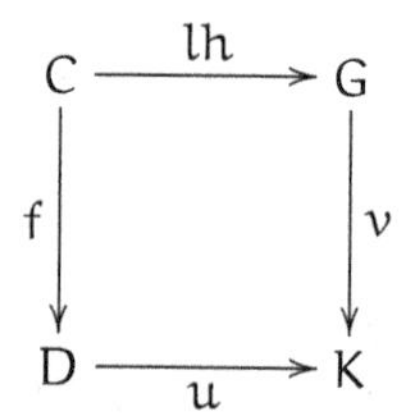

By the pushout property for the first square, there is a unique morphism $w : F \to K$ such that

$$wg = u, \qquad wk = vl.$$

The two maps w and v determine, by the pushout property of the second square, a unique map $x : H \to K$ such that

$$xn = w, \qquad xm = v;$$

it follows that

$$xng = u, \qquad xm = v.$$

To complete the proof we must show that x is the only morphism satisfying these last equations.

Suppose that x' satisfies

$$x'ng = u, \qquad x'm = v.$$

Then $x'ng = u$, $x'nk = x'ml = vl$. By uniqueness of the construction of w, $x'n = w$. Further, by uniqueness of the construction of x, $x' = x$. □

Suppose, in particular, that $\mathcal{C}$ is the category Top and that $f : C \to D$ is the inclusion of the closed subspace C of D. Then F, H are adjunction spaces and 6.6.5 can be stated as

$$G\ {}_{l}{\sqcup}\ (E\ {}_{h}{\sqcup}\ D) = G\ {}_{lh}{\sqcup}\ D.$$

Other applications of 6.6.5 (mainly to groupoids) will occur later.

Our next result says, roughly, that a retract of a pushout is a pushout. Now the term retract is meaningful in any category, so in order to give meaning to the last sentence it is enough to define morphisms of commutative squares.

Consider the following cubical diagram of morphisms of $\mathcal{C}$.

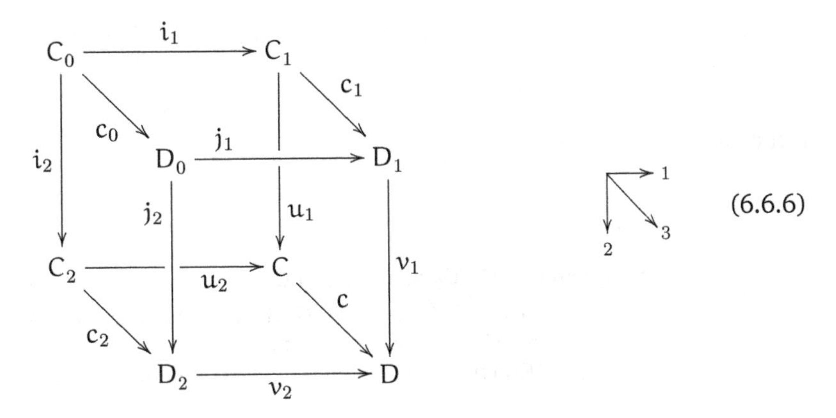

(6.6.6)

Let us write $\mathbf{C}$ for the back square (in direction 3) and $\mathbf{D}$ for the front square (in direction 3), both of which we suppose commutative. If the whole diagram is commutative, then it is called a *morphism* $\mathbf{c} : \mathbf{C} \to \mathbf{D}$. Clearly we have an identity morphism $\mathbf{C} \to \mathbf{C}$, and the composite of morphisms is again a morphism.

So we have a category $\mathcal{C}_{\square}$ of commutative squares and morphisms of squares and in this category the notion of retraction is well-defined.

6.6.7 *Let* $\mathbf{C}, \mathbf{D}$ *be commutative squares in* $\mathcal{C}$ *such that* $\mathbf{D}$ *is a pushout. If there is a retraction* $\mathbf{D} \to \mathbf{C}$ *then* $\mathbf{C}$ *is a pushout.*

Proof Let $\mathbf{c} : \mathbf{C} \to \mathbf{D}$, $\mathbf{d} : \mathbf{D} \to \mathbf{C}$ be morphisms such that $\mathbf{dc} = 1_{\mathbf{C}}$. Suppose also (referring to the diagram (6.6.6)) that we are given morphisms

$$C_1 \xrightarrow{w_1} C' \xleftarrow{w_2} C_2$$

such that $w_1 i_1 = w_2 i_2$. Consider the morphisms

$$D_1 \xrightarrow{w_1 d_1} C' \xleftarrow{w_2 d_2} D_2.$$

Since $\mathbf{d}$ is a map of squares, $d_1 j_1 = i_1 d_1$, $d_2 j_2 = i_2 d_0$. Hence

$$(w_1 d_1) j_1 = (w_2 d_2) j_2.$$

Since $\mathbf{D}$ is a pushout, there is a unique morphism $x : D \to C'$ such that

$$x v_1 = w_1 d_1, \qquad x v_2 = w_2 d_2.$$

Let $w = xc : C \to C'$. Then for $\alpha = 1, 2$

$$w u_\alpha = x c u_\alpha = x v_\alpha c_\alpha = w_\alpha d_\alpha c_\alpha = w_\alpha$$

as we required.

Suppose $w' : C \to C'$ also satisfied $w'_\alpha u_\alpha = w_\alpha$, $\alpha = 1, 2$. Then

$$w' d v_\alpha = w' u_\alpha d_\alpha = w_\alpha d_\alpha = x v_\alpha, \quad \alpha = 1, 2.$$

Hence $w' d = x$ and so $w' = w' dc = xc = w$. □

EXERCISES

1. Let $\mathcal{C}$ be a category and let C_1, C_2 be objects of $\mathcal{C}$. Let $\mathcal{C}$ be the category whose objects are diagrams $C_1 \xrightarrow{i_1} C \xleftarrow{i_2} C_2$ (written (i_1, i_2, C)) and whose morphisms (i'_1, i'_2, d') are morphisms $f : C \to C$ of $\mathcal{C}$ such that $f i_\alpha = i'_\alpha$, $\alpha = 1, 2$. Prove that an initial object of $\mathcal{C}'$ [Exercise 7 of Section 6.1] is a coproduct diagram $C_1 \to C_1 \sqcup C_2 \leftarrow C_2$.
2. Let $\mathcal{C}$ be a category and let C_1, C_2 be objects of $\mathcal{C}$. A *product* of C_1, C_2 is a point in the category $\mathcal{C}''$ of diagrams (p_1, p_2, C) (where $p_1 : C \to C_1$, $p_2 : C \to C_2$) and morphisms of such diagrams. Prove that the category of groups admits products.
3. Prove that if a category has an initial object and arbitrary pushouts, then it has coproducts.
4. Let $f, g : A \to B$ be morphisms in $\mathcal{C}$. A *coequaliser* of f, g is a morphism $c : B \to C$ for some C such that c is φ-universal for the property $cf = cg$. Prove that the categories Set, Top, Grp all admit arbitrary coequalisers.

5. Prove that if a category $\mathcal{C}$ has coproducts and coequalisers, then it has arbitrary pushouts.
6. Let $f : A \to B$, $g : B \to C$ be morphisms in $\mathcal{C}$. Prove that the square

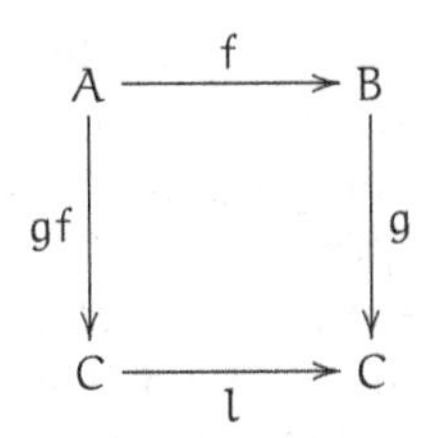

is a pushout if f is epic.
7. Prove that if in diagram (6.6.4) the composite square is a pushout, and g is epic, then the right-hand square is a pushout.
8. Prove that the category of pointed spaces has a coproduct. [This coproduct is called the *wedge* and is written $X \vee Y$.]
9. The *square category* $\square$ has four objects 0, 1, 2, 3 and it has non-identity morphisms $i_1 : 0 \to 1$, $i_2 : 0 \to 2$, $u_1 : 1 \to 3$, $u_2 : 2 \to 3$, $u : 0 \to 3$ with the rule $u_1 i_1 = u_2 i_2 = u$. Prove that this does specify a category, that an object of $\mathcal{C}_\square$ corresponds to a functor $C : \square \to \mathcal{C}$, and that a morphism of squares corresponds to a morphism of functors $\square \to \mathcal{C}$.
10. The *wedge category* $\vee$ has three objects 0, 1, 2 and two non-identity morphisms $i_1 : 0 \to 1$, $i_2 : 0 \to 2$. Prove that a commutative square in a category $\mathcal{C}$ can be regarded as a morphism $\Gamma \to \Gamma'$ of functors $\vee \to \mathcal{C}$ such that Γ' is a constant functor.
11. Let $\Gamma : \mathcal{C} \to \mathcal{D}$ be a functor. A morphism $\eta : \Gamma \to \Gamma'$ of functors is said to be a *colimit* of Γ if (i) $\Gamma' : \mathcal{C} \to \mathcal{D}$ is a constant functor, (ii) η is φ-universal for morphisms of Γ to a constant functor: that is, if $\delta : \Gamma \to \Delta$ is a morphism such that Δ is constant then there is a unique morphism $\delta' : \Gamma' \to \Delta$ such that $\delta'\eta = \delta$. Prove that a pushout square in $\mathcal{C}$ can be regarded as a colimit of a functor $\vee \to \mathcal{C}$.
12. Let $\mathcal{U}$ be an open cover of a space X. Regard $\mathcal{U}$ as a subcategory of Top with objects the elements of $\mathcal{U}$ and morphisms all inclusions $U \to V$ such that $U, V \in \mathcal{U}$. Prove that the inclusion functor $\mathcal{U} \to \mathsf{Top}$ has as colimit a morphism to the constant functor with value X.
13. Given categories $\mathcal{C}$ and $\mathcal{D}$ there is a category whose objects are the morphisms of functors $\mathcal{C} \to \mathcal{D}$—this category is essentially $\mathsf{Fun}(\mathbb{I}^{\to}, \mathsf{Fun}(\mathcal{C}, \mathcal{D}))$ and it is isomorphic to $\mathsf{Fun}(\mathcal{C} \times \mathrm{I}^{\to}, \mathcal{D})$. In any case, the notion of a retract of a morphism of functors is well defined. Prove that a retract of a colimit is again a colimit. [This generalises 6.6.7.]
14. The *dual* of a construction for categories is obtained in a category by carrying out this construction in the opposite category $\mathcal{C}^{op}$ and the transferring the construction to $\mathcal{C}$ by means of the obvious contravariant functor $D : \mathcal{C}^{op} \to \mathcal{C}$. In this way a coproduct $C_1 \sqcup C_2$ in $\mathcal{C}^{op}$ becomes a *product* $C_1 \sqcap C_2$ in $\mathcal{C}$; a pushout in $\mathcal{C}^{op}$ becomes a *pullback* in $\mathcal{C}$; coequaliser becomes *equaliser*; colimit becomes *limit*; initial object becomes *terminal object* and *vice versa*. Write out the definitions in $\mathcal{C}$ of these constructions and discuss their properties.

6.7 The fundamental groupoid of a union of spaces

Throughout this section let X be a topological space and let X_0, X_1, X_2 be subspaces of X such that $X_0 = X_1 \cap X_2$ and the interiors of X_1, X_2 cover X. Our aim is to determine the groupoid πX, and also certain of its full subgroupoids, in terms of the groupoids πX_i, $i = 0, 1, 2$ and the morphisms induced by inclusions. The general interpretation of the theorem which we prove must, however, wait until chapter 9 when the necessary algebraic theory of groupoids has been developed. It is hoped that the reader will by now be so familiar with the concept of pushout that the theorem will have its appeal even with only one example of its use in computations. This example is in any case a crucial example, namely the computation of the fundamental group of the circle.

By 2.5.11, the following square of inclusions

$$\begin{array}{ccc} X_0 & \xrightarrow{i_1} & X_1 \\ {\scriptstyle i_2}\downarrow & & \downarrow{\scriptstyle u_1} \\ X_2 & \xrightarrow[u_2]{} & X \end{array} \qquad (6.7.1)$$

is a pushout in the category of topological spaces. We denote this square by **X**.

Let A be any set. If $i : Y \to X$ is the inclusion of a subspace Y of X, then i induces a morphism

$$\pi Y A \to \pi X A$$

which should be denoted by $\pi i A$, but which we shall denote simply by i. We shall also denote $Pi : PY \to PX$ by i'.

A set A is called *representative* in X if A meets each path-component of X. Our main result is the following.

6.7.2 *If* A *is representative in* X_0, X_1, X_2, *then the square*

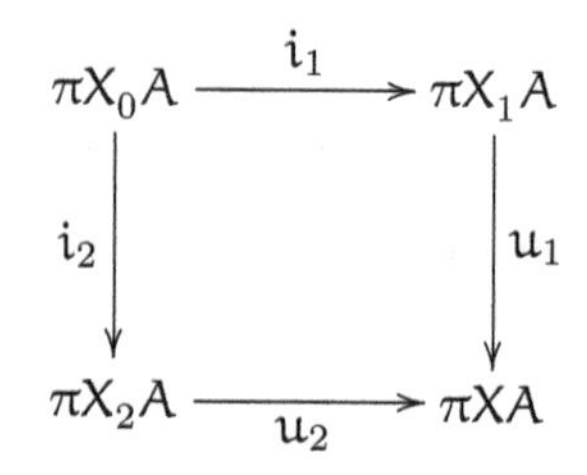

is a pushout in the category of groupoids.

We write the above square of groupoid morphisms as $\pi\mathbf{X}A$.

In using this theorem, it is usual to take X to be path-connected—otherwise the result is used on each path-component of X at a time. Also, we shall assume that A is a subset of X, since any points of A not in X are irrelevant to the theorem. The sort of picture for X_1, X_2 is the following, in which shading is used to distinguish X_1, X_2; points of A are denoted by dots; and the components of X_1, X_2 should be thought of as complicated spaces (for example, a real projective space) rather than the simple subsets of $\mathbb{R}^2$ shown.

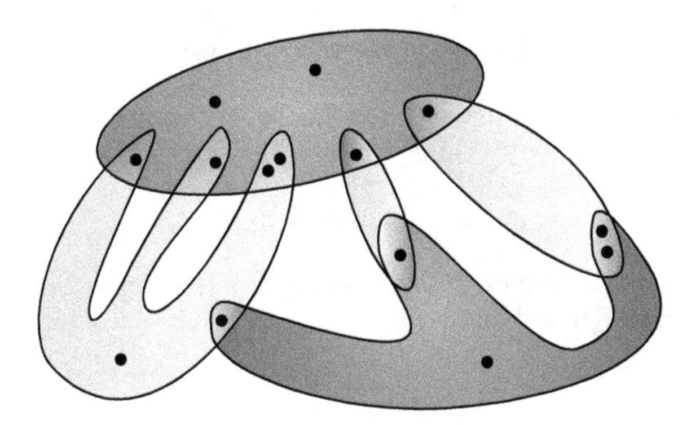

Fig. 6.10

The proof of 6.7.2 is in two parts—the case $A = X$, and the general case. The first case is the only one involving topology.

Proof of 6.7.2—*the case* $A = X$

Consider the following three diagrams:

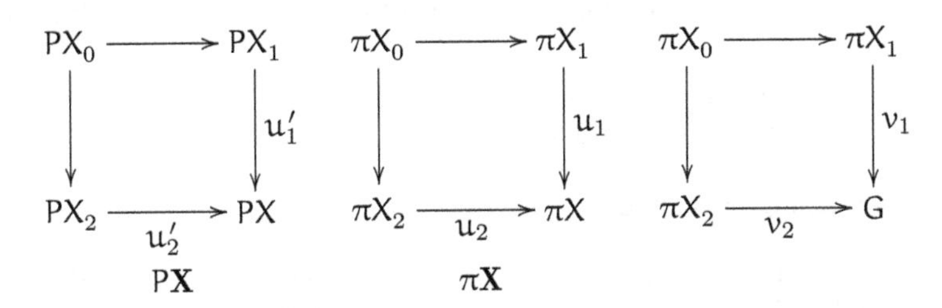

We know that the first two diagrams are commutative. Suppose also that the third is commutative. We wish to prove that there is a unique morphism

$v : \pi X \to G$ such that $vu_1 = v_1$, $vu_2 = u_2$.

Step 1. We know that the projections $p_\lambda : PX_\lambda \to \pi X_\lambda$ $(\lambda = 0, 1, 2)$ and $p : PX \to \pi X$ are morphisms (a word we use here for functor). Let $w_\lambda = v_\lambda p_\lambda$ $(\lambda = 1, 2)$. We use w_1, w_2 to construct a morphism $w : PX \to G$.

Let a be a path of X and suppose first of all that $\mathrm{Im}\, a$ is contained in one or other of X_1, X_2. Then a is $u_1' b_1$ or $u_2' b_2$ for a unique path b_λ in X_λ and we define

$$wa = w_\lambda b_\lambda.$$

This definition is sensible because if $\mathrm{Im}\, a$ is contained in $X_1 \cap X_2 = X_0$, then $b_1 = i_1 b$, $b_2 = i_2 b$ for a path b in X_0 and

$$w_1 b_1 = w_1 i_1 b = w_2 i_2 b = w_2 b_2.$$

Next, suppose a is any path in X. By a corollary of the Lebesgue covering theorem [3.6.4 (Corollary 1)] there is a subdivision

$$a = a_n + \cdots + a_1$$

such that $\mathrm{Im}\, a_i$ is contained in one or other of X_1, X_2 for each $i = 1, \ldots, n$. Then wa_i is well-defined and we set

$$wa = wa_n + \cdots + wa_1.$$

The usual arguments of superimposing subdivisions show that wa is independent of the subdivision, and that if a, b are paths such that $b + a$ is defined then

$$w(b + a) = wb + wa.$$

Clearly $w : PX \to G$ is a morphism, in fact, the only morphism such that

$$wu_1' = w_1, \qquad wu_2' = w_2.$$

It is also clear that w maps constant paths to zeros of G, since both w_1 and w_2 do so.

(This part of the proof does not use the inverse of G. In effect we have proved that $P\mathbf{X}$ is a pushout in Cat.)

We next show that w maps equivalent paths to the same element of G. We know that this is true for w_1, w_2.

Step 2. Consider the rectangle R in $\mathbb{R}^2$ and a map $F : R \to X$ such that $\mathrm{Im}\, F$ is contained in X_λ $(\lambda = 1$ or $2)$. The map F determines on the sides of R paths a, b, c, d [Fig. 6.11], such that $a = u_\lambda' a_\lambda, \ldots, d = u_\lambda' d_\lambda$.

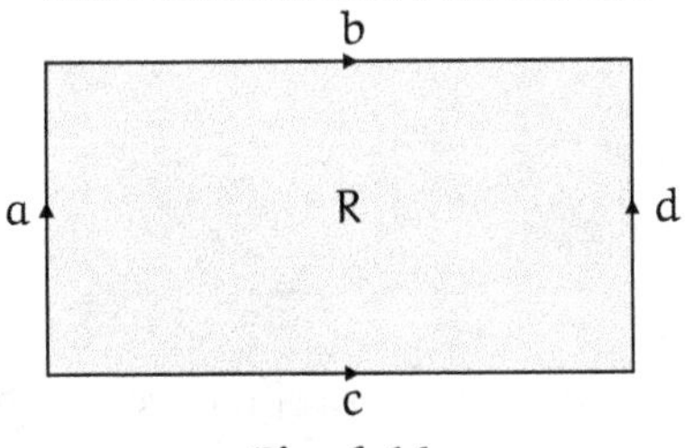

Fig. 6.11

Since R is convex

$$b_\lambda + a_\lambda \sim d_\lambda + c_\lambda$$

whence

$$wb + wa = w_\lambda(b_\lambda + a_\lambda) = w_\lambda(d_\lambda + c_\lambda) = wd + wc.$$

Step 3. This step is a quite simple cancellation argument.

Let a, b be paths in X and let $F : [0, r] \times \mathbb{I} \to X$ be a homotopy $a \sim b$. The Lebesgue covering lemma shows that we can, by a grid composed of lines

$$\{(ri/n, t) : t \in \mathbb{I}\}, \quad i = 0, \dots, n$$
$$\{(s, j/n) : s \in [0, r]\}, \quad j = 0, \dots, n$$

subdivide $[0, r] \times \mathbb{I}$ into rectangles so small that each is mapped by F into X_1 or X_2. Let a_j be the path $s \mapsto F(s, j/n)$

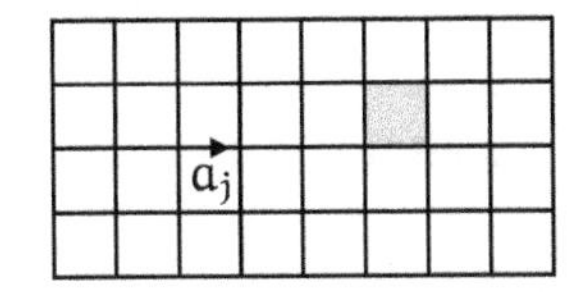

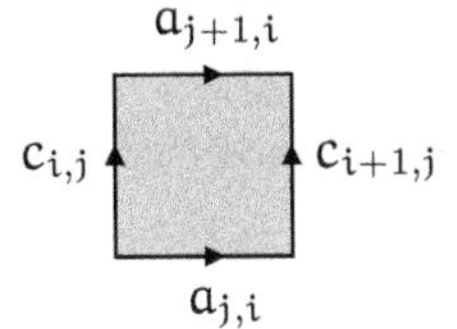

Fig. 6.12

so that $a_0 = a$, $a_n = b$. The vertical lines determine a subdivision

$$a_j = a_{j,n-1} + \cdots + a_{j,0}$$

and also paths $c_{i,j} : t \mapsto F(ri/n, t + j/n)$ such that

$$a_{j+1,i} + c_{i,j} \sim c_{i+1,j} + a_{j,i}.$$

Therefore

$$
\begin{aligned}
wa_j &= \sum_{i=0}^{n-1} wa_{j,i} \\
&= \sum_{i=0}^{n-1} \{-wc_{i+1,j} + wa_{j+1,i} + wc_{i,j}\} \quad \text{by Step 2} \\
&= -wc_{n,j} + \sum_{i=0}^{n-1} wa_{j+1,i} + w_{c_{0,j}} \\
&= wa_{j+1}
\end{aligned}
$$

the first and last terms being zero since $c_{0,j}, c_{n,j}$ are constant paths. Hence by induction $wa = wb$.

Step 4. Let a, b be equivalent paths in X. Then there are constant paths r, s such that $r + a$ is homotopic to $s + b$. It follows that

$$wa = w(r + a) = w(s + b) = wb.$$

Thus w maps equivalent paths to the same element of G and so defines a morphism $v : \pi X \to G$ such that $vp = w$. Thus for any path a in X_λ $(\lambda = 1, 2)$

$$
\begin{aligned}
vu_\lambda(\operatorname{cls} a) &= v \operatorname{cls} u'_\lambda a = wu'_\lambda a = w_\lambda a \\
&= v_\lambda(\operatorname{cls} a)
\end{aligned}
$$

whence

$$vu_\lambda = v_\lambda.$$

Step 5. Let $v' : \pi X \to G$ be any morphism such that

$$v' u_\lambda = v_\lambda, \quad \lambda = 1, 2.$$

Then

$$
\begin{aligned}
v' p u'_\lambda &= v' u_\lambda p_\lambda \\
&= v_\lambda p_\lambda \\
&= w_\lambda .
\end{aligned}
$$

Since $v'p$ is a morphism it follows that

$$v'p = w = vp.$$

Since $p : PX(x, y) \to \pi X(x, y)$ is surjective for each x, y in X, it follows that

$$\nu' = \nu.$$

□

Proof of 6.7.2—*the general case*

For this it is sufficient, by 6.6.7, to prove that the square $\pi\mathbf{X}A$ is a retract of $\pi\mathbf{X}$. We do this by constructing retractions $r_\lambda : \pi X_\lambda \to \pi X_\lambda A$ ($\lambda = 0, 1, 2$ or) consistent with the various morphisms induced by inclusion.

Since A is representative in X_0 we can choose for each x in X_0 a path class $\theta_0 x$ in πX_0 from some point of $A \cap X_0$ to x. Since A is representative in X_1 and in X_2 we can choose for each x in $X_\lambda \setminus X_0$ ($\lambda = 1, 2$) a path class $\theta_\lambda x$ in πX_λ from some point of $A \cap X_\lambda$ to x, and we extend θ_λ over all of X_λ by setting $\theta_\lambda x = i_\lambda \theta_0 x$ for x in X_0. Finally, for any x in X we define

$$\theta x = \begin{cases} u_1\theta_1 x, & \text{if } x \in X_1 \\ u_2\theta_2 x, & \text{if } x \in X_2. \end{cases}$$

Then θ is well-defined. Further, θ_λ ($\lambda = 0, 1, 2,$) defines, by 6.5.13, a retraction $r_\lambda : \pi X_\lambda \to \pi X_\lambda A$; these morphisms define a retraction $r : \pi\mathbf{X} \to \pi\mathbf{X}A$. □

As will be clear from chapter 9, 6.7.2 can often be interpreted to give information on the fundamental groups themselves. For example, if X_0, X_1, X_2 are path-connected, then we can take $A = \{x_0\}$ where x_0 is a point of X_0, and then 6.7.2 determines completely the fundamental group $\pi(X, x_0)$. But we shall want to use 6.7.2 when X_0 at least is not path-connected, and in this case it is useful to carry out a further retraction. The main lemma for this purpose is the following.

6.7.3 *Let $\mathcal{C}, \mathcal{D}$ be categories and $f : \mathcal{C} \to \mathcal{D}$ a functor such that* $\mathrm{Ob}(f)$ *is injective. Then any full representative subcategory $\mathcal{C}'$ of $\mathcal{C}$ gives rise to a pushout square*

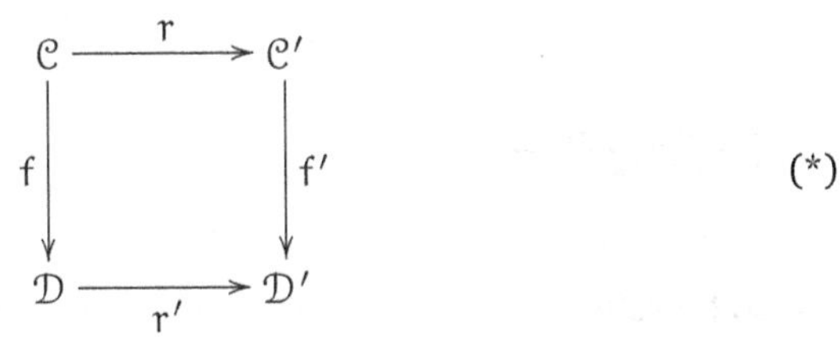

in which f' is a restriction of f and r, r' are deformation retractions.

Proof Let $\mathcal{D}'$ be the full subcategory of $\mathcal{D}$ whose objects are those equal to fx for x in $\mathrm{Ob}(\mathcal{C}')$, and those not equal to fx for any x in $\mathrm{Ob}(\mathcal{C})$; that is

$$\mathrm{Ob}(\mathcal{D}') = f[\mathrm{Ob}(\mathcal{C}')] \cup (\mathrm{Ob}(\mathcal{D}) \setminus f[\mathrm{Ob}(\mathcal{C})]).$$

Then we have a commutative diagram

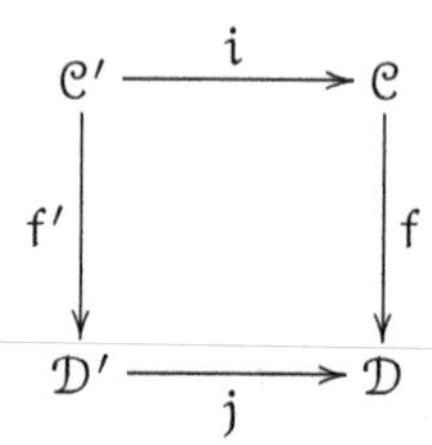

in which i, j are inclusions and f' is a restriction of f. We choose a deformation retraction $r : \mathcal{C} \to \mathcal{C}'$ and a homotopy function $\theta : ir \simeq 1 \operatorname{rel} \mathcal{C}'$ as is possible by 6.5.13.

For each y in $\mathrm{Ob}(\mathcal{D})$ let $\varphi y = f\theta x$ if $y = fx$ and otherwise let $\varphi y = 1_y$. Thus $\varphi f = f\theta$; φy is invertible, has initial point in $\mathrm{Ob}(\mathcal{D}')$ and has final point y; further, $\varphi y = 1_y$ if $y \in \mathrm{Ob}(\mathcal{D}')$. It follows that φ is a homotopy function $jr' \simeq 1 \operatorname{rel} \mathcal{D}'$ where $r' : \mathcal{D} \to \mathcal{D}'$ is a deformation retraction.

If $a \in \mathcal{C}(x, x')$ then

$$\begin{aligned} f'r(a) &= f((\theta x')^{-1} a(\theta x)) \\ &= (\varphi f x')^{-1} f a(\varphi f x) \\ &= r' f(a). \end{aligned}$$

Therefore $f'r = r'f$ and the square (*) commutes.

To prove (*) to be a pushout suppose there is given a commutative diagram

$$\begin{array}{ccc} \mathcal{C} & \xrightarrow{r} & \mathcal{C}' \\ {\scriptstyle f}\downarrow & & \downarrow{\scriptstyle u} \\ \mathcal{D} & \xrightarrow[v]{} & \mathcal{E} \end{array}$$

If there is a functor $w : \mathcal{D}' \to \mathcal{E}$ such that $wr' = v$ then

$$w = wr'j = vj$$

and so there is at most one such w. On the other hand, let $w = vj$. Then

$$wf' = vjf' = vfi = uri = u.$$

Further φy, and hence also $\nu\varphi y$, is the identity if $y \neq fx$ for any x, while if $y = fx$ then

$$\nu\varphi f(x) = \nu f\theta(x) = ur\theta(x) = u(1) = 1.$$

It follows that $\nu\varphi$ is the constant homotopy function and so

$$wr' = \nu jr' = \nu.$$

This proves that the square is a pushout. □

We apply this result to the situation at the beginning of this section. Thus we are given subspaces X_1, X_2 of X whose interiors cover X. Further $X_0 = X_1 \cap X_2$, and A is a subset of X representative in X_0, X_1, X_2. Note that if X is path-connected then each path component of one of the spaces X_1, X_2 meets the other space [Exercise 8 of Section 3.4]—it is then common and convenient to take A as one point in each path-component of X_0, but this assumption is not essential. The main point is that A may be chosen in a way appropriate to the geometry of the situation. This is the advantage of using the groupoid πXA rather than just the fundamental group at some point.

6.7.4 *Let A' be a subset of $A \cap X_1$ representative in X_1 and let $A_1 = A' \cup (A \setminus X_1)$. Then there is a pushout diagram*

$$\begin{array}{ccccc} \pi X_0 A & \xrightarrow{i_1} & \pi X_1 A & \xrightarrow{r} & \pi X_1 A_1 \\ \downarrow{\scriptstyle i_2} & & & & \downarrow{\scriptstyle u_1} \\ \pi X_2 A & \xrightarrow[u_2]{} & \pi XA & \xrightarrow[r']{} & \pi XA_1 \end{array} \qquad (*)$$

in which r, r' are deformation retractions.

Proof If we insert the induced morphism $\pi X_1 A \to \pi XA$ in (*) we obtain two squares. The left-hand square is a pushout by 6.7.2. To prove the right-hand square a pushout we use 6.7.3 with the substitutions $\mathcal{C} = \pi X_1 A$, $\mathcal{D} = \pi XA$, $\mathcal{C}' = \pi X_1 A' = \pi X_1 A_1$. Then we have

$$\mathrm{Ob}(\mathcal{D}') = A' \cup (A \setminus (A \cap X_1)) = A' \cup (A \setminus X_1) = A_1.$$

Finally, (*) itself is a pushout, since it is the composite of two pushouts. □

The result will be used in chapter 9 to prove the Van Kampen theorem on the fundamental group of an adjunction space. The interpretation of that theorem in general requires a lot of preliminary algebra. Here we finish this chapter by showing that 6.7.4 enables us to determine the fundamental group of a circle—this gives us our first simple example of a path-connected but not simply-connected space.

6.7.5 *There is an isomorphism*

$$\pi(\mathbb{S}^1, 1) \simeq \mathbb{Z}.$$

Proof

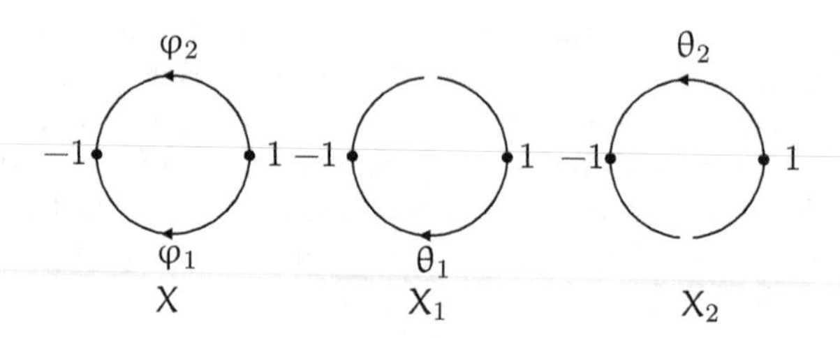

Fig. 6.13

We use complex number notation. Let $X_1 = \mathbb{S}^1 \setminus \{i\}$, $X_2 = \mathbb{S}^1 \setminus \{-i\}$, $A = \{-1, 1\}$, $A_1 = \{1\}$. Then X_1, X_2 are simply connected (they are both homeomorphic to $]0, 1[$) while X_0 is the topological sum of two simply-connected components. Therefore $\pi X_2 A$ is isomorphic to the groupoid $\mathbf{I}$ while $\pi X_0 A$ is isomorphic to the discrete groupoid $\{0, 1\}$. Thus by 6.7.4 we have a pushout

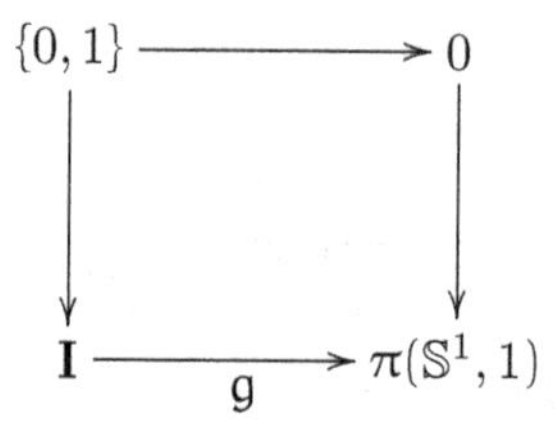

in which 0 is the trivial group and g is specified completely as a morphism by the fact that

$$g(\iota) = \varphi, \qquad \text{say.}$$

The above pushout implies the following: *if* $f : \mathbf{I} \to K$ *is any morphism to a group* K, *then there is a unique morphism* $h : \pi(\mathbb{S}^1, 1) \to K$ *such that* $hg = f$. In particular, letting $f : \mathbf{I} \to \mathbb{Z}$ be the morphism such that $f(\iota)$ is the element 1 of $\mathbb{Z}$, there is a unique morphism $h : \pi(\mathbb{S}^1, 1) \to \mathbb{Z}$ such that $h(\varphi) = 1$.

Let $k : \mathbb{Z} \to \pi(\mathbb{S}^1, 1)$ be the morphism $n \mapsto n\varphi$. Clearly $hk(1) = 1$ and so $hk = 1 : \mathbb{Z} \to \mathbb{Z}$. On the other hand,

$$khg(\iota) = kf(\iota) = k(1) = \varphi = g(\iota).$$

Therefore $khg = f$ and so, by the uniqueness part of the universal property, $kh = 1 : \pi(\mathbb{S}^1, 1) \to \pi(\mathbb{S}^1, 1)$. □

The proof also shows that φ is a generator of $\pi(\mathbb{S}^1, 1)$. In order to determine φ, let θ_i be the unique element of $\pi X_i(1, -1)$ $(i = 1, 2)$ and let $\varphi_i = u_i \theta_i$ in $\pi\mathbb{S}^1(1, -1)$. The retraction r' satisfies

$$r'\varphi_2 = -\varphi_1 + \varphi_2$$

and so if we take the isomorphism $\pi X_2 A \to \mathbf{I}$ to be that which sends θ_2 to ι we deduce that

$$\varphi = -\varphi_1 + \varphi_2.$$

Clearly φ is the class of the path $t \mapsto e^{2\pi i t}$ of length 1.

6.7.5 *(Corollary 1)* $\mathbb{S}^1$ is not a retract of $\mathbb{E}^2$.

Proof If $r : \mathbb{E}^2 \to \mathbb{S}^1$ were a retraction, then so also would be $r : \pi\mathbb{E}^2(1) \to \pi\mathbb{S}^1(1)$. Since a retraction is surjective this is clearly impossible. □

EXERCISES

1. Prove that the fundamental group of the torus $\mathbb{S}^1 \times \mathbb{S}^1$ is isomorphic to $\mathbb{Z} \times \mathbb{Z}$. Prove also that the two maps $\mathbb{S}^1 \to \mathbb{S}^1 \times \mathbb{S}^1$ which send $x \mapsto (x, a)$, $x \mapsto (a, x)$ respectively are not homotopic.
2. The coproduct in the category of groups exists [cf. chapter 8]—it is called the free product and it written $*$. Prove that the fundamental group of $\mathbb{S}^1 \vee \mathbb{S}^1$ is isomorphic to $\mathbb{Z} * \mathbb{Z}$.
3. Suppose there is given the commutative square of 6.7.3 in which r, s are deformation retractions and f' is the restriction of f. Suppose also that θ, φ are homotopy functions $ir \simeq 1 \operatorname{rel} \mathcal{C}'$, $js \simeq 1 \operatorname{rel} \mathcal{D}'$ respectively. Prove that the square is a pushout if the following condition holds: if $y \in \mathrm{Ob}(\mathcal{D})$ and $\varphi y \neq 1$, then there are objects $x_1, \dots x_n$ of $\mathcal{C}$ such that $\varphi y = (f\theta x_n) \cdots (f\theta x_1)$.
4. Suppose that in the situation of 6.7.2, A_1, A_2 are subsets of A representative in X_1, X_2 respectively, and that $A \cap X_0 = A_1 \cap A_2$. Prove that there is a pushout square

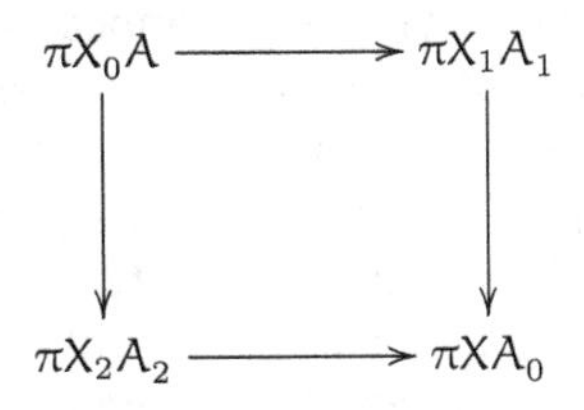

5. Prove that $\mathbb{S}^{n-1}$ is a retract of $\mathbb{E}^n$ if and only if there is a map $\mathbb{E}^n \to \mathbb{E}^n$ without fixed points.

6. Let the open cover $\mathcal{U}$ of the space X be regarded as a category as in Exercise 12 of Section 6.6. The fundamental groupoid functor restricts to a functor $\pi \mid \mathcal{U} : \mathcal{U} \to \mathsf{Grpd}$. Prove that if $\mathcal{U}$ has a subcover $\mathcal{V}$ such that $\mathcal{U}$ consists of all intersections $V \cap V'$ for V, V' in $\mathcal{V}$, then the functor $\pi \mid \mathcal{U}$ has as colimit a morphism to the constant functor with value πX.

7. [This exercise requires knowledge of ordinals and transfinite induction.] Let $\mathcal{U}$ be an open cover of X such that the intersection of two elements of $\mathcal{U}$ belongs to $\mathcal{U}$. We say $\mathcal{U}$ is *stratified* if there is a function f from $\mathcal{U}$ to the ordinals such that $fU < fU'$ whenever U is a proper subset of U'. Prove $\mathcal{U}$ is stratified if $\mathcal{U}$ is finite or is well-ordered by inclusion. Let $\mathcal{U}$ be stratified, and let A be a subset of X representative in each U of $\mathcal{U}$. Prove that the functor $\mathcal{U} \to \mathsf{Grpd}$, $U \mapsto \pi UA$ has as colimit a morphism to the constant functor with value πXA. [Use the previous exercise and Exercises 12, 13 of Section 6.6. See [BR84] for a better result.]

8. Let σX be the semi-fundamental groupoid of X as constructed in Exercise 7 of Section 6.2. Investigate whether or not theorems corresponding to 6.7.2, 6.7.4 hold for σ.

9. Prove that the coproduct of trivial groups is trivial. Deduce that in the situation of 6.7.2, if X_0, X_1, X_2 are 1-connected, then so also is $X = X_1 \cup X_2$.

10. Formulate the notions of *commutative square* in a category or groupoid, and of the compositions of commutative squares in two directions. Prove that any composition of commutative squares is commutative. Use this to give another exposition of Step 3 of the proof of 6.7.2.

NOTES

The Bibliography gives a number of books on category theory which will enhance the introductory account given here. The reader is urged to look at the original paper defining categories ([EM45]), and to read the Presidential address [Mac65].

Groupoids have been known since [Bra26]. The paper [Bro87] gives a survey of their widespread use in mathematics, which includes group theory, differential topology, ergodic theory, differential topology, algebraic geometry, functional analysis, homotopy theory, Galois theory, and others areas. Thus there is evidence that the groupoid concept is an extension of the notion of group which gives considerably more flexibility and power without any loss. Such a view has not yet appeared in many texts.

The fundamental group of a union $K \cup L$ of simplicial complexes K, L such that K, L and $K \cap L$ are all connected was described in terms of generators and relations in [Sei31]. The fundamental group of a kind of adjunction space was given by [Kam33] (apparently not knowing of Seifert's paper); Van Kampen stated results more general than those of Seifert, since he dealt also with the case of non-connected intersection, but the name Van Kampen, or Seifert-Van Kampen, has usually been applied to any theorem on the fundamental group of a union of spaces. The proofs of the results in Van Kampen's paper are difficult to follow, partly because he did not have the language to describe properly the situations in which he was interested, and to use their formal properties.

The modern version of the theorem and proof for the fundamental group was given in [Cro59]. By this time, the notion of universal property was well used and understood. A version in terms of non-abelian cohomology was given in [Olu58], and this was generalised to the non-connected case in [Bro65]. Another cohomological formulation for a general non-connected union is given in [Wei61]. The paper [BHK83] derives the main result of section 6.7 using non-abelian cohomology with coefficients in a groupoid.

The results of section 6.7 and the general Van Kampen theorem given in section 9.1 are taken from [Bro65]. A general formulation of the theorem for the fundamental groupoid of infinite unions is given in [Raz76] and with a different proof in [BR84]. This proof goes back in spirit to the proof of [Cro59].

As explained in the Preface to the second edition of this book, writing up this material on the fundamental groupoid suggested that there should be a higher dimensional version of Van Kampen's theorem. The intuitive basis for this idea is explained in [Bro82]. To express this theorem, and to prove it, one needs various notions of *higher homotopy groupoids.* One form of these is the ω-groupoids of [BH78] and [BH81]; another is the cat^n-groups of [Lod82] which were used by [BL87] to formulate and prove a Van Kampen theorem for cubical diagrams of spaces, with powerful applications. It was possibly a general reluctance to make the extension from groups to groupoids that prevented models of these kinds being discovered earlier. On the other hand, whereas the proof of the Van Kampen theorem for filtered spaces in [BH81] does follow the intuitive lines of the proof in dimension 1, a number of new ideas are needed to make the methods work in all dimensions. The main problem is that whereas it is clear what is meant by the composite of the edges of a square, it is by no means clear what is meant by the composite of the faces of a cube. That is, the problem is to generalise Steps 2 and 3 of the proof of 6.7.2. For this reason, the general case needs considerable algebraic development of the theory of ω-groupoids. The more powerful theorem for n-cubes of spaces given in [BL87] uses in its proof a gamut of techniques from algebraic topology.

Chapter 7

Cofibrations

In the last chapter, we introduced the notion of homotopy type of spaces. This is a coarser notion than homeomorphism type—more spaces are of the same homotopy type than are of the same homeomorphism type. Hence in a given class of spaces there are fewer homotopy types than homeomorphism types and so the determination of homotopy types can be expected to be easier. In rough terms, although we will not be able to justify this statement here, by passing to homotopy classes the classification problem is in many cases reduced from an uncountable problem to a countable one. Thus there is a countable set of homotopy types of (finite) 2-dimensional cell complexes but an uncountable set of homeomorphism types.

The obvious question arises: how do we determine whether or not two spaces are of the same homotopy type? This is a difficult problem, and even for simply connected cell complexes of dimension 5 it is a problem of current research (cf. [Bau88]). There are two useful techniques.

First, to prove X, Y are *not* of the same homotopy type we construct homotopy type invariants. For example $\mathbb{S}^1$ is not of the homotopy type of a point because $\mathbb{S}^1$ has fundamental group $\mathbb{Z}$, which is not isomorphic to the trivial group. On the other hand $\mathbb{S}^2$ and $\mathbb{S}^3$ both have trivial fundamental group, yet are not of the same homotopy type. The proof of this requires other invariants, for example homology groups, with which we do not deal in this book.

Second, to prove X and Y are of the same homotopy type, we must construct a homotopy equivalence $X \to Y$. For this, we need to know something about X and Y. A common case is that X and Y are given as repeated adjunction spaces, and because of this we take as our main aim the discussion of the homotopy type of adjunction spaces. The main tool used is a *gluing theorem* for homotopy equivalences. To prove this theorem we need the notion of *cofibration*, and to conceptualise the proof it is convenient to use the *track groupoid* and the *operations* of this groupoid. These operations are important in their own right. We obtain these operations from the more general situation of a *fibration of groupoids*. This has the advantage of applicability in other situations.

7.1 The track groupoid

For those who have studied section 5.9, which topologises the space Y^X of continuous functions $X \to Y$, we can define the track groupoid πY^X simply as the fundamental groupoid of the space Y^X, and this gives us a theory for the category of k-spaces equivalent to that exposed below. However, to give an independent version, we define the track groupoid analogously to the fundamental groupoid, but replacing paths by homotopies.

We refer the reader back to section 6.5 for the notion of a homotopy $F : f \simeq g$ of length q, where f and g are maps $X \to Y$ and $q \geqslant 0$. Then we define $-F : g \simeq f$ by $(-F)(x,t) = F(x, q-t)$. If further $G : g \simeq h$ is a homotopy of length r, then we define $G + F : f \simeq h$ of length $q + r$ by

$$(G+F)(x,t) = \begin{cases} F(x,t) & \text{if } 0 \leqslant t \leqslant q, \\ G(x, t-q) & \text{if } q \leqslant t \leqslant q+r. \end{cases}$$

Note that $G+F$ is continuous by 2.5.12 since $X \times [0,q]$ and $X \times [q, q+r]$ are closed subspaces of $X \times [0, q+r]$. There is also a constant homotopy $f \simeq f$ of length 0 for any map f. So, analogously to the case of paths, we obtain a category PY^X whose objects are the maps $X \to Y$ and whose morphisms from f to g are the homotopies, of various lengths, from f to g.

We now wish to construct from this category a groupoid. In the case of paths we used the notion of 'homotopic rel end points'. In the present case, we need the notion of two homotopies $F, G : f \simeq g$ of the same length q being *homotopic rel end maps*, which means that there is a map

$$\mathcal{H} : X \times [0,q] \times \mathbb{I} \to Y$$

such that for all x in X, s in $[0,q]$ and t in $\mathbb{I}$

$$\mathcal{H}(x,s,0) = F(x,s) \qquad \mathcal{H}(x,s,1) = G(x,s)$$
$$\mathcal{H}(x,0,t) = fx \qquad \mathcal{H}(x,q,t) = gx.$$

This amounts to saying that if $\mathcal{H}_t$ denotes the map $(x,s) \mapsto \mathcal{H}(x,s,t)$, then $\mathcal{H}_0 = F$, $\mathcal{H}_1 = G$ and each $\mathcal{H}_t$ is a homotopy $f \simeq g$ of length q.

We now define a *constant homotopy of length* q to be a map $C : X \times [0, q] \to Y$ such that $C(x, 0) = C(x, t)$ for all x in X, t in $[0, q]$. Such a homotopy is denoted ambiguously by q. Then two homotopies F and G are *equivalent* if there are real numbers q, r such that $q + F$, $r + G$ are homotopic rel end maps. This is an equivalence relation, and it is shown by methods only formally different from those of chapter 6 that these equivalence classes form a groupoid. This groupoid we call a *track groupoid* and write it

$$\pi Y^X.$$

The components of this groupoid are the homotopy classes of maps $X \to Y$. The set of these homotopy classes is written

$$[X, Y].$$

We shall make essential use of the fact that these objects depend *functorially* on X and on Y. This means that if $h : Y \to Z$ and $i : A \to X$ are maps, then there are *induced* morphisms of groupoids

$$h_* : \pi Y^X \to \pi Z^X, \qquad i^* : \pi Y^X \to \pi Y^A$$

defined as follows. If $f : X \to Y$ is a map, and so an object of πY^X, then

$$h_*(f) = hf, \qquad i^*(f) = fi.$$

If $F : X \times [0, q] \to Y$ is a homotopy representing an element $\operatorname{cls} F$ of πY^X, then we set

$$h_*(\operatorname{cls} F) = \operatorname{cls} hF, \qquad i^*(\operatorname{cls} F) = \operatorname{cls} F(i \times 1),$$

where $i \times 1 : A \times [0, q] \to X \times [0, q]$ is given by $(a, t) \mapsto (ia, t)$. It is easy to check that h_* and i^* are well-defined morphisms of groupoids. Further, the following functorial rules are satisfied: if also $g : Z \to W$ and $j : B \to A$, then

$$(gh)_* = g_* h_*, \qquad (ij)^* = j^* i^*,$$

while if h is the identity, so also is h_*, and if i is the identity, so also is i^*.

Spaces and groupoids each have a notion of homotopy. It is natural therefore to ask how the above constructions of h_* and i^* behave with regard to homotopies.

7.1.1 *If* $h \simeq k : Y \to Z$, *then* $h_* \simeq k_* : \pi Y^X \to \pi Z^X$. *If* $i \simeq j : A \to X$, *then* $i^* \simeq j^* : \pi Y^X \to \pi Y^A$.

Proof Let $F : h \simeq k$, $G : i \simeq j$ be homotopies, and suppose for convenience that both homotopies are of length 1. The homotopy $F_* : h_* \simeq k_*$ is defined on an object f of πY^X, i.e. for a map $f : X \to Y$, to be the class of the homotopy $(x, t) \mapsto F(f \times 1)$. That is,

$$F_*(f) = \mathrm{cls}\, F(f \times 1).$$

To prove that F_* is a homotopy of morphisms of groupoids, we have to show that if H is a homotopy $f \simeq g$, then the homotopies $kH + F(f \times 1)$ and $F(g \times 1) + hH$ are homotopic rel end maps.

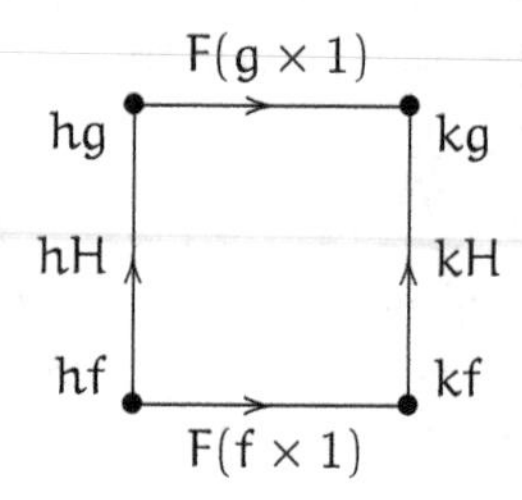

Define $\mathcal{G} : X \times \mathbb{I} \times \mathbb{I} \to Z$ by $\mathcal{G}(x, s, t) = F(H(x, s), t)$. Then

$$\mathcal{G}(x, s, 0) = h(H(x, s)), \qquad \mathcal{G}(x, s, 1) = k(H(x, s)),$$
$$\mathcal{G}(x, 0, t) = F(fx, t), \qquad \mathcal{G}(x, 1, t) = F(gx, t).$$

Of course $\mathcal{G}$ itself is not the required homotopy, but we can easily manufacture the right one from $\mathcal{G}$. We subdivide a square in two ways as shown in Fig. 7.1 and define a map $\varphi : \mathbb{I} \times \mathbb{I} \to \mathbb{I} \times \mathbb{I}$ by mapping the two parts of the left hand square to the two triangles of the right hand square so that the solid thick lines are shrunk to the two vertices, while the middle vertical line is mapped to the diagonal. Let $\mathcal{H} = \mathcal{G}(1 \times \varphi)$. Then $\mathcal{H}$ is the required homotopy.

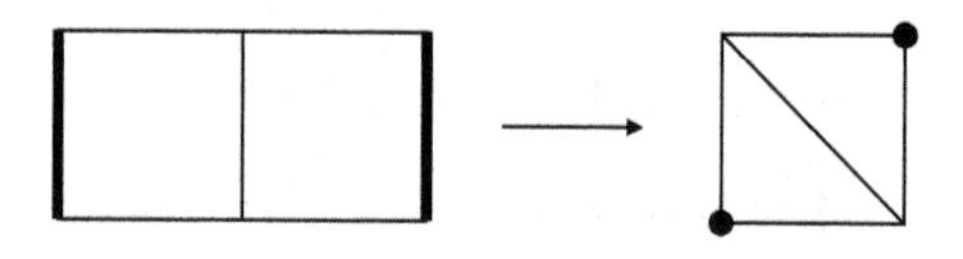

Fig. 7.1

The proof that $i^* \simeq j^*$ is similar and is left to the reader. □

We now give some motivation for our next constructions and definitions.

Many geometric problems can be translated into *factorisation problems for maps*. A typical such factorisation problem is the following. Suppose given three spaces A, X, Y and two out of three of the maps in the diagram:

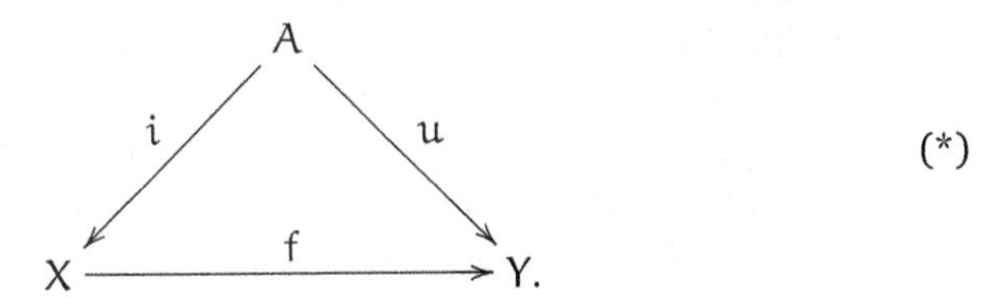

The problem is to decide whether or not the third map can exist.

An example of this type of problem is to decide if a subspace A of X is a retract of X. Then in diagram (*) we take i to be the inclusion, $Y = A$, and u to be the identity. A retraction $X \to A$ is simply a map f such that $fi = 1$. For example, we can ask: is the n-sphere $\mathbb{S}^n$ a retract of the n-cell $\mathbb{E}^{n+1}$? The answer is: no. For $n = 0$, the proof is easy, since $\mathbb{E}^1$ is connected and $\mathbb{S}^0$ is not. For $n = 1$, a proof can be given by considering the diagram

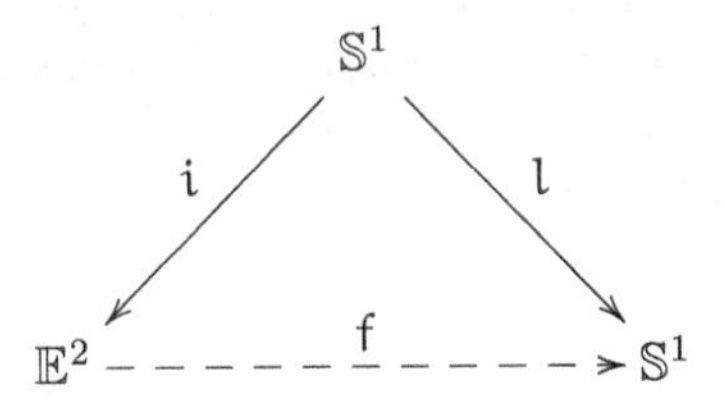

and applying the fundamental group functor for some base point. This leads to a diagram of morphisms of groups

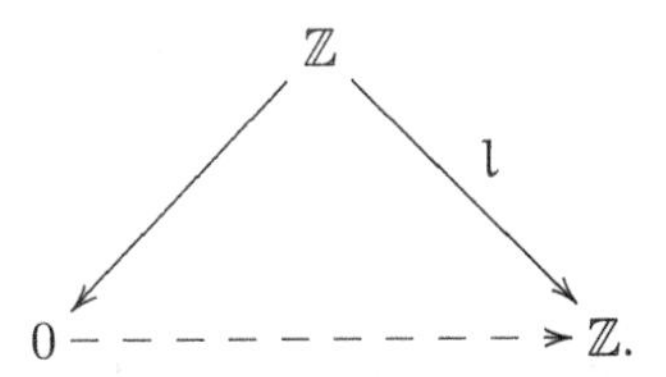

Clearly there is no dotted morphism making this diagram commute, and so the retraction $f : \mathbb{E}^2 \to \mathbb{S}^1$ cannot exist. There is a similar proof, using the nth homology group, for the general case—for details, see any book on the homology theory of spaces, for example [HW60], [Mas67], [Mun75]. Note that the intuitive interpretation of this result for $n = 1$ is that the skin of a drum cannot be deformed to the rim of the drum, without breaking the skin.

It is a cliché that such a negative result, namely the non-existence of a retraction, can be used to deduce positive results. In this case we can deduce the *Brouwer fixed point theorem: Every map* $g : \mathbb{E}^{n+1} \to \mathbb{E}^{n+1}$ *has a fixed point.* For suppose g does not have a fixed point, that is, there is no x in $\mathbb{E}^{n+1}$ such that $gx = x$. For each x in $\mathbb{E}^{n+1}$ let fx be the point in $\mathbb{S}^n$ such that fx, x and gx lie in that order on a line. We leave the reader to prove that f is well defined and continuous. But then f is a retraction $\mathbb{E}^{n+1} \to \mathbb{S}^n$, which we already know cannot exist. So g must have a fixed point. This proves the Brouwer Fixed Point Theorem. This theorem has a range of applications in mathematics, because it is possible to transform many useful existence problems into the problem of the existence of a fixed point of some map.

Now we get back to our main aims. We wish to study in diagram (*) the case where i and u are given, and the problem is the existence of the map f. This is in general a difficult problem, and we follow the standard practice in mathematics of introducing some simplifying procedure. In this case, we fix i, and then ask: *is the existence of* f *independent of the choice of* u *in its homotopy class?* That is, if $fi = u$ and $u \simeq v$, must there be a g such that $gi = v$? In order to study this question, and to use the relevant condition on i, it is convenient to introduce some broader considerations.

Let Top denote the category of topological spaces and continuous maps. Let A be a topological space. The category Top^A of *topological spaces under* A has objects the maps $i : A \to X$ for all topological spaces X. Since we keep A fixed in what follows, we can where convenient write such a map as (X, i). In the category Top^A a *morphism* $f : (X, i) \to (Y, u)$ consists of the pairs $(X, i), (Y, u)$ and a map $f : X \to Y$ such that $fi = u$. Thus it is common to write a morphism in Top^A as a commutative diagram

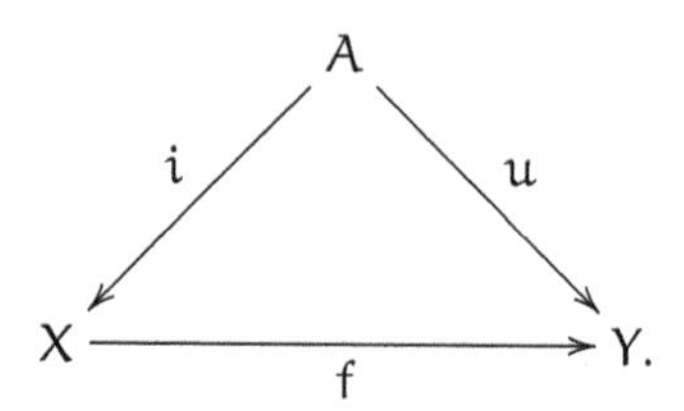

We refer to f as a *map under* A. The composition in Top^A is the obvious one: the composite of $f : (X, i) \to (Y, u)$ and $g : (Y, u) \to (Z, w)$ is $gf : (X, i) \to (Z, w)$. It is clear that Top^A becomes in this way a category.

We now define homotopy in Top^A. Let $f, g : (X, i) \to (Y, u)$ be maps under A. A *homotopy under* A is a homotopy $F : f \simeq g$ of maps of spaces such that $F(ia, t) = ua$ for all $a \in A$ and $t \in \mathbb{I}$. If i is an inclusion, it is common in the literature to refer to F also as a homotopy rel A, and we shall often use this terminology.

Many of the previous constructions on homotopies can also be carried out for homotopies under A. In particular, homotopy under A is an equivalence relation on the set of maps $(X, i) \to (Y, u)$ under A; the set of homotopy classes is written

$$[(X, i), (Y, u)].$$

As suggested above, one of our major tools will be an analysis of how the set $[(X, i), (Y, u)]$ depends on the homotopy class of u. For example, it will be useful to know whether a homotopy $\alpha : u \simeq v$ determines a bijection

$$\alpha_{\#} : [(X, i), (Y, u)] \to [(X, i), (Y, v)].$$

If this is so, then the domain of $\alpha_{\#}$ is non-empty if and only if its range is non-empty, and so the existence of a factorisation of u through i will depend only on the homotopy class of u. We will impose on i the condition of being a *cofibration*, and this will imply the existence of the bijection as above. In order to apply the cofibration condition, it is convenient to define a particular kind of morphism of groupoids called a *fibration*. These will be studied in the next section.

We can use homotopies under A to define a *track groupoid under* A which we shall write

$$\pi(Y, u)^{(X,i)}.$$

We shall not make too much use of this groupoid, but there is one special case which it is worth relating to a body of standard work in homotopy theory.

We write a singleton space as $\cdot$. Let X and Y be pointed spaces. Then we have unique pointed maps $i : \cdot \to X$, $u : \cdot \to Y$ mapping the element of $\cdot$ to the respective base points. A map $X \to Y$ under $\cdot$ is simply a pointed map. A homotopy under A of pointed maps is called a *pointed homotopy*. We write

$$\pi^{\cdot} Y^X$$

for the track groupoid $\pi(Y, u)^{(X,i)}$. We abbreviate $[(X, i), (Y, u)]$ to $[X, Y]_{\cdot}$. The constant pointed map $X \to Y$ is also written $\cdot$.

Recall that ΣX is the reduced suspension of a pointed space X (cf. Section 5.8).

7.1.2 *There is a bijection*

$$(\pi^{\cdot} Y^X)(\cdot) \to [\Sigma X, Y]_{\cdot}.$$

Hence $[\Sigma X, Y]_{\cdot}$ *obtains a group structure from that on the object group* $(\pi^{\cdot} Y^X)(\cdot)$.

Proof The elements of $G = (\pi^{\cdot} Y^X)(\cdot)$ can be represented by homotopies of length 1, i.e. by maps $f : X \times \mathbb{I} \to Y$. The condition that f is a homotopy under $\cdot$ is that $f[\cdot \times \mathbb{I}] = \cdot$, where we ambiguously write $\cdot$ for any base point. The condition that f is a homotopy of the constant map $\cdot$ to itself is that $f[X \times \dot{\mathbb{I}}] = \cdot$. But ΣX is the space obtained from $X \times \mathbb{I}$ by identifying

$$\cdot \times \mathbb{I} \cup X \times \dot{\mathbb{I}}$$

to a point. So the representatives f of length 1 of elements of G are bijective with the pointed maps $\bar{f} : \Sigma X \to Y$. Under this correspondence, a homotopy f_t of such representatives corresponds to a homotopy $\bar{f}_t$ of pointed maps $\Sigma X \to Y$, by 4.3.2, and since $\mathbb{I}$ is locally compact. So the required bijection is constructed.

The transfer of the group structure from $(\pi^{\boldsymbol{\cdot}} Y^X)(\cdot)$ to $[\Sigma X, Y]_{\boldsymbol{\cdot}}$ is immediate. □

It is convenient to identify these groups by means of this bijection. The track group $[\Sigma X, Y]_{\boldsymbol{\cdot}}$ plays an important role in homotopy theory. In particular, if $X = \mathbb{S}^{n-1}$, we obtain the *nth homotopy group* $\pi_n(Y, \cdot)$, so that $\pi_1(Y, \cdot)$ is essentially the fundamental group of Y at the base point. The calculation of these homotopy groups, even for the case $Y = \mathbb{S}^2$, is very difficult and has not been completed. For further information, see [Whi78], and other books on homotopy theory.

It is a striking fact that the group $[\Sigma\Sigma X, Y]_{\boldsymbol{\cdot}}$ is abelian, and in particular the higher homotopy groups $\pi_n(Y, \cdot)$ are abelian for $n \geqslant 2$ (see Exercise 12). See Chapter 12 for more comments on this area.

EXERCISES

1. Let $i : X \to \Gamma X$ be the embedding $x \mapsto (x, 1)$ from the pointed space X to the subset $X \times 1$ of the reduced cone ΓX. Let $u : X \to Y$ be a pointed map. Prove that the set $S = [(\Gamma X, i), (Y, u)]$ may be identified with $(\pi^{\boldsymbol{\cdot}} Y^X)(\cdot, u)$. Let β'', β', β be elements of S. By using this identification, and a description of ΣX as $\Gamma^+ X \cup \Gamma^- X$ corresponding to the description of the unreduced suspension SX given on p. 115–117, show how the element $-\beta' + \beta$ becomes an element $d(\beta', \beta)$ of $[\Sigma X, Y]_{\boldsymbol{\cdot}}$, called the *difference element*. Prove that $d(\beta'', \beta') + d(\beta', \beta) = d(\beta'', \beta)$.
2. Suppose further to the previous exercise, that

$$\alpha', \alpha \in [\Sigma X, Y]_{\boldsymbol{\cdot}} = (\pi^{\boldsymbol{\cdot}} Y^X)(\cdot).$$

Then with these identifications, $\beta + \alpha$ can be regarded as an element of $[(\Gamma X, i), (Y, u)]$, and is then written $\beta \perp \alpha$. Prove that (i) $(\beta \perp \alpha) \perp \alpha' = \beta \perp (\alpha + \alpha')$, (ii) $\beta \perp 0 = \beta$, (iii) $d(\beta, \beta \perp \alpha) = \alpha$.
3. Let $\mathcal{C}$ be a category which admits coproducts. Prove that for any objects C_1, C_2, C_3 of $\mathcal{C}$ there are isomorphisms

$$C_1 \sqcup C_2 \to C_2 \sqcup C_1, \qquad C_1 \sqcup (C_2 \sqcup C_3) \to (C_1 \sqcup C_2) \sqcup C_3.$$

Prove also that if 0 is an initial object of $\mathcal{C}$ then there are isomorphisms

$$C_1 \to C_1 \sqcup 0, \qquad C_1 \to 0 \sqcup C_1.$$

4. Suppose further to the previous exercise that for each pair (C_1, C_2) of $\mathcal{C}$ we choose a specific coproduct and write this coproduct $C_1 \sqcup C_2$. Prove that $\sqcup$ is a functor $\mathcal{C} \times \mathcal{C} \to \mathcal{C}$, and interpret the isomorphisms of the previous exercises as natural equivalences of functors.
5. Let A, B, C be objects of the category $\mathcal{C}$. Let $\mathcal{C}^{A,B}$ denote the functor $X \mapsto \mathcal{C}(A, X) \times \mathcal{C}(B, X)$ from $\mathcal{C}$ to Set, and let $\mathcal{C}^C$ denote as usual the functor $X \mapsto \mathcal{C}(C, X)$. Prove that C is a coproduct of A and B if and only if there is a natural equivalence of functors $\mathcal{C}^C \simeq \mathcal{C}^{A,B}$. Use this and similar results to give another solution to Exercise 3.
6. A *comultiplication* on an object A of $\mathcal{C}$ is a morphism $c : A \to A \sqcup A$. Prove that a comultiplication on A induces a natural transformation $\mathcal{C}^{A,A} \to \mathcal{C}^A$. On the other hand, any such natural transformation is called a *natural multiplication* on the sets $\mathcal{C}(A, X)$, $X \in \mathrm{Ob}(\mathcal{C})$. Prove, conversely, that any natural multiplication on these sets is induced by a comultiplication $A \to A \sqcup A$.
7. Let $c : A \to A \sqcup A$ be a comultiplication on the object A of $\mathcal{C}$. We say c is *associative* if the following diagram commutes

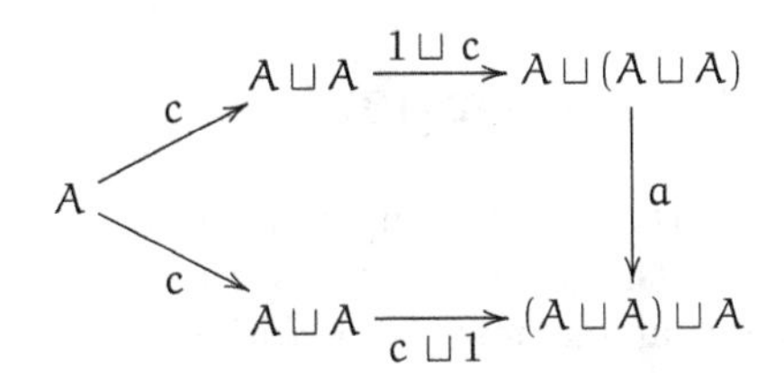

where a is the isomorphism given by Exercise 3. Prove that c is associative if and only if the multiplication induced on each $\mathcal{C}(A, X)$, for $X \in \mathrm{Ob}(\mathcal{C})$, is associative.
8. Let $c : A \to A \sqcup A$ be a comultiplication on the object A of $\mathcal{C}$, and for each X in $\mathcal{C}$ let $c_X : \mathcal{C}(A, X) \times \mathcal{C}(A, X) \to \mathcal{C}(A, X)$ be the induced multiplication. A *natural identity* for c_X is for each X an element e_X of $\mathcal{C}(A, X)$ which is an identity for the multiplication c_X and which is natural, i.e., for any morphism $f : X \to Y$ the induced function $f_* : \mathcal{C}(A, X) \to \mathcal{C}(A, Y)$ maps e_X to e_Y. Find necessary and sufficient conditions on c for the induced natural multiplications to have a natural identity. Further, supposing that c_X has a natural identity, define a *natural inverse* on $\mathcal{C}(A, X)$, $X \in \mathrm{Ob}(\mathcal{C})$ and find necessary and sufficient conditions on c for c_X to have a natural inverse.
9. Prove that, in the category HTop. of pointed spaces and homotopy classes of pointed maps, the set $[\Sigma A, X]$. for all pointed spaces X, has a natural group structure (that is, a natural multiplication which is associative and has natural identity and inverse). Prove that the wedge of pointed spaces is a coproduct in HTop.. Give an explicit formula for a comultiplication $\Sigma A \to \Sigma A \vee \Sigma A$ which induces the above natural group structure.
10. Dualise (in the sense of Exercise 14 of Section 6.6) Exercises 3–8. [Here the dual of a comultiplication $A \to A \sqcup A$ is a multiplication $A \sqcap A \to A$.]

11. Let S be a set with two functions $m_0, m_1 : S \times S \to S$; we write $m_i(x,y) = x +_i y$, $x, y \in S$. These additions induce additions in $S \times S$ be the usual rule

$$(x,y) +_i (x',y') = (x +_i x', y +_i y').$$

We say that m_i is a *morphism* for m_{1-i} if $m_i(z +_{1-i} w) = m_i(z) +_{1-i} m_i(w)$ for all $z, w \in S \times S$. Prove that m_0 is a morphism for m_1 if and only if

$$(x +_1 x') +_0 (y +_1 y') = (x +_0 y) +_1 (x' +_0 y')$$

for all $x, x', y, y' \in S$ and that this is equivalent to m_1 is a morphism for m_0. Suppose that these equivalent conditions hold. Suppose further that there are elements 0_0 and 0_1 in S which are zeros for $+_0$ and $+_1$ respectively. Prove that $0_0 = 0_1$. Prove also that $+_0 = +_1$, and that both additions are associative and commutative.
12. Prove that the group structure on $[\Sigma A, X]_\bullet$ is natural with respect to maps of A in the sense that a pointed map $f : A \to B$ induces a morphism of groups $f^* : [\Sigma B, X]_\bullet \to [\Sigma A, X]_\bullet$, $\operatorname{cls} g \mapsto \operatorname{cls}(g(\Sigma f))$. Deduce that $[\Sigma^2 A, X]_\bullet$ has two multiplications each of which is a morphism for the other; hence, show that these two multiplications coincide and are commutative. Express these facts as statements about $\Sigma^2 A$ and its coproducts.

7.2 Fibrations of groupoids

Let $p : E \to B$ be a morphism of groupoids. We say p is a *fibration* if the following condition holds: for all objects x of E and elements b in B with initial point px, there is an element e of E with initial point x and such that $pe = b$.

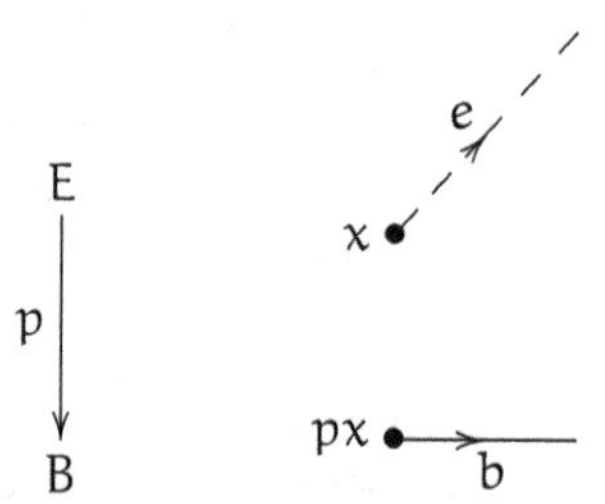

Recall that we use additive notation in groupoids, so that $a : x \to y$ and $b : y \to z$ have a sum $b + a : x \to z$.

EXAMPLES
1. An isomorphism $E \to B$ is a fibration. In particular, an identity morphism $B \to B$ is a fibration. The projection $p_1 : B \times F \to B$ from a product is a fibration. The proof is easy. Let $x = (y, z)$, so that $p_1 x = y$. If b in B has initial point y, then $e = (b, 0_z)$ is an element of $B \times F$ with initial point x and such that $pe = b$.

2. Let $\mathbb{Z}_n$ be the cyclic group of order n, with generator written t, so that the elements of $\mathbb{Z}_n$ are $0, t, 2t, \ldots, (n-1)t$. Define a morphism $p : \mathbf{I} \to \mathbb{Z}_n$ by its value on $\iota \in \mathbf{I}(0,1)$, namely $p\iota = t$. Then $p(-\iota) = (n-1)t$. It follows that p is a fibration if and only if $n = 1$ or 2. Note that p is surjective if $n = 3$.
3. Let E and B be groups and let $p : E \to B$ be a morphism. Then p is a fibration if and only if p is surjective. Thus groupoids have a richer theory than that of groups since there is a greater variety of morphisms of groupoids than there is of groups. More examples of this variety will arise in the next two chapters.
4. We will later give useful conditions on a map $i : A \to X$ of spaces for the induced morphism

$$i^* : \pi Y^X \to \pi Y^A$$

to be a fibration of groupoids.

Let $p : E \to B$ be a morphism of groupoids. If u is an object of B, we write $p^{-1}[u]$ for the subgroupoid of E with objects those x in $\mathrm{Ob}(E)$ such that $px = u$, and with elements those e in E such that $pe = 0_u$. The following result contains our basic construction of an *operation*, or *action*, arising from a fibration of groupoids.

7.2.1 *Let $p : E \to B$ be a fibration of groupoids. Then there is an assignment to each $b \in B(u,v)$ of a function*

$$b_{\#} : \pi_0 p^{-1}[u] \to \pi_0 p^{-1}[v]$$

with the properties that
(a) *if b is a zero, then $b_{\#}$ is the identity;*
(b) *if $b \in B(u,v)$, $c \in B(v,w)$, then $(c+b)_{\#} = c_{\#} b_{\#}$;*
(c) *each function $b_{\#}$ is a bijection.*

Proof The construction of $b_{\#}$ is as follows.

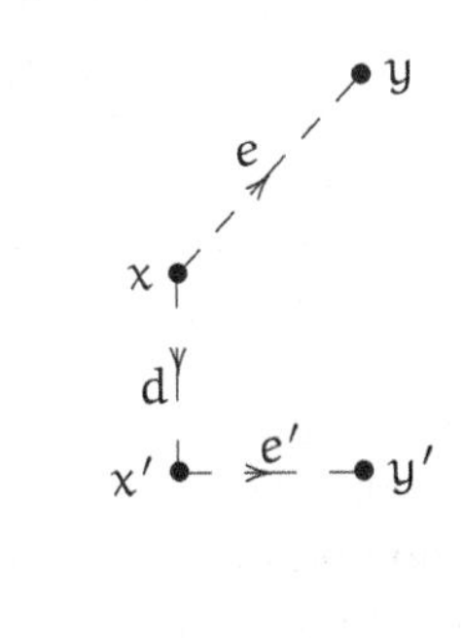

Let x be an object of E such that $px = u$, so that x represents an element of $\pi_0 p^{-1}[u]$. By the fibration condition, there is an element e in E with initial point x and such that $pe = b$. Let y be the final point of e. We claim that the class $\operatorname{cls} y$ of y in $\pi_0 p^{-1}[v]$ is independent of the choices that have been made.

To this end, suppose x' is another object of E such that $px' = u$ and also $\operatorname{cls} x' = \operatorname{cls} x$ as elements of $\pi_0 p^{-1}[u]$. Then there is an element $d \in E(x, x')$ such that $pd = 0_u$. Choose an element e' in E with initial point x' and such that $pe' = b$. Let y' be the final point of e'. Then

$$p(e' + d - e) = b + 0_u - b = 0_v.$$

Hence $\operatorname{cls} y = \operatorname{cls} y'$ in $\pi_0 p^{-1}[v]$. This shows that $b_\#$ is well defined.

If $b = 0_u$ and $px = u$, then $p(0_x) = b$. This proves (a).

If further $c \in B(v, w)$, $px = u$, and $e \in E(x, y)$ satisfies $pe = b$, choose f in E with initial point y and such that $pf = c$. Then $p(f + e) = c + b$, and if z is the final point of f, then

$$c_\# b_\# (\operatorname{cls} x) = c_\# (\operatorname{cls} y) = \operatorname{cls} z = (c + b)_\# (\operatorname{cls} x).$$

This proves (b).

The proof of (c) is now easy, since (a) and (b) imply that $(b_\#)^{-1} = (-b)_\#$. □

We express 7.2.1 by saying that the groupoid B *acts,* or *operates*, on the family of sets $\pi_0 p^{-1}[u]$, $u \in \operatorname{Ob}(B)$. We will use actions of groupoids on families of sets in a different context in section 10.4.

We now come to a topological application of fibrations of groupoids. Let $i : A \to X$ be a map of spaces. We say i is a *cofibration* if the following condition holds for all spaces Y: *if* $X : A \to Y$ *is a map,* $q \geqslant 0$, *and* $G : A \times [0, q] \to Y$ *is a homotopy with initial map* fi, *then there is a homotopy* $H : X \times [0, q] \to Y$ *with initial map* f *such that* $H(i \times 1) = G$. We will give many examples of cofibrations in the next section. In this section we shall show that a cofibration $i : A \to X$ is an embedding, with closed image if X is Hausdorff. In the case where i is an inclusion we say that $f : X \to Y$ *extends* fi, and that a homotopy H on X *extends* the homotopy $H(i \times 1)$. For this reason, it is also common to say that if an inclusion $A \to X$ is a cofibration, then the pair (X, A) has the *homotopy extension property,* which is often abbreviated to HEP.

Here is an example of an inclusion which is not a cofibration. Let $X = \mathbb{L} \cup [-1, 0]$ and let $A = [-1, 0]$. The inclusion $i : A \to X$ extends to the identity $1 : X \to X$. But the homotopy $(a, t) \mapsto a - ta - t$ (which deforms i to the constant map at -1) does not extend to a homotopy of $1 : X \to X$.

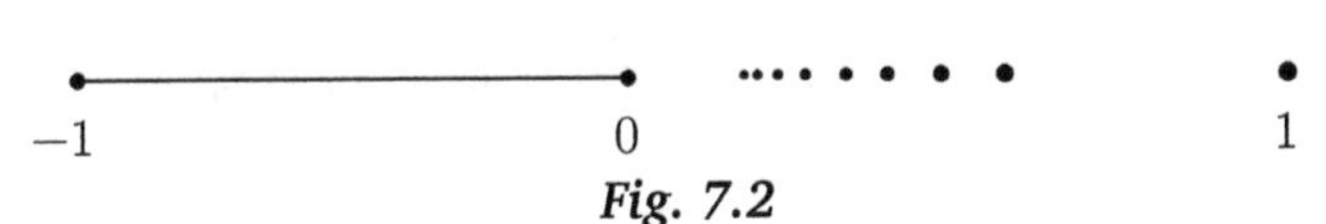

Fig. 7.2

7.2.2 *Let $i : A \to X$ be a cofibration. Then the induced morphism*

$$i^* : \pi Y^X \to \pi Y^A$$

is a fibration of groupoids.

Proof Let $f : X \to Y$ be a map and let b be an element of πY^A with initial point fi. Then b has a representative G which is a homotopy of fi. By the cofibration condition, there is a homotopy H of f such that $H(i \times 1) = G$. The class e of H in πY^X satisfies $i^*(e) = b$. □

The reader will have noted that the cofibration condition is stronger than is required to obtain i^* is a fibration. The exercises explore various weakenings of the cofibration condition, although only one of these weakenings has been found to be generally useful.

In the definition of cofibration we use homotopies of arbitrary length q. But the condition for a cofibration is satisfied for homotopies of any length if it is satisfied for homotopies of length 1. So it is usually convenient to restrict attention to such homotopies.

Let $i : A \to X$ be a map. Consider the diagram

$$\begin{array}{ccc} A \times 0 & \xrightarrow{i \times 1} & X \times 0 \\ \downarrow{\scriptstyle \varepsilon_0} & & \downarrow{\scriptstyle \varepsilon_0} \\ A \times \mathbb{I} & \xrightarrow[i \times 1]{} & X \times \mathbb{I} \end{array} \qquad (7.2.3)$$

where as often in this chapter we find it convenient to denote a space $\{x\}$ by x, and where ε_0 denotes the inclusion. A map $f : X \to Y$ is determined by the map $f' : X \times 0 \to Y$ sending $(x, 0) \mapsto fx$; similarly, $u : A \to Y$ is determined by $u' : A \times 0 \to Y$. A homotopy U of $u : A \to Y$ is a map $U : A \times \mathbb{I} \to Y$ whose restriction to $A \times 0$ is u'. Thus the condition that $i : A \to X$ is a cofibration is equivalent to the condition that (7.2.3) is a *weak pushout*, a condition which is the same as the pushout condition but without the uniqueness [cf. p. 121].

Let $M(i)$ be the mapping cylinder of i; that is, $M(i)$ is the adjunction space $(X \times 0)\ {}_{i\times 1}\sqcup\ (A \times \mathbb{I})$, and is given by the pushout diagram

$$\begin{array}{ccc} A \times 0 & \xrightarrow{i \times 1} & X \times 0 \\ {\scriptstyle \varepsilon_0}\downarrow & & \downarrow \\ A \times \mathbb{I} & \longrightarrow & M(i). \end{array}$$

Then diagram (7.2.3) determines a map $\mu : M(i) \to X \times \mathbb{I}$.

7.2.4 *Let $i : A \to X$ be a map. The following conditions are equivalent.*
(a) *The map i is a cofibration.*
(b) *If Y is a space and $G : M(i) \to Y$ is a map, then there is a map $H : X \times \mathbb{I} \to Y$ such that $H\mu = G$.*
(c) *There is a map $\rho : X \times \mathbb{I} \to M(i)$ such that $\rho\mu = 1$—that is, μ is a coretraction.*

Proof (a) $\Leftrightarrow$ (b) This is a restatement of the cofibration condition.

(b) $\Leftrightarrow$ (c) This is a special case of the following proposition: *If $j : C \to Z$ is a map, then j is a coretraction if and only if for any map $g : C \to Y$, there is a map $f : Z \to Y$ such that $fj = g$.* The proof is as follows. If $r : Z \to C$ is a map such that $rj = 1$, then $gr : Z \to Y$ satisfies $grj = g$. For the converse, take $Y = C$ and $g = 1$. Then there is a map $r : Z \to C$ such that $rj = 1$. □

For part of our next corollary, we need a definition. Let C and D be subspaces of a space W. We say C and D satisfy the *gluing property* if for any space Y, any maps $C \to Y$ and $D \to Y$ which agree on $C \cap D$ define a map on $C \cup D$; this is equivalent to saying that the square of inclusions

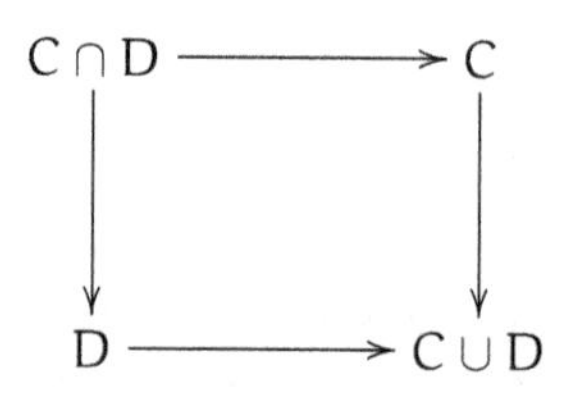

is a pushout.

7.2.4 *(Corollary 1) Let $i : A \to X$ be a cofibration. Then*
(a) *i is an embedding;*
(b) *the pair $i[A] \times \mathbb{I}$, $X \times 0$ of subspaces of $X \times \mathbb{I}$ satisfies the gluing property; and*

(c) *if further* X *is Hausdorff, then* i *is a closed map.*

Proof Condition (b) of 7.2.4 implies that $\mu : M(i) \to X \times \mathbb{I}$ is an embedding. Since $M(i)$ is defined as a pushout, this proves (b). Also the composite

$$A \to A \times 1 \to M(i) \to X \times \mathbb{I}$$

is an embedding. Since the map $X \to X \times 1 \to X \times \mathbb{I}$ is an embedding, it follows that $i : A \to X$ is an embedding, which proves (a).

If X is Hausdorff then so also is $X \times \mathbb{I}$, and hence $\mu[M(i)]$, which is the subset of $X \times \mathbb{I}$ on which the identity and $\mu\rho$ agree, is closed in $X \times \mathbb{I}$. This proves (c). □

Because of this last result, in the case that i is a cofibration we think of $M(i)$ as a subspace of $X \times \mathbb{I}$ and picture $M(i)$ as in Fig. 7.3.

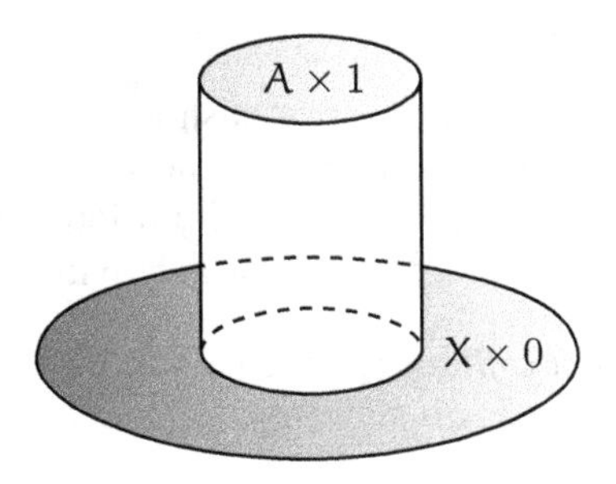

Fig. 7.3

7.2.4 *(Corollary 2) Let* $i : A \to X$ *be a cofibration and let* B *be a space. If* i *is closed, or* B *is locally compact, then* $1 \times i : B \times A \to B \times X$ *is a cofibration.*

Proof Since i is a cofibration, the map $\mu : M(i) \to X \times \mathbb{I}$ is a coretraction with retract ρ, say. Then

$$1 \times \mu : B \times M(i) \to B \times X \times \mathbb{I}$$

is a coretraction with retract $1 \times \rho$. Now comes the subtle point, namely that we need the natural map

$$M(1 \times i) \to B \times M(i)$$

to be a homeomorphism, that is we need that the diagram

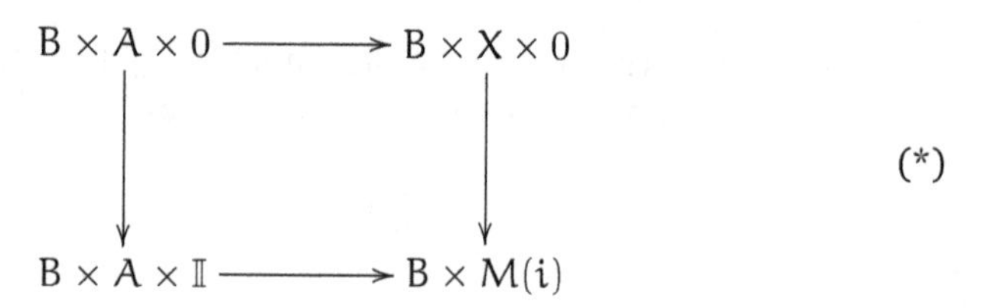

is a pushout. This is true if B is locally compact, by 4.6.6. On the other hand, if i is a closed cofibration, then by the gluing rule 2.5.12 for continuous functions, the diagram

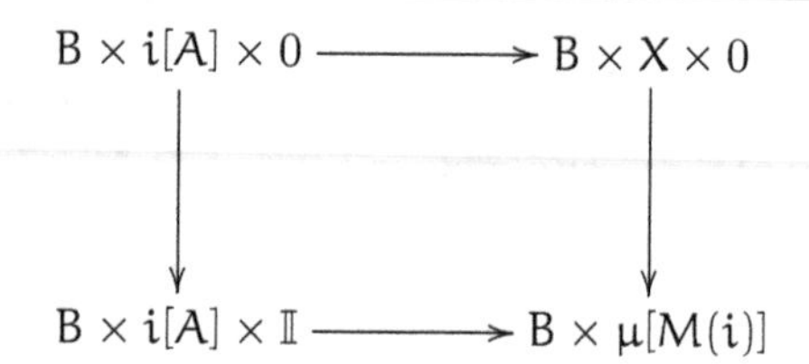

is also a pushout. It follows that (*) is a pushout. □

In order to apply the consequences of the fact that if $i : A \to X$ is a cofibration, then $p = i^* : \pi Y^X \to \pi Y^A$ is a fibration of groupoids, we need to identify $\pi_0 p^{-1}[u]$. The main result is the following.

7.2.5 *Let* $i : A \to X$ *be a cofibration. Let* $u : A \to Y$ *be a map. Let* $p = i^* : \pi Y^X \to \pi Y^A$. *Then there is a canonical bijection*

$$\pi_0 p^{-1}[u] \cong [(X, i), (Y, u)].$$

Proof Both sets are sets of equivalence classes of maps $f : X \to Y$ such that $fi = u$. On the left, the equivalence relation is *homotopic by homotopies* H *such that* $H(i \times 1)$ *is homotopic rel end maps to the constant homotopy.* On the right, the equivalence relation is *homotopic under* A, that is, by homotopies H' such that $H'(i \times 1)$ is the constant homotopy. The following lemma shows that these two equivalence relations are the same.

7.2.5 *(Lemma) Let* $i : A \to X$ *be a cofibration. Let* $H : X \times \mathbb{I} \to Y$ *be a homotopy* $f \simeq g$, *and let* $\mathcal{G}$ *be a homotopy rel end maps* $G = H(i \times 1)$ *to a homotopy* $G' : u \simeq v$. *Then* H *is homotopic rel end maps to a homotopy* H' *such that* $H'(i \times 1) = G'$.

Proof Since i is a cofibration, we may assume that i is an inclusion of a subspace, and that $M(i) = A \times \mathbb{I} \cup X \times 0$. We next prove that

$$W = A \times \mathbb{I} \times \mathbb{I} \cup X \times \dot{\mathbb{I}} \times \mathbb{I} \cup X \times \mathbb{I} \times 0$$

is a retract of $Z = X \times \mathbb{I} \times \mathbb{I}$. To prove this, choose a homeomorphism $\varphi : \mathbb{I} \times \mathbb{I} \to \mathbb{I} \times \mathbb{I}$ such that (see Fig. 7.4)

$$\varphi[\dot{\mathbb{I}} \times \mathbb{I} \cup \mathbb{I} \times 0] = \mathbb{I} \times 0.$$

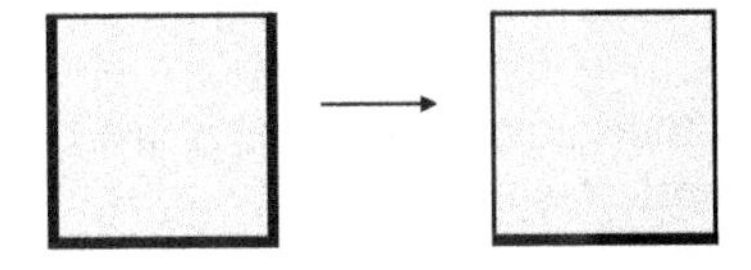

Fig. 7.4

Then $1 \times \varphi : X \times \mathbb{I} \times \mathbb{I} \to X \times \mathbb{I} \times \mathbb{I}$ is a homeomorphism mapping W to $A \times \mathbb{I} \times \mathbb{I} \cup X \times \mathbb{I} \times 0$. But $i \times 1 : A \times \mathbb{I} \to X \times \mathbb{I}$ is a cofibration, by 7.2.4 (Corollary 2), since $\mathbb{I}$ is locally compact, and this implies that any map on $A \times \mathbb{I} \times \mathbb{I} \cup X \times \mathbb{I} \times 0$ extends over Z. Hence any map on W extends over Z.

To apply this last fact, we need to specify a map on W. Define

$$H' : X \times \mathbb{I} \times 0 \to Y \text{ by } (x, s, 0) \mapsto H(x, s);$$
$$f' : X \times 0 \times \mathbb{I} \to Y \text{ by } (x, 0, t) \mapsto f(x);$$
$$g' : X \times 1 \times \mathbb{I} \to Y \text{ by } (x, 1, t) \mapsto gx.$$

These are all continuous, agree on their common domains, and are defined on closed subspaces of their union which we write T, say, which is considered as a subspace of $X \times \mathbb{I} \times \mathbb{I}$. Hence they define a map K on T. Now $1 \times \varphi$ maps T homeomorphically to $X \times \mathbb{I} \times 0$, a space which with $A \times \mathbb{I} \times \mathbb{I}$ has the gluing property. Hence T and $A \times \mathbb{I} \times \mathbb{I}$ have the gluing property. So H', f', g' and $\mathcal{G}$ define a map on W. This extends to a map $\mathcal{H}$ on $X \times \mathbb{I} \times \mathbb{I}$, and $\mathcal{H}$ is a homotopy of H rel end maps as required. □

To go back to the main result, the lemma with $u = v$ and G' the constant homotopy shows that these two equivalence relations are identical. □

The following is immediate from 7.2.1, 7.2.2, and 7.2.5.

7.2.5 *(Corollary 1) Let $i : A \to X$ be a cofibration and let Y be a topological space. Then the track groupoid πY^A acts on the family of sets $[(X, i), (Y, u)]$, for maps $u : A \to Y$, in the sense that if $\alpha \in \pi Y^A(u, v)$, then there is a bijection*

$$\alpha_{\#} : [(X, i), (Y, u)] \to [(X, i), (Y, v)]$$

such that $0_{\#} = 1$ and $\beta_{\#}\alpha_{\#} = (\beta + \alpha)_{\#}$ if further $\beta \in \pi Y^A(v, w)$.

Also, if $\alpha_{\#}(\text{cls}\, f) = \text{cls}\, g$, then any representative of α extends to a homotopy $f \simeq g$. □

Now we exploit the 'functorial' dependence of the sets $[(X, i), (Y, u)]$ on the space-under-A (Y, u). Suppose that $f : Y \to Z$ is a map of spaces. Then f determines by composition a function

$$\begin{aligned} f_* : [(X, i), (Y, u)] &\to [(X, i), (Z, fu)] \\ \operatorname{cls} g &\mapsto \operatorname{cls} fg. \end{aligned}$$

It is clear that if f is a homeomorphism, then f_* is a bijection. Our next aim is the following more subtle result.

7.2.6 *Let $i : A \to X$ be a cofibration. Let $f : Y \to Z$ be a homotopy equivalence. Then for each map $u : A \to Y$ the function*

$$f_* : [(X, i), (Y, u)] \to [(X, i), (Z, fu)]$$

is a bijection.

Proof Let $g : Z \to Y$ be a homotopy inverse of f, so that $gf \simeq 1$, $fg \simeq 1$. Consider the functions

$$\begin{aligned} [(X, i), (Y, u)] \xrightarrow{f_*} [(X, i), (Z, fu)] \xrightarrow{g_*} [(X, i), (Y, gfu)] \\ \xrightarrow{f_*} [(X, i), (Y, fgfu)]. \end{aligned}$$

We shall prove below that the composites of the first two functions, and of the last two functions, are bijections. It follows easily that each of the functions is a bijection (cf. Exercise 2 of Section 6.1).

In order to prove that the composites $g_* f_*$ and $f_* g_*$ are bijections, we use the following two lemmas.

7.2.6 *(Lemma 1) Let $i : A \to X$ be a cofibration, and let $f, f' : Y \to Z$ be homotopic maps. Then for each map $u : A \to Y$ there is an element $\theta \in \pi Z^A(fu, f'u)$ such that the following diagram commutes:*

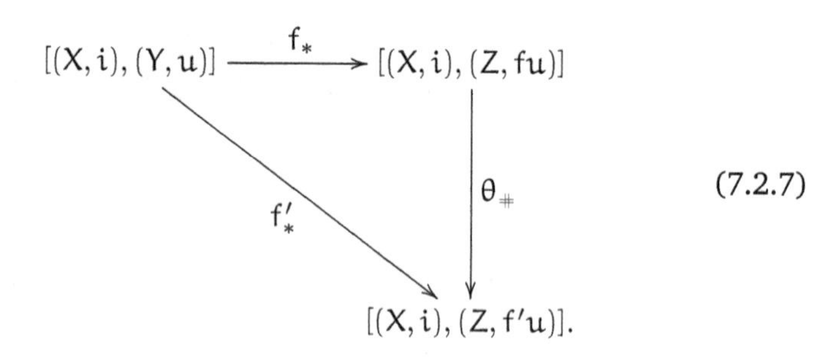

(7.2.7)

Proof Let $F_t : Y \to Z$ be a homotopy $f \simeq f'$, and let θ be the class in πZ^A of the homotopy $F_t u$. Then for each $g : (X, i) \to (Y, u)$ the homotopy $F_t g$ is a homotopy $fg \simeq f'g$ which extends the homotopy $F_t u : A \to Z$. It follows that $\theta_{\#}(\text{cls}\, fg) = \text{cls}\, f'g$. □

7.2.6 *(Lemma 2) If $f \simeq 1 : Y \to Y$, then for each map $u : A \to Y$ the function*

$$f_* : [(X, i), (Y, u)] \to [(X, i), (Y, fu)]$$

is a bijection.

Proof In diagram (7.2.7) we take $Y = Z$ and $f' = 1$. The result follows since $f'_* = 1$ and $\theta_{\#}$ is a bijection. □

This now completes the proof of 7.2.6. □

Note that this proof is an analogue of the proof of 6.5.12. The idea for the proof of 7.2.6 arose in the following way.

In texts on homotopy theory, one studies the homotopy groups $\pi_n(Y, y)$. Here $\pi_1(Y, y)$ is the fundamental group of Y based at y, while for all $n \geqslant 1$, the set $\pi_n(Y, y)$ can be identified with the set of homotopy classes $[(\mathbb{S}^n, i), (Y, j)]$ of maps under A where A is a singleton $\{a\}$, ia is a given base point x, say, of the n-sphere $\mathbb{S}^n$, and $ja = y$. This set of homotopy classes is more clearly written as $[(\mathbb{S}^n, x), (Y, y)]$. One of the early results is usually that a homotopy equivalence $f : Y \to Z$ of spaces induces an isomorphism $f_n : \pi_n(Y, y) \to \pi_n(Z, fy)$ of the nth homotopy groups for all $n \geqslant 1$ and all $y \in Y$. This result is proved by showing that the fundamental groupoid πY acts on the family of groups $\pi_n(Y, y)$.

There were many developments in homotopy theory which involved replacing the sphere $\mathbb{S}^n$ in the definition of homotopy groups by a general space X with a chosen base point x, say. The above result on the effect of a homotopy equivalence on homotopy groups then straightforwardly generalises to give that a homotopy equivalence $f : Y \to Z$ induces a bijection of sets

$$f_* : [(X, x), (Y, y)] \to [(X, x), (Z, fy)]$$

if the inclusion $\{x\} \to X$ is a cofibration.

The final generalisation is to replace the inclusion $\{x\} \to X$ by a general cofibration $A \to X$. Essentially the same proof then gives 7.2.6; all that is needed is to set up an appropriate notation which will allow the proof to be transcribed. The advantage of the new format is that it is more symmetrical—there are two variables X and Y of roughly similar status. It often happens in mathematics that reformulating a result to make it look more pleasing can also make the result more powerful and useful. Our present example will be exploited in section 7.4 to prove the gluing theorem for homotopy equivalences. We can also prove immediately the following result.

7.2.8 *Let* $f : (X, i) \to (Y, u)$ *be a map under* A *and suppose that* $i : A \to X$ *and* $u : A \to Y$ *are cofibrations. Let* $f : X \to Y$ *be a homotopy equivalence of spaces. Then* $f : (X, i) \to (Y, u)$ *is a homotopy equivalence under* A.

Proof By 7.2.6, but with some roles reversed, the map

$$f_* : [(Y, u), (X, i)] \to [(Y, u), (Y, u)]$$

is a bijection. Hence there is a map $g : (Y, u) \to (X, i)$ such that $f_*(\operatorname{cls} g) = \operatorname{cls} 1_Y$. This says that $fg \simeq 1$ under A. But $f : X \to Y$ is a homotopy equivalence. Let g' be a homotopy inverse of f. Then

$$gf = 1_X gf \simeq (g'f)(gf) = g'(fg)f \simeq g'1_Y f = g'f \simeq 1_X.$$

So we have proved that g also is a homotopy inverse of f as maps of spaces; but we have not yet proved that $gf \simeq 1$ *under* A.

However we can now apply the same argument to $g : (Y, u) \to (X, i)$ to find a map $f' : (X, i) \to (Y, u)$ such that $gf' \simeq 1$ under A. We then obtain homotopies under A

$$gf \simeq gfgf' \simeq g1_Y f' = gf' \simeq 1_X.$$

This completes the proof. □

Here is an application of the last result which will be useful later in section 7.4. Let $i : A \to X$ be an inclusion. We say A is a *deformation retract* of X if there is a retraction $r : X \to A$ such that $ir : X \to X$ is homotopic to 1_X under A. In this case, a homotopy $1_X \simeq ir$ under A is called a *retracting homotopy* of X into A and r is called a *deformation retraction.* Another way of stating these ideas is to say that A is a deformation retract of X if the inclusion i gives a homotopy equivalence $i : (A, 1_A) \to (X, i)$ of spaces under A.

Clearly, to prove A is a deformation retract of X it is sufficient to exhibit such a retracting homotopy, that is, to exhibit a homotopy $R_t : X \to X$ under A such that

$$R_0 = 1_X, \qquad R_1[X] \subseteq A.$$

For example, 0 is a deformation retract of $\mathbb{I}$. More generally, if a is a point of the convex set X in a normed space, then a is a deformation retract of X since the homotopy $(x, t) \mapsto (1 - t)x + ta$ is a retracting homotopy of X onto a.

If A is a deformation retract of X, then of course the inclusion $i : A \to X$ is a homotopy equivalence. The converse is false. For example, let $X = C\mathbb{L}$ (Fig. 7.5). In section 7.5 it is proved that CY is contractible for any space Y. So X is contractible and hence the inclusion $\{(0,0)\} \to X$ is a homotopy equivalence. But $(0,0)$ is not a deformation retract of X. The reason, expressed intuitively, is that in deforming X to $(0,0)$, all points $(n^{-1}, 0)$ have to move at least to the vertex of the cone, and so, by continuity, the point $(0,0)$ also cannot remain stationary in such a deformation.

However, the converse is true if i is a cofibration.

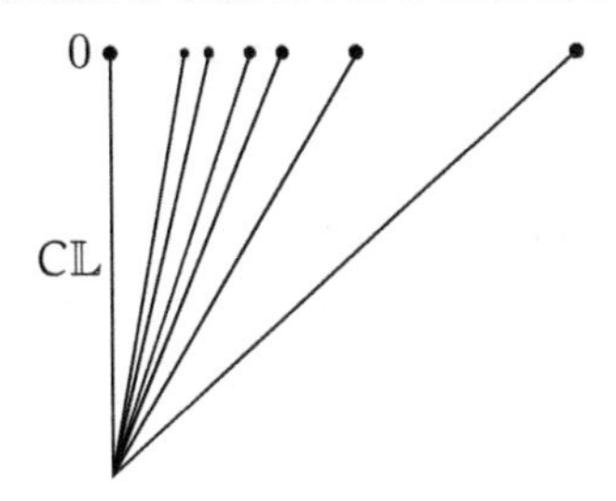

Fig. 7.5

7.2.8 (Corollary 1) *Let $i : A \to X$ be an inclusion which is both a cofibration and a homotopy equivalence. Then A is a deformation retract of X.*

Proof This is immediate from 7.2.8 and the fact that i defines a map $(A, 1_A) \to (X, i)$ of spaces under A. □

7.2.8 (Corollary 2) *Let $i : A \to X$ be an inclusion which is a cofibration. Then $X \times 0 \cup A \times \mathbb{I}$ is a deformation retract of $X \times \mathbb{I}$.*

Proof The inclusion $j : X \times 0 \cup A \times \mathbb{I} \to X \times \mathbb{I}$ is a homotopy equivalence since both spaces contain $X \times 0$ as a deformation retract. So the result follows from the previous corollary if we can prove j is a cofibration. For this it is sufficient to show that $B = (X \times \mathbb{I} \times 0) \cup (X \times 0 \cup A \times \mathbb{I}) \times \mathbb{I}$ is a retract of $D = X \times \mathbb{I} \times \mathbb{I}$.

There is a homeomorphism $\psi : \mathbb{I} \times \mathbb{I} \to \mathbb{I} \times \mathbb{I}$ which maps $\mathbb{I} \times 0 \cup 0 \times \mathbb{I}$ to $\mathbb{I} \times 0$. Then $1 \times \psi : D \to D$ maps B to $C = X \times \mathbb{I} \times 0 \cup A \times \mathbb{I} \times \mathbb{I}$. But $i \times 1 : A \times \mathbb{I} \to X \times \mathbb{I}$ is a cofibration (7.2.4 (Corollary 2)), and so C is a retract of D. Hence B is a retract of D. □

There is a more direct proof of this corollary [Exercise 6], but the above proof is an instructive use of the previous ideas.

We now return to fibrations of groupoids to explain the *exact sequences* of a fibration. We give some applications of these to homotopy of spaces, but in fact the idea is purely algebraic. There are a number of algebraic applications, some of which we indicate in the Exercises. One reason for giving some results in this area is that the idea of exact sequence is so pervasive in algebraic topology, and some other areas of mathematics, that it is worthwhile giving an impression of the method.

The statement of the next result is long since it contains a lot of small individual points. The proofs are quite simple. The value of the result is in systematising a number of related facts. Also, it suggest that whenever you come across a fibration of groupoids, it is useful to consider the family of exact sequences and to see what consequences can be drawn from it.

7.2.9 (The exact sequences of a fibration of groupoids) *Let* $p : E \to B$ *be a fibration of groupoids. Let* x *be an object of* E, *and let* F_x *be the fibre* $p^{-1}[0_y]$ *of* p *over* $y = px$. *Then there is a sequence of maps of three groups and three pointed sets*

$$F_x(x) \xrightarrow{i'} E(x) \xrightarrow{p'} B(y) \xrightarrow{\partial} \pi_0 F_x \xrightarrow{i_*} \pi_0 E \xrightarrow{p_*} \pi_0 B$$

in which i' *is the restriction of the inclusion* $i : F_x \to E$, p' *is the restriction of* p, i_* *and* p_* *are the induced maps, and* ∂ *is given by* $\partial\alpha = \alpha_{\#}\bar{x}$, *where* $\bar{x}$ *denotes the component of* x *in* F_x *and* $\alpha_{\#}$ *is given by the operation of* $B(px)$ *on* $\pi_0 F_x$ *of 7.2.1. The above sequence has the following properties:*

(a) i' *is injective and* $i'[F_x(x)] = p'^{-1}[0_y]$;
(b) $\partial\alpha = \partial\beta$ *if and only if there is a* $\gamma \in E(x)$ *such that* $p'\gamma = -\beta + \alpha$;
(c) *if* $\bar{u}$ *denotes the component in* F_x *of an object* u *of* F_x, *then* $i_*\bar{u} = i_*\bar{v}$ *if and only if there is an* $\alpha \in B(y)$ *such that* $\alpha_{\#}\bar{u} = \bar{v}$;
(d) *if* $\hat{y}$ *denotes the component of* y *in* B *then*

$$i_*[\pi_0 F_x] = p_*^{-1}[\hat{y}].$$

Proof (a) The injectivity of i' is clear, since i is injective. Also, by definition of F_x, $pi[F_x] = \{0_x\}$, and so $i'[F_x(x)] \subseteq p'^{-1}[0_y]$. We have to prove the reverse inclusion. But this is clear, since if $\gamma \in E(x)$ and $p\gamma = 0_y$, then $\gamma \in F_x$.

(b) Let $\partial\alpha = \partial\beta$. Then $\alpha_{\#}\bar{x} = \beta_{\#}\bar{x}$. This means that if $\tilde{\alpha} \in E(x, u)$, $\tilde{\beta} \in E(x, v)$ satisfy $p\tilde{\alpha} = \alpha$, $p\tilde{\beta} = \beta$, then u and v lie in the same component of F_x, that is there is an element $\delta \in F_x(v, u)$. Let $\gamma = -\tilde{\alpha} + \delta + \tilde{\beta} \in E(x)$. Then $p\gamma = -\alpha + \beta$.

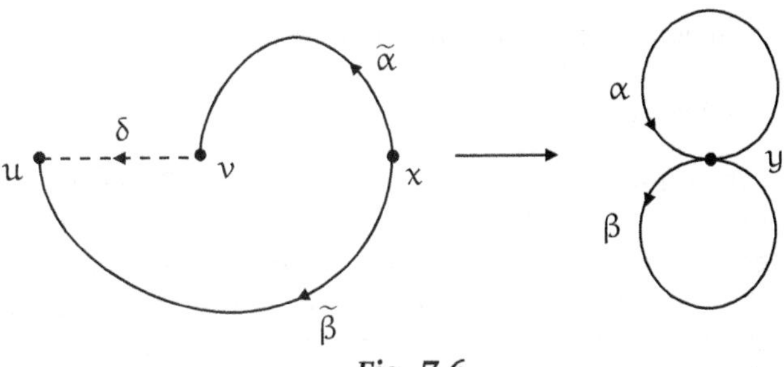

Fig. 7.6

Conversely, if $-\alpha+\beta = p\gamma$ where $\gamma \in E(x)$, then from $(p\gamma)_{\#}\bar{x} = \bar{x}$, we deduce $(-\alpha)_{\#}\beta_{\#}(\bar{x}) = \bar{x}$. So $\alpha_{\#}\bar{x} = \beta_{\#}\bar{x}$, and hence $\partial\alpha = \partial\beta$.
(c) Let u and v be objects of F_x, so that $pu = pv = y$. If $i_*\bar{u} = i_*\bar{v}$, then there is an $\tilde{\alpha} \in E(u,v)$, and so $\alpha = p\tilde{\alpha} \in B(y)$ satisfies $\alpha_{\#}\bar{u} = \bar{v}$. Conversely, if there is an $\alpha \in B(y)$ such that $\alpha_{\#}\bar{u} = \bar{v}$, then by definition of the operation $\alpha_{\#}$ there is an element $\tilde{\alpha} \in E(u,v)$ and so $i_*\bar{u} = i_*\bar{v}$.
(d) Let u be an object of F_x. Then $pu = y$ and so $i_*\bar{u} \in p_*^{-1}[\hat{y}]$. Suppose conversely that v is an object of E whose component v' in E satisfies $p_*v' = \hat{y}$. This means that pv and y belong to the same component of B and so there is an element $\alpha \in B(pv,y)$. Since p is a fibration, there is an element $\tilde{\alpha}$ in E starting at v and such that $p\tilde{\alpha} = \alpha$. Let u be the final point of $\tilde{\alpha}$. Then u is an object of F_x and $i_*\bar{u} = v'$. □

Here are some applications of the exact sequence to homotopy theory. The second of these will be used in section 7.6.

7.2.10 *Let the inclusion $i : A \to X$ of A in X be a cofibration, and suppose that r and s are two retractions $X \to A$ which are homotopic. Then r and s are homotopic under A.*

Proof Consider the fibration of groupoids $i^* : \pi A^X \to \pi A^A$ with base point in πX^A the retraction $r : X \to A$. Part of the exact sequence of the fibration is

$$\pi A^X(r) \xrightarrow{i^*} \pi A^A(1) \xrightarrow{\partial} [(X,i),(A,1)].$$

Let H_t be a homotopy $r \simeq s$. Let α be the class in $\pi A^A(1_A)$ of the homotopy $G_t = H_t \mid A$. Then $\alpha_{\#}(\operatorname{cls} r) = \operatorname{cls} s$. But $\alpha = i^*\gamma$ where γ is the class in $\pi A^X(r)$ of the homotopy $x \mapsto G_t rx$. So $\alpha_{\#}(\operatorname{cls} r) = \operatorname{cls} r$. Hence $r \simeq s$ under A. □

Another way of expressing the conclusion of 7.2.10 is that the natural map $[(X,i),(A,1)] \to [X,A]$ is injective.

7.2.11 *Let* a *be a point of a space* X *and suppose that the inclusion* $i : \{a\} \to X$ *is a cofibration. Let* $f : X \to Y$ *be an inessential map. Then there is a homotopy under* $\{a\}$ *of* f *to a constant map.*

Proof Let H_t be a homotopy of f to a constant map. Then the sum of H_t and the homotopy $(x, t) \mapsto H_{1-t}(a)$ is a homotopy $K_t : f \simeq g$ such that g is constant and $f(a) = g(a) = y$, say. Consider the fibration of groupoids $i^* : \pi Y^X \to \pi Y^{\{a\}}$ and its fibre F_g determined by the base point g of πY^X. Part of the exact sequence of the fibration at this base point is

$$\pi Y^X(g) \xrightarrow{i^*} \pi Y^{\{a\}}(h) \xrightarrow{\partial} [(X, a), (Y, y)]$$

where h is the restriction of g, i.e. $h(a) = y$. Let α be the class in $\pi Y^{\{a\}}(h)$ of the homotopy $(a, t) \mapsto K_t(a)$. Then $\alpha_{\#}(\operatorname{cls} f) = \operatorname{cls} g$. But $\alpha = i^*\gamma$ where γ is the class of the homotopy $(x, t) \mapsto K_t(a)$. Hence $\alpha_{\#}(\operatorname{cls} f) = \operatorname{cls} f$. So $\operatorname{cls} f = \operatorname{cls} g$ and hence f is homotopic to g under $\{a\}$. □

7.2.12 *Let* X, Y *be pointed spaces, regarded as spaces under* $\{a\}$. *Suppose that the inclusion* $i : \{a\} \to X$ *is a cofibration. Then the group* $\pi(Y, \cdot)$ *operates on the set* $[X, Y]_{\cdot}$ *so that if* Y *is path-connected then the map* $[X, Y]_{\cdot} \to [X, Y]$ *is surjective and two elements* α, β *of* $[X, Y]_{\cdot}$ *have the same image in* $[X, Y]$ *if and only if there is an element* $\gamma \in \pi(Y, \cdot)$ *such that* $\gamma \cdot \alpha = \beta$.

Proof Since the map i is a cofibration, the induced morphism $i^* : \pi Y^X \to \pi Y^{\cdot}$ is a fibration of groupoids. Part of the exact sequence of this fibration is

$$\pi Y^{\cdot}(\cdot) \to [(X, \cdot), (Y, \cdot)] \to [X, Y] \to [\cdot, Y].$$

We can identify $\pi Y^{\cdot}(\cdot)$ with the fundamental group $\pi(Y, \cdot)$ of Y at the base point $\cdot$. The result now follows from 7.2.9, since if Y is path-connected then $[\cdot, Y]$ is a singleton. □

Another application of the exact sequence is for the case of pointed spaces X, Y with subspace $i : A \to X$ such that i is a cofibration. As base point of Y^X we take the constant pointed map $X \to Y$. The identifications of section 7.1 then yield the exact sequence of sets of pointed homotopy classes:

$$[\Sigma(X/A), Y]_{\cdot} \to [\Sigma X, Y]_{\cdot} \to [\Sigma A, Y]_{\cdot} \to [X/A, Y]_{\cdot} \to [X, Y]_{\cdot} \to [A, Y]_{\cdot}$$

in which the exactness is as given in 7.2.9.

Exercises

1. (i) Prove that if A is a deformation retract of B, and B is a deformation retract of C, then A is a deformation retract of C.
(ii) Prove that if A is a deformation retract of X and B is a deformation retract of Y, then $A \times B$ is a deformation retract of $X \times Y$.
2. Let $(X, i), (X', i')$ be spaces under A. Let $f : X \to X'$ be a map under A. Prove that if f is a homotopy equivalence under A, then f induces a bijection $[(X', i'), (Y, u)] \to [(X, i), (Y, u)]$. Use such a homotopy equivalence to construct an action of πY^A on the family of sets $[(X', i'), (Y, u)]$ in the case $i : A \to X$ is a cofibration and f is a homotopy equivalence under A.
3. Prove that $\pi(Y \times Z)^A$ is isomorphic to $\pi Y^A \times \pi Z^A$. Use this fact to prove that if $f \simeq f' : Y \to W$, then $f_* \simeq f'_* : \pi Y^A \to \pi W^A$.
4. A map $i : A \to X$ is said to be a *weak cofibration*, or to have the *weak homotopy extension property* (written WHEP), if for any map $f : X \to Y$ and any F of $u = fi : A \to Y$, there is a real number $r \geqslant 0$ such that the homotopy $F + r$ of u extends to a homotopy of f. Prove that if $j : A \to Z$ is a cofibration, and there are maps $g : X \to Z$, $h : Z \to X$ under A such that $hg \simeq 1_X$ under A, then i is a weak cofibration.
5. Prove that if $i : A \to X$ is a map, then $i^* : \pi Y^X \to \pi Y^A$ is a fibration of groupoids if and only if for any map $f : X \to Y$ and any homotopy H of $u = fi$, then H is homotopic rel end map to a homotopy K such that K extends to a homotopy of f. [If this property holds for all Y and all maps f, then we say i has the *rather weak homotopy extension property*. This property, which was introduced in the exercises of the first edition of this book, has been further studied by [Kie77]. See also Exercise 7 of Section 7.5.]
6. Let A be a subspace of X and let $\rho : X \times \mathbb{I} \to X \times 0 \cup A \times \mathbb{I}$ be a retraction. Let ρ_1 and ρ_2 be the two components of ρ. Define $R_t : X \times \mathbb{I} \to X \times \mathbb{I}$ by $(x, s) \mapsto (\rho_1(x, st), s(1-t) + t\rho_2(x, s))$. Prove that R_t is a homotopy rel $X \times 0 \cup A \times \mathbb{I}$ such that $R_0 = 1$, $R_t : (x, s) \mapsto \rho(x, s)$.
7. Compare the 5-lemma for a map of fibrations of groupoids given in [Bro70] with treatments of the 5-lemma in books on algebraic topology or homological algebra.
8. Let C and D be subspaces of a topological space W. Prove that C and D have the gluing property if and only if for all subsets U of $C \cup D$, U is open in $C \cup D$ if and only if $U \cap C$ and $U \cap D$ are open in C and D respectively.

7.3 Examples

It is convenient to define a *cofibred pair* (X, A) to be a space X and subspace A of X such that the inclusion of A into X is a cofibration. The pair is *closed* if A is closed in X. In much of the literature, a cofibred pair is called a pair with the homotopy extension property. We shall derive our main examples of cofibrations from the trivial examples that for any space X the pairs (X, X) and $(X, \varnothing)$ are both cofibred, together with the following basic example.

7.3.1 *The pair* $(\mathbb{I}, \{0,1\})$ *is cofibred.*

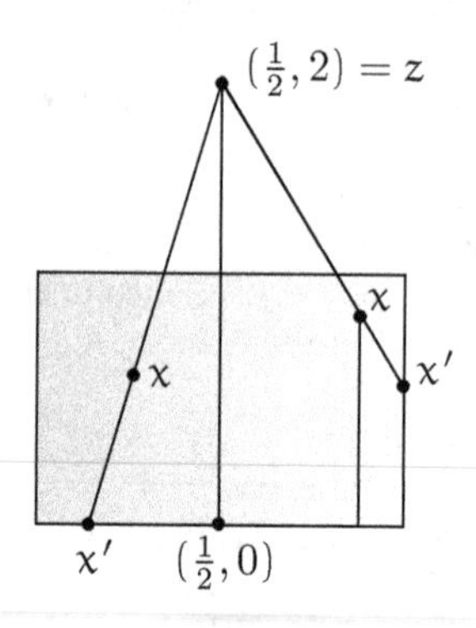

Fig. 7.7

Proof Recall that we write $\dot{\mathbb{I}}$ for $\{0,1\}$. We have to construct a retraction $\mathbb{I} \times \mathbb{I} \to \mathbb{I} \times 0 \cup \dot{\mathbb{I}} \times \mathbb{I}$. The latter space is shown in thick lines in Fig. 7.7. Let z be the point $(\frac{1}{2}, 2)$ in $\mathbb{R}^2$, and regard $\mathbb{I} \times \mathbb{I}$ as a subset of $\mathbb{R}^2$. For each x in $\mathbb{I} \times \mathbb{I}$ let x' be the unique point of $\mathbb{I} \times 0 \cup \dot{\mathbb{I}} \times \mathbb{I}$ such that z, x, x' are collinear. Then $\rho : x \mapsto x'$ is the required retraction. We leave the reader to work out a formula for ρ (e.g. by writing $x = (\frac{1}{2} + s, 2 - t)$ where $|s| \leqslant \frac{1}{2}$ and $1 \leqslant t \leqslant 2$) and so to prove that ρ is continuous. $\square$

In 7.2.4 (Corollary 2) we gave one way of constructing a new cofibration from an old one. Here are some other ways of generating cofibrations.

7.3.2 *The composite of cofibrations is a cofibration.*

7.3.3 *If* (X, A) *and* (Y, B) *are closed cofibred pairs, then so also is* $(X \times Y, A \times B)$.

7.3.4 *Let* (X, A) *be a closed cofibred pair, and let* $f : A \to B$ *be a map. Then* $(B\ {}_f\!\sqcup\ X, B)$ *is a closed cofibred pair.*

7.3.5 *For any spaces* X, B*, the pair* $(B \sqcup X, B)$ *is a closed cofibred pair.*

Proof of 7.3.2 Let $i : A \to X$ and $j : X \to U$ be cofibrations. Let $f : U \to Y$ be a map and let $G : A \times \mathbb{I} \to Y$ be a homotopy with initial map fji. Since i is a cofibration, there is a homotopy H of fj extending G, and since j is a cofibration there is a homotopy K of f extending K. Then K extends G and so ji is a cofibration. □

Proof of 7.3.3 By 7.2.4 (Corollary 2) the maps $i \times 1 : A \times B \to X \times B$ and $1 \times j : X \times B \to X \times Y$ are cofibrations and hence by the previous result $i \times j = (1 \times j)(i \times 1)$ is a cofibration. Clearly $i \times j$ is also closed. □

Remark 7.3.6 In order to prove 7.3.4 we need a method of constructing homotopies on adjunction spaces. First, it is convenient to stick to homotopies of length 1 and to write a homotopy $H : X \times \mathbb{I} \to Y$ as $H_t : X \to Y$ (it being understood that $0 \leqslant t \leqslant 1$), and to say H_t is a homotopy $H_t : H_0 \simeq H_1$. Second, suppose there is given a pushout

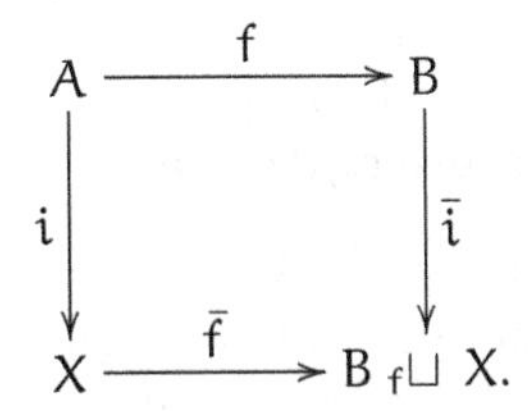

Since $\mathbb{I}$ is locally compact we can apply 4.6.6 which, with the above notation, becomes: *if we are given a commutative square*

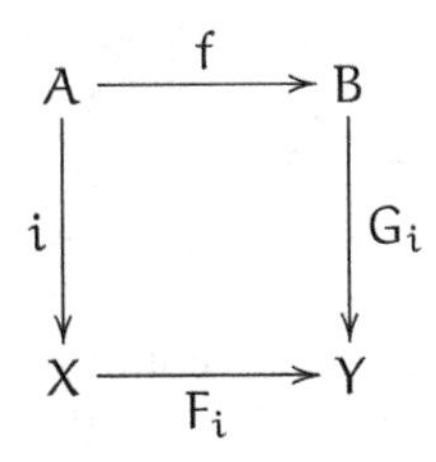

in which F_t and G_t are homotopies, then there is a unique homotopy H_t : $B \,_f\!\sqcup X \to Y$ such that

$$H_t\bar{f} = F_t, \qquad H_t\bar{i} = G_t.$$

That is, the pushout property allows us to define not only maps but also homotopies.

Proof of 7.3.4 Let $g : B \,_f\!\sqcup X \to Y$ be a map and let K_t be a homotopy of $g\bar{i}$. Then $K_t f$ is a homotopy of $g\bar{f}i$ which, since i is a cofibration, extends to a homotopy H_t of $g\bar{f}$. By the previous remark, K and H_t define a homotopy

G_t of g as required. □

Proof of 7.3.5 This follows from 7.3.4 on taking $A = \varnothing$. A direct proof is easy. □

EXAMPLES

1. For any A, the inclusion $A \times 0 \to A \times \mathbb{I}$ is a cofibration.

Proof We know that the inclusion $\dot{\mathbb{I}} \to \mathbb{I}$ is a cofibration. By 7.3.5, the inclusion $0 \to \dot{\mathbb{I}}$ is a cofibration. By 7.3.3 the inclusion $0 \to \mathbb{I}$ is a cofibration. The identity $A \to A$ is a cofibration. So the result follows from 7.2.4 (Corollary 2). □

2. Let $f : A \to B$ be a map. Regard the mapping cylinder $M(f)$ of f [Example 5 of Section 4.6] as $(A \sqcup B)\ {}_{f'}\!\sqcup\ (A \times \mathbb{I})$ where $f' : A \times \dot{\mathbb{I}} \to B$ maps $(a, 0) \mapsto fa$, $(a, 1) \mapsto a$. We claim that: *the inclusions* $A \sqcup B \to M(f)$, $A \to M(f)$, $B \to M(f)$ *are closed cofibrations.*

Proof It follows from 7.3.1 and 7.3.3 that the inclusion $A \times \dot{\mathbb{I}} \to A \times \mathbb{I}$ is a closed cofibration. The result now follows from 7.3.4 and 7.3.2. □

3. For any A the inclusion $A \to CA$ of A into the cone on A, which identifies A with the subset $A \times 1$ of CA, is a closed cofibration.

Proof This follows from the previous example since $CA = M(f)$ where $f : A \to \{v\}$ is the constant map. □

4. Two maps $i : A \to X$ and $j : B \to Y$ are called *homeomorphic* if there are homeomorphisms $h : A \to B$ and $k : X \to Y$ such that $jh = ki$. It is clear that in such cases, i is a cofibration if and only if j is a cofibration, and i is closed if and only if j is closed. Now the inclusion $\mathbb{S}^{n-1} \to \mathbb{E}^n$ is homeomorphic to the inclusion $\mathbb{S}^{n-1} \to C\mathbb{S}^{n-1}$. By the previous example, the inclusion $\mathbb{S}^{n-1} \to \mathbb{E}^n$ is a closed cofibration. This, with 7.3.4, shows that for any B and any map $f : \mathbb{S}^{n-1} \to B$, the inclusion $B \to B\ {}_f\!\sqcup\ \mathbb{E}^n$ is a closed cofibration.

5. Let X be a cell complex and let A be a subcomplex of X. Then the inclusion $A \to X$ is a closed cofibration.

Proof The cell complex X is obtained from A by adjoining in order of increasing dimension those cells of X which are not already in A. So this example follows from the previous example and the fact that a composite of closed cofibrations is a closed cofibration. □

6. However, not all inclusions of closed subspaces are closed cofibrations, as we showed earlier by an example.

We now derive a useful condition for a pair (X, A) to be closed cofibred. This will imply a useful product rule for cofibrations. However, this rule will not be used until the last part of section 7.5, and so the reader may therefore wish to turn directly to applications of cofibrations in the next section.

Consider a pair (X, A). By a *Strøm structure on* (X, A) is meant a pair (w, h) consisting of a map $w : X \to \mathbb{I}$ such that $w[A] = \{0\}$, and a homotopy $h : X \times \mathbb{I} \to X$ rel A of the identity 1_X such that $h(x, t) \in A$ whenever $t > w(x)$. (See [Str66], [Str69]).

7.3.7 *Let* (X, A) *be a pair where* A *is closed in* X*. Then* (X, A) *is cofibred if and only if* (X, A) *admits a Strøm structure.*

Proof Suppose (X, A) is cofibred, so that there is a retraction

$$\rho : X \times \mathbb{I} \to X \times \{0\} \cup A \times \mathbb{I}.$$

We write $\rho = (\rho_1, \rho_2)$ where ρ_1 and ρ_2 are the components of ρ. Define the map

$$w : X \to \mathbb{I}, \qquad x \mapsto \sup_{t \in \mathbb{I}}(t - \rho_2(x, t)).$$

We postpone till the end of this section the proof that w is continuous. That $w(x) \geqslant 0$ follows from $\rho_2(x, 0) = 0$. Clearly $w[A] = \{0\}$. Suppose $w(x) = 0$; then $\rho_2(x, t) \geqslant t$ for all t whence $\rho(x, t) \in A \times \mathbb{I}$ for all $t > 0$. Since $A \times \mathbb{I}$ is closed, this implies that $\rho(x, t) \in A \times \mathbb{I}$ for $t = 0$; but $\rho(x, 0) = (x, 0)$, so that $x \in A$.

Let $h = \rho_1$. Then h is a homotopy rel A of the identity map on X. Suppose $t > w(x)$. Then $\rho_2(x, t) > 0$, and so $\rho_1(x, t) \in A$.

Suppose conversely that (w, h) is a Strøm structure on (X, A). Then a retraction $\rho : X \times \mathbb{I} \to X \times \{0\} \cup A \times \mathbb{I}$ is defined by

$$\rho(x, t) = \begin{cases} (0, h(x, t)) & t \leqslant w(x) \\ (t - w(x), h(x, t)) & t \geqslant w(x). \end{cases}$$

Hence (X, A) is cofibred. □

The following result is a simple corollary of the last characterisation.

7.3.8 *Let* (X, A) *and* (Y, B) *be closed cofibred pairs. Then the pair*

$$(X \times Y, A \times X \cup X \times B)$$

is also a closed cofibred pair.

Proof Let (w, h) and (u, k) be Strøm structures for the pairs (X, A) and (Y, B) respectively. Define $\nu : X \times Y \to \mathbb{I}$ by

$$\nu(x, y) = \min(w(x), \nu(y)).$$

Define $l : X \times Y \times \mathbb{I} \to X \times Y$ by

$$l(x, y, t) = (h(x, \min(u(y), t)), k(y, \min(w(x), t)).$$

Then (l, v) is a Strøm structure for the pair $(X \times Y, A \times X \cup X \times B)$. □

Remark Let $w : X \to \mathbb{I}$ be a map and let $A = w^{-1}[0]$. Then A is the intersection of the sets $w^{-1}[0, n^{-1}[$ for all $0 \neq n \in \mathbb{N}$. Thus A is a G_δ-set (the intersection of a countable family of open sets). It may be proved [cf. Exercise 16 of Section 3.6] that if A is a closed G_δ-set in a normal space X, then there is a map $w : X \to \mathbb{I}$ such that $A = w^{-1}[0]$.

Another general result which implies the product rule 7.3.8 is the following result due to [Lil73], following a suggestion in an exercise of the first edition that a result with similar assumptions should be true. We leave the reader to go through the proof (cf. [Jam84, Proposition 6.14]).

7.3.9 *Let (X, A), (X, B) and $(X, A \cap B)$ be closed cofibred pairs. Then $(X, A \cup B)$ is a closed cofibred pair.* □

In order to complete the proof of 7.3.7, and so of 7.3.8, we have a continuity rule to prove. This follows from the following more general result.

7.3.10 *Let $\varphi : X \times C \to \mathbb{R}$ be a map, let C be compact and define*

$$\begin{aligned} w : X &\to \mathbb{R} \\ x &\mapsto \sup_{c \in C} \varphi(x, c). \end{aligned}$$

Then w is well defined and continuous.

Proof For each x in X, $\varphi[x \times C]$ is a compact and hence bounded subset of $\mathbb{R}$. Therefore w is well defined.

Suppose that $wx = r$ and that $N = [r - \varepsilon, r + \varepsilon]$ is a neighbourhood of r. By definition of r, $c \in C \Rightarrow \varphi(x, c) \leqslant r$, and so

$$x \times C \subseteq \varphi^{-1}]\leftarrow, r + \varepsilon[.$$

By 3.5.6 (Corollary 2) there is an open neighbourhood U_1 of x such that

$$U_1 \times C \subseteq \varphi^{-1}]\leftarrow, r + \varepsilon[$$

and this implies that

$$w[U_1] \subseteq \,]\leftarrow, r + \varepsilon[. \qquad (*)$$

However, there is a c in C such that $\varphi(x,c) \in \operatorname{Int} N$ and so there is a neighbourhood U_2 of x such that

$$\varphi[U_2 \times c] \subseteq N. \qquad (**)$$

So, if $y \in U_1 \cap U_2$, then (*) implies $w(y) \leqslant r + \varepsilon$ while (**) implies $w(y) \geqslant r - \varepsilon$; hence $w[U_1 \cap U_2] \subseteq N$. □

Exercises

1. Give an example of an inclusion $i : A \to X$ and maps $u, v : A \to Y$ such that $u \simeq v$ and u extends over X, but v does not extend over X.
2. Let A, B be disjoint closed subsets of X such that there are maps $\lambda, \mu : X \to [0,1]$ with $A = \lambda^{-1}[0]$, $B = \mu^{-1}[0]$. Prove that there is a map $\nu : X \to [0,1]$ such that $A = \nu^{-1}[0]$, $B = \nu^{-1}[1]$.
3. Suppose given the pushout square of remark 7.3.6. Let $u : B \to Y$ be a map. Prove that the function

$$\begin{aligned} \bar{f}^* : [(B\ {}_f\!\sqcup X, \bar{i}), (Y,u)] &\to [(X,i),(Y,uf)] \\ \operatorname{cls} g &\mapsto \operatorname{cls} g\bar{f} \end{aligned}$$

is a bijection.
4. Give all the details of the proof that in the proof of 7.3.8, (l, ν) is a Strøm structure as stated.
5. Let (w, h) be a Strøm structure for (X, A). Prove that if R_t is a retracting homotopy (rel A) of X onto A, then it may be assumed that w is everywhere less than 1. Hence show that if (X, A) and (Y, B) are closed cofibred pairs and B is a deformation retract of Y, then $X \times B \cup A \times Y$ is a deformation retract of $X \times Y$.

7.4 The gluing theorem for homotopy equivalences of closed unions

In 2.7.3 we proved a rather trivial gluing theorem for homeomorphisms. The main objective of this section is a *gluing theorem for homotopy equivalences*. This theorem is useful for constructing homotopy equivalences, since it is often easy to recognise that 'pieces' of a map are homotopy equivalences, while, as we shall see, the process of constructing the homotopy inverse of the map obtained by gluing the pieces is not so straightforward.

In order to explain the proof, it is useful to introduce the category Top^2 whose objects are the maps $i : X_0 \to X$ of spaces and whose maps $i \to j$, where $j : Y_0 \to Y$, are the commutative diagrams of maps

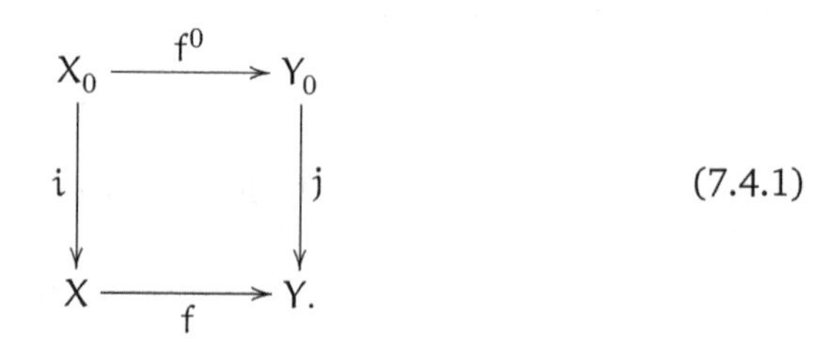

(7.4.1)

Such a diagram is sometimes written $(f^0, f) : i \to j$. The composition in the category Top^2 is the 'horizontal' composition:

$$(g^0, g)(f^0, f) = (g^0 f^0, gf).$$

In this category a map (f^0, f) is an isomorphism if and only if it is a homeomorphism of maps in the sense defined in Example 4 of Section 7.3.

As usual in topological situations, we have a notion not only of maps (in this case, maps of maps) but also of homotopies of maps (again, of maps of maps). Let (f^0, f) and (g^0, g) be maps $i \to j$ of maps. A *homotopy* $(h, k) : (f^0, f) \simeq (g^0, g)$ is a pair of homotopies $h_t : f^0 \simeq g^0$ and $k_t : f \simeq g$ such that $jh_t = k_t i$ for all $t \in \mathbb{I}$. This means that each (h_t, k_t) is a map of maps $i \to j$.

Again, as is usual when homotopy has been defined, we have the notions of *homotopic* maps of maps, *homotopy equivalence* of maps, *domination,* and so on. We leave the reader to write down the definitions and basic properties of these notions following the scheme of chapter 6.

The notion of homotopy equivalence of maps is more restrictive than the notion of just a pair of homotopy equivalences. That is, in the diagram (7.4.1), if we are given that both f^0 and f are homotopy equivalences, it does not follow that (f^0, f) is a homotopy equivalence of maps. Of course we know that f^0 has a homotopy inverse g^0 and that f has a homotopy inverse g, but there is no reason why (g^0, g) should be a map $j \to i$, and even if this is so, there is still no reason why there should be homotopies of maps of the required form.

However this does follow if i and j are cofibrations, and this is our first main result, and a crucial step in the proof of the main gluing theorem.

7.4.2 *Suppose in the diagram* (7.4.1) *that* f^0 *and* f *are homotopy equivalences and that* i *and* j *are cofibrations. Then* $(f^0, f) : i \to j$ *is a homotopy equivalence of maps.*

Actually our proof will give more information on the homotopies involved. This information is used in the proof of the gluing theorem 7.4.3, and it also implies that 7.4.2, with the next addendum, does generalise 7.2.8, which dealt with homotopy equivalences under A.

7.4.2 *(Addendum) Let* $g^0 : Y_0 \to X_0$ *be any homotopy inverse of* f^0 *and let* $H_t^0 : f^0 g^0 \simeq 1$, $K_t^0 : g^0 f^0 \simeq 1$ *be homotopies. Then* g^0 *extends to a homotopy inverse* g *of* f *such that the homotopy* $fg \simeq 1$ *extends* H_t^0 *while the homotopy* $gf \simeq 1$ *extends the sum*

$$K^0 + g^0 H^0 f^0 - g^0 f^0 K^0$$

of the homotopies

$$g^0 f^0 = g^0 f^0 1_{X_0} \simeq g^0 f^0 g^0 f^0 \simeq g^0 1_{Y_0} f^0 \simeq 1_{X_0}$$

determined by H_t^0 *and* K_t^0.

Proof Consider the diagram

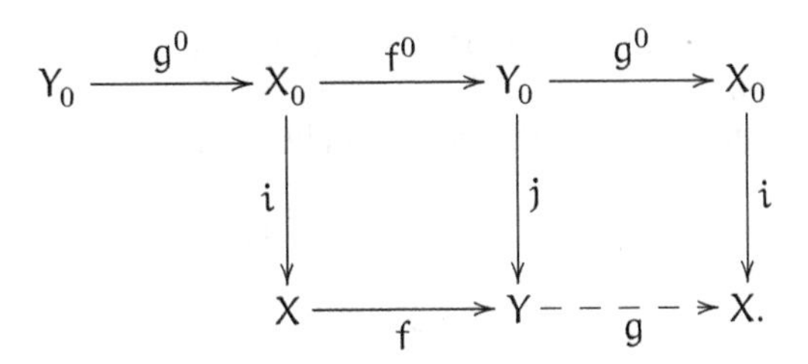

We first work in the category of maps under Y_0, and consider the induced maps of sets of homotopy classes

$$[(Y,j),(X,ig^0)] \xrightarrow{f*} [(Y,j),(Y,fig^0)] \xrightarrow{\alpha_\#} [(Y,j),(Y,j)]$$

where α is the class in πY^{Y_0} of the homotopy $jH_t^0 : jf^0 g^0 \simeq j$. Since f is a homotopy equivalence, f_* is a bijection [(7.2.7)]. Also $\alpha_\#$ is a bijection. So there is a homotopy class $\operatorname{cls} g$ in $[(Y,j),(X,ig^0)]$ such that

$$\alpha_\# f_*(\operatorname{cls} g) = \operatorname{cls} 1_Y. \qquad (*)$$

Then in the first place $gj = ig^0$. Also (*) in conjunction with 7.2.6 shows that there is a homotopy $fg \simeq 1$ which agrees on Y_0 with the given homotopy $H_t^0 : f^0 g^0 \simeq 1$.

We next prove that g is a homotopy equivalence. This follows from the fact that f is a homotopy equivalence and $fg \simeq 1$, since if g' is any homotopy inverse of f then

$$gf = (gf)1_X \simeq (gf)g'f = g(fg')f \simeq g1_Y f \simeq 1_X.$$

(This argument is really the same as that in 6.1.1 which proves that any right inverse of an isomorphism is itself an isomorphism. At that stage, the argument seems trivial. At this stage, it acquires some real power.)

Now we can apply to g instead of f the argument which produced g. This shows that there is a map $f' : X \to Y$ which agrees with f^0 on X_0, and that there is a homotopy $K_t : gf' \simeq 1$ which agrees on X_0 with the given homotopy $K_t^0 : g^0 f^0 \simeq 1$.

Finally we can apply again the kind of trick which proved g is a homotopy inverse of f. That is we have homotopies determined by H_t and K_t

$$gf1_X \simeq gfgf' \simeq gf' \simeq 1_X.$$

This proves 7.4.2 and the addendum. □

We leave it as an exercise for the reader to deduce 7.2.8.

Our main gluing theorem is an easy consequence of 7.4.2, and in fact the statement of the theorem is almost as long as the proof.

7.4.3 (The gluing theorem 1: closed subspaces) *Let $f : X \to Y$ be a map of spaces. Suppose that:*
(a) *X and Y are each given as the union of closed subspaces*

$$X = X_1 \cup X_2, \qquad Y = Y_1 \cup Y_2$$

where $X_1 \cap X_2 = X_0$, $Y_1 \cap Y_2 = Y_0$, say.
(b) *$f(X_n) \subseteq Y_n$ for $n = 0, 1, 2$.*
(c) *The restrictions of f*

$$f^0 : X_0 \to Y_0, \qquad f^1 : X_1 \to Y_1, \qquad f^2 : X_2 \to Y_2$$

are homotopy equivalences.
(d) *Each inclusion $X_0 \to X_1$, $X_0 \to X_2$, $Y_0 \to Y_1$, $Y_0 \to Y_2$ is a cofibration.*
Then f is a homotopy equivalence.

Proof The main point is that we apply 7.4.2 and its addendum, with X and Y replaced first by X_1 and Y_1, and then by X_2 and Y_2. This allows us to start with a homotopy inverse g^0 of f^0 and to construct homotopy inverses g^1, g^2 of f^1, f^2 extending g^0. Since all the subspaces are closed, the maps g^1 and g^2 define a map $g : Y \to X$.

Further, if we choose homotopies $f^0 g^0 \simeq 1$, $g^0 f^0 \simeq 1$, then for $n = 1, 2$ the homotopies $f^n g^n \simeq 1$, $g^n f^n \simeq 1$ may be chosen to agree with each other on the subspaces Y_0 and X_0 respectively, because of the specific formulae given in the addendum to 7.4.2. So the homotopies on the subspaces define also homotopies $fg \simeq 1$, $gf \simeq 1$, and the gluing theorem is proved. □

Note that the proof of the theorem gives also information similar to that in the addendum to 7.4.2. It would be possible to make this clearer by setting up a category of *triads* $(X; X_1, X_2)$ and their maps and homotopies. However, for the purposes of this book, such formality would take us too far afield, and would not bring too much benefit.

For an application of the gluing theorem, we first recall the suspension SX of a space X as defined in section 4.4. We shall need the fact that SX can be written as the union of two closed subspaces C^+X and C^-X, with intersection which can be identified with X. Further, each of C^+X, C^-X is homeomorphic to the cone CX. In effect, we shall regard SX as given by a pushout diagram

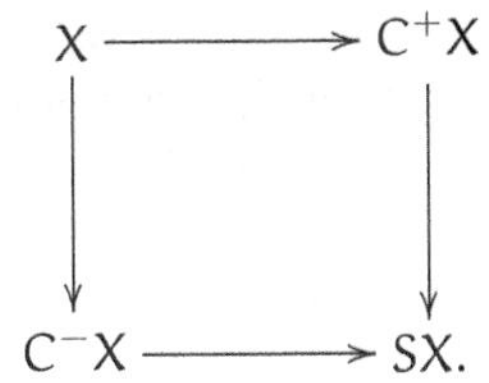

The key property which we need of CX, and so of C^+X and C^-X, is that *if* $f : X \to Y$ *is an inessential map, then* f *extends to a map* $CX \to Y$. The proof of this is easy: Let $H : X \to Y$ be a null-homotopy of f, that is $H : g \simeq f$ where g is constant. Then H is a map $X \times \mathbb{I} \to Y$ such that $H[X \times 0]$ is a singleton. Hence H determines a map $CX \to Y$ which agrees with f on $X = X \times 1$.

We shall also need the fact that CX is contractible, and that hence so also are C^-X and C^+X. This will be proved carefully in the next section, so for the moment we just assume it. The two properties of contractible spaces that we shall use are that any map to a contractible space is inessential, and that any map between contractible spaces is a homotopy equivalence.

7.4.3 *(Corollary) Let the space* X *be the union of closed subspaces* X_1 *and* X_2 *with intersection* X_0 *and such that the inclusions* $X_0 \to X_1$, $X_0 \to X_2$ *are cofibrations. Suppose that* X_1 *and* X_2 *are contractible. Then* X *is of the homotopy type of the suspension* SX_0.

Proof Let $i_n : X_0 \to X_n$ be the inclusion for $n = 1, 2$. Since X_n is contractible, i_n is inessential, that is, i_n is homotopic to a constant map. Hence i_1 extends to a map $f_1 : C^-X_0 \to X_1$ and i_2 extends to a map $f_2 : C^+X_0 \to X_2$. These together define a map $f : SX_0 \to X$. Since X_1 and X_2 are contractible, as are C^-X_0 and C^+X_0, the maps f_1 and f_2 are homotopy equivalences, as is their restriction to X_0, namely the identity $X_0 \to X_0$. All the inclusions under consideration are closed cofibrations. Hence by the gluing theorem, f is a homotopy equivalence. □

In the next section we apply the gluing theorem for unions of closed subspaces to prove a gluing theorem on the homotopy type of adjunction spaces. This is a more general and often more immediately applicable result.

We conclude this section with a result of a somewhat technical character but which will be used in section 7.6. It is not an application of the gluing theorem, but is placed here because it requires the notion of maps and homotopies of maps which were defined at the beginning of the section.

Let $i : A \to X$ and $j : B \to Y$ be two inclusions of subspaces. By a *map of pairs* $\mathbf{f} : (X, A) \to (Y, B)$ is meant a map of maps $(f^0, f) : i \to j$, that is a map $f : X \to Y$ such that $f[A] \subseteq B$, and where f^0 denotes the restriction $f \mid A, B$. By a *homotopy* $\mathbf{f} \simeq \mathbf{g}$ of such maps we mean simply a homotopy of maps as defined previously, and this is just a homotopy f_t such that $f_0 = f$, $f_1 = g$, and $f_t[A] \subseteq B$ for all t. We say $\mathbf{f}$ is *deformable into* B if there is a homotopy $\mathbf{f} \simeq \mathbf{g}$ such that $g[X] \subseteq B$. If also the homotopy is under A then we say that $\mathbf{f}$ is *deformable into* B *under* A.

7.4.4 *Let* $\mathbf{f} : (X, A) \to (Y, B)$ *be a map of pairs such that* $\mathbf{f}$ *is deformable into* B. *Assume that the inclusion* $i : A \to X$ *is a cofibration. Then* $\mathbf{f}$ *is deformable into* B *under* A.

Proof Let $j : B \to Y$ denote the inclusion. Let f_t be a homotopy $(f^0, f) \simeq (g^0, g)$ such that $g[A] \subseteq B$. Let α denote the class in πB^A of the restriction of the homotopy f_t to a homotopy $f^0 \simeq g^0$. Let $\beta = j_* \alpha$ in πY^A. We consider the diagram

$$\begin{array}{ccccc} & & [(X,i),(B,g^0)] & \xrightarrow{(-\alpha)_\#} & [(X,i),(B,jf^0)] \\ & & \downarrow j_* & & \downarrow j_* \\ [(X,i),(Y,jf^0)] & \xrightarrow{\beta_\#} & [(X,i),(Y,jg^0)] & \xrightarrow{(-\beta)_\#} & [(X,i),(Y,jf^0)]. \end{array}$$

Let θ in $[(X,i),(Y,jf^0)]$ be the class of $f : X \to Y$ so that

$$\beta_\#(\theta) = \operatorname{cls} g.$$

But $g[X] \subseteq B$ and so $g = jg'$, where g' is the restriction of g to a map $X \to B$. Let $\varphi = \operatorname{cls} g'$. Then $\beta_\#(\theta) = j_* \varphi$ and so since the above diagram is commutative

$$\begin{aligned}\theta &= (-\beta)_{\#}\beta_{\#}(\theta)\\ &= (-\beta)_{\#}j_*(\varphi)\\ &= j_*(-\alpha)_{\#}(\varphi).\end{aligned}$$

So if $h : X \to B$ represents $(-\alpha)_{\#}(\varphi)$ then jh represents $\theta = \mathrm{cls}\, f$. Hence f is homotopic under A to jh. □

It should be confessed that a simpler proof of this result is possible [Exercise 5], but the above proof is given here as another illustration of the use of the operations.

EXERCISES

1. Generalise 7.4.3 to the case where X is the union of closed subspaces $X_1, \ldots, X_n$ such that if X_0 is the intersection of the X_i then
(i) each inclusion $X_0 \to X_i$ is a cofibration,
(ii) $X_i \cap X_j = X_0$ if $i \neq j$,
and similar conditions hold for Y.
2. Let $f : (X, A) \to (Y, B)$ be a map of pairs such that B is a deformation retract of Y. Prove that f is deformable into B under A.
3. Let the inclusion $A \to X$ be a cofibration, let $B \subseteq Y$ and let F_t be a homotopy such that (i) $F_1[X] \subseteq B$, (ii) F_0 and F_1 agree on A, (iii) F_0 maps A homeomorphically to B. Prove that F_0 is homotopic to F_1 rel A. [Model the proof of 7.4.4.]
4. Prove 7.2.8 (Corollary 1), 7.2.10 and 7.2.11 using the previous exercise.
5. Let $R_t : X \times \mathbb{I} \to X \times \mathbb{I}$ be a retracting homotopy of $X \times \mathbb{I}$ onto $W = X \times 1 \cup A \times \mathbb{I}$, so that $R_0 = 1$, $R_t[X \times \mathbb{I}] \subseteq W$ and R_t is a homotopy rel W. Let $F_t : (X, A) \to (Y, B)$ be a homotopy deforming F_0 into B. Prove that the homotopy $(x, t) \mapsto FR_t(x, 0)$ deforms F_0 into B rel A.
6. Let $f : (X, A) \to (Y, B)$ be a map of cofibred pairs. Prove that f has a right homotopy inverse if (i) $f : X \to Y$ is a homotopy equivalence and the restriction $f^0 : A \to B$ of f has a right homotopy inverse, or (ii) there is a map $g : (Y, B) \to (X, A)$ such that $fg \simeq 1_Y$ and $f^0 g^0 \simeq 1_B$.
7. Prove that $\{0\}$ is not a deformation retract of $C\mathbb{L}$.

7.5 The homotopy type of adjunction spaces

Suppose we are given an adjunction space $B\ {}_f\!\sqcup X$ so that we have a pushout

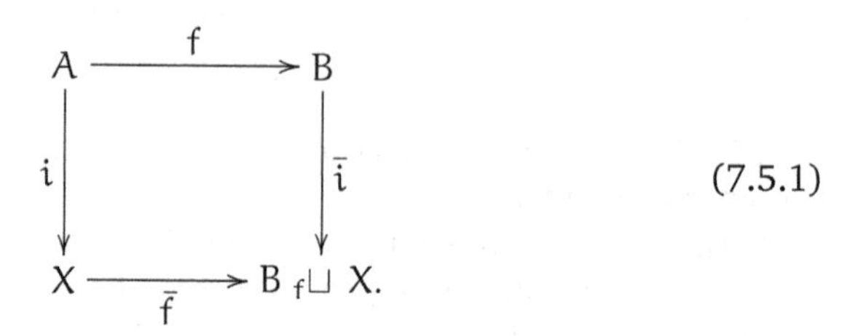

(7.5.1)

In this section we shall apply the gluing theorem from the previous section to show that if i is a cofibration then the homotopy type of $B\ {}_f\!\sqcup X$ depends only on the homotopy class of f and on the homotopy types of B and of the map i.

We first need some simple lemmas on deformation retracts.

7.5.2 *Let* $A \subseteq D \subseteq X$ *and let* D *be a deformation retract of* X*. Then* $B\ {}_f\!\sqcup D$ *is a deformation retract of* $B\ {}_f\!\sqcup X$.

Proof Let $R_t : X \to X$ be a retracting homotopy of X onto D. Let

$$F_t = \bar{f}R_t : X \to B\ {}_f\!\sqcup X, \ G_t = \bar{i} : B \to B\ {}_f\!\sqcup X.$$

Then $F_t i = G_t f$ (since D contains A); by Remark 7.3.6, F_t and G_t define a homotopy $H_t : B\ {}_f\!\sqcup X \to B\ {}_f\!\sqcup X$. It is easily checked that H_t is a retracting homotopy of $B\ {}_f\!\sqcup X$ onto $B\ {}_f\!\sqcup D$. □

Intuitively, 7.5.2 is 'obvious', since the homotopy which retracts X down onto D also retracts $B\ {}_f\!\sqcup X$ onto $B\ {}_f\!\sqcup D$. But this last statement, although it contains the essential idea, is not accurate and also does not indicate why the resulting homotopy is continuous.

7.5.2 *(Corollary 1) If* A *is a deformation retract of* X*, then* B *is a deformation retract of* $B\ {}_f\!\sqcup X$.

Proof Take $A = D$ in (7.5.1). □

We recall that $\{0\}$ is a deformation retract of $\mathbb{I}$. It follows easily that, for any A, $A \times 0$ is a deformation retract of $A \times \mathbb{I}$; so the following result is immediate.

7.5.2 *(Corollary 2) For any* A*, the vertex of the cone* CA *is a deformation retract of* CA.

A similar argument shows that if $f : A \to B$ is a map then B is a deformation retract of the mapping cylinder [Example 2 of Section 7.3]

$$M(f) = B\ {}_{f'}\sqcup\ (A \to \mathbb{I})$$

where $f' : A \times 0 \to B$ is $(a, 0) \mapsto fa$. In fact, if $r : A \times \mathbb{I} \to A \times 0$ is the map $(a, t) \mapsto (a, 0)$, then r is a deformation retraction, and therefore so also is the map $q : M(f) \to B$ which is defined by the maps 1_B and $f'r$. Let $j : A \to M(f)$ be the map $a \mapsto (a, 1)$ by which we identify A as a subspace of $M(f)$. Then we have a commutative diagram,

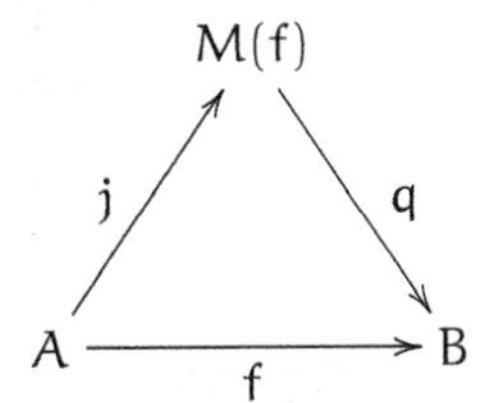

in which j is an inclusion and q is a deformation retraction. This 'factorisation' of f is very useful.

7.5.3 *The map* $f : A \to B$ *is a homotopy equivalence if and only if* A *is a deformation retract of* $M(f)$.

Proof We know that q is a homotopy equivalence. Therefore, j is a homotopy equivalence if and only if f is a homotopy equivalence. But $(M(f), A)$ is cofibred [Section 7.3]. By 7.2.8 (Corollary 1) j is a homotopy equivalence if and only if A is a deformation retract of $M(f)$. □

This result shows that two spaces are of the same homotopy type if and only if there is a third space containing each as a deformation retract.

Consider again the adjunction space $B\ {}_f\sqcup\ X$ of diagram (7.5.1). In order to study the homotopy type of this space it is convenient to use some auxiliary space, namely, $M(f)$, $M(f) \cup X$ and $M(\bar{f})$; these spaces are represented in the following figure.

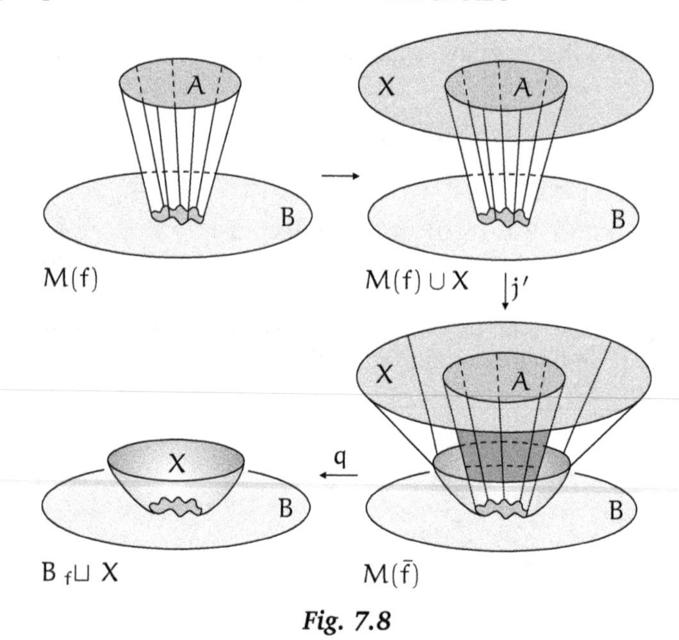

Fig. 7.8

By 4.6.1 $M(f) \cup X$ is a subspace of $M(\bar{f})$—we write j' for the inclusion map. Let $q : M(\bar{f}) \to B\ {}_f\sqcup\ X$ be the standard deformation retraction. It turns out that if (X, A) is cofibred then qj' is a homotopy equivalence, and therefore, in this case, $B\ {}_f\sqcup\ X$ can be replaced by $M(f) \cup X$. The gluing theorem 7.4.3 may then be conveniently applied to the union $M(f) \cup X$.

7.5.4 *If* (X, A) *is cofibred, then*

$$p = qj' : M(f) \cup X \to B\ {}_f\sqcup\ X$$

is a homotopy equivalence rel B.

Proof We know q is a homotopy equivalence $\operatorname{rel} B$. It follows from 7.2.8 (Corollary 2) that $X \times 1 \cup A \times \mathbb{I}$ is a deformation retract of $X \times \mathbb{I}$. By 7.5.2 with A, D, X replaced by $A \times 1$, $X \times 1 \cup A \times \mathbb{I}$, $X \times \mathbb{I}$ respectively (and with $X \times 1$ identified with X), $M(f) \cup X$ is a deformation retract of $M(\bar{f})$. Thus j' also is a homotopy equivalence $\operatorname{rel} B$. □

As a simple example of the previous result, let $X = \mathbb{S}^2$, let A consist of the North and South Poles of $\mathbb{S}^2$, and let B consist of a single point. Then the adjunction space $B\ {}_f\sqcup\ X$ is simply $\mathbb{S}^2$ with North and South Poles identified. On the other hand, $M(f) \cup X$ is homeomorphic to $\mathbb{S}^2$ with an arc C joining the North to the South Pole. 7.5.4 shows that the map $M(f) \cup X \to B\ {}_f\sqcup\ X$ which shrinks C to a point is a homotopy equivalence.

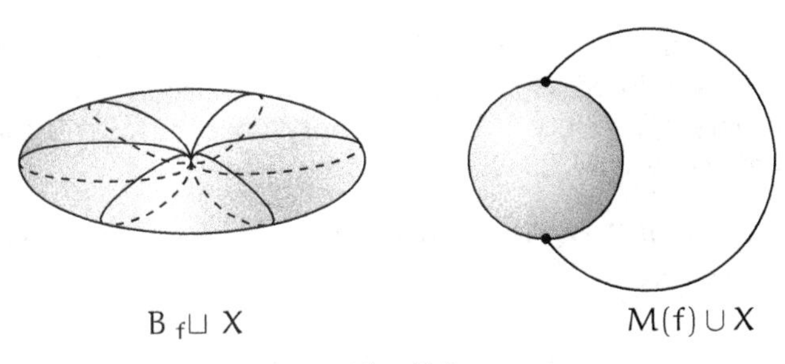

Fig. 7.9

The preceding result is false without some assumptions on the pair (X, A) [cf. the Example on p. 297].

We now show that the homotopy type of $M(f) \cup X$ depends only on the homotopy class of f.

Suppose that $f_t : f_0 \simeq f_1$ is a homotopy of maps $A \to B$. Let

$$F : M(f_0) \cup X \to M(f_1) \cup X$$

be the identity on B and on X, and on the part $A \times]0, 1]$ of the mapping cylinder be given by

$$F(a, t) = \begin{cases} f_{2t}a, & 0 < t \leqslant \frac{1}{2} \\ (a, 2t-1), & \frac{1}{2} \leqslant t \leqslant 1 \end{cases}$$

(where $(a, 1)$ is identified with a). The map F is illustrated for the case $X = A = \{a\}$ in Fig. 7.8 which amalgamates the pictures of $M(f_0)$, $M(f_1)$. The proof of the continuity of F is left as an exercise.

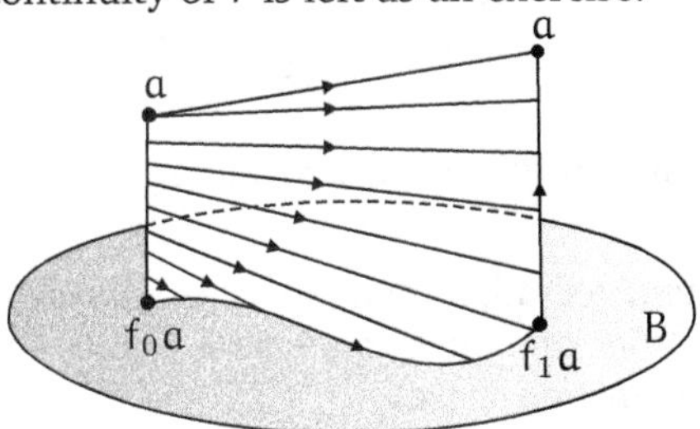

Fig. 7.10

7.5.5 *The above map* F *is a homotopy equivalence* $\mathrm{rel}(B \cup X)$.

Proof Let $G = F \mid M(f_0), M(f_1)$, and let $i_\varepsilon : B \to M(f_\varepsilon)$ $(\varepsilon = 0, 1)$ be the inclusion. Since G is the identity on B, $Gi_0 = i_1$. Since i_0, i_1 are homotopy equivalences, so also is G. Since $G \mid B \cup A$, $B \cup A$ is the identity, and $(M(f_\varepsilon), B \cup A)$ is cofibred, G is actually a homotopy equivalence $\mathrm{rel}(B \cup A)$ [7.2.8 (Corollary 1)]. Since F is the identity on X, it follows that F is a homotopy equivalence $\mathrm{rel}(B \cup X)$. □

7.5.5 *(Corollary 1) If* (X, A) *is cofibred and* $f_0 \simeq f_1 : A \to B$*, then there is a homotopy equivalence*

$$B\ {}_{f_0}\!\sqcup X \to B\ {}_{f_1}\!\sqcup X \,\mathrm{rel}\, B.$$

Proof This follows from 7.5.5 and 7.5.4. □

As a simple application of this fact, note that if B is path-connected then $B\ {}_f\!\sqcup \mathbb{E}^1$ (where $f : \mathbb{S}^0 \to B$) is always of the homotopy type of $B \vee \mathbb{S}^1$.

We now show the dependence of the homotopy type of $B\ {}_f\!\sqcup X$ on the homotopy types of B and of the map $i : A \to X$. Suppose given a commutative diagram of maps

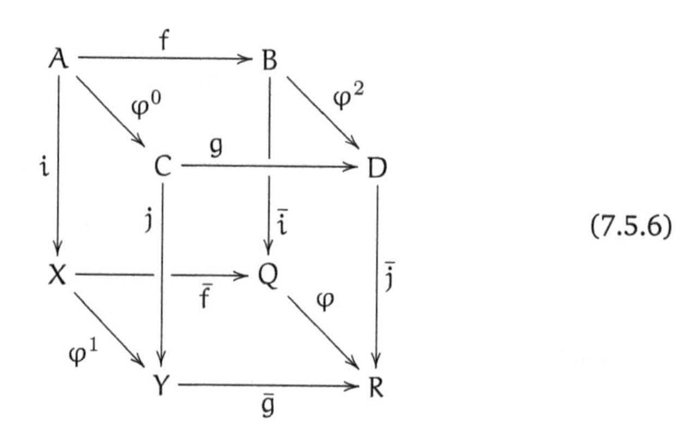

where i and j are inclusions of closed subspaces.

7.5.7 (Gluing theorem for adjunction spaces) *Suppose* i *and* j *are closed cofibrations and the front and back squares of* (7.5.6) *are pushouts, determining* Q *and* R *as adjunction spaces*

$$Q = B\ {}_f\!\sqcup X, \qquad R = D\ {}_g\!\sqcup Y.$$

Suppose that φ^0, φ^1, *and* φ^2 *are homotopy equivalences. Then the map*

$$\varphi : B\ {}_f\!\sqcup X \to D\ {}_g\!\sqcup Y$$

determined by $\varphi^0, \varphi^1, \varphi^2$ *is a homotopy equivalence.*

Proof The main point of the proof is to replace the adjunction spaces $Q = B\ {}_f\!\sqcup X$, $R = D\ {}_g\!\sqcup Y$ by the corresponding homotopy pushouts $Q' = M(f) \cup X$, $R' = M(g) \cup Y$. The diagram (7.5.6) determines a map $\psi : Q' \to R'$ which agrees with φ^1 on X, with φ^2 on B, and with $\varphi^0 \times 1$ on $A \times]0, 1[$. There is a commutative diagram

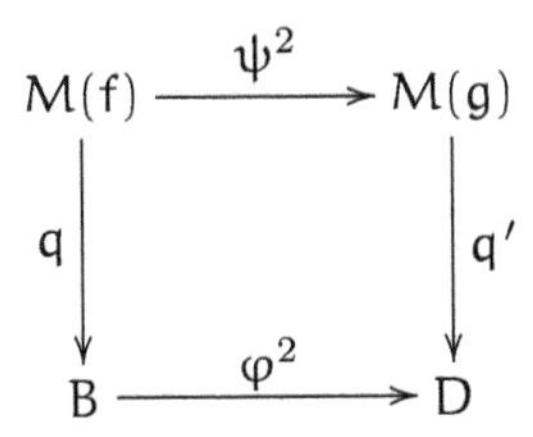

in which ψ^2 is the restriction of ψ and the vertical maps are the standard deformation retractions. Since φ^2, q and q' are homotopy equivalences, it follows that ψ^2 is a homotopy equivalence.

We now apply the gluing theorem 7.4.3 with

$$\begin{aligned} X_0 &= A, & X_1 &= M(f), & X_2 &= B, \\ Y_0 &= C, & Y_1 &= M(g), & Y_2 &= D \end{aligned}$$

to the map ψ. The cofibration and homotopy equivalence conditions for 7.4.3 are satisfied. It follows that ψ is a homotopy equivalence.

However the following diagram is commutative:

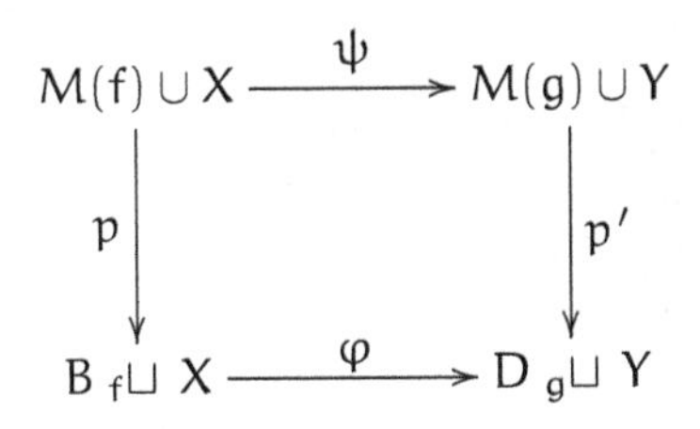

where the vertical maps are the homotopy equivalences of 7.5.4. Since also ψ is a homotopy equivalence, it follows that φ is a homotopy equivalence. □

7.5.7 *(Corollary 1) Let* $(\varphi^0, \varphi^1) : i \to j$ *be a homotopy equivalence of maps where* $i : A \to X$ *and* $j : C \to Y$ *are closed cofibrations. Let* $g : C \to D$ *be a map. Then the induced map*

$$\varphi : D\ {}_{g\varphi^0}\!\sqcup X \to D\ {}_g\!\sqcup Y$$

determined by (φ^0, φ^1) *is a homotopy equivalence.*

Proof This is the case $B = D$, $\varphi^2 = 1$ of 7.5.7. □

7.5.7 *(Corollary 2) Let* $i : A \to X$ *be a closed cofibration and let* $f : A \to B$ *be a homotopy equivalence. Then the induced map*

$$\bar{f} : X \to B \;{}_f\!\sqcup X$$

is a homotopy equivalence.

Proof This is the following special case of 7.5.7:

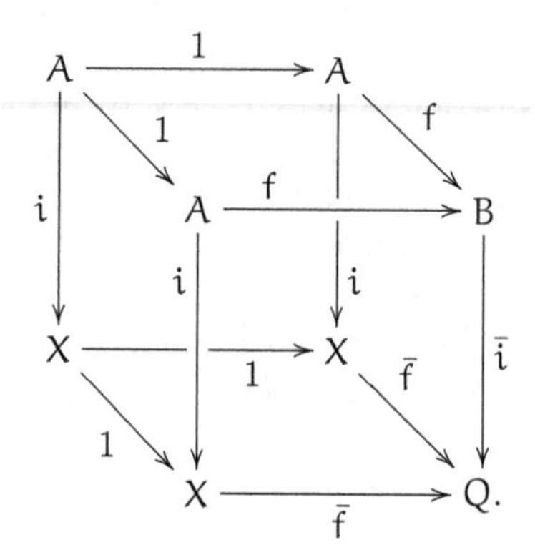

Since f and the identity maps $A \to A$ and $X \to X$ are homotopy equivalences, it follows that $\bar{f}$ is a homotopy equivalence. □

7.5.7 *(Corollary 3) If* $i : A \to X$ *is a closed cofibration and* A *is contractible, then the identification map* $p : X \to X/A$ *is a homotopy equivalence.*

Proof This is the case of the previous corollary when B consists of a single point. □

This last corollary has the following simple proof. Let $F_t : A \to A$ be a homotopy such that $F_0 = 1$ and F_1 is constant. Let $i : A \to X$ be the inclusion. Then iF_t extends to a homotopy $G_t : X \to X$ such that $G_0 = 1$ and $G_1[A]$ is a single point of A. Therefore, G_1 defines a map $g : X/A \to X$ such that $gp = G_1$—whence $gp \simeq 1_X$. Also, since $G_t[A] \subseteq A$, G_t defines a homotopy $H_t : X/A \to X/A$ such that $H_1 p = pG_t$. Here $H_0 = 1_{X/A}$ and $H_1 p = pG_1 = pgp$, whence $H_1 = pg$. This proves that g is a homotopy inverse of p.

Similar proofs can be given for some other cases of 7.5.7 (Corollary 2) (this remark is useful in the solution of Exercise 8).

7.5.7 (Corollary 2) is often useful when combined with 4.5.8 as follows. Suppose given a Hausdorff space Q, a closed subspace B of Q and a closed subspace A of the compact space X. Let $h : X \to Q$ be a map such that $h[A] \subseteq B$ and $h \mid X \setminus A, Q \setminus B$ is defined and is a bijection. Let $f = h \mid A, B$. Finally, let (X, A) be cofibred.

7.5.8 *Under these conditions, if* $f : A \to B$ *is a homotopy equivalence, then so also is* $h : X \to Q$.

Proof By 4.5.8 there is a homeomorphism $g : B \,{}_f\!\sqcup X \to Q$ such that $g\bar{f} = h$. By 7.5.7 (Corollary 2) $\bar{f}$ is a homotopy equivalence. Therefore h is a homotopy equivalence. □

The previous results are false without some conditions on the pair (X, A).

EXAMPLE

1. Consider the subspaces of $\mathbb{R}$

$$Y = \{0\} \cup \bigcup_{n \geqslant 1} \left[\frac{1}{2n}, \frac{1}{2n-1}\right], \quad B = \{0, 1\} \cup \bigcup_{n \geqslant 1} \left[\frac{1}{2n+1}, \frac{1}{2n}\right].$$

Let $i : \mathbb{L} \to B$ be the inclusion, let $f = q : M(i) \to B$, $A = M(i)$, $X = A \cup Y$. Then $B \,{}_f\!\sqcup X \cong B \,{}_i\!\sqcup Y \cong \mathbb{I}$. But X is not path-connected (draw the picture and compare with Example 2, p. 80). So $\bar{f}$ is not a homotopy equivalence even though f is.

We now give some applications of the preceding results, first of all using only 7.5.5 (Corollary 1).

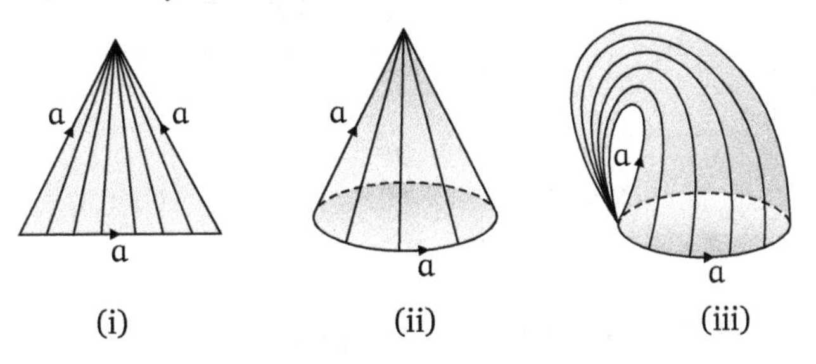

Fig. 7.11

The *dunce's hat* is obtained from a triangle by identifying the sides according to the pattern shown in the above figure. Two steps in the identification are shown; the final step is the identification of the two thickly drawn circles in (iii). The following, somewhat intuitive, discussion shows that the dunce's hat is contractible.

We can think of the dunce's hat as a adjunction space $\mathbb{S}^1 {}_f\sqcup \Delta^2$ where $f : \dot{\Delta}^2 \to \mathbb{S}^1$ is determined by the identification shown in (i). In fact, regarding each side a of Δ^2 as a map of a line segment onto $\mathbb{S}^1$, we can write

$$f = a + a - a.$$

It is clear that f is homotopic to the obvious homeomorphism $g : \dot{\Delta}^2 \to \mathbb{S}^1$. But $\mathbb{S}^1 {}_g\sqcup \Delta^2$ is homeomorphic to $\mathbb{E}^2$. By 7.5.5 (Corollary 1), the dunce's hat is of the homotopy type of $\mathbb{E}^2$ and so is contractible.

Our next applications are to joins and smash products. For the remainder of this section *all spaces will be assumed pointed, compact, and Hausdorff.* The base points of X, Y are to be x_0, y_0 respectively. The base point of $X * Y$ will then be $e = \frac{1}{2}x_0 + \frac{1}{2}y_0$; the base point of SX will be the top vertex v_1 (i.e., the set $X \times 1$). We will also identify $X \vee Y$ with the space obtained from $X \sqcup Y$ by identifying the base point of X with that of Y. In CZ we identify Z with the subspace$Z \times 1$ by the map $z \mapsto (z, 1)$.

7.5.9 *There is a homotopy equivalence*

$$S(X \times Y) \to (X * Y) \vee SX \vee SY.$$

Proof

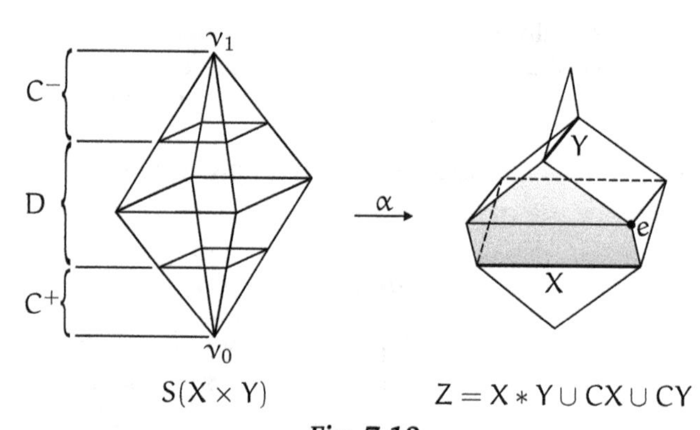

Fig. 7.12

The points of $S(X \times Y)$ we write as (z, t) for z in $X \times Y$ and t in $\mathbb{I}$, so that $(z, 0) = v_0$, $(z, 1) = v_1$, for any z in $X \times Y$. Let C^-, C^+, D be respectively the set of points (z, t) such that $t \geqslant \frac{3}{4}$, $t \leqslant \frac{1}{4}$, $\frac{1}{4} \leqslant t \leqslant \frac{3}{4}$. Obviously, we have a homeomorphism from $S(X \times Y)$ to

$$W = (X \times Y \times \mathbb{I}) \cup C(X \times Y \times 0) \cup C(X \times Y \times 1)$$

in which, C^-, C^+, D are mapped to $C(X \times Y \times 1)$, $C(X \times Y \times 0)$ and $X \times Y \times \mathbb{I}$ respectively. So we now replace $S(X \times Y)$ by W.

Consider the maps

$$p : C(X \times Y \times 0) \to CX \qquad p' : C(X \times Y \times 1) \to CY$$
$$(x, y, 0, t) \mapsto (x, t), \qquad (x, y, 1, t) \mapsto (y, t),$$
$$q : X \times Y \times \mathbb{I} \to X * Y$$
$$(x, y, s) \mapsto (1 - s)x + sy.$$

Clearly p, p' agree with q on $X \times Y \times \dot{\mathbb{I}}$; therefore, these maps define a map

$$\alpha : W \to Z = X * Y \cup CX \cup CY.$$

Now p, p' are homotopy equivalences, since they each are maps from a contractible space to a contractible space. By 7.5.8 α is a homotopy equivalence (in 7.5.8 take $X = W$, $A = C(X \times Y \times 0) \cup C(X \times Y \times 1)$; the fact that (X, A) is cofibred is an easy consequence of results of section 7.3). We complete the proof by showing that Z is of the homotopy type of $(X * Y) \vee SX \vee SY$.

The inclusion $\iota : X \to X * Y$ is homotopic to the constant map c with value y_0, since a homotopy $i \simeq c$ is given by

$$(x, t) \mapsto (1 - t)x + ty_0$$

(in Fig. 7.12, the shaded face of $X * Y$ is homeomorphic to CX, and this homotopy is simply sliding X up the cone). Also, c itself is homotopic to the constant map with value e (such a homotopy is given by $(x, t) \mapsto (1 - t)y_0 + te$). Since i is the attaching map of the cone CX and CX/X is homeomorphic to SX, it follows from 7.5.5 (Corollary 1) that Z is of the homotopy type of

$$((X * Y) \vee SX) \cup CY.$$

A similar argument shows that this space is of the homotopy type of $(X * Y) \vee SX \vee SY$. □

A pointed space X is called *well-pointed* if $(X, \{x_0\})$ is cofibred. We recall that all spaces are assumed compact and Hausdorff.

7.5.10 *There is an identification map*

$$X * Y \to \Sigma(X \divideontimes Y)$$

which is a homotopy equivalence if X and Y are well-pointed.

Proof We proved in chapter 5 that $\Sigma(X \divideontimes Y)$ is homeomorphic to an identification space $(X * Y)/R$. Thus, to complete the proof we have only to show that if X, Y are well-pointed then R is contractible and $(X * Y, R)$ is cofibred.

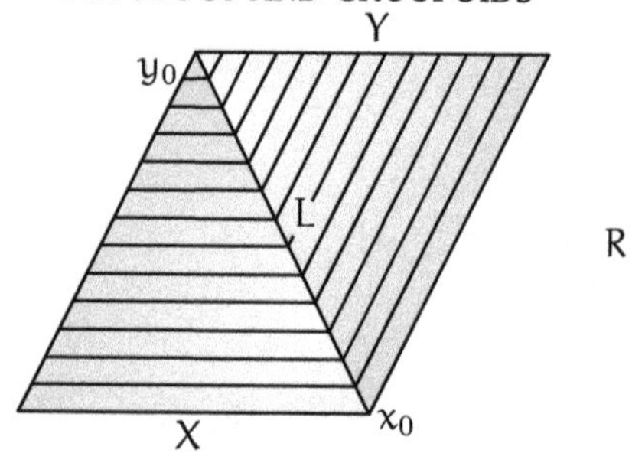

Fig. 7.13

Let L be the subspace of R of points rx_0+sy_0. Then R is the union of two cones with a common 'generator' L. The inclusion $L \to CX$ is a homotopy equivalence, since both L and CX are contractible. Also $(X \times \mathbb{I}, X \times 0 \cup x_0 \times \mathbb{I})$ is cofibred by 7.3.8 and hence so also does $(CX, Cx_0) = (CX, L)$. By 7.2.8 (Corollary 1) L is a deformation retract of CX. Similarly, L is a deformation retract of CY and it follows that L is a deformation retract of R. Therefore, R is contractible, since L is contractible.

Let $J : X \times Y \times \Delta^1 \to X * Y$ be the identification map of 5.7.2 (Corollary 1). Then

$$J^{-1}[R] = x_0 \times Y \times \Delta^1 \cup X \times y_0 \times \Delta^1 \cup X \times Y \times \dot{\Delta}^1.$$

Now 7.3.7 applied to the product of the three pairs (X, x_0), (Y, y_0) and $(\Delta^1, \dot{\Delta}^1)$ show that $(X\times Y\times\Delta^1, J^{-1}[R])$ is cofibred. Since J is an identification map, it follows that $(X * Y, R)$ is cofibred, and the proof is complete. □

7.5.11 *If X is well-pointed then the identification map*

$$SX \to \Sigma X = SX/Sx_0$$

is a homotopy equivalence.

Proof By an argument similar to that in 7.5.10 the pair (SX, Sx_0) is cofibred. Since Sx_0 is contractible, the result follows. □

7.5.12 *If X and Y are well-pointed then there is a homotopy equivalence*

$$\Sigma(X \times Y) \to \Sigma(X \divideontimes Y) \vee \Sigma X \vee \Sigma Y.$$

Proof As base point of $X \times Y$ we take (x_0, y_0) where x_0, y_0 are the base points of X, Y respectively. By 7.3.2, $X \times Y$ is well-pointed and so $\Sigma(X \times Y)$ is of the same homotopy type as $S(X \times Y)$. Similarly, we have a map

$$(X * Y) \vee SX \vee SY \to \Sigma(X \divideontimes Y) \vee \Sigma X \vee \Sigma Y$$

which is a homotopy equivalence on each 'summand' of the wedge. By two applications of the gluing theorem 7.4.3, this map is a homotopy equivalence (we leave the reader to verify the necessary cofibration conditions). The result now follows from 7.5.9. □

7.5.12 *(Corollary 1) There is a homotopy equivalence*

$$\Sigma(\mathbb{S}^m \times \mathbb{S}^n) \to \mathbb{S}^{m+n+1} \vee \mathbb{S}^{m+1} \vee \mathbb{S}^{n+1}.$$

Proof This is the case $X = \mathbb{S}^m$, $Y = \mathbb{S}^n$ of 7.5.12. □

EXERCISES

1. Let X, Y, Z be spaces and $h : X \to Z$ a map. (i) Let $f : X \to Y$ be a map; prove that there is a map $g : Y \to Z$ such that $gf \simeq h$ if and only if h extends over $M(f)$. (ii) Let $g : Y \to Z$; prove that there is a map $f : X \to Y$ such that $gf \simeq h$ if and only if $jh : X \to M(g)$ is deformable into Y (where $j : Z \to M(g)$ is the inclusion). Interpret these statements in the case $X = Z$, $h = 1$.
2. Let $\mathbb{L}$ have base point 0, and let $C\mathbb{L}$ have the same base point as $\mathbb{L}$. Prove that $C\mathbb{L} \vee C\mathbb{L}$ is not contractible. [Let $Z = C\mathbb{L} \vee C\mathbb{L}$ and let ν_1, ν_{-1} be the vertices of the two cones of Z. Let F_t be a homotopy $Z \to Z$ such that $F_0 = 1$, F_1 is constant. Without loss of generality it may be assumed that $F_t 0$ takes the values ν_1, ν_{-1}. Let $\alpha_\varepsilon = \inf\{t : F_t(0) = \nu_\varepsilon\}$ for $\varepsilon = 1, -1$. Show that either assumption $\alpha_1 < \alpha_{-1}$, $\alpha_{-1} < \alpha_1$ leads to a contradiction.]
3. (i) Let $f : \mathbb{L} \to \mathbb{L} \times C\mathbb{L}$ be the map $x \mapsto (x, 0)$. Prove that $\bar{f} : \mathbb{I} \to (\mathbb{L} \times C\mathbb{L}) \,{}_f\!\sqcup \mathbb{I}$ is not a homotopy equivalence. (ii) Prove that if $e \in \mathbb{S}^n$, then $\mathbb{S}^n/(\mathbb{S}^n \setminus \{e\})$ is contractible.
4. Prove that if X, Y are compact and Hausdorff then there is a homotopy equivalence
$$SX \times Y/\nu_1 \times Y \to (X * Y) \vee SX.$$
Deduce that if, further, X, Y are well-pointed, then there is a homotopy equivalence
$$\Sigma X \times Y/\cdot \times Y \to \Sigma(X \circledast Y) \vee \Sigma X.$$
5. Let $X_1, \dots, X_n$ be compact, Hausdorff, well-pointed spaces. For any non-empty subset N of $\{1, \dots, n\}$ let Z_n denote the smash product of the spaces X_i for i in N. Prove that there is a homotopy equivalence from $\Sigma(X_1 \times \cdots \times X_n)$ to the wedge of the spaces ΣZ_N for all non-empty subsets N of $\{1, \dots, n\}$.
6. Let A be a set of m points in $\mathbb{S}^n$ $(n \geqslant 1)$. Prove that $\mathbb{S}^n/A$ has the homotopy type of the wedge of $\mathbb{S}^n$ with $(m-1)$ circles.
7. Let A be a closed subspace of X. Prove that the following conditions are equivalent: (i) There is a map $\nu : X \to \mathbb{I}$ and a homotopy $\psi_t : 1_X \simeq \psi_1 \operatorname{rel} A$ such that $\nu[A] = \{0\}$, and for all $x \in X$, $\nu x < 1$ implies $\psi_1 x \in A$. (ii) There is a neighbourhood V of A in X, a map $u : X \to \mathbb{I}$ such that $u[X \setminus V] = \{1\}$ and $u[A] = \{0\}$, and a homotopy $\varphi_t : V \to X \operatorname{rel} A$ such that φ_0 is the inclusion $V \to X$ and $\varphi_1[V] \subseteq A$. [If these equivalent conditions hold, we say A is a WNDR (*weak neighbourhood deformation retract*) of X.]

8. Let $i : A \to X$ be the inclusion of the closed subspace A of X and let A be identified as usual with the subspace $A \times 1$ of $M(i)$. Let $p : M(i) \to X$ be the projection. Prove that the following conditions are equivalent: (i) A is a WNDR of X, (ii) there is a homotopy $h_t : M(i) \to M(i)$ of the identity rel A such that

$$h_1[A \times \mathbb{I}] \subseteq A, \qquad h_t[a \times \mathbb{I}] \subseteq a \times \mathbb{I}, \qquad a \in A,$$

(iii) $p : M(i) \to X$ is a homotopy equivalence rel A, (iv) there is a homotopy equivalence $M(i) \to X$ rel A, (v) (X, A) has the WHEP. [cf. [Pup67].]
9. Consider the following properties of a pair (X, A), which we label as the *rather*, *very* and *completely* WHEP. (RWHEP) If $f : X \to Y$ is a map and $u_t : A \to Y$ is a homotopy of $f \mid A$, then u_t is homotopic rel end maps to a homotopy which extends to a homotopy of f. (VWHEP) If $f : X \to Y$ is a map, $u = f \mid A$ and $u \simeq v$, then some homotopy $u \simeq v$ extends to a homotopy of f. (CWHEP) If $u, v : A \to Y$ are maps and $u \simeq v$, then u extends over X if and only if v extends over X. Prove that WHEP $\Rightarrow$ RWHEP $\Rightarrow$ VWHEP $\Rightarrow$ CWHEP. Prove also that CWHEP is not a property invariant under homotopy equivalence of pairs.
10. Show that 7.5.7 (Corollary 3) is true if (X, A) satisfies only the VWHEP.
11. Let $\mathbf{f} : (X, A) \to (Y, B)$ have homotopy inverse $\mathbf{g}$ such that $\mathbf{fg} \simeq 1$ rel B. Prove that if (Y, B) has the VWHEP, then so also does (X, A).
12. Let $(X, A), (Y, B)$ have the WHEP, and let $Z = (X \times Y)/(X \times B \cup A \times Y)$, with the usual base point. Prove that $(Z, \cdot)$ has the WHEP.
13. If $f : X \to Y$ is a map, then there is an inclusion map $f' : Y \to C(f)$ where $C(f)$ is the mapping cone [Example 5 of Section 4.5]. The following sequence of maps

$$X \xrightarrow{f} Y \xrightarrow{f'} C(f) \xrightarrow{f''} C(f') \xrightarrow{f'''} C(f'') \longrightarrow \cdots \xrightarrow{f^{(n)}} C(f^{(n-1)})$$

is called the (*unpointed*, or *free*) *Puppe sequence*; here $f^{(n)}$ is defined inductively by $f^{(n)} = (f^{(n-1)})'$. Prove that there is a diagram, commutative up to homotopy

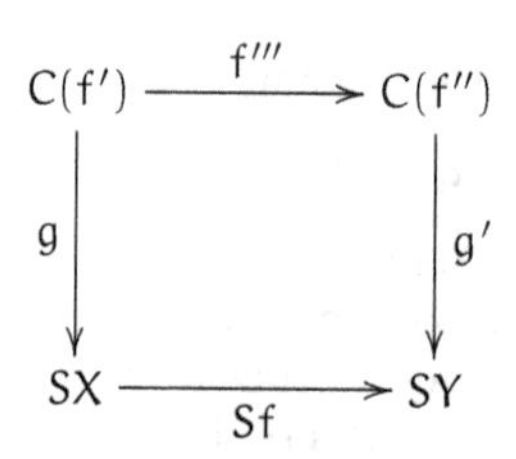

in which g, g' are homotopy equivalences.
14. Prove a result similar to that of the previous exercise for the pointed case, in which C is replaced by Γ and S by Σ.
15. A sequence $A_1 \xleftarrow{a_1} A_2 \xleftarrow{a_2} A_3 \longleftarrow \cdots$ of pointed sets and pointed functions is called *exact* if for each $i \geqslant 2$, $\operatorname{Im} a_i = a_{i-1}^{-1}[\cdot]$. A sequence $X_1 \xrightarrow{f_1} X_2 \xrightarrow{f_2} X_3 \longrightarrow \cdots$ of pointed maps of pointed spaces is called *exact* if the induced sequence of sets

$$[X_1, Z]_\cdot \xleftarrow{f_1^*} [X_2, Z]_\cdot \xleftarrow{f_2^*} [X_3, Z]_\cdot \longleftarrow \cdots$$

is exact for any pointed space—here the base point of $[X_n, Z]\cdot$ is the class of the constant map and f_i^* is $\mathrm{cls}\, g \mapsto \mathrm{cls}\, gf_i$. Prove that the (pointed) Puppe sequence is exact. Deduce that for any pointed map $f : X \to Y$ there is an exact sequence

$$X \xrightarrow{f} Y \xrightarrow{f'} \Gamma(f) \longrightarrow \Sigma X \xrightarrow{\Sigma f} \Sigma Y \xrightarrow{\Sigma f'} \Sigma\Gamma(f) \longrightarrow \cdots .$$

Prove further that if X is a closed subspace of Y, $f : X \to Y$ is the inclusion and (Y, X) is cofibred, then there is an exact sequence

$$X \xrightarrow{f} Y \xrightarrow{p} Y/X \longrightarrow \Sigma X \xrightarrow{\Sigma f} \Sigma Y \longrightarrow \Sigma(Y/X) \longrightarrow \cdots .$$

16. Compare the last sequence with the sequence of sets in the proof of 7.2.12.
17. Let $i : X \to Y$ be the inclusion of the closed subspace X of Y. Let (Y, X) be a pointed pair cofibred. We say X is *retractile* in Y if $\Sigma i : \Sigma X \to \Sigma Y$ has a left-homotopy inverse. Prove that the following conditions are equivalent: (i) X is retractile in Y, (ii) ΣX is a retract of ΣY, (iii) for each Z the following sequence is exact

$$0 \to [\Sigma(Y/X), Z]\cdot \xrightarrow{p^*} [\Sigma Y, Z]\cdot \xrightarrow{i^*} [\Sigma X, Z]\cdot \to 0$$

where 0 denotes a trivial group [cf. [Jam84]].
18. Prove that if X, Y are well-pointed, then $X \vee Y$ is retractile in $X \times Y$.
19. Let $f : X \to Y$ be a pointed map. Prove that Y is retractile in $\Gamma(f)$ if and only if Σf is inessential.
20. Let X, Y be well-pointed, compact, and Hausdorff. The homotopy equivalences $X * Y \to \Sigma(X \divideontimes Y)$, $SX \to \Sigma X$, $SY \to \Sigma Y$ and the Whitehead product map $w : X * Y \to SX \vee SY$ determine a Whitehead product map $w' : \Sigma(X \divideontimes Y) \to \Sigma X \vee \Sigma Y$ uniquely up to homotopy. Prove that $\Sigma w'$ is inessential [cf. [Ark62]].
21. Let X, Y be well-pointed, and let $Z = \Sigma X \vee \Sigma Y$. Let $\rho_1 : \Sigma(X \times Y) \to Z$ be $i_1(\Sigma p_1)$ where $p_1 : X \times Y \to X$ is the projection and $i_1 : \Sigma X \to Z$ is the inclusion. Let $\rho_2 : \Sigma(X \times Y) \to Z$ be defined similarly as $i_2(\Sigma p_2)$. Prove that the following sequence

$$0 \to [\Sigma(X \divideontimes Y), Z]\cdot \xrightarrow{p^*} [\Sigma(X \times Y), Z]\cdot \xrightarrow{i^*} [Z, Z]\cdot \to 0$$

is exact, and that if w' is defined as in the previous exercise then

$$p^*(\mathrm{cls}\, w') = -\sigma_2 - \sigma_1 + \sigma_2 + \sigma_1 \text{ where } \sigma_i = \mathrm{cls}\, \rho_i .$$

7.6 The cellular approximation theorem

As motivation for the work of this section we consider the following question: suppose X, B are cell complexes, A is a subcomplex of X and $f : A \to B$ is a map; is then $B\, {}_f\!\sqcup\, X$ of the homotopy type of a cell complex? We know that $B\, {}_f\!\sqcup\, X$ is a cell complex if f is a cellular map. Also we know that the homotopy type of $B\, {}_f\!\sqcup\, X$ depends only on the homotopy class of f (by 7.5.5 (Corollary 1) and since (X, A) is cofibred). Our question is thus answered by the *cellular approximation theorem*, a special case of which asserts that any map $f : A \to B$ is homotopic to a cellular map. The word approximation here is used in a rough sense only—we will not be concerned with questions of metrics nor with a real number $\varepsilon > 0$.

The main technical work is in the following result; the elegant formulation of the proof is due to J. F. Adams.

7.6.1 *The following statements are true for each* $n \geqslant 1$.

$\alpha(n)$ *Any map* $\mathbb{S}^r \to \mathbb{S}^n$ *with* $r < n$ *is inessential.*

$\beta(n)$ *Any map* $\mathbb{S}^r \to \mathbb{S}^n$ *with* $r < n$ *extends over* $\mathbb{E}^{r+1}$.

$\gamma(n)$ *Let* B *be path-connected and let* Q *be formed by attaching a finite number of* n*-cells to* B*. Then any map*

$$(\mathbb{E}^r, \mathbb{S}^{r-1}) \to (Q, B)$$

with $r < n$ *is deformable into* B.

The proof is by induction by means of the implications

$$\gamma(n) \Rightarrow \alpha(n) \Leftrightarrow \beta(n) \Rightarrow \gamma(n+1)$$

the only difficult step being the proof of $\beta(n) \Rightarrow \gamma(n+1)$. The start of the induction—the proof of $\gamma(1)$—is easy; in fact, since $\mathbb{E}^0$ consists of a single point and $\mathbb{S}^{-1}$ is the empty set, $\gamma(1)$ is equivalent to the statement that Q is path-connected, and this is a special case of 4.6.3.

Proof of $\gamma(n) \Rightarrow \alpha(n)$

Let $f : \mathbb{S}^r \to \mathbb{S}^n$ be a map such that $r < n$. Let $p : \mathbb{E}^r \to \mathbb{S}^r$ be an identification map which shrinks the boundary $\mathbb{S}^{r-1}$ of $\mathbb{E}^r$ to a point x of $\mathbb{S}^r$, and let $e^0 = \{fx\}$. Then $\mathbb{S}^n$ has a cell structure

$$\mathbb{S}^n = e^0 \cup e^n$$

and so $fp : \mathbb{E}^r \to \mathbb{S}^n$ defines a map

$$g : (\mathbb{E}^r, \mathbb{S}^{r-1}) \to (e^0 \cup e^n, e^0).$$

The hypothesis $\gamma(n)$ implies that g is deformable into e^0. By 7.4.2 g is deformable into e^0 rel $\mathbb{S}^{r-1}$. Throughout this homotopy $\mathbb{S}^{r-1}$ is mapped to the point fx, and so, by Remark 7.3.6, this homotopy defines a homotopy $f \simeq f' : \mathbb{S}^r \to \mathbb{S}^n$ rel$\{x\}$ such that f' is constant. Thus f is inessential. □

Proof of $\alpha(n) \Leftrightarrow \beta(n)$

The proof depends on the fact that there is a homeomorphism $\mathbb{E}^{r+1} \to C\mathbb{S}^r$ which is the identity on $\mathbb{S}^r$ (where $\mathbb{S}^r$ is identified with the subset $\mathbb{S}^r \times 1$ of

$C\mathbb{S}^r$). Let $p : \mathbb{S}^r \times \mathbb{I} \to C\mathbb{S}^r$ be the identification map and let $g : \mathbb{S}^r \to X$ denote a constant map.

A homotopy $g \simeq f : \mathbb{S}^r \to X$ is a map $\mathbb{S}^r \times \mathbb{I} \to X$ which is $(x, 0) \mapsto gx$ on $\mathbb{S}^r \times 0$ and $(x, 1) \mapsto fx$ on $\mathbb{S}^r \times 1$. Since g is constant, such a homotopy defines an extension $C\mathbb{S}^r \to X$ of f. Conversely, if $f : \mathbb{S}^r \to X$ is a map and $F : C\mathbb{S}^r \to X$ is an extension of f, then the composite Fp is a homotopy $g \simeq f$ such that g is constant. □

Proof of $\beta(n) \Rightarrow \gamma(n+1)$

Let $\dot{\mathbb{I}}^r$ denote the boundary of the r-cube $\mathbb{I}^r$, that is, the set of points x of $\mathbb{I}^r$ which have at least one coordinate with a value 0 or 1. The pairs $(\mathbb{E}^r, \mathbb{S}^{r-1})$ and $(\mathbb{I}^r, \dot{\mathbb{I}}^r)$ are homeomorphic and so we can assume $\beta(n)$ in the form that *any map* $\dot{\mathbb{I}}^{r+1} \to \mathbb{S}^n$ *with* $r < n$ *extends over* $\mathbb{I}^{r+1}$. Further, since $(\mathbb{I}^{r+1}, \dot{\mathbb{I}}^{r+1})$ is cofibred, we can by (7.2.7) assume that this is true not only for $\mathbb{S}^n$ but also for any space of the homotopy type of $\mathbb{S}^n$.

We assume that

$$Q = B \,{}_{k_1}\!\sqcup\, \mathbb{E}^{n+1} \,{}_{k_2}\!\sqcup\, \mathbb{E}^{n+1} \cdots {}_{k_m}\!\sqcup\, \mathbb{E}^{n+1}.$$

Let $\overline{k_i} : \mathbb{E}^{n+1} \to Q$ be the usual extension of $k_i : \mathbb{S}^n \to B$. We use these maps to define an open cover of Q.

Let U_i be the image under $\overline{k_i}$ of the set $\{x \in \mathbb{E}^{n+1} : |x| < \frac{2}{3}\}$, and let U_i' be the image under $\overline{k_i}$ of the set $\{x \in \mathbb{E}^{n+1} : |x| > \frac{1}{3}\}$. Let $U = B \cup U_1' \cup \cdots \cup U_m'$. Then U_i and U are open in Q and we set

$$\mathcal{U} = \{U, U_1, \ldots, U_m\}.$$

Notice that $U \cap U_i$ is homeomorphic to the space of points x in $\mathbb{E}^{n+1}$ such that $\frac{1}{3} < |x| < \frac{2}{3}$, and this space is homeomorphic to $\mathbb{S}^n \times]\frac{1}{3}, \frac{2}{3}[$. Hence $U \cap U_i$ is of the homotopy type of $\mathbb{S}^n$, and so we have
(*): $\beta(n)$ *can be applied for maps into* $U \cap U_i$ *rather than* $\mathbb{S}^n$.

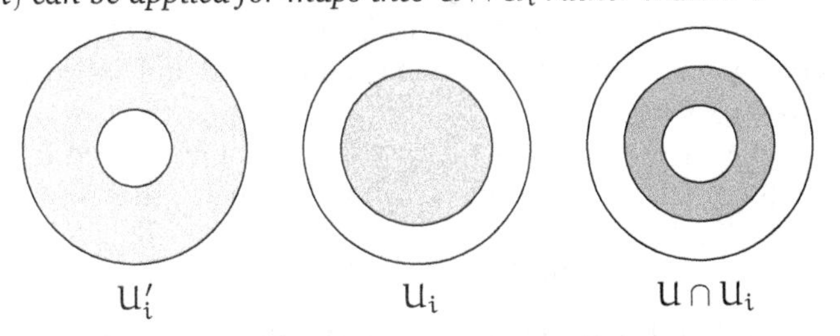

Fig. 7.14

Suppose now we are given a map

$$f : (\mathbb{I}^r, \dot{\mathbb{I}}^r) \to (Q, B)$$

where $r < n + 1$. We must prove that **f** is deformable into B.

By means of hyperplanes in $\mathbb{R}^r$ with equations of the the form

$$x_j = s/N, \qquad s = 1, \ldots, N-1, \qquad j = 1, \ldots, r$$

we may subdivide $\mathbb{I}^r$ into cubes of diameter $\leqslant \sqrt{r}/N$. By the Lebesgue covering lemma, N may be chosen so large that each such cube is mapped by f into some set of $\mathcal{U}$.

Let A be the union of all cubes J of all dimensions in this subdivision such that $f[J] \subseteq U$. Let K^q be the union of all cubes J of dimension $\leqslant q$ (so that $K^{-1} = \emptyset$, K^0 consists of isolated points, and $K^r = \mathbb{I}^r$), and let $K_q = K^q \cup A$.

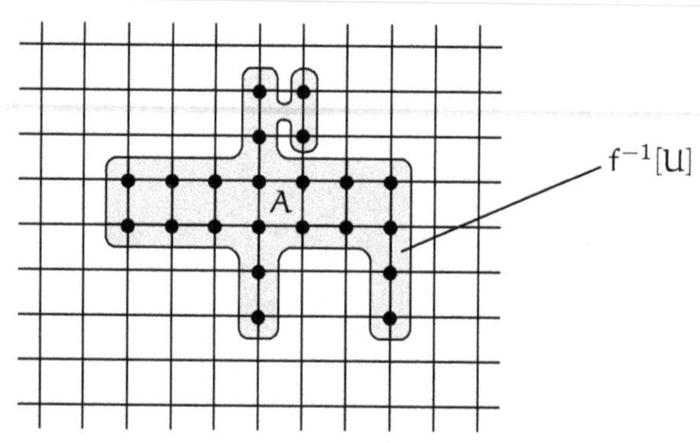

Fig. 7.15

We construct maps $g_q : K_q \to Q$ by induction on q so that the following conditions are fulfilled:

1_q) g_q agrees with f on A, and $g_q \mid K_{q-1} = g_{q-1}$,

2_q) if $x \in K_q$ and $fx \in U_i$ then $g_q x \in U \cap U_i$.

To start the induction we define g_{-1} to be $f \mid A$—clearly conditions 1_{-1}) and 2_{-1}) are satisfied. Suppose g_q has been defined and satisfies 1_q) and 2_q); we extend g_q over K_{q+1}.

Let J^{q+1} be a $(q+1)$-cube of K_{q+1} which is not contained in A. Then for some unique i, $f[J^{q+1}] \subseteq U_i$ and it follows from 2_q) that

$$g_q[\dot{J}^{q+1}] \subseteq U \cap U_i.$$

But $q + 1 \leqslant r < n + 1$, whence $q < n$. By (*) above, g_q extends to a map $J^{q+1} \to U \cap U_i$, and we define $g_{q+1} : K_{q+1} \to Q$ to agree on J^{q+1} with this map. Clearly, conditions 1_{q+1}) and 2_{q+1}) are satisfied, so the induction is complete.

Let $g = g_r : \mathbb{I}^r \to Q$. We prove that $f \simeq g$.

The map $\overline{k_i} : \mathbb{E}^{n+1} \to Q$ maps the set of points x with $|x| < \frac{2}{3}$ bijectively onto U_i: we suppose there is given a linear structure on U_i by means of this bijection. We then define a homotopy $h_t : \mathbb{I}^r \to Q$ by

$$h_t x = \begin{cases} fx, & fx \in \text{ no } U_i \\ (1-t)fx + tgx, & fx \in \text{ some } U_i. \end{cases}$$

Each r-cube J^r in the given subdivision of $\mathbb{I}^r$ is mapped by f into U or into some U_i; thus the formula for h_t shows that h_t is continuous on J^r. Since there cubes from a cover of $\mathbb{I}^r$ by closed subsets, this implies the continuity of h_t, as a function $\mathbb{I}^r \times \mathbb{I} \to Q$.

Clearly $h_0 = f$. We prove that $h_1 = g$.

Suppose J is an r-cube of the subdivision and $x \in J$. If fx belongs to some U_i, then the formula for h_1 shows that $h_1 x = gx$. Suppose fx belongs to no U_i (in which case $h_1 x = fx$). Then $f[J]$ is contained in no U_i and so $f[J] \subseteq U$. Hence $x \in A$ and so $h_1 x = fx = gx$.

Since $\dot{\mathbb{I}}^r \subseteq A$ the homotopy h_t is $\operatorname{rel} \dot{\mathbb{I}}^r$; hence h_t defines a homotopy $\mathbf{f} \simeq \mathbf{g}$ where $\mathbf{g} : (\mathbb{I}^r, \dot{\mathbb{I}}^r) \to (Q, B)$ is the map defined by g.

Finally, $\operatorname{Im} g \subseteq U$ and B is a deformation retract of U. Therefore, $\mathbf{g}$ is deformable into B [cf. Exercise 2 of Section 7.4]. □

The cellular approximation theorem itself is a consequence of the following deformation theorem. (We recall now that the r-skeleton of a cell complex K is written K^r.)

7.6.2 *Let L be a space and $(L_r)_{r\geqslant 0}$ a sequence of subspaces of L such that for all $r \geqslant 0$*

(a) $L_r \subseteq L_{r+1}$,

(b) *any map $(\mathbb{E}^{r+1}, \mathbb{S}^r) \to (L, L_r)$ is deformable into L_{r+1}.*

Let K be a cell complex, A a subcomplex of K and $f : K \to L$ a map such that

$$f[A^r] \subseteq L_r, \qquad r = 0, 1, \ldots.$$

Then f is homotopic $\operatorname{rel} A$ to a map $g : K \to L$ such that

$$g[K^r] \subseteq L_r, \qquad r = 0, 1, \ldots.$$

Proof Let $K_r = A \cup K^r$. We construct a sequence of maps $f^r : K \to L$ and homotopies $f^{r-1} \simeq f^r$ such that

1_r), $f^r[K_s] \subseteq L_s$, $0 \leqslant s \leqslant r$,

2_r), f^r agrees with f^{r-1} on K^{r-1},

3_r), $f^{r-1} \simeq f^r \operatorname{rel} K^{r-1}$.

The induction is started with $f^{-1} = f$ when the above conditions are vacuously satisfied.

By condition (b) any map $(\mathbb{E}^{r+1}, \mathbb{S}^r) \to (L, L_r)$ is deformable into $L_{r+1} \operatorname{rel} \mathbb{S}^r$. It follows easily, using Remark 7.3.6, that if e^{r+1} is any $(r+1)$-cell of $K \setminus A$ then $f^r \mid K^r \cup e^{r+1}$ is deformable into $L_{r+1} \operatorname{rel} K_r$. By applying this to each $(r+1)$-cell of $K \setminus A$ in turn we obtain a homotopy $f^r \mid K_{r+1} \simeq f' \operatorname{rel} K_r$ such that $f[K_{r+1}] \subseteq L_{r+1}$. Since (K, K_{r+1}) is cofibred this homotopy extends to a homotopy $f^r \simeq f^{r+1} \operatorname{rel} K_r$. Clearly, f^{r+1} satisfies $1_{r+1}), 2_{r+1}), 3_{r+1})$.

Since $K = K^N$ for some N, the map $g = f^N$ is the required map. □

7.6.2 *(Corollary 1) (the cellular approximation theorem). Let* K, L *be complexes and* A *a subcomplex of* K. *If* $f : K \to L$ *is a map such that* $f \mid A$ *is cellular, then* f *is homotopic* rel A *to a cellular map.*

Proof This follows from 7.6.1 $\gamma(n)$ and 7.6.2 with $L_r = L^r$. □

Exercises

1. Let L_0 be a subcomplex of L such that for all $r \geqslant 0$ any map $(\mathbb{E}^{r+1}, \mathbb{S}^r) \to (L, L_0)$ is deformable into L_0. Prove that L_0 is a deformation retract of L.
2. Let $f_0, f_1 : K \to L$ be cellular maps and A a subcomplex of K. Suppose that $F : f_0 \simeq f_1$ is a homotopy such that for all $t \in \mathbb{I}$, $F_t[A^r] \subseteq L^{r+1}$, $r = 0, 1, \ldots$. Prove that F is homotopic rel $A \times \mathbb{I} \cup K \times \dot{\mathbb{I}}$ to a homotopy $G : f_0 \simeq f_1$ such that for all $t \in \mathbb{I}$, $G_t[K^r] \subset L^{r+1}$, $r = 0, 1, \ldots$.
3. Let $i : K^2 \to K$ be the inclusion of the 2-skeleton of the cell complex K. Prove that for any subset A of K^2, i induces an isomorphism $\pi K^2 A \to \pi K A$.
4. Prove that $\mathbb{S}^n$ is simply connected for $n > 1$. Prove also that for any point x of $\mathbb{S}^n$ the ith homotopy group $\pi_i(\mathbb{S}^n, x)$ [cf. Section 7.1] is 0 for $i < n$.
5. Prove the cellular approximation theorem for (infinite) CW-complexes.
6. Let K be a connected subcomplex of the complex L. Let $j : K \to L$ be the inclusion and let $x \in K^0$, $e \in \mathbb{S}^r$. Prove that any map $(\mathbb{E}^{r+1}, \mathbb{S}^r) \to (L, K)$ is homotopic to a map h such that $he = x$.
7. Continuing the notation of the previous exercise, suppose that $j_* : \pi_r(K, x) \to \pi_r(L, x)$ is injective. Let $\mathbf{h} : (\mathbb{E}^{r+1}, \mathbb{S}^r) \to (L, K)$ be a map such that $he = x$. Prove that $h^1 : \mathbb{S}^r \to K$ is inessential rel e, and hence show that **h** is homotopic rel e to a map **k** such that $k[\mathbb{S}^r] = \{x\}$.
8. Suppose further to the previous exercise, that $j_* : \pi_{r+1}(K, x) \to \pi_{r+1}(L, x)$ is surjective. Prove that **h** is deformable into K.
9. Let $j : K \to L$ be the inclusion map of the subcomplex K of L. Let K, L be connected, and suppose $j_* : \pi_r(K, x) \to \pi_r(L, x)$ is an isomorphism for all $r \geqslant 1$ (where $x \in K^0$). Prove that K is a deformation retract of L.
10. Let $f : K \to K'$ be a map of connected complexes such that $f_* : \pi_r(K, x) \to \pi_r(K', fx)$ is an isomorphism for all $r \geqslant 1$ (where $x \in K^0$). Prove that f is a homotopy equivalence.

NOTES

The notion of cofibration of spaces has a dual version, that of *fibration.* For accounts of both of these, see many books on algebraic topology or homotopy theory, for example [May99, Hat02]. The notion of fibration of groupoids [Bro70] was borrowed from the notion of fibration of spaces, using the idea that the groupoid **I** is for groupoids a kind of model of the unit interval. Fibrations of spaces and of groupoids generalise the covering maps of spaces and covering morphisms of groupoids which will be treated in chapter 10. The notion of action of a groupoid on sets generalises the notion of action of a group on a set, which is important in many branches of mathematics.

The way in which the notion of homotopy occurs in different contexts has led to a number of abstract homotopy theories (for example [Qui67], [Kam72a], [Mey84], [Bau88]). These have the usual advantages of abstractions in mathematics, namely of (i) covering several examples at the same time, (ii) suggesting the possibility of similar methods covering new examples, and (iii) simplifying proofs, by allowing the concentration on essential features. In particular, the gluing theorem for homotopy equivalences has been proved in these theories (cf. [Shi84]).

The dual of the gluing theorem is called the *cogluing theorem* in [BH70]. It is shown in [Hea70] that the gluing theorem follows from the cogluing theorem. See [Hea78] for more information on the use of groupoid methods in homotopy theory.

The papers [Rut72], [Rut74] give further results on joins and homotopy types.

The notion of double mapping cylinder is an example of a *homotopy pushout*. The general theory of constructions (particularly limits and colimits) 'up to homotopy' is playing an important role in a number of areas of mathematics, such as the applications of algebraic topology to algebraic geometry (cf. [BK72], [Vog73], [CP86] for some basic results and ideas in this area). For other proofs of the gluing lemma, but with different assumptions, see [SW57], [Die71], [Fuc83], [Jam84]. A generalisation of 7.2.8 which includes a result both for fibrations and cofibrations is given in [Lam80].

Track groups (i.e. the object groups of πY^X) were first studied extensively in [Bar55a], [Bar55b], who gave the remarkable calculation that if $X = Y = \mathbb{S}^r \cup e^{r+1}$, $(r \geqslant 2)$, where the cell is attached by a map $\mathbb{S}^r \to \mathbb{S}^r$ of degree 2, then the group of homotopy classes of pointed maps $X \to Y$, with addition given by the representation of X as $\Sigma(\mathbb{S}^{r-1} \cup e^r)$, is cyclic of order 4. The easier part of this work, showing that the group concerned was an extension of $\mathbb{Z}_2$ by $\mathbb{Z}_2$, involved the construction of exact sequences which were later generalised to the relative case by [SW57], and also given an elegant formulation by [Pup58] (cf. Exercise 15 of Section 7.5).

The results of section 7.3 have a long history—see the references in [Pup58] and the papers of Strøm ([Str66], [Str69], [Str72]).

I am indebted to conversations with D. Grayson on the notation and style of the revision of this chapter.

Chapter 8

Some combinatorial groupoid theory

In order to describe in the next chapter various of the groupoids which arise as πXA (for 'nice' spaces X) by application of the van Kampen theorem 6.7.2. To do this, we must build up some of the constructions on groupoids which are needed for this.

Here we encounter the fact that in the first instance groupoids are more complicated than groups, so this seems to make for extra difficulties. On the other hand, we find that we can use one construction on groupoids to describe several constructions on groups, such as free groups and free products of groups. The extra 'spatial component' to groupoids given by the objects of a groupoid, allows for constructions on groupoids which model the geometry, and make clearer the analogies between constructions on spaces and constructions on groupoids.

A key construction on groupoids is to take a groupoid G with object set X and change the object set X by a function $\sigma : X \to Y$. We will obtain a new groupoid $U_\sigma(G)$ and a morphism $\overline{\sigma} : G \to U_\sigma(G)$ which has a φ-universal property, i.e. we can construct morphisms from $U_\sigma(G)$. This construction will turn out to imply constructions of free groupoids and of free products of groupoids, and will also usefully model, as we show in the next chapter, certain constructions on spaces.

Another construction we need is the analogue of quotient groups by a normal subgroup. This construction is again more interesting than in the case of groups, and a generalisation of it will be applied in Chapter 11 in connection with orbit spaces under the action of a group. The underlying graph structure of a groupoid allows groupoids better to model the geometry under consideration than groups alone.

In essence, this chapter sets up some of the basic combinatorial groupoid theory required for applications.

All these constructions are used in section 9.1 for computing fundamental groups of adjunction spaces, and then for more geometric applications.

8.1 Universal morphisms

We have seen that groupoids model well the product and sum of topological spaces. In this section, we construct what in full could be called φ-universal groupoids with respect to Ob, but which we shall find convenient to call simply *universal groupoids* and *universal morphisms*. The universal properties developed here are a special case of a more general theory involving what are called *opfibrations of categories*, which is important in many areas of mathematics, and which tends to have a different terminology from that we use here. This is discussed in the Notes at the end of the chapter. Certainly the analogy should be borne in mind with the situation for final topologies, where the forgetful functor giving the underlying set $\mathsf{Top} \to \mathsf{Set}$ is an opfibration. However, in an algebraic category such as that of groupoids the construction of the universal objects is radically different from that for spaces.

Let G be a groupoid, X a set and

$$\sigma : \mathrm{Ob}(G) \to X$$

a function. We shall construct from G and σ a groupoid U and morphism $\overline{\sigma} : G \to U$ such that (i) $\mathrm{Ob}(U) = X$ and (ii) $\overline{\sigma}$ is exactly σ on objects.

Our exposition gives the construction first and the universal property afterwards, although there is a case for taking the opposite order!

The idea of this construction is as follows (we use multiplicative notation for groupoids throughout most of this chapter). Let $a_1 \in G(x_1, y_1)$, $a_2 \in G(x_2, y_2)$. Then $a_2 a_1$ is defined if and only if $y_1 = x_2$; in such a case, a morphism $\overline{\sigma} : G \to U$ will satisfy

$$\overline{\sigma} a_2 \overline{\sigma} a_1 = \overline{\sigma}(a_2 a_1).$$

Suppose, however, that $y_1 \neq x_2$ but $\sigma y_1 = \sigma x_2$. Then $\overline{\sigma} a_2 \overline{\sigma} a_1$ will be defined in U but will not necessarily be $\overline{\sigma} b$ for any b. So, in order to construct U from G and σ, we must have a method of constructing elements such as $\overline{\sigma} a_2 \overline{\sigma} a_1$. This is done by means of 'words' in the elements of G.

Let $x, x' \in X$. A *word of length* n ($n \geqslant 1$) from x to x' is a sequence

$$a = (a_n, \ldots, a_1)$$

of elements of G such that, if $a_i \in G(x_i, x'_i)$, $i = 1, \ldots, n$, then
(a) $x'_i \neq x_{i+1}$, $i = 1, \ldots, n-1$,
(b) $\sigma x'_i = \sigma x_{i+1}$, $i = 1, \ldots, n-1$,
(c) $\sigma x_1 = x$, $\sigma x'_n = x'$,
(d) no a_i is the identity.
These conditions ensure that no product $a_{i+1}a_i$ is defined in G, but that $\bar{\sigma}a_{i+1}\bar{\sigma}a_i$ will be defined in U.

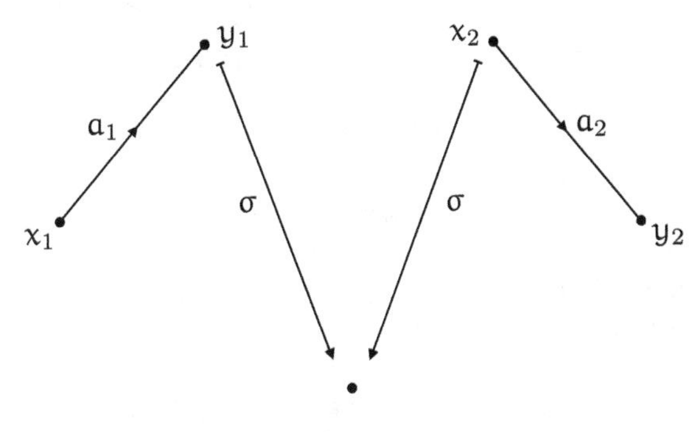

Fig. 8.1

The set of all words from x to x' of length $\geqslant 1$ is written $U(x, x')$; except that, if $x = x'$, we also include in $U(x, x)$ the empty word of length 0 which we write $(\ )_x$ (and thus suppose that $(\ )_x \neq (\ )_y$ when $x \neq y$).

We now define by induction on length a multiplication

$$U(x', x'') \times U(x, x') \to U(x, x'').$$

The empty word in $U(x, x)$ is to act as identity; that is, if $a \in U(x, x')$ then

$$(\ )_{x'}a = a, \ a(\ )_x = a.$$

Suppose that a is as above and $b = (b_m, \ldots, b_1) \in U(x', x'')$, where $b_j \in G(y_j, y'_j)$. We define ba to be, in the various cases:

$(b_m, \ldots, b_1, a_n, \ldots, a_1)$ if $y_1 \neq x'_n$,

$(b_m, \ldots, b_2, b_1 a_n, a_{n-1}, \ldots, a_1)$ if $y_1 = x'_n$ but $b_1 a_n \neq 1$,

$(b_m, \ldots, b_2)(a_{n-1}, \ldots, a_1)$ by induction if $y_1 = x'_n$ and $b_1 a_n = 1$.

This can be expressed as: multiply the two words by putting them end to end, computing in G and cancelling identities where possible.

The definition of multiplication shows that

$$(a_n, \ldots, a_1)(a_1^{-1}, \ldots, a_n^{-1}) = (\ )_{x'},$$
$$(a_1^{-1}, \ldots, a_n^{-1})(a_n, \ldots, a_1) = (\ )_x.$$

Thus each word has a left and right inverse.

We now show that multiplication is associative.

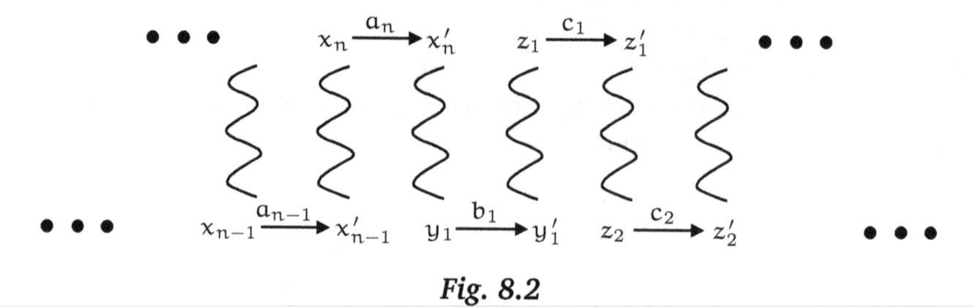

Fig. 8.2

Let $c = (c_r, \ldots, c_1) \in U(x'', x''')$ where $c_k \in G(z_k, z'_k)$. If b is of length 0 then certainly $c(ba) = ca = (cb)a$; this is true, similarly, if a or c is of length 0. Suppose that $m = 1$; we check the value of $c(ba)$ and $(cb)a$ in each case that can arise. These values are

$$
\begin{array}{ll}
(c_r, \ldots, c_1, b_1, a_n, \ldots, a_1) & \text{if } x'_n \neq y_1,\, y'_1 \neq z_1, \\
(c_r, \ldots, c_1, b_1 a_n, a_{n-1} \ldots, a_1) & \text{if } y'_1 \neq z_1,\, x'_n = y_1,\, b_1 a_n \neq 1, \\
(c_r, \ldots, c_1 b_1, a_n, \ldots, a_1) & \text{if } y'_1 = z_1,\, c_1 b_1 \neq 1,\, x'_n \neq y_1, \\
(c_r, \ldots, c_1)(a_{n-1}, \ldots, a_1) & \text{if } y'_1 \neq z_1,\, b_1 a_n = 1, \\
(c_r, \ldots, c_2)(a_n, \ldots, a_1) & \text{if } x'_n \neq y_1,\, c_1 b_1 = 1, \\
(c_r, \ldots, c_1 b_1 a_n, a_{n-1}, \ldots, a_1) & \text{if } y'_1 = z_1,\, x'_n = y_1,\, c_1 b_1 a_n \neq 1, \\
(c_r, \ldots, c_2)(a_{n-1}, \ldots, a_1) & \text{if } y'_1 = z_1,\, x'_n = y_1,\, c_1 b_1 a_n = 1.
\end{array}
$$

If $m > 1$ we proceed by induction. Write $b = b''b'$ with b', b'' shorter than b. Then

$$
\begin{aligned}
c(ba) &= c((b''b')a) = c(b''(b'a)) = (cb'')(b'a) \\
&= ((cb''b')a = (cb)a
\end{aligned}
$$

as was to be shown.

So if we take X as the set of objects of U we have proved that U is a groupoid.

Notice that if x of x', does not belong to $\operatorname{Im}\sigma$, then $U(x, x')$ is either empty (if $x \neq x'$), or contains the identity alone (if $x = x'$). Thus $U = U' \sqcup U''$ where U' is the full subgroupoid on $\operatorname{Im}\sigma$ and U'' is the discrete groupoid on $X \setminus \operatorname{Im}\sigma$.

The groupoid U depends on G and on σ, and we therefore write $U_\sigma(G)$ for U.

We define a morphism $\overline{\sigma} : G \to U_\sigma(G)$ to be σ on objects and to be defined for a in $G(x_1, x_1')$ by

$$\overline{\sigma}a = \begin{cases} (a), & a \neq 1 \\ (\ )_{\sigma x_1}, & x_1 = x_1' \text{ and } a = 1. \end{cases}$$

Thus $\overline{\sigma}$ is injective on the set of non-identities of G and also maps no non-identity to an identity. It is immediate from the definition of the multiplication of $U_\sigma(G)$ that $\overline{\sigma}$ is a morphism.

Although the above construction of $U_\sigma(G)$ gives a good idea of its structure, the important property of $U_\sigma(G)$ from a general point of view is that $\overline{\sigma}$ satisfies a φ-universal property.

Let us, temporarily, identify the set of objects $\mathrm{Ob}(G)$ of a groupoid G with the wide subgroupoid of G whose elements are all identities: thus we regard $\mathrm{Ob}(G)$ as a subgroupoid of G. (A reader who dislikes this convention may instead call this wide subgroupoid $\mathrm{Id}(G)$, and replace suitably Ob by Id in what follows.)

A morphism $f : G \to H$ of groupoids is said to be *universal*, or simply to be a *universal morphism* if the following square:

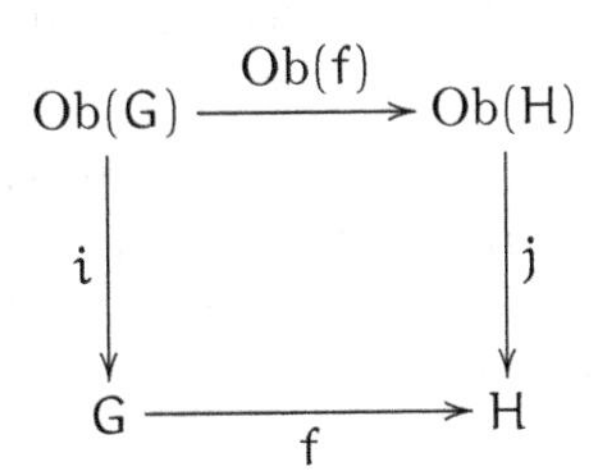

in which the vertical morphisms are inclusions, is a pushout (in the category of groupoids). A useful restatement of this definition is:

8.1.1 *A morphism* $f : G \to H$ *of groupoids is universal with respect to* $\mathrm{Ob}(f)$ *if and only if for any morphism* $g : G \to K$ *of groupoids and any function* $\tau : \mathrm{Ob}(H) \to \mathrm{Ob}(K)$ *such that*

$$\mathrm{Ob}(g) = \tau\,\mathrm{Ob}(f)$$

there is a unique morphism $g^* : H \to K$ *such that* $\mathrm{Ob}(g^*) = \tau$ *and* $g^*f = g$, *as shown in the diagram:*

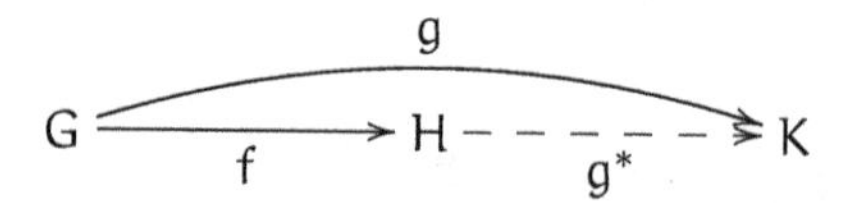

$$\mathrm{Ob}(G) \xrightarrow[\mathrm{Ob}(f)]{} \mathrm{Ob}(H) \xrightarrow[\tau]{} \mathrm{Ob}(K)$$

Proof Let k be the inclusion $\mathrm{Ob}(K) \to K$. Clearly the function $\tau : \mathrm{Ob}(H) \to \mathrm{Ob}(K)$ and the morphism $\tau' : \mathrm{Ob}(H) \to K$ determine each other by the rule $\tau' = k\tau$. The condition $\mathrm{Ob}(g) = \tau\,\mathrm{Ob}(f)$ is equivalent to $gi = \tau'\,\mathrm{Ob}(f)$, and the condition $\mathrm{Ob}(g^*) = \tau$ is equivalent to $g^*j = \tau'$. So the result follows from the definition of pushouts. □

From now on we will often abbreviate 'universal with respect to $\mathrm{Ob}(f)$' simply to 'universal'.

The pushout property ensures that if $f : G \to H$ is universal, then H is determined up to isomorphism by $\mathrm{Ob}(f)$ and G. On the other hand, given G and $\sigma : \mathrm{Ob}(G) \to X$, there is a universal morphism $f : G \to H$ with $\mathrm{Ob}(f) = \sigma$. This is a consequence of:

8.1.2 *Let G be a groupoid and $\sigma : \mathrm{Ob}(G) \to X$ a function. The morphism $\overline{\sigma} : G \to U_\sigma(G)$ is universal.*

Proof We use 8.1.1. Let $g : G \to K$ be a morphism and $\tau : X \to \mathrm{Ob}(K)$ a function such that $\mathrm{Ob}(g) = \tau\sigma$. Let $U = U_\sigma(G)$.

We wish a morphism $g^* : U \to K$ such that $\mathrm{Ob}(g^*) = \tau$. Thus g is determined on the identities of U by τ. An element of U which is not an identity is a word

$$a = (a_n, \dots, a_1), \quad a_i \in G(x_i, x_i')$$

such that $\sigma x_i' = \sigma x_{i+1}$, $i = 1, \dots, n-1$. If $n = 1$, then we set

$$g^*(a_1) = g a_1;$$

this is the only definition consistent with $g^*\sigma^* = g$. If $n > 1$, then

$$g x_i' = \tau\sigma x_i' = \tau\sigma x_{i+1} = g x_{i+1};$$

therefore we can set

$$\begin{aligned} g^* a &= g a_n \dots g a_1 \\ &= g^*(a_n) \dots g^*(a_1) \end{aligned}$$

the product on the right being defined in K. Clearly, this is the only possible definition of a morphism g^* consistent with $g^*\overline{\sigma} = g$, $\mathrm{Ob}(g^*) = \tau$. To complete this proof we verify that this g^* is a morphism.

Suppose a is as above, b is a word $(b_m, \ldots, b_1)$ where $b_j \in G(y_j, y_j')$ and ba is defined. The equation $g^*(ba) = g^*bg^*a$ is clearly true if m or n is 0. Suppose it is true for $m = p - 1$, $n = q - 1$. Then for $m = p$, $n = q$, referring to the definition of ba, we find that $g^*(ba)$ is

$$\begin{array}{ll} gb_m \ldots gb_1 ga_n \ldots ga_1 & \text{if } y_1 \neq x_n', \\ gb_m \ldots gb_2(b_1a_n)ga_{n-1} \ldots ga_1 & \text{if } y_1 = x_n',\ b_1a_n \neq 1, \\ g^*((b_m, \ldots, b_2)(a_{n-1}, \ldots, a_1)) & \text{if } y_1 = x_n',\ b_1a_n = 1. \end{array}$$

This is the same as g^*bg^*a by the inductive hypothesis and the fact that g is a morphism. □

From now on, we identify each non-identity a_1 of G with its image (a_1) in $U_\sigma(G)$.

Let $f : G \to H$ be a morphism. We say f is *strictly universal* if for any morphism $g : G \to K$ such that $\mathrm{Ob}(g)$ factors through $\mathrm{Ob}(f)$, there is a unique morphism $g^* : H \to K$ such that $g^*f = g$ (this is the same as the condition in 8.1.1 except that we drop the requirement that $\mathrm{Ob}(g^*) = \tau$).

8.1.3 *A morphism* $f : G \to H$ *is strictly universal if and only if* f *is universal and* $\mathrm{Ob}(f)$ *is surjective.*

Proof Suppose $f : G \to H$ is a strictly universal. Let $\sigma = \mathrm{Ob}(f)$. The given universal properties imply that there is an isomorphism $f^* : H \to U_\sigma(G)$ such that $f^*f = \overline{\sigma}$, $\mathrm{Ob}(f^*) = 1$ the identity on $\mathrm{Ob}(H)$. It follows easily that $\overline{\sigma} : G \to U_\sigma(G)$ is a strictly universal. This implies that $\sigma = \mathrm{Ob}(f)$ is surjective, for otherwise the condition $g^*\overline{\sigma} = g$ does not always determine g^* on the identities of $U_\sigma(G)$ which are not images by $\overline{\sigma}$ of identities of G.

Suppose given $g : G \to K$ and $\tau : \mathrm{Ob}(H) \to \mathrm{Ob}(K)$ such that $\mathrm{Ob}(g) = \tau\,\mathrm{Ob}(f)$. Since f is a strictly universal there is a unique morphism $g^* : H \to K$ such that $g^*f = g$ and $\mathrm{Ob}(g^*) = \tau$. But the last condition is redundant, since $g^*f = g$ implies $\mathrm{Ob}(g^*)\,\mathrm{Ob}(f) = \mathrm{Ob}(g) = \tau\,\mathrm{Ob}(f)$ and so, since $\mathrm{Ob}(f)$ is surjective, that $\mathrm{Ob}(g^*) = \tau$. □

This shows that a strictly universal morphism is analogous to an identification map of topological spaces.

EXAMPLES

1. Let G be a groupoid and σ the unique function from $\mathrm{Ob}(G)$ to a single point set $\{x\}$. Then $U_\sigma(G)$ is a groupoid with only one object, that is, $U_\sigma(G)$

is a group. This group is called the *universal group* of G and is written $\mathsf{U}G$. The morphism $\overline{\sigma} : G \to \mathsf{U}G$ is universal for morphisms from G to groups, i.e., if $g : G \to K$ is a morphism to a group K, then there is a unique morphism $g^* : \mathsf{U}G \to K$ such that $g^*\overline{\sigma} = g$. By the construction of $\mathsf{U}G$, the elements of this group are the identity and also all words $(a_n, \ldots, a_1)$ such that $a_i \in G$, no a_i is the identity and no $a_{i+1}a_i$ is defined in G.

2. In particular, consider the groupoid **I** with two objects $0, 1$ and two non-identities ι, ι^{-1} from 0 to 1, and 1 to 0 respectively. A word of length n in $\mathsf{U}\mathbf{I}$ is

$$\iota\iota\ldots\iota \quad \text{or} \quad \iota^{-1}\iota^{-1}\ldots\iota^{-1}$$

and these we can write ι^n and ι^{-n} respectively. Clearly, there is an isomorphism $\mathsf{U}\mathbf{I} \to \mathbb{Z}$ which sends $\iota^{\pm n} \mapsto n$.

Notice that the computation of $\pi(\mathbb{S}^1, 1)$ in 6.7.5 is now easy—it is immediate from 6.7.4 that there is a strictly universal morphism $\mathbf{I} \to \pi(\mathbb{S}^1, 1)$.

3. Let A be a set. A *free group on* A is a group FA and a function $\lambda : A \to FA$ which is universal for functions from A into groups, that is, if $\mu : A \to G$ is any function to a group G then there is a unique morphism $\mu^* : FA \to G$ such that $\mu^*\lambda = \mu$. We prove that *a free group on* A *always exists.*

Proof Let A be regarded as a discrete groupoid, let FA be the universal group of $A \times \mathbf{I}$. In the following diagram

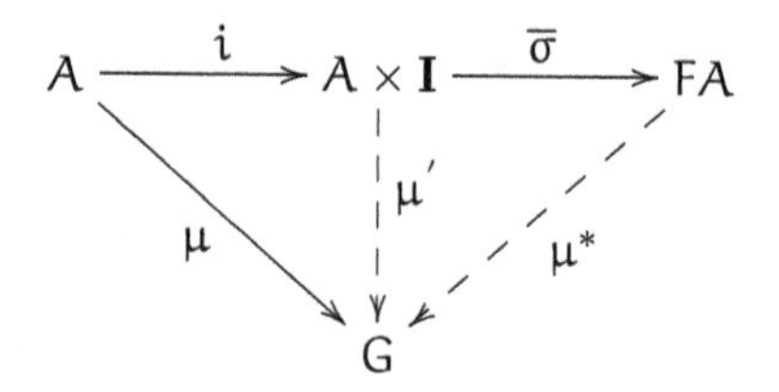

i is the function $a \mapsto (a, \iota)$, $\overline{\sigma}$ is the strictly universal morphism and μ is a given function to a group G. The elements μa of G define uniquely a morphism $\mu' : A \times \mathbf{I} \to G$ which sends $(a, \iota) \mapsto \mu a$ (so that $\mu' i = \mu$). Since μ' is a morphism to a group, μ' defines uniquely $\mu^* : FA \to G$ such that $\mu^*\overline{\sigma} = \mu'$. Hence $\mu^*\overline{\sigma}i = \mu' i = \mu$.

Suppose $\overline{\mu} : FA \to G$ is any morphism such that $\overline{\mu}\overline{\sigma}i = \mu$. Then $\overline{\mu}\overline{\sigma} = \mu'$ whence $\overline{\mu} = \mu^*$. This shows that FA with $\overline{\sigma}i : A \to FA$ is a free group on A. (We generalise this later to the existence of free groupoids on a graph.) □

In the above proof, note that each (a, ι) is a non-identity of $A \times \mathbf{I}$; so $\overline{\sigma}$ is injective on $\operatorname{Im} i$, whence $\lambda = \overline{\sigma}i$ is injective. Let us identify each a in A with λa, so that $(\lambda a)^{-1}$ is written a^{-1} (note that $a^{-1} = \overline{\sigma}(a, \iota^{-1})$). Then the non-identity elements of FA are uniquely written as products

$$a_n^{\varepsilon_n} \ldots a_1^{\varepsilon_1}, \qquad a_i \in A, \quad \varepsilon_i = \pm 1$$

such that for no i is it true that both $a_i = a_{i+1}$ and $\varepsilon_{i+1} = -\varepsilon_i$.

4. Let G, H be groupoids, and let

$$G \xrightarrow{j_1} K \xleftarrow{j_2} H$$

be morphisms. We say these morphisms present K as the *free product* of G and H if the following property is satisfied: if $g : G \to L$, $h : H \to L$ are any morphisms which agree on $\mathrm{Ob}(G) \cap \mathrm{Ob}(H)$, then there is a unique morphism $k : K \to L$ such that $kj_1 = g$, $kj_2 = h$. We prove that *such a free product always exists.*

Proof Let $X = \mathrm{Ob}(G) \cup \mathrm{Ob}(H)$ and let

$$\sigma : \mathrm{Ob}(G) \sqcup \mathrm{Ob}(H) \to X$$

be the function defined by the two inclusions into X. Note that σ is always a surjection, and σ is a bijection if and only if $\mathrm{Ob}(G)$, $\mathrm{Ob}(H)$ are disjoint.

In the following diagram

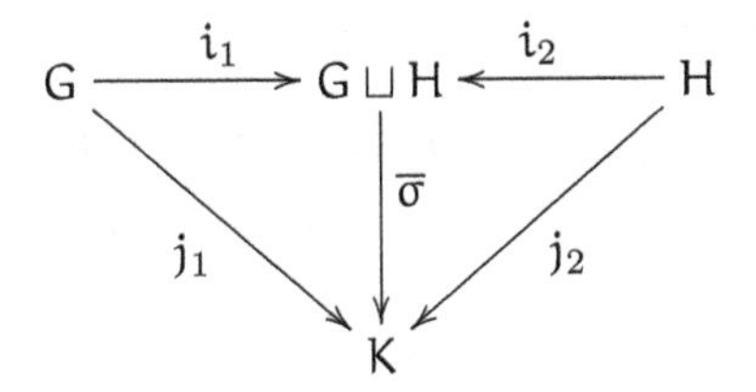

i_1, i_2 are the injections of the coproduct, $K = U_\sigma(G \sqcup H)$ and $j_1 = \overline{\sigma} i_1$, $j_2 = \overline{\sigma} i_2$. The universal property for j_1, j_2 is trivial to verify. □

Notice that the universal property of the free product can also be expressed by saying that the following square is a pushout

$$\begin{array}{ccc} \mathrm{Ob}(G) \cap \mathrm{Ob}(H) & \xrightarrow{i_2} & H \\ {\scriptstyle i_1}\downarrow & & \downarrow{\scriptstyle j_2} \\ G & \xrightarrow[j_1]{} & K \end{array}$$

where i_1, i_2 are the inclusions.

The free product of G and H is usually written $G * H$. In particular, let G, H be groups (supposed to have the same object). Then it is clear that the *injections into the free product* $G * H$ *form a coproduct of groups.* If G, H

have no common elements, then the elements of $G * H$ are the identity and all products

$$k_n \dots k_2 k_1$$

where (i) each k_i belongs to one or other of G, H, (ii) no k_i is an identity, (iii) for no i do k_i, k_{i+1} belong to the same group. (When we write $G * H$ for groups G, H we will always assume that this is the coproduct of groups in the above sense.)

The aim of the rest of this section is to determine the universal group of any connected groupoid. This is useful for the topological applications in Chapter 9.

8.1.4 *The composite of universal morphisms is universal.*

Proof This is immediate from the definition and the fact that a composite of pushouts is a pushout 6.6.5. □

8.1.4 *(Corollary 1) Let G be a groupoid and $\sigma : Ob(G) \to X$, $\tau : X \to Y$ functions. Then $U_\tau U_\sigma(G)$ is isomorphic to $U_{\tau\sigma}(G)$.*

Proof This is clear from previous results. □

8.1.4 *(Corollary 2) Let G, H be groupoids. Then the groups*

$$U(G \sqcup H), \quad U(G * H), \quad UG * UH$$

are all isomorphic.

Proof Let X be a set with one object. Consider the commutative diagram

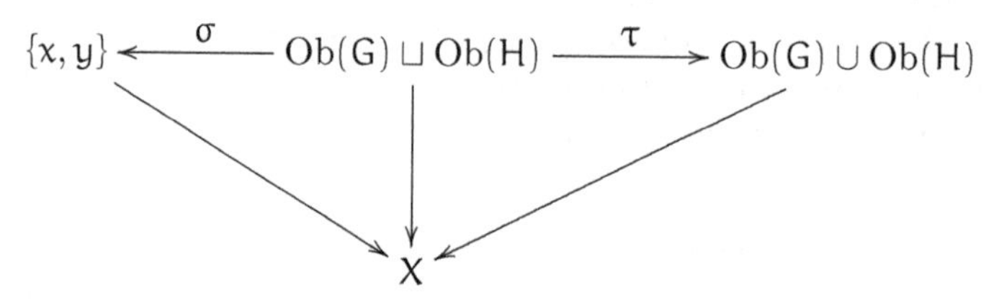

in which the downward functions are constant, τ is the inclusion on $Ob(G)$ and on $Ob(H)$, and σ maps $Ob(G)$ to x and $Ob(H)$ to y (where $x \neq y$). Then we obtain a diagram of strictly universal morphisms.

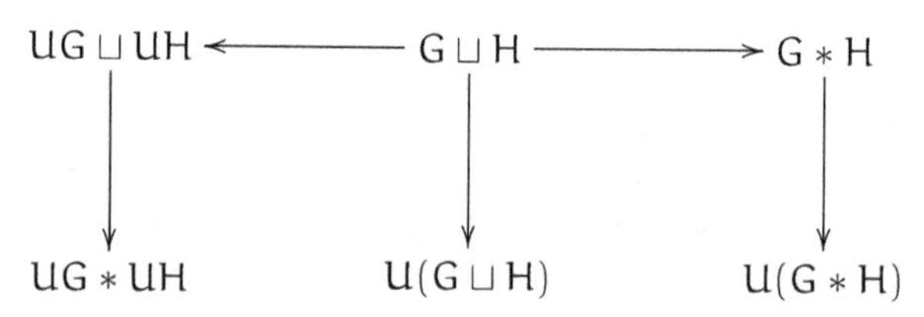

By 8.1.4 (Corollary 1), all of the bottom groups in this diagram are isomorphic. □

We now show how to determine UG for any connected groupoid G. First we prove:

8.1.5 *Let G be a connected groupoid and T a wide, tree subgroupoid of G. Then for any object x_0 of G, the canonical morphism*

$$G(x_0) * T \to G$$

determined by the inclusions, is an isomorphism.

Proof Let $j_i : G(x_0) \to G$, $j_2 : T \to G$ be the two inclusions. Each element a of $G(x, y)$ can be written uniquely as

$$\tau_y a' \tau_x^{-1}$$

for $a' \in G(x_0)$ and $\tau_y, \tau_x \in T$. Therefore, morphisms $f_1 : G(x_0) \to K$, $f_2 : T \to K$ which agree on x_0 define a morphism $f : G \to K$ by

$$fa = f_2(\tau_y) f_1(a') f_2(\tau_x^{-1})$$

and f is the only morphism such that $fj_1 = f_1$, $fj_2 = f_2$. □

The proof of 8.1.5 shows that the isomorphism $G \to G(x_0) * T$ is given by $a \mapsto \tau_y a' \tau_x^{-1}$ $(a \in G(x, y))$.

8.1.6 *If T is a tree groupoid, then UT is a free group. If further T has n objects, then UT is a free group on $(n-1)$ elements.*

Proof Let x_0 be an object of T and for each object x of T let τ_x be the unique element of $T(x_0, x)$. Let A be the set of these τ_x for all $x \neq x_0$, and in the groupoid $A \times \mathbf{I}$ let ι_x, ι_x^{-1} denote respectively (τ_x, ι) and its inverse (τ_x, ι^{-1}).

Let $f : A \times \mathbf{I} \to T$ be the morphism which sends $\iota_x \mapsto \tau_x$, so that $\sigma = \mathrm{Ob}(f)$ simply identifies all $(\tau_x, 0)$ to x_0. In $U_\sigma(A \times \mathbf{I})$ the only non-identity words are $\iota_x, \iota_x^{-1}, \iota_y \iota_x^{-1}$ $(x \neq y)$. So the morphism $\varphi : U_\sigma(A \times \mathbf{I}) \to T$ which sends $\iota_x \mapsto \tau_x$, $\iota_x^{-1} \to \tau_x^{-1}$, $\iota_y \iota_x^{-1} \mapsto \tau_y \tau_x^{-1}$ is an isomorphism such that $\varphi\bar{\sigma} = f$. Therefore f is a strictly universal. Hence the composite $A \times \mathbf{I} \xrightarrow{f} T \to UT$ is strictly universal and so UT is isomorphic to FA, the free group on A. Finally, if T has n objects, then A has $(n-1)$ elements.

□

8.1.6 *(Corollary 1) If G is a connected groupoid and x_0 is an object of G, then UG is isomorphic to $G(x_0) * F$ where F is a free group.*

Proof By 8.1.4 (Corollary 2), 8.1.5 and 8.1.6 it is enough to find a wide tree subgroupoid T of G. This can be done by choosing for each object

$x \neq x_0$ of G an element τ_x of $G(x_0, x)$ and defining T to have all elements τ_x, their inverses and their products. The only element of $T(x, y)$ is then $\tau_y \tau_x^{-1}$ and so T is a wide, tree subgroupoid of G. □

EXERCISES

1. Define the coproduct $\sqcup_{\alpha \in A} G_\alpha$ for an arbitrary family in (i) the category of groupoids, (ii) the category of groups. Prove that if $(G_\alpha)_{\alpha \in A}$ is a family of groupoids, then the universal group of their coproduct is isomorphic to the coproduct (i.e., free product) of their universal groups. Hence show that, if G is a groupoid, then UG is isomorphic to the free product of the groups UG_α for all components G_α of G.
2. Let $\mathcal{C}$ be a category, let X be a set and $\sigma : \mathrm{Ob}(\mathcal{C}) \to X$ a surjection. Prove that there is a category $\mathcal{U}$ and a functor $\overline{\sigma} : \mathcal{C} \to \mathcal{U}$ such that (i) $\mathrm{Ob}(\mathcal{U}) = X$, $\mathrm{Ob}(\overline{\sigma}) = \sigma$, (ii) if $\tau : \mathcal{C} \to \mathcal{D}$ is any functor such that $\mathrm{Ob}(\tau)$ factors through σ, then there is a unique functor $\overline{\tau} : \mathcal{U} \to \mathcal{D}$ such that $\tau^* \overline{\sigma} = \tau$.
3. Let $f : G \to H$ be a morphism of groupoids such that $\mathrm{Ob}(G) = \mathrm{Ob}(H) = X$ and $\mathrm{Ob}(f) = 1_X$. Let $\sigma : X \to Y$ be a surjection. Prove that there is a unique morphism $f' : U_\sigma(G) \to U_\sigma(H)$ such that $f'\overline{\sigma} = \overline{\sigma} f$, and that f' is injective (on elements) if f is.
4. Prove that if $f : G \to H$ is a strictly universal morphism then f is epic in the category of groupoids.
5. Prove that if $f : G \to H$, $g : H \to K$ are morphisms of groupoids such that gf is strictly universal and f is epic, then g is strictly universal.
6. A groupoid G is called the *internal free product* of subgroupoids G_λ if (i) each identity of G lies in some G_λ, (ii) for each non-identity element x of G there is a unique sequence $\lambda_1, \dots, \lambda_n$ $(n \geqslant 1)$ with $\lambda_i \neq \lambda_{i+1}$, and unique non-identity elements x_i in G_{λ_i} such that $x = x_n x_{n-1} \dots x_1$. Prove that G is the internal free product of the G_λ if and only if the canonical morphism $\sqcup_\lambda G_\lambda \to G$ is strictly universal. Prove that in such a case, any two distinct G_λ meet in a discrete groupoid.
7. Let (G_λ) be a family of subgroups of the groupoid G such that any element x of G is a product of elements of various G_λ. Prove that G is the internal free product of the G_λ if and only if the following condition is satisfied: if $x_i \in G_{\lambda_i}$ $(i = 1, \dots, n;$ $n \geqslant 1)$ with $\lambda_i \neq \lambda_{i+1}$ $(i = 1, \dots, n-1)$ and if $x_n \dots x_1$ is defined in G and is an identity element, then at least one of the x_i is an identity element.
8. Suppose there is given a square of groupoid morphisms which is a pushout in the category of groupoids.

$$\begin{array}{ccc} G_0 & \xrightarrow{i_1} & G_1 \\ {\scriptstyle i_2}\downarrow & & \downarrow{\scriptstyle u_1} \\ G_2 & \xrightarrow[u_2]{} & G \end{array}$$

Suppose also that $\sigma_1 = \mathrm{Ob}(u_1)$ and $\sigma_2 = \mathrm{Ob}(u_2)$ are surjective. Prove that the following square in which $j_1 = \sigma_1^* i_1$, $j_2 = \sigma_2^* i_2$, $\nu_1 = u_1^*$, $\nu_2 = u_2^*$, is also a pushout.

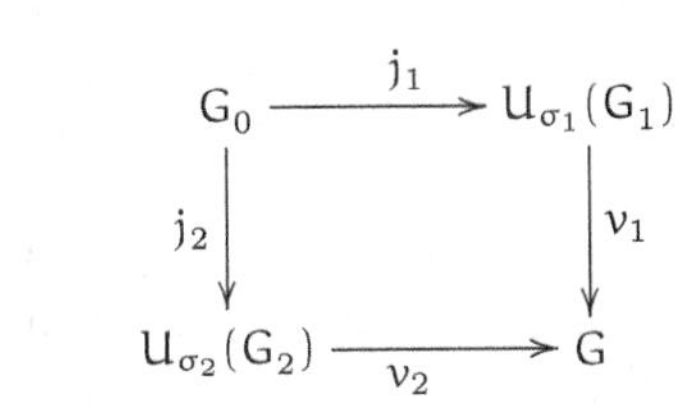

9. Suppose that the first square of the previous exercise is a pushout, that G_0, G_1 are discrete and that $\mathrm{Ob}(i_2)$ is surjective. Prove that $u_2 : G_2 \to G$ is a universal morphism.
10. Let G' be a subgroupoid of G and let $f : G \to H$ be a universal morphism. Prove that there is a subgroupoid H' of H such that f restricts to a universal morphism $G' \to H'$.
11. Compare our exposition of the construction of $U_\sigma(G)$ with that in [Hig05], and with the several proofs of normal form theorems for free groups and free products of groups in [Coh89].

8.2 Free groupoids

The free groupoids generalise the free groups. As the reader will by now have expected, free groupoids are defined by means of a universal property—to express this we need a little of the language of graph theory.

A *graph* Γ consists of a set $\mathrm{Ob}(\Gamma)$ of *objects* (or *vertices*) and for each x, y in $\mathrm{Ob}(\Gamma)$ a set $\Gamma(x,y)$ (often called the set of *edges* from x to y). As usual we write $\gamma : x \to y$ for $\gamma \in \Gamma(x,y)$, and x is the initial, y is the final point of γ. The sets $\Gamma(x,y)$ for various x, y in $\mathrm{Ob}(\Gamma)$ are supposed disjoint. In particular, if $x \neq y$ then $\Gamma(x,y)$ does not meet $\Gamma(y,x)$, so that the graphs we are concerned with are often called *oriented* or *directed* graphs. As we did for categories, we shall usually write Γ for the union of the sets $\Gamma(x,y)$ for any x,y in $\mathrm{Ob}(\Gamma)$, so that $a \in \Gamma$, or a is an element of Γ, means $a \in \Gamma(x,y)$ for some x,y in $\mathrm{Ob}(\Gamma)$.

An object x of a graph Γ is called *discrete* (in Γ) if there are no elements of Γ with initial or final point x. The graph is discrete if all its objects are discrete.

Let Γ, Δ be graphs. A *graph morphism* $f : \Gamma \to \Delta$ assigns to x in $\mathrm{Ob}(\Gamma)$ and object fx of $\mathrm{Ob}(\Delta)$ and to each a in $\Gamma(x,y)$ an element fa in $\Delta(fx, fy)$. It is easy to verify that the graphs and graph morphisms form a category.

If $\mathrm{Ob}(\Gamma) \subseteq \mathrm{Ob}(\Delta)$, $\Gamma \subseteq \Delta$ and the inclusion $\Gamma \to \Delta$ is a graph morphism, then we say Γ is a *subgraph* of Δ: further, Γ is *wide* in Δ if $\mathrm{Ob}(\Gamma) = \mathrm{Ob}(\Delta)$; and Γ is *full* in Δ if $\Gamma(x,y) = \Delta(x,y)$ for all x,y in $\mathrm{Ob}(\Gamma)$.

Suppose Γ' is a subgraph of Γ and $f : \Gamma \to \Delta$ is a graph morphism. Then by $f[\Gamma']$ we mean the subgraph Δ' of Δ whose objects are fx for x in $\mathrm{Ob}(\Gamma')$, and such that $\Delta'(w, z)$ is the union of the sets $f[\Gamma'(x, y)]$ for all x, y in $\mathrm{Ob}(\Gamma')$ such that $fx = w$, $fy = z$. In particular, $\mathrm{Im}\, f$ is defined to be the graph $f[\Gamma]$.

The *coproduct* in the category of graphs is clearly disjoint union. There is a graph 2 which has two objects $0, 1$ and one element $\iota : 0 \to 1$. If γ is an element of a graph Γ then there is a unique morphism of graphs $\hat{\gamma} : 2 \to \Gamma$ such that $\hat{\gamma}(\iota) = \gamma$.

Given a graph Γ, its *dispersion* $D(\Gamma)$ is obtained as a disjoint union Δ of copies of the graph 2, one for each element of Γ, and of a discrete graph on the discrete objects of Γ. The following picture gives an example of this construction:

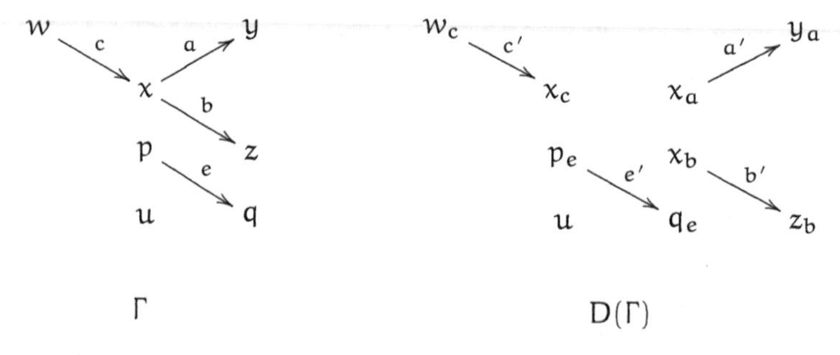

There is a graph morphism $\varphi : D(\Gamma) \to \Gamma$ which re-identifies the objects of $D(\Gamma)$ to give Γ again: thus $D(\Gamma)$ has the same discrete objects as Γ but for each element $a : x \to y$ of Γ there is an element $a' : x_a \to y_a$ of $D(\Gamma)$, and φ sends $x_a \mapsto x$, $a' \mapsto a$.

If $\mathcal{C}$ is a category then $\mathcal{C}$ defines a graph, also written $\mathcal{C}$, simply by forgetting about the composition of elements of $\mathcal{C}$. If $f : \mathcal{C} \to \mathcal{D}$ is a functor, then f defines also a graph morphism $\mathcal{C} \to \mathcal{D}$—the converse, of course, is false. Since groupoids are special kinds of categories, these remarks apply also to groupoids and morphisms of groupoids. In particular, we can talk about subgraphs of a groupoid. Notice that if $f : G \to H$ is a morphism of groupoids then $\mathrm{Im}\, f$ is a subgraph of H but not usually a subgroupoid. For example, if $f : \mathbf{I} \to \mathbb{Z}$ is the strictly universal morphism, then $\mathrm{Im}\, f$ has only three elements $0, 1, -1$.

Let Γ be as above a graph in a groupoid G. The subgroupoid of G *generated* by Γ is the intersection of all subgroupoids of G which contain Γ: it is thus the smallest subgroupoid of G containing Γ, and its elements are clearly all identities at points of $\mathrm{Ob}(\Gamma)$ and all products

$$a_n \dots a_1$$

which are well defined in G and for which $a_i \in \Gamma$ or $a_i^{-1} \in \Gamma$.

We come now to free groupoids. Let Γ be a graph in a groupoid G. We say G is *free on* Γ if Γ is wide in G and for any groupoid H, any graph morphism $f : \Gamma \to H$ extends uniquely to a morphism $G \to H$; and if such Γ exists, we say G is a *free groupoid.*

The free groupoid on the graph 2 is clearly the unit interval groupoid **I**, and the free groupoid on a discrete graph Δ is the discrete groupoid on the objects of Δ. The free groupoid on a coproduct of graphs is the coproduct of the free groupoids on each.

8.2.1 *Let Γ be a graph in a groupoid G. The following conditions are equivalent:*
(a) G *is free on* Γ.
(b) *If* $\varphi : D(\Gamma) \to \Gamma$ *is the dispersion of* Γ, *then the induced morphism* $\bar{\varphi} : F(D(\Gamma)) \to G$ *is strictly universal.*
(c) Γ *generates* G *and the non identity elements of* G *can be written uniquely as products*

$$a_n^{\varepsilon_n} \dots a_1^{\varepsilon_1}$$

such that $a_i \in \Gamma$, $\varepsilon_i = \pm 1$ *and for no* i *is it true that both* $a_i = a_{i+1}$ *and* $\varepsilon_i = -\varepsilon_{i+1}$.

Proof We prove (a) $\Leftrightarrow$ (b) $\Leftrightarrow$ (c).
(a) $\Rightarrow$ (b) Let $\sigma = \mathrm{Ob}(\varphi)$. Suppose $h : F(D(\Gamma)) \to H$ is a morphism and $\tau : \mathrm{Ob}(G) \to \mathrm{Ob}(H)$ is a function such that

$$\mathrm{Ob}(h) = \tau\sigma.$$

We first construct from h and τ a graph morphism $f : \Gamma \to H$.

On objects, f is to be τ; on elements $fa = h(a')$. Clearly, f is a graph morphism. By the assumption that G is free on Γ, f extends uniquely to a morphism $h^* : G \to H$. If $a \in \Gamma$ then

$$h^*\bar{\varphi}(a') = h^*a = fa = h(a');$$

it follows that $h^*\bar{\varphi} = h$. Since Γ is wide in G we must also have $\mathrm{Ob}(h^*) = \tau$.

If $\bar{h} : G \to H$ is any morphism such that $\bar{h}\bar{\varphi} = h$, then $\bar{h}$ extends f and so $\bar{h} = h^*$. This proves that $\bar{\varphi}$ is strictly universal.
(b) $\Rightarrow$ (a) Let $f : \Gamma \to H$ be a graph morphism and let $\tau = \mathrm{Ob}(f)$. We define a morphism $h : F(D(\Gamma)) \to H$ by $a' \mapsto fa$. Then $\mathrm{Ob}(h) = \tau\sigma$ and so there is a unique morphism $h^* : G \to H$ such that $\mathrm{Ob}(h^*) = \tau$ and $h^*\bar{\varphi} = h$. Thus $\mathrm{Ob}(h^*) = \mathrm{Ob}(f)$ and for each element a of Γ, $h^*a = h(a') = fa$. This shows that h^* extends f.

Let $\bar{h} : G \to H$ be any morphism extending f. Since Γ is wide in G, we have $\mathrm{Ob}(\bar{h}) = \mathrm{Ob}(f)$. Further, if $a \in \Gamma$, then

$$\bar{h}\bar{\varphi}(a') = \bar{h}a = fa$$

whence $\bar{h}\bar{\varphi} = h$. If follows that $\bar{h} = h^*$.
(b) $\Leftrightarrow$ (c) This follows from the explicit construction of $U_\sigma(A \times \mathbf{I})$. □

8.2.1 *(Corollary 1) Let* G *be a free groupoid on* Γ. *If* $f : G \to H$ *is strictly universal, then* H *is free on* $f[\Gamma]$. *In particular,* UG *is a free group.*

Proof This follows from 8.2.1 and 8.1.4. □

8.2.1 *(Corollary 2) Let* G *be a free groupoid on* Γ *and let* Δ *be a subgraph of* Γ. *If* H *is the subgroupoid of* G *generated by* Δ, *then* H *is free on* Δ.

Proof We use 8.2.1(c). First of all, Δ is certainly wide in H. Next, if a is a non-identity of H, then, since Δ generates H, a is a product $a_n^{\varepsilon_n} \dots a_1^{\varepsilon_1}$ such that $a_i \in \Delta$ and $\varepsilon_i = \pm 1$.

If for some i we have $a_i = a_{i+1}$, $\varepsilon_i = -\varepsilon_{i+1}$, then we can cancel $a_{i+1}^{\varepsilon_{i+1}} a_i^{\varepsilon_i}$. Further, we can repeat such cancellations until no relation of this form holds. However, since $\Delta \subseteq \Gamma$ and G is free on Γ, the resulting expression for a is unique (by 8.2.1(c)). It follows again from 8.2.1(c) that H is free on Δ. □

In fact, *any* subgroupoid of a free groupoid is free, but this is more difficult to prove [cf. Section 10.8].

Let G be a free groupoid on Γ. The cardinality of the set of elements of Γ is called the *rank* of G. Now it is easy to see from 8.2.1(b) or (c) that no element of Γ is an identity of G; so, if $f : G \to H$ is strictly universal, then f is injective on the elements of Γ and hence G and H have the same rank. But it may be proved that two free groups are isomorphic if and only if they have the same rank [CF63, p. 48]. So the rank of G depends only on G and not on the particular choice of Γ freely generating G.

If G, G' are two free groupoids on a graph Γ, then the universal property shows that there is a unique isomorphism $G \to G'$ which is the identity of Γ. On the other hand, given a graph Γ, then 8.2.1(b) shows how to construct a free groupoid $F\Gamma$ on Γ, while 8.2.1(c) exhibits the elements of $F\Gamma$.

Because of the form of elements given in 8.2.1(c), elements of the free groupoid $F\Gamma$ on Γ are called *paths* in Γ, and if $w \in F\Gamma(x, y)$, then w is called a path from x to y. The graph Γ is called *circuit free* (or a *forest*) if all the object groups of $F\Gamma$ are trivial, Γ is called a *tree* if $F\Gamma$ is a tree groupoid; and Γ is *connected* if $F\Gamma$ is connected.

We shall need the following result which is basic in graph theory, but whose proof is expressed nicely in groupoid language.

8.2.2 *If Γ is a connected graph, then Γ contains a tree T which is wide in Γ.*

Proof The result is trivial if $\mathrm{Ob}(\Gamma)$ is empty, so we suppose $\mathrm{Ob}(\Gamma)$ non-empty. Let $\mathcal{T}$ be the set of all trees contained in Γ; $\mathcal{T}$ is non-empty since it contains for example a tree with one object and no edges.

Clearly $\mathcal{T}$ is partially ordered by inclusion. We claim that an element of $\mathcal{T}$ is maximal in $\mathcal{T}$ if and only if it is wide in Γ.

For the proof of the claim, suppose first that T is an element of $\mathcal{T}$ which is wide in Γ. Suppose a is an element of $\Gamma(x, y)$ such that a does not belong to T.

Let w be the unique element of $FT(y, x)$. Then wa is a non-trivial element of the group $F\Gamma(x, x)$, and so $T \cup \{a\}$ is not a tree. This contradiction shows that T is maximal.

Suppose, conversely, that T is a tree in Γ which is not wide in Γ. Let $x \in \mathrm{Ob}(T)$, $y \in \mathrm{Ob}(\Gamma) \setminus \mathrm{Ob}(T)$. Since Γ is connected, there is an element $w \in F\Gamma(x, y)$. Write $w = a_n^{\varepsilon_n} \dots a_1^{\varepsilon_1}$ as in 8.2.1(c), so that w is a path traversing objects $x = x_1, x_2, \dots, x_n, x_{n+1} = y$. Let x_i be the first of these objects which does not lie in T. Let T' be obtained from T by adjoining the object x_{i+1} and the edge a_i. Then FT' is connected. Also FT' is a tree groupoid since if $u \in FT'(x)$, then 8.2.1(c) shows that $u \in FT(x)$, and hence $u = 1$, since T is a tree. Since T' contains T, it follows that T is not maximal.

We now show that the set $\mathcal{T}$ has a maximal element. If Γ is finite then the set $\mathcal{T}$ is finite, and so has a maximal element. If Γ is not finite, we must apply Zorn's Lemma (cf. Glossary).

Let $\mathcal{C}$ be any non-empty ordered subset of $\mathcal{T}$, and let C be the union of the elements of $\mathcal{C}$. If $x, y \in \mathrm{Ob}(C)$, then $x, y \in \mathrm{Ob}(C')$ for some C' in $\mathcal{C}$. Hence $FC'(x, y)$, and so $FC(x, y)$, is non-empty. If $u, v \in FC(x, y)$, then by 8.2.1(c), $u, v \in FC''(x, y)$ for some C'' in $\mathcal{C}$ containing C'. Hence $u = v$. So FC is a tree groupoid, and so $C \in \mathcal{T}$.

Hence any non-empty ordered subset of $\mathcal{T}$ has an upper bound in $\mathcal{T}$. By Zorn's Lemma, $\mathcal{T}$ has a maximal element T. Then T is wide in Γ. □

8.2.3 *If G is a connected, free groupoid then each object group $G(x)$ of G is free. If further G is of rank n_1 and has n_0 objects, then $G(x)$ is of rank $n_1 - n_0 + 1$.*

Proof Let $x \in \mathrm{Ob}(G)$ and let Γ be a graph freely generating G. It follows from 8.2.2 that G contains a tree groupoid T such that $T \cap \Gamma$ generates T. For each y in $\mathrm{Ob}(G)$ there is a unique element ty of $T(x, y)$. By 6.5.11 (see also 6.7.3), these elements define retractions $r : T \to \{x\}$, $r' : G \to G(x)$.

Let Δ be the wide subgraph of Γ whose elements are those a in Γ which do not lie in T. Let H be the (free) subgroupoid of G generated by Δ. We claim that G is the free product $T * H$.

Let us grant this claim for the moment. Consider the diagram

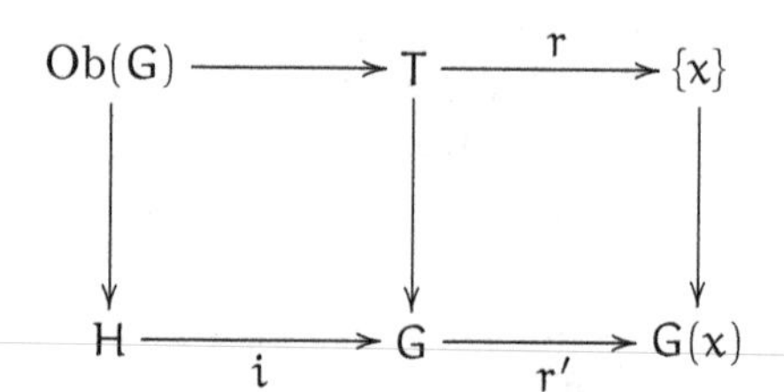

in which the left-hand square of inclusions is a pushout, since $G = T * H$, and the right-hand square is a pushout, by 6.7.3. Therefore the composite square is a pushout. It follows that $r'i : H \to G(x)$ is strictly universal, and so $G(x)$ is a free group on $r'[\Delta]$.

If G is of rank m then Γ has m elements. If G has n objects then $\Gamma \cap T$ has $n-1$ elements. Then Δ, and so also $r'[\Delta]$, has $m-(n-1)$ elements—hence $G(x)$ is of rank $m - n + 1$.

To complete the proof we must show that $G = T * H$. Let $t : T \to K$, $h : H \to K$ be morphisms which agree on $\mathrm{Ob}(G)$. Let $\Gamma' = \Gamma \cap T$, so that T is free on Γ'. Then t, h restrict to graph morphisms $t' : \Gamma' \to K$, $h' : \Delta \to K$ which (since they agree on objects and $\Gamma' \cap \Delta$ has no elements) together define a graph morphism $f : \Gamma \to K$. Since G is free on Γ, f extends uniquely to a morphism $k : G \to K$. However $k \mid T$ extends t', $k \mid H$ extends h'. Therefore $k \mid T = t$, $k \mid H = h$ and this proves that t and h extend to a morphism $k : G \to H$.

If $\bar{k} : G \to H$ also extends t and h, then $\bar{k}\mid_\Gamma = f$ and it follows that $\bar{k} = k$.

□

EXERCISES

1. Prove that a groupoid G is free if and only if each component of G is free.
2. Prove that a simply-connected groupoid is free.
3. Prove that the coproduct, and free product of free groupoids is free.
4. Give another proof of 8.2.3 using 8.2.1(b), Exercise 3 of Section 6.7, and Exercise 9 of Section 8.1.
5. Give another proof of 8.2.1 (Corollary 2) using Exercise 10 of Section 8.1.

8.3 Quotient groupoids

Let G be a groupoid. A subgroupoid N of G is called *normal* if N is wide in G and, for any objects x, y of G and G in $G(x, y)$,

$$aN(x)a^{-1} \subseteq N(y)$$

that is,

$$a_x[N(x)] \subseteq N(y).$$

This last condition implies that $(a^{-1})_x[N(y)] \subseteq N(x)$ and hence, since $(a^{-1})_x = (a_x)^{-1}$, we will in fact have

$$a_x[N(x)] = N(y).$$

EXAMPLE

1. Let $f : G \to H$ be a morphism. Then $\operatorname{Ker} f$, the wide subgroupoid of G whose elements are all k in G such that fk is an identity of H, is a normal subgroupoid of G. In fact, it is obvious that $\operatorname{Ker} f$ is wide in G, and normality follows from

$$f(aka^{-1}) = fafkfa^{-1} = fafa^{-1} = 1, \qquad k \in N(x), \quad a \in G(x, y).$$

We note also that if $\operatorname{Ob}(f)$ is injective then $\operatorname{Ker} f$ is totally disconnected.

A morphism $f : G \to H$ is said to *annihilate* a subgraph Γ of G if $f[\Gamma]$ is a discrete subgroupoid of H. Thus $\operatorname{Ker} f$ is the largest subgroupoid annihilated by f.

8.3.1 *Let* N *be a totally disconnected, normal subgroupoid of* G. *Then there is a groupoid* G/N *and a morphism* $p : G \to G/N$ *such that* p *annihilates* N *and is universal for morphisms from* G *which annihilate* N.

Proof We define $\operatorname{Ob}(G/N) = \operatorname{Ob}(G)$. If $x, y \in \operatorname{Ob}(G)$ we define $G/N(x, y)$ to consist of all cosets

$$aN(x), \quad a \in G(x, y).$$

If $a \in G(x, y)$, $b \in G(y, z)$ then, by normality,

$$\begin{aligned} bN(y)aN(x) &= baN(x)N(x) \\ &= baN(x). \end{aligned}$$

Therefore, multiplication of cosets again gives a coset. The associativity of multiplication is obvious. The identity element of $G/N(x, x)$ is the coset

$N(x)$ and the inverse of $aN(x)$ is $a^{-1}N(y)$ ($a \in G(x, y)$). So G/N is a groupoid.

The morphism $p : G \to G/N$ is the identity on objects, and on elements is defined by $a \mapsto aN(x)$—clearly p is a morphism and $\operatorname{Ker} p = N$.

The universal property of p is that if $f : G \to H$ is any morphism which annihilates N, then there exists a unique morphism $f^* : G/N \to H$ such that $f^*p = f$. Now the cosets of N are exactly the equivalence classes of the elements of G under the relation $a \sim b \Leftrightarrow ab^{-1}$ is defined and belongs to N; so $a \sim b \Leftrightarrow pa = pb$. The universal property follows easily from A.4.6. □

Remark 8.3.1 is true on the assumption only that N is normal [cf. Exercise 2]; we need this more complicated construction of G/N in chapter 11, and give it there. We call G/N a *quotient groupoid* of G.

The usual homomorphism theorem for groups (that if $f : G \to H$ is a morphism then $\operatorname{Im} f$ is isomorphic to $G/\operatorname{Ker} f$) is false for groupoids, one reason being that $\operatorname{Im} f$ need not be a subgroupoid of H; for example, the strictly universal morphism $f : \mathbf{I} \to \mathbb{Z}$ has $\mathbf{I}/\operatorname{Ker} f$ isomorphic to $\mathbf{I}$. However we do have:

8.3.2 *Let* $f : G \to H$ *be a morphism such that* $\operatorname{Ob}(f)$ *is injective. Then* $\operatorname{Im} f$ *is a subgroupoid of* H *and the canonical morphism*

$$G/\operatorname{Ker} f \to \operatorname{Im} f$$

is an isomorphism.

Proof To prove that $\operatorname{Im} f$ is a subgroupoid of H, it is sufficient to prove that if $c, d \in \operatorname{Im} f$ and $d^{-1}c$ is defined in H, then $d^{-1}c \in \operatorname{Im} f$.

Suppose $c = fa$, $d = fb$ where $a \in G(x, y)$, $b \in G(z, w)$. Since $d^{-1}c$ is defined, $fy = fw$, which implies (since $\operatorname{Ob}(f)$ is injective) that $y = w$. Hence, $b^{-1}a$ is defined and $d^{-1}c = f(b^{-1}a)$ which belongs to $\operatorname{Im} f$.

Since f annihilates $\operatorname{Ker} f$, which is a wide, totally disconnected and normal subgroupoid of G, there is a canonical morphism $f' : G/\operatorname{Ker} f \to H$ such that $f'p = f$. Since f' is defined by $f'(a \operatorname{Ker} f) = fa$, it is clear that $\operatorname{Im} f' = \operatorname{Im} f$. Let $f'' : G/\operatorname{Ker} f \to \operatorname{Im} f$ be the restriction of f'. Then $\operatorname{Ob}(f'')$ is bijective, and for each $a, b \in G$

$$\begin{aligned} f''(a \operatorname{Ker} f) = f''(b \operatorname{Ker} f) &\Leftrightarrow fa = fb \\ &\Leftrightarrow a \operatorname{Ker} f = b \operatorname{Ker} f. \end{aligned}$$

It follows from 6.4.3 that f'' is an isomorphism. □

We now consider relations in a groupoid. Suppose given for each object x of the groupoid G a set $R(x)$ of elements of $G(x)$—thus R can be regarded as a wide, totally disconnected subgraph of G. The *normal closure* $N(R)$ of R is the smallest wide normal subgroupoid of G which contains R. This obviously exists since the intersection of any family of normal subgroupoids of G is again a normal subgroupoid of G. Further, $N(R)$ is totally disconnected since the family of object groups of any normal subgroupoid N of G is again a normal subgroupoid of G.

Alternatively, $N = N(R)$ can be constructed explicitly. Let x be an object of G. By a *consequence* of R at x is meant either the identity of G at x, or any product

$$\rho = a_n^{-1}\rho_n a_n \dots a_1^{-1}\rho_1 a_1 \qquad (*)$$

for which $a_i \in G(x, x_i)$ and ρ_i, or ρ_i^{-1}, is an element of $R(x_i)$. Clearly, $N(x)$, the set of consequences of R at x, is a subgroup of $G(x)$ and the family N of these groups is a wide totally disconnected subgroupoid of G containing R. Also N is normal, since if $a \in G(y, x)$ then

$$a^{-1}\rho a = (a_n a)^{-1}\rho_n(a_n a)\dots(a_1 a)^{-1}\rho_1 a_1 a$$

is an element of $N(y)$. On the other hand, any normal, wide subgroupoid of G which contains R must clearly contain all products such as (*) and so must contain N. Hence $N = N(R)$.

The projection $p : G \to G/N(R)$ clearly has the universal property : *if* $f : G \to H$ *is any morphism which annihilates* R *then there is a unique morphism* $f' : G/N(R) \to H$ *such that* $f'p = f$. We call $G/N(R)$ the groupoid G *with the relations* $\rho = 1$, $\rho \in R$.

In applications, we are often given G, R as above and wish to describe the object groups of $G/N(R)$. These are determined by the following result.

8.3.3 *Let G be connected, let* $x \in \mathrm{Ob}(G)$ *and let* $r : G \to G(x)$ *be a deformation retraction. Let* $H = G/N(R)$. *Then* $H(X)$ *is isomorphic to the group* $G(x)$ *with the relations*

$$r(\rho) = 1, \quad \rho \in R.$$

Proof The deformation retraction r and the morphism $p : G \to H$ determine as in 6.7.3 a deformation retraction $s : H \to H(x)$ such that the following square is a pushout

$$\begin{array}{ccc} G & \xrightarrow{r} & G(x) \\ \Big\downarrow{\scriptstyle p} & & \Big\downarrow{\scriptstyle p'} \\ H & \xrightarrow[s]{} & H(x) \end{array} \qquad (**)$$

where p' is the restriction of p. We verify that p' satisfies the required universal property.

Let $f : G(x) \to K$ be a morphism such that f annihilates $r[R]$, i.e. fr annihilates R. Then there is a unique morphism $g : H \to K$ such that $gp = fr$. Since (**) is a pushout, there is a unique morphism $f' : H(x) \to K$ such that $f'p' = f$, $f's = g$. To complete the proof we must show that the last condition is redundant as far as the uniqueness of f' is concerned.

Let $f'' : H(X) \to K$ be any morphism such that $f''p' = f$. Then $f''sp = f''p'r = fr = gp$. But p is surjective, so that $f''s = g$. The pushout property of (**) now implies that $f'' = f'$. □

EXERCISES

1. Prove that there are morphisms $f : \mathbf{I} \to \mathbb{Z}$, $g : \mathbf{I} \to \mathbb{Z}_2$ such that $\operatorname{Ker} f = \operatorname{Ker} g$ and $\operatorname{Im} f$ is not (graph) isomorphic to $\operatorname{Im} g$.
2. Let N be any normal subgroupoid of the groupoid G. Prove that there is a groupoid G/N and morphism $p : G \to G/N$ such that p annihilates N and is universal for this property. [The elements of G/N are equivalence classes of elements of G under the relation $a \sim b$ if and only if $a = xby$ for some x, y in N.]
3. Prove that if $f : G \to H$ is a morphism with kernel N, then $f = pf'$ where $p : G \to G/N$ is the projection and $f' : G/N \to H$ has discrete kernel. If f' is an isomorphism onto $\operatorname{Im} f$, then we shall call f a *projection*.
4. Prove that any morphism f can be factored as $f = f_2 f_1$ where $\operatorname{Ob}(f_1)$ is the identity and f_2 is faithful [for faithful, cf. Exercise 14 of Section 6.5].
5. Prove that any $f : G \to H$ can be factored as $f = gr$ where r is a deformation retraction and $\operatorname{Ker} g$ is totally disconnected.
6. Prove that any deformation retraction is a projection.
7. Prove that if $f : G \to H$ is faithful, then $f = gr$ where r is a deformation retraction and $\operatorname{Ker} g$ is discrete.
8. Prove that any projection is a deformation retraction followed by projection g with $\operatorname{Ob}(g) = 1$.
9. Prove that the category of groupoids admits coequalisers and deduce that any morphisms $G_0 \to G_1$, $G_0 \to G_2$ have a pushout.

8.4 Some computations

We conclude this chapter with a computation of a pushout of groupoids which arises in the determination of the fundamental group of the union of two connected spaces with non connected intersection, for example as shown in Fig. 8.3.

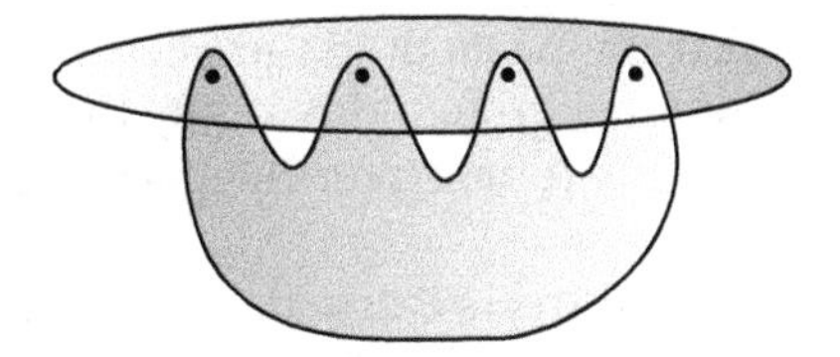

Fig. 8.3

Note that one of the problems of attempting to solve this question with the fundamental group alone is that a choice of base point is then required. It is clear that a base point should be in a component of the intersection of the two pieces, but it is not clear which component to choose. We therefore avoid a decision by choosing a *set* of base points, one in each component of the intersection, and so work with groupoids. This strategy of avoidance of decision seems to be optimal.

The formulae that we obtain are stated in Van Kampen's paper on the fundamental group ([Kam33]) but his proof even of the connected case is difficult to follow.

8.4.1 *Suppose given a pushout of groupoids*

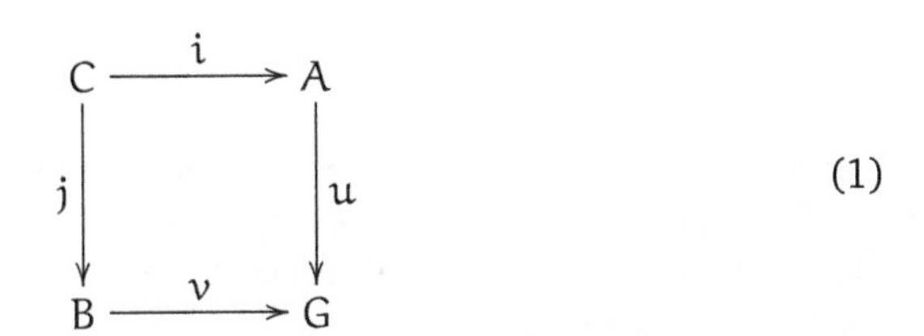

such that (a) *all the groupoids have the same set* J *of objects,* (b) i, j, u, v *are the identity on objects, and* (c) A *and* B *are connected groupoids, and* C *is totally disconnected. Let* p *be a chosen element of* $J = \mathrm{Ob}(C)$. *Let* $r : A \to A(p)$, $s : B \to B(p)$ *be retractions obtained by choosing elements* $\alpha_x \in A(p, x)$, $\beta_x \in B(p, x)$, *for all* $x \in J$, *with* $\alpha_p = 1$, $\beta_p \in 1$. *Let* $f_x = (u\alpha_x)^{-1}(v\beta_x)$ *in* $G(p)$, *and let* F *be the free group on the elements* f_x, $x \in J$, *with the relation* $f_p = 1$. *Then the object group* $G(p)$ *is isomorphic to the quotient of the free product group*

$$A(p) * B(p) * F$$

by the relations

$$(ri\gamma)f_x(sj\gamma)^{-1}f_x^{-1} = 1 \tag{2}$$

for all $x \in J$ *and all* $\gamma \in C(x, x)$.

Proof We first remark that the pushout (1) implies that the groupoid C is the quotient of the free product groupoid $A * B$ by the relations $(i\gamma)(j\gamma)^{-1}$ for all $\gamma \in C$. The problem is to interpret this fact in terms of the object group at p of G.

To this end, let T, S be the tree subgroupoids of A, B respectively generated by the elements α_x, β_x, $x \in J$. The elements α_x, β_x, $x \in J$, define isomorphisms

$$\varphi : A \to A(p) * T, \qquad \psi : B \to B(p) * S$$

where if $g \in G(x, y)$ then

$$\varphi g = \alpha_y (rg)\alpha_x^{-1}, \qquad \psi g = \beta_y (sg)\beta_x^{-1}.$$

So G is isomorphic to the quotient of the groupoid

$$H = A(p) * T * B(p) * S$$

by the relations

$$(\varphi i\gamma)(\psi j\gamma)^{-1} = 1$$

for all $\gamma \in C$. By 8.3.3, the object group $G(p)$ is isomorphic to the quotient of the group $H(p)$ by the relations

$$(r\varphi i\gamma)(r\psi j\gamma)^{-1} = 1$$

for all $\gamma \in C$.

Now if $J' = J \setminus \{p\}$, then T, S are free groupoids on the elements α_x, β_x, $x \in J'$, respectively. By 8.2.1 (Corollary 1), $T * S$ is the free groupoid on all the elements α_x, β_x, $x \in J'$. It follows from 8.2.3, that $(T * S)(p)$ is the free group on the elements $r\beta_x = \alpha_x^{-1}\beta_x = f_x$, $x \in J'$. Let $f_p = 1 \in F$. Since

$$r\varphi i\gamma = ri\gamma, \qquad r\psi j\gamma = f_x(sj\gamma)f_x^{-1},$$

the result follows. □

There is a consequence of the above computation which we shall use in the next chapter in proving the Jordan Curve theorem. First, if F and H are groups, recall that we say that F is a *retract* of H if there are morphisms $\imath : F \to H$, $\rho : H \to F$ such that $\rho\imath = 1$. This implies that F is isomorphic to a subgroup of H.

8.4.1 *(Corollary) Under the situation of* 8.4.1, *the free group* F *is a retract of* $G(p)$. *Hence if* $J = \mathrm{Ob}(C)$ *has more than one element, then the group* $G(p)$ *is not trivial, and if* J *has more than two elements, then* $G(p)$ *is not abelian.*

Proof Let $M = A(p) * B(p) * F$, and let $\imath' : F \to M$ be the inclusion. Let $\rho' : M \to F$ be the retraction which is trivial on $A(p)$ and $B(p)$ and is the identity on F. Let $q : M \to G(p)$ be the quotient morphism. Then it is clear that ρ' preserves the relations (2), and so ρ' defines uniquely a morphism $\rho : G(p) \to F$ such that $\rho q = \rho'$. Let $\iota = qi'$. Then $\rho\iota = \rho' i' = 1$. So F is a retract of $G(p)$.

The concluding statements are clear. □

Here is another application of the Van Kampen theorem for the fundamental groupoid, namely to HNN-extensions of groups. This construction is named after the initial letters of the authors of the paper which introduced the construction ([HNN49]).

Our approach also shows the advantage of the use of groupoids: because there is a homotopy theory of groupoids, we can make analogues in groupoid theory of constructions for spaces, particularly double mapping cylinders.

Let G be a group and let $i_A : A \to G$, $i_B : B \to G$ be inclusions of two subgroups of G. Suppose given an isomorphism $\theta : A \to B$. The aim is to construct a new group $H = G*_\theta$ which contains A, B as subgroups and in which there is an element $t \in H$ which conjugates A to B, that is for all $a \in A$ we have $tat^{-1} = \theta a$ in H.

To this end we form the pushout of groupoids

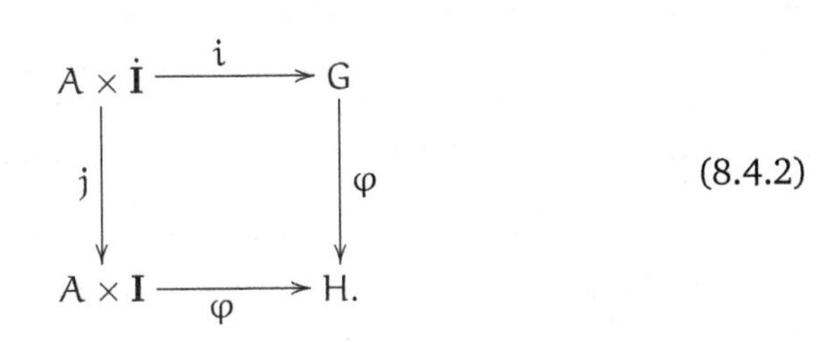

(8.4.2)

Here j is the inclusion; i is the morphism defined by $(a, 0) \mapsto i_A a$, $(a, 1) \mapsto i_B \theta a$; and φ, ψ are defined by the pushout, so that $\varphi i = \psi j$. Hence for $a \in A$, $\psi(a, 0) = \varphi i(a, 0) = a$, $\psi(a, 1) = \varphi i(a, 1) = \varphi\theta a$.

Note that if e denotes the identity element of the group G, and $a \in A$, then, in the multiplication in $A \times \mathbf{I}$, we have $(a, 1)(e, \iota) = (e, \iota)(a, 0)$. Let $t = \psi(e, \iota) \in H$. So in H

$$\begin{aligned} t(\varphi a) &= (\psi(e, \iota)(\varphi i(a, 0)) \\ &= \psi((e, \iota)(a, 0)) \\ &= \psi((a, 1)(e, \iota)) \\ &= (\psi(a, 1))t \\ &= \varphi(\theta a)t. \end{aligned}$$

Hence $t(\varphi a)t^{-1} = \varphi(\theta a)$ in H.

Note that this result and construction makes sense whether or not $i_A : A \to G$ is injective and whether or not θ is an isomorphism. However it is a standard result of combinatorial group theory, see for example [LS77], that if i_A is injective and θ is an isomorphism, then φ is also injective. It is in these circumstances that H is called an HNN-extension and written $G*_\theta$.

This is a corollary of a more general result on graphs of groups [Ser80], which has been put in a powerful way in terms of a normal form for the elements of the fundamental groupoid of a graph of groups in [Hig76]. This itself has been generalised to the fundamental groupoid of a *graph of groupoids* in Emma Moore's Bangor thesis [EJM01] (which is available for download). A consequence of her results is that in the situation of 8.4.1, if i, j are injective, then so also are u, v.

Notes

Most of the results on groupoids are due to Higgins ([Hig64], [Hig66], [Hig05]) but the proof of 8.4.1 is new. The proof of the associativity of the multiplication of the words of $U_\sigma(G)$ is borrowed from the treatment of free products of groups in [Lan65]. The construction of arbitrary colimits of categories and groupoids is given in [Hig05], and indeed is a special case of results of [Hig63]. The reader is invited to compare these approaches to the construction of $U_\sigma(G)$ with the standard constructions in combinatorial group theory of free groups and free products, and the derivation of normal forms. For example [Coh66] discusses four methods for obtaining these. The paper [Hig76] has a very elegant treatment of normal forms for the fundamental groupoid of graph of groups, and a computational form of this is in Emma Moore's thesis [EJM01].

The term 'combinatorial groupoid theory' seems appropriate though is non standard, and it is an area which needs development. It certainly includes methods of covering groupoids as presented in Chapter 9, and in [Hig05]. Other contributions are [Bra04, Hum94]. The latter paper, and [EJM01], also deal with computational aspects.

There is an important general theory of fibred, opfibred and bifibred categories which was initiated by Grothendieck and developed strongly by Jean Benabou. These notions are discussed in for example [Bor94], [Joh02], [Tay99]. In particular, the functor Ob from groupoids to sets is, in these terms, a bifibration (both fibration and opfibration) of categories. Unfortunately, the terminology is not entirely stable (the last cited books use the terms 'prone, supine' for what was previously called 'cartesian, co-cartesian'), so we have decided to stick with the terminology of Higgins which dates back to 1963. However, similar ideas occur in module theory, with notions of 'change of ring', and in the theory of Mackey functors, but there again with a different terminology. The work of Higgins on universal morphisms of groupoids was a stimulus for later work on 'induced crossed modules', which are structures which carry homotopical information in dimensions 0,1 and 2. The general problem is to develop language and modes of calculation to describe how mathematical structures with information at various levels are changed in high levels when changes are made in a low level.

Chapter 9

Computation of the fundamental groupoid

In this chapter we compute the fundamental groupoid of some useful adjunction spaces, and hence of cell complexes, by applying the methods developed in chapter 8. We then develop some geometric applications, to knots, the Phragmen-Brouwer property, and the Jordan Curve Theorem.

9.1 The Van Kampen theorem for adjunction spaces

In Section 6.7 we proved a van Kampen theorem for the fundamental groupoid πXA when X is given as a union of open subsets.

Suppose given an adjunction space $W\,{}_f\!\sqcup\, Z$ as in the pushout square

$$\begin{array}{ccc} Y & \xrightarrow{f} & W \\ {\scriptstyle i}\downarrow & & \downarrow{\scriptstyle \bar{\imath}} \\ Z & \xrightarrow[\bar{f}]{} & W\,{}_f\!\sqcup\, Z. \end{array} \tag{9.1.1}$$

Our object is to determine the groupoid

$$\pi(W\,{}_f\!\sqcup\, Z)B$$

for certain (useful) B.

In chapter 7, we studied the homotopy type of $W\,{}_f\!\sqcup Z$ and showed its dependence on the homotopy types of W and (Z, Y) if (Z, Y) is cofibred. This is a local condition on Y in Z. To determine the fundamental groupoid $\pi(W\,{}_f\!\sqcup Z)$ as a pushout, we also need some local conditions—these conditions are essentially in dimensions 0 and 1, and are described in terms of the natural map

$$p : M(f) \cup Z \to W\,{}_f\!\sqcup Z$$

and its induced morphism of fundamental groupoids.

Suppose that C is a subset of Z representative in Z and in Y, that D is a subset of W representative in W, and that $f[C] \subseteq D$. Let

$$g = f \mid C \cap Y, D, \qquad B = D\,{}_g\!\sqcup C.$$

Under these conditions we have:

9.1.2 *(The Van Kampen theorem) The following square*

$$\begin{array}{ccc} \pi YC & \xrightarrow{f} & \pi WD \\ {\scriptstyle i}\downarrow & & \downarrow{\scriptstyle \bar{i}} \\ \pi ZC & \xrightarrow[\bar{f}]{} & \pi(W\,{}_f\!\sqcup Z)B \end{array} \qquad (9.1.3)$$

is a pushout if and only if the morphism

$$p : \pi(M(f) \cup Z)A \to \pi(W\,{}_f\!\sqcup Z)B$$

in which $A = D \cup C$, is a homotopy equivalence of groupoids. In particular, (9.1.3) is a pushout if (Z, Y) is cofibred.

Proof Let $X = M(f) \cup Z$, $X_2 = X \setminus W$, $X_1 = M(f)$, $X_0 = X_1 \cap X_2$.

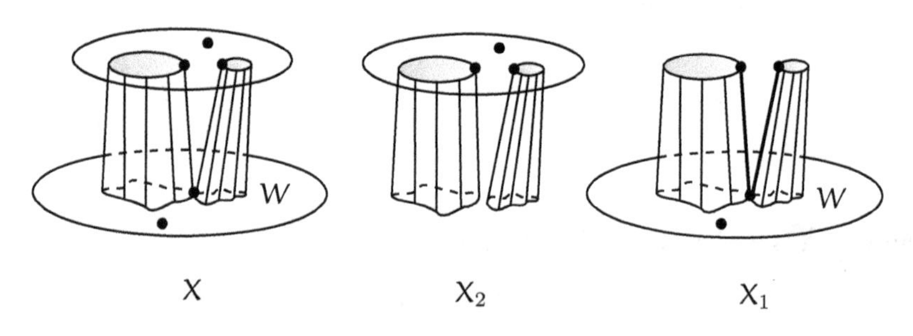

Fig. 9.1

The interiors of X_1, X_2 cover X and so we are in the position to apply 6.7.4 to give us a pushout isomorphic to (9.1.3).

Let $A' = D$, so that in the notation of 6.7.4, $A_1 = D \cup (C \setminus Y)$. Then $p : M(f) \cup Z \to W\, {}_f\!\sqcup\, Z$ maps A_1 bijectively onto B. Further A', and hence also A_1, is representative in X_1. Indeed, A_1 meets each path component of W because D is representative in W. Also $C \cap Y$ is representative in Y and each point c of $C \cap Y$ can be joined by the path down the mapping cylinder of fc, which belongs to D; this path is shown as a thick line in X_1 in Fig. 9.1. Notice also that if θc is the class in $\pi X_1 A$ of this path, then $p(\theta c)$ is the identity at fc in πW.

Consider the following diagram in which (i) $Q = W\, {}_f\!\sqcup\, Z$, (ii) the front square is the pushout determined by A_1 and the above elements θc as in 6.7.4, (iii) the back square is (9.1.3).

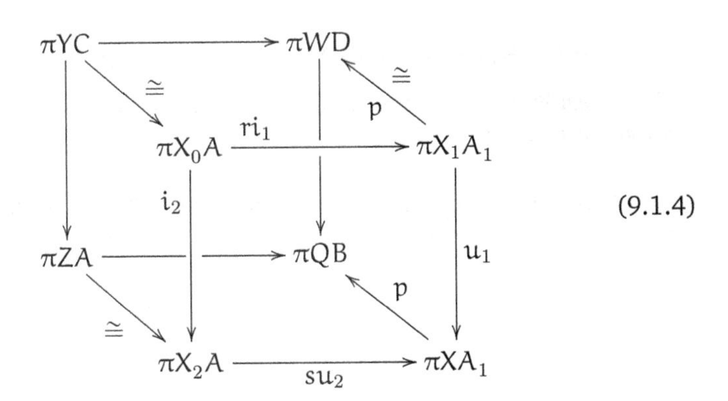

The left-hand square is induced by inclusions and so is commutative. The right-hand square is induced by p and its restrictions, so the right-hand square is commutative. The commutativity of the top and bottom squares is a consequence of $p(\theta c) = 1$ $(c \in C \cap Y)$. Thus (9.1.4) is commutative.

Each morphism marked $\cong$ is induced by a homotopy equivalence and is bijective on objects. Therefore these morphisms are isomorphisms. Hence, each of the following statements is equivalent to its successor: (a) (9.1.3) is a pushout, (b) (9.1.4) determines an isomorphism of its front square to its back square, (c) $p : \pi XA_1 \to \pi QB$ is an isomorphism.

However, the last morphism is bijective on objects so (c) is equivalent to (d) $p : \pi XA_1 \to \pi QB$ is a homotopy equivalence. Since πXA_1 is a deformation retract of πXA, (d) itself is equivalent to (e) $p : \pi XA \to \pi QB$ is a homotopy equivalence.

This proves the main part of 9.1.2. The last statement of 9.1.2 follows from 7.5.4. □

9.1.2 *(Corollary 1) Let* W, Z *be closed in* $W \cup Z$*, let* $(Z, W \cap Z)$ *be cofibred and let* B *be a set representative in* $W \cap Z$, W, Z*. Then the square of morphisms induced by inclusions*

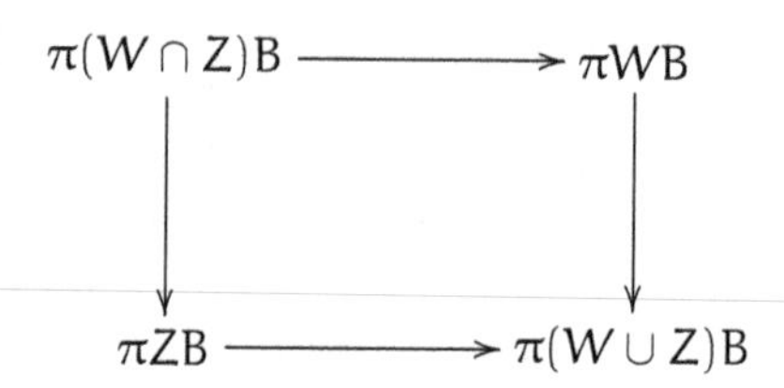

is a pushout.

Proof This is a consequence of 9.1.2 with $f : Y \to W$ the inclusion. □

This result is, of course, similar to 6.7.2, and is in many cases more convenient to use than the earlier result.

9.1.2 *(Corollary 2) Suppose the assumptions of* 9.1.2 (Corollary 1) *hold and also* $\pi(W \cap Z)B$ *is discrete. Then* $\pi(W \cup Z)B$ *is isomorphic to the free product of groupoids*

$$\pi ZB * \pi WB.$$

Proof From the pushout square of 9.1.2 (Corollary 1) it is easy to deduce the morphism $\pi ZB \sqcup \pi WB \to \pi(W \cup Z)B$ determined by the two inclusions of Z, W into $W \cup Z$ is a 0-identification. □

Remark Even this corollary is false without some local assumptions on Y in Z (or in W). For example, let H be the subspace of $\mathbb{R}^2$ which is the union of all circles centre $(1/n, 0)$ for n a positive integer—this space has been called the 'Hawaiian earring'. Let $0 = (0, 0)$ be the base point of H. The space CH is contractible and so the group $\pi(CH, 0)$ is trivial. However, [Gri54], [Gri56] has shown that $\pi(CH \vee CH, 0)$ is non-trivial and in fact can be generated only by an uncountable number of elements. Again, the fundamental group of $H \vee H$ is not the free product $\pi(H, 0) * \pi(H, 0)$.

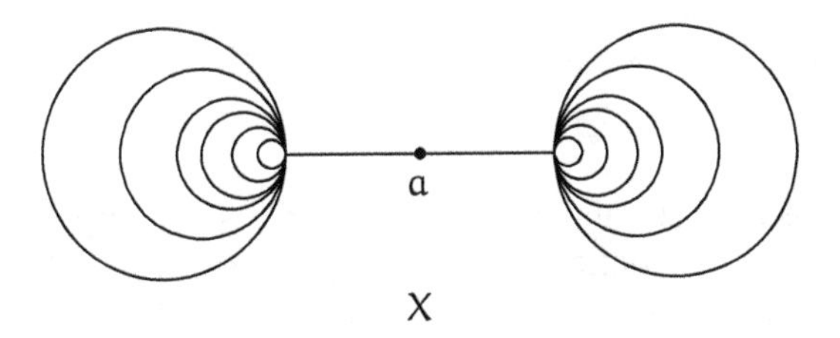

X

Fig. 9.2

However the space X of Fig. 9.2 formed by joining two Hawaiian earrings together does have has its fundamental group isomorphic to $\pi(H, 0) * \pi(H, 0)$—this is easy to prove from 9.1.2 (Corollary 2) by taking W, Z to be left- and right-hand halves of X meeting in $\{a\}$. In this case, the obvious map $X \to H \vee H$ induces a morphism $\pi(X, a) \to \pi(H \vee H, 0)$ which is injective but not surjective (the proof of this statement is not easy—cf. [Gri54] and [MM86]).

Suppose now that we are in the situation of 9.1.2, that (Z, Y) is cofibred, that $C \subseteq Y$ and $B = D = f[C]$.

9.1.2 *(Corollary 3) If further πYC, πWD are discrete then*

$$\bar{f} : \pi ZC \to \pi(W \,{}_f\!\sqcup\, Z)D$$

is a 0-identification morphism.

Proof This follows from 9.1.2 and the definition of 0-identification morphism. □

We can derive a number of useful results from this. For example, if D consists of a single point d (and the other assumptions of 9.1.2 (Corollary 3) hold) then the fundamental group $\pi(W \,{}_f\!\sqcup\, Z, d)$ is isomorphic to $U(\pi ZC)$, the universal group of πZC. In particular, if Z is path-connected and $c_0 \in C$, then

$$\pi(W \,{}_f\!\sqcup\, Z, d) \cong \pi(Z, c_0) * F$$

where F is a free group with one generator for each element of C other than c_0.

We now derive the fundamental group of a cell complex, first dealing with the 1-dimensional case.

9.1.5 *If K is a connected cell complex and $v \in K^0$, then the groupoid $\pi K^1 K^0$ is a free groupoid and the fundamental group $\pi(K^1, v)$ is a free group on $r_1 - r_0 + 1$ generators where r_n is the number of n-cells of K, $n = 0, 1$.*

Proof K^1 is obtained by adjoining 1-cells to K^0, that is,

$$K^1 = K^0 \,{}_f\sqcup\, (\Lambda \times \mathbb{E}^1)$$

where Λ is a discrete set and $f : \Lambda \times \mathbb{S}^0 \to K^0$ is the attaching map. Let $C = \Lambda \times \mathbb{S}^0$; since K^1 is connected, $f[C] = K^0$. Since $\pi(\Lambda \times \mathbb{S}^0)C$ and $\pi K^0 K^0$ are discrete groupoids (and also $(\Lambda \times \mathbb{E}^1, \Lambda \times \mathbb{S}^0)$ is cofibred) the morphism

$$\bar{f} : \pi(\Lambda \times \mathbb{E}^1)C \to \pi K^1 K^0$$

is a 0-identification. But $\pi(\Lambda \times \mathbb{E}^1)C$ is isomorphic to $\Lambda \times \mathbf{I}$. So the result follows from the discussion of free groupoids in section 8.2. □

Notice that 9.1.5 also gives the generators of $\pi K^1 K^0$ as follows. For each λ in Λ let ι_λ denote the unique path class in $\pi(\lambda \times \mathbb{E}^1)C$ from $(\lambda, -1)$ to $(\lambda, +1)$. Then the generators of $\pi K^1 K^0$ are the elements $\bar{f}(\iota_\lambda)$, λ in Λ; thus if v_0, v_1 are vertices of K joined by a 1-cell, then the path class in $\pi K(v_0, v_1)$ determined by the characteristic map of this 1-cell is one of these generators.

9.1.5 *(Corollary 1) The fundamental group of the circle, $\pi(\mathbb{S}^1, 1)$ is isomorphic to $\mathbb{Z}$, with generator the class of the path*

$$\begin{aligned} \mathbb{I} &\to \mathbb{S}^1 \\ t &\mapsto e^{2\pi i t}. \end{aligned}$$

Proof This is immediate from 9.1.2 and the previous remark, since the given path is a characteristic map for the 1-cell of $\mathbb{S}^1$. □

We now show that $\pi(K^2, v)$ is isomorphic to $\pi(K^1, v)$ with relations for each 2-cell. Let us suppose

$$K^2 = K^1 \,{}_g\sqcup\, (M \times \mathbb{E}^2)$$

where $g : M \times \mathbb{S}^1 \to K^1$. Suppose also that K^1 is connected. For each m in M, let $v_m = g(m, e)$ where $e = (1, 0)$ and let

$$\rho_m = g(\iota_m) \in \pi(K^1, v_m)$$

where ι_m is a generator of the fundamental group of $M \times \mathbb{S}^1$ at (m, e). Let v be an element of K^1 and let a_m be an assigned element of $\pi K^1(v, v_m)$ (with $a_m = 1$ if perchance $v_m = v$).

9.1.6 *The fundamental group $\pi(K^2, v)$ is isomorphic to the free group $\pi(K^1, v)$ with the relations*

$$a_m^{-1} \rho_m a_m = 1, \quad m \in M.$$

Proof We first show that if $V = \{v_m : m \in M\} \cup \{v\}$ then $\pi K^2 V$ is the groupoid $\pi K^1 V$ with the relations $\rho_m = 1$, $m \in M$. Let $C = M \times \{e\}$. We have a pushout square

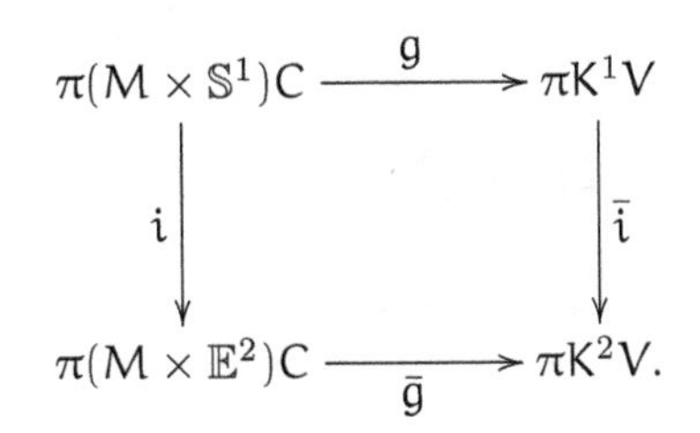

$$\begin{array}{ccc} \pi(M \times \mathbb{S}^1)C & \xrightarrow{g} & \pi K^1 V \\ \downarrow i & & \downarrow \bar{i} \\ \pi(M \times \mathbb{E}^2)C & \xrightarrow[\bar{g}]{} & \pi K^2 V. \end{array}$$

Suppose $f : \pi K^1 V \to F$ is any morphism such that $f\rho_m = 1$, $m \in M$. Then $\mathrm{Im}(fg)$ is discrete. Since $\pi(M \times \mathbb{E}^2)C$ is a discrete groupoid on C, f defines a morphism $\bar{f} : \pi(M \times \mathbb{E}^2)C \to F$ such that $\bar{f}i = fg$. So there is a unique morphism $f' : \pi K^2 V \to F$ such that $f'\bar{i} = f$, $f'\bar{g} = \bar{f}$. The last condition is redundant, since $\pi(M \times \mathbb{E}^2)C$ is discrete and so $\bar{g}$ is determined by $\mathrm{Ob}(\bar{g}) = \mathrm{Ob}(g) : C \to V$, a surjective function.

This proves that $\pi K^2 V$ is $\pi K^1 V$ with relations $\rho_m = 1$, $m \in M$. The conclusion of 9.1.6 follows from 8.3.3. □

9.1.7 *If K is a cell complex and A a subset of K^2, then the inclusion $K^2 \to K$ induces an isomorphism $\pi K^2 A \to \pi K A$.*

Proof We first prove that $\mathbb{S}^n$ is simply-connected for $n > 1$. Let e be a point of $\mathbb{S}^{n-1} = E^n_+ \cap E^n_-$. By 9.1.2 (Corollary 1) we have a pushout square

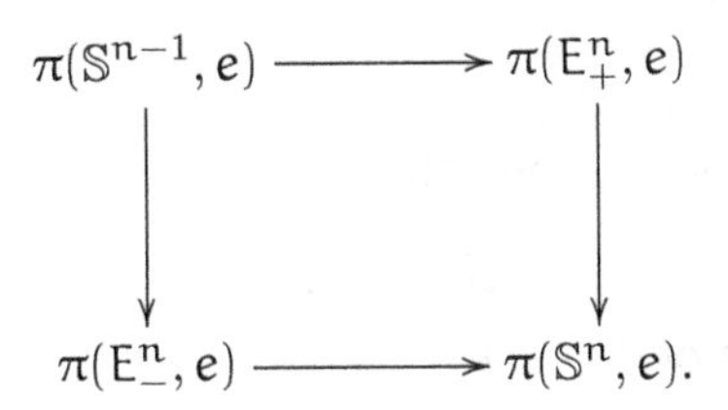

$$\begin{array}{ccc} \pi(\mathbb{S}^{n-1}, e) & \longrightarrow & \pi(E^n_+, e) \\ \downarrow & & \downarrow \\ \pi(E^n_-, e) & \longrightarrow & \pi(\mathbb{S}^n, e). \end{array}$$

But E^n_+, E^n_- are homeomorphic to $\mathbb{E}^n$ and so are simply-connected. Hence $\pi(E^n_+, e), \pi(E^n_-, e)$ are trivial groups and therefore $\pi(\mathbb{S}^n, e)$ is trivial.

Now consider any adjunction space $W \,{}_f\!\sqcup\, \mathbb{E}^{n+1}$ where $f : \mathbb{S}^n \to W$ and $n > 1$. Let $e \in \mathbb{S}^n$ and suppose W is path-connected. There is a pushout

square

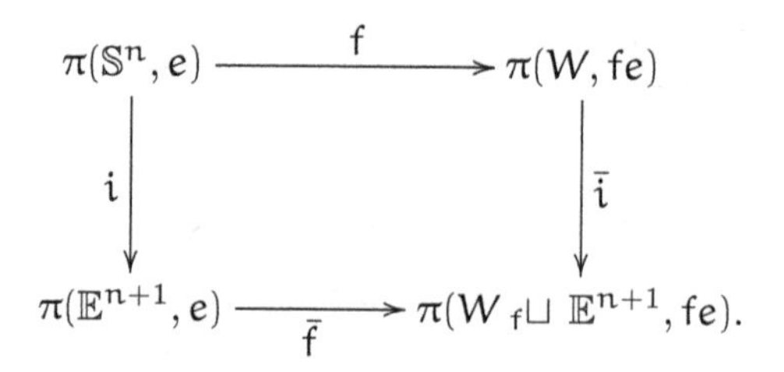

Since the two left-hand groups are trivial, $\bar{\imath}$ is an isomorphism. □

In order to compute the fundamental group of spaces, it is clearly necessary to compute maps $\pi(\mathbb{S}^1, e) \to \pi(K^1, fe)$. The following result is crucial.

9.1.8 *Let* $f : \mathbb{S}^1 \to \mathbb{S}^1$ *be the map* $z \mapsto z^n$, *n an integer. Then the induced morphism* $f : \pi(\mathbb{S}^1, e) \to \pi(\mathbb{S}^1, e)$ *of additive groups is multiplication by n.*

Proof The result is clearly true if $n = 0$, since f is then constant, or if $n = 1$, since f is then the identity. Suppose $n > 1$; let w be the complex number $e^{2\pi i/n}$ and let $w^r = e^{2\pi ir/n}$, $r = 0, 1, \dots, n-1$. Let X_r be the subset of $\mathbb{S}^1$ of points $e^{2\pi i\theta}$, $r/n \leqslant \theta \leqslant (r+1)/n$, let C_r consist solely of w^r, w^{r+1} and let $C = \{w^r : 0 \leqslant r < n\}$. Since X_r is simply-connected there is a unique element ι_r in $\pi X_r(w^r, w^{r+1})$. The morphism $\pi X_r C_r \to \pi\mathbb{S}^1 C$ induced by inclusion is injective and so we regard $\pi X_r C_r$ as a subgroupoid of $\pi\mathbb{S}^1 C$. A generator a of $\pi(\mathbb{S}^1, e)$ where $e = w^0$, is then given by

$$a = \iota_{n-1} + \dots + \iota_0.$$

The map $f : \mathbb{S}^1 \to \mathbb{S}^1$ determines by restriction $f' : X_r \to \mathbb{S}^1$; clearly $f'\iota_r = a$. Therefore,

$$fa = f(\iota_{n-1} + \dots + \iota_0) = a + \dots + a = na.$$

If $n < 0$, let $m = -n$. The $z \mapsto z^n$ is the composite of $g : z \mapsto z^{-1}$ and $z \mapsto z^m$. But if $b : \mathbb{I} \to \mathbb{S}^1$ is the path $t \mapsto e^{2\pi it}$, then gb is $t \mapsto e^{-2\pi it}$, that is, $gb = -b$. Hence, in $\pi(\mathbb{S}^1, e)$, $ga - fa$; therefore $fa = -ma = na$. □

If $f : \mathbb{S}^1 \to \mathbb{S}^1$ is a map such that $f : \pi(\mathbb{S}^1, 1) \to \pi(\mathbb{S}^1, 1)$ is multiplication by n, then we say f is of *degree* n.

EXAMPLES

1. Let $K = \mathbb{S}^1 \vee \dots \vee \mathbb{S}^1$ be a wedge of n circles, with the cell structure $e^0 \cup e^1_1 \cup \dots \cup e^1_n$. Let v be the vertex of K. Then $\pi(K, v)$ is a free group on n-generators, the generators being the classes of the loops which pass once round one of the circles.

2. The fundamental group of the real projective plane $P^2(\mathbb{R})$ and the real projective n-space $P^n(\mathbb{R})$ $(n > 1)$ are the same, by 9.1.7 and the fact that $P^2(\mathbb{R})$ can be identified with the 2-skeleton of $P^n(\mathbb{R})$. Also, $P^2(\mathbb{R}) = \mathbb{S}^1 {}_f\sqcup \mathbb{E}^2$ where $f : \mathbb{S}^1 \to \mathbb{S}^1$ is of degree 2 [Section 5.3]. It follows that the fundamental group of $P^2(\mathbb{R})$ is the group $\mathbb{Z}/2\mathbb{Z} = \mathbb{Z}_2$.
3. We can also state that the fundamental groups of $\mathbb{S}^2$ and $\mathbb{S}^1 \times \mathbb{S}^1$ are 0 and $\mathbb{Z} \times \mathbb{Z}$ respectively. It follows that no two of the spaces $\mathbb{S}^2$, $\mathbb{S}^1 \times \mathbb{S}^1$, $P^n(\mathbb{R})$ are of the same homotopy type; *a fortiori*, no two of these spaces are homeomorphic.
4. The Klein bottle has a cell structure $K = e^0 \cup e^1_1 \cup e^1_2 \cup e^2$. From Fig. 4.4, p. 98, it is clear that, if $\{a, b\}$ is a set of generators of $\pi(K^1, v)$ as given in 9.1.5, then the relation determined by the 2-cell of K is $abab^{-1}$. Thus $\pi(K, v)$ is a free group on two generators a, b with the relation $abab^{-1} = 1$.

It is a simple consequence of 9.1.2 that if we form a pushout of spaces $Q = B {}_f\sqcup (X \times \mathbb{I})$ by attaching a cylinder $X \times \mathbb{I}$ to B by means of a map $f : X \times \dot{\mathbb{I}} \to B$ then the fundamental groupoid of Q, on an appropriate set, may be described as a pushout of groupoids in a manner analogous to (8.4.2). We leave the reader to describe this precisely.

The results on HNN-extensions in section 8.4, and our earlier proofs that the fundamental group of a circle is infinite cyclic, show that some groups are well described as constructed from groupoids. On the other hand, it is sometimes convenient to regard groups as object groups of groupoids. As an example, consider the trefoil group $\mathrm{Tr} = \mathrm{gp}\langle x, y \mid x^3y^{-2}\rangle$. This is known to be an infinite group, but from this viewpoint it is not so easy to find a normal form for its elements.

A different way into its structure comes from seeing the trefoil group as a fundamental group of a cell complex given as a double mapping cylinder. Let the unit circle $\mathbb{S}^1$ have base point e, say, and consider the double mapping cylinder $M = M(3, 2)$ of the maps $\mathbb{S}^1 \to \mathbb{S}^1$ given by $z \mapsto z^3$, $z \mapsto z^2$ respectively. This space M contains two copies of $\mathbb{S}^1$ with base points e_3, e_2, say. Let E be the set of these base points. Then the fundamental groupoid $\widehat{T} = \pi(M, E)$ has a presentation with three generators $\hat{x} \in \widehat{T}(e_3)$, $\hat{y} \in \widehat{T}(e_2)$, $w \in \widehat{T}(e_3, e_2)$ with the relation $w\hat{x}^3 = \hat{y}^2 w$. An advantage of this 'Trefoil groupoid' $\widehat{T}$ over the Trefoil group Tr is that the generator w acts as a kind of 'separator' of the generators $\hat{x}, \hat{y}$ for any well defined word in these generators not containing consecutive symbols u, u^{-1}, for $u = x, y, w$ or their inverses. For this reason, it is easy to give a normal form for words, and we leave this as an exercise for the reader.

Notice also that the double mapping cylinder M as above is a cell complex. By contrast, the pushout P of the two maps $\mathbb{S}^1 \to \mathbb{S}^1$ as above is not even Hausdorff, and it is not clear what might be its fundamental groupoid.

By contrast, the groupoid $\widehat{T}$ is easily understood: it is the double mapping cylinder in the category of groupoids of the two morphisms $\mathbb{Z} \to \mathbb{Z}$ given by multiplication by 2 and by 3 respectively.

We now give an intuitive account of the computation of the fundamental group of the complement of a graph embedded in Euclidean space $\mathbb{R}^3$. This computation is important in knot theory.

By a *graph* in $\mathbb{R}^3$ we mean a cell complex K of dimension 1 which is embedded in $\mathbb{R}^3$. Such a graph is usually represented by a diagram with vertices, overpasses, edges and crossings as in Fig. 9.3.

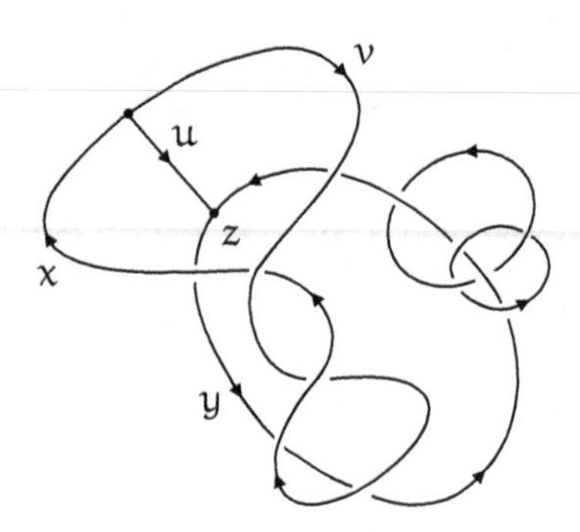

Fig. 9.3

In order to avoid *wildness* problems, for example edges which wind infinitely often round others (see [FA49]) we suppose given a 1-dimensional complex K and an embedding $i : K \times \mathbb{E}^2 \to \mathbb{R}^3$. That is, the edges of the graph are taken to have a certain thickness.

The diagram D of the embedding can be regarded as the projection of $i[K \times \mathbb{E}^2]$ onto $\mathbb{R}^2$ by the mapping $(x, y, z) \mapsto (x, y)$, and it is supposed that the embedding is arranged so that the diagram has edges crossing only at double points. This conforms with our picture. The diagram allows us to divide the embedded graph into *vertices*, namely the images of $v \times \mathbb{E}^2$ where v is a vertex of K at which more than two edges meet, and *overpasses*, namely the edges between vertices, between crossings, or between vertices and crossings. Thus the overpasses are labelled by letters in Fig. 9.3. We also orient the graph by choosing a direction for each overpass as shown.

9.1.9 *Under the above circumstances, the fundamental group $\pi(\mathbb{R}^3 \setminus i[K \times \mathbb{E}^2], p)$ has a presentation with a generator for each overpass and relations of two types:*

(a) *at each vertex v there is a relation $x_1^{\varepsilon_1} x_2^{\varepsilon_2} \dots x_r^{\varepsilon_r} = 1$, where the x_i are the edges at the vertex and the sign ε_i is $+1$ if the arrow for x_i points towards the vertex, and -1 otherwise;*

(b) *at each crossing with overpass* x *crossing* y *and* z *as shown in* Fig. 9.4 (ii), *there is a relation* $y = xzx^{-1}$.

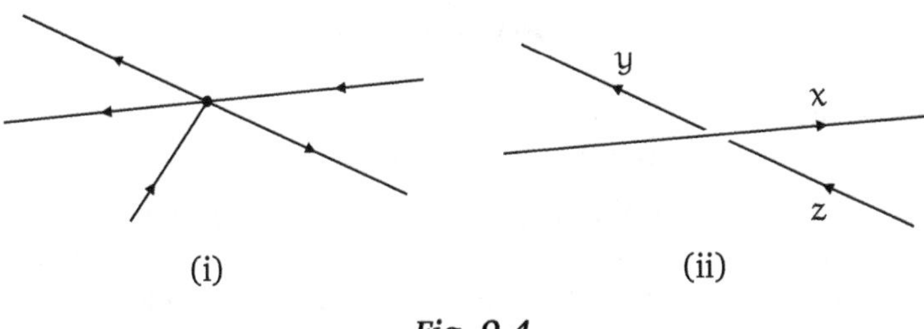

Fig. 9.4

A proof goes roughly as follows. Imagine the graph $i[K \times \mathbb{E}^2]$ as part of a slice $S = \mathbb{R}^2 \times [-\varepsilon, \varepsilon]$ of $\mathbb{R}^3$, while S is regarded as the intersection of two half spaces $H_+ = \mathbb{R}^2 \times [-\varepsilon, \rightarrow[$ and $H_- = \mathbb{R}^2 \times]\leftarrow, \varepsilon]$. Surround each vertex v by a solid ball E_v and each crossing c by a solid ball D_c as in Fig. 9.5.

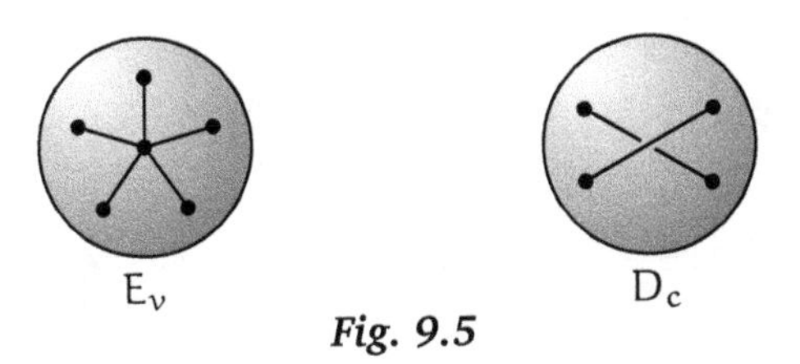

Fig. 9.5

Let X be the union of $i[K \times \mathbb{E}^2]$ and all the balls E_v and D_c. Arrange these so that $H_+ \setminus X$ and $H_- \setminus X$ intersect in a non connected space with one component for each region into which the diagram of the graph divides the plane. By 9.1.8, the fundamental group $\mathbb{R}^3 \setminus X$ is a free group with one generator for each edge e_i of the diagram between crossings or vertices. Now replace the balls E_v and D_c with their intersection with $i[K \times \mathbb{E}^2]$ excised. It is easy to see that E_v contributes a relation as given at each vertex. The balls D_c contribute two relations. One of these 'continues' an overpass while the other is the relation we want; that is in the situation shown in Fig. 9.6,

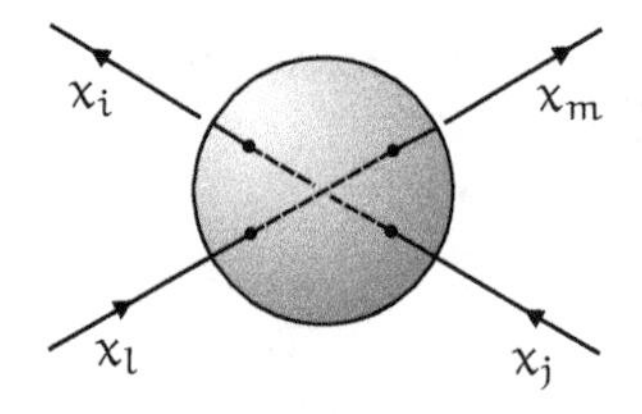

Fig. 9.6

the relations are

$$x_l = x_m, \quad x_i = x_l x_j x_k^{-1}.$$

Another way of seeing the intuitive basis of these relations is given in Fig. 9.7. The thick lines denote parts of the embedded graph, and the thin lines denote various positions of a deformation of part of a path. The reader is also urged to demonstrate these relations with string and wire models.

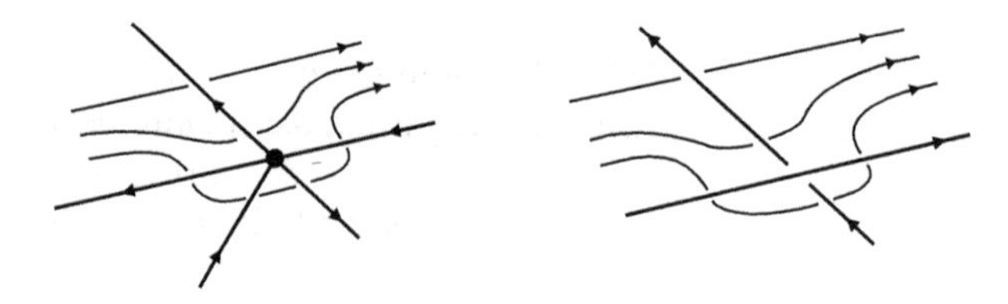

Fig. 9.7

As one application, for the pentoil shown in Fig. 9.8

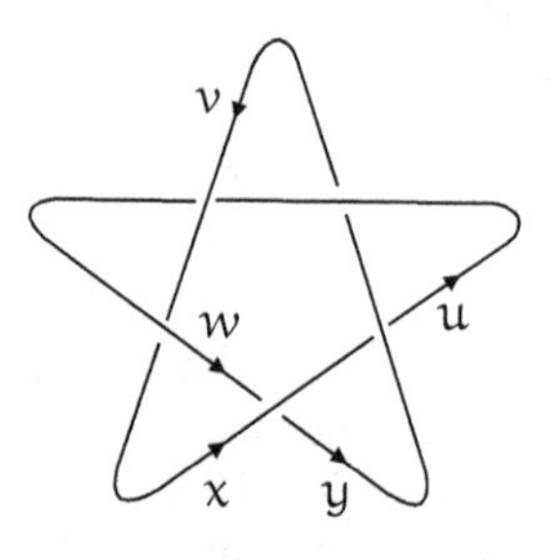

Fig. 9.8

we obtain a presentation for the fundamental group of the complement as having generators x, y, u, v, w with relations $w = xyx^{-1}$, $x = yuy^{-1}$, $y = uvu^{-1}$ and $v = wxw^{-1}$. By elimination of u, v, and w we may obtain the presentation with generators x and y and one relation

$$xyxyxy^{-1}x^{-1}y^{-1}x^{-1}y^{-1} = 1.$$

This relation corresponds to wrapping string around a part of knot model as shown in Fig. 9.9.

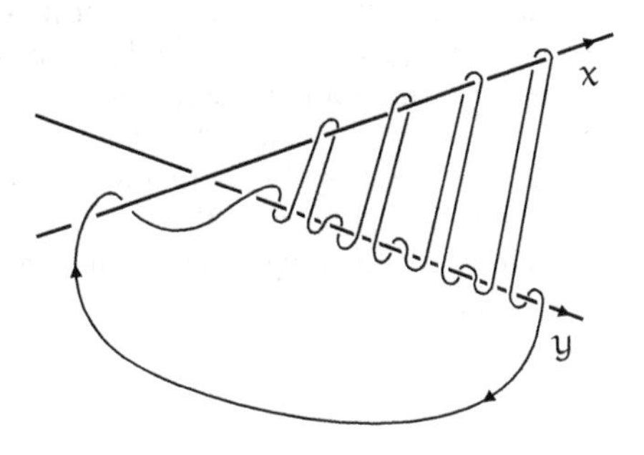

Fig. 9.9

If you wrap string around a model of the pentoil in precisely this way, and then tie the ends together, the loop will disentangle itself from the knot, thus demonstrating the calculation. (See [Bro88].)

For further information on knots and links, consult also [Kau87]. For relations of the fundamental group to the important area of configuration spaces (these are spaces of n distinct points in a space X) see also [Bir75].

EXERCISES

1. Prove that 9.1.7 is a consequence of the cellular approximation theorem.
2. Prove that the spaces $P^n(\mathbb{H})$ are simply-connected.
3. Prove that $\mathbb{S}^1$ is not a retract of $P^n(\mathbb{R})$, $n > 1$.
4. Let $X = Y \cup Z$ where Z, Y are path-connected and (X, Y) is cofibred. Let $a_0, a_1, \dots, a_n$ be points one in each path-component of $Y \cap Z$. Let i, j be the inclusions of $Y \cap Z$ into Y, Z respectively. Let $\alpha_r \in \pi Y(a_0, a_r)$, $\beta_r \in \pi Z(a_0, a_r)$, $r = 0, \dots, n$, with $\alpha_0 = 1$, $\beta_0 = 1$. Let F be a free group on elements γ_r, $r = 0, \dots, n$ with the relation $\gamma_0 = 1$. Prove that $\pi(X, a_0)$ is isomorphic to the free product of the groups $\pi(Y, a_0)$, $\pi(Z, a_0)$ and F with the relations

$$\alpha_r^{-1}(i\rho_r)\alpha_r = \gamma_r(\beta_r^{-1}(j\rho_r)\beta_r)\gamma_r^{-1}$$

for all $\rho_r \in \pi(Y \cap Z, a_r)$ and $r = 0, \dots, n$. [Here γ_r corresponds to the element $(u\beta_r^{-1})(v\alpha_r)$ of $\pi(X, a_0)$ where u, v are the inclusions of Y, Z respectively into X.]
5. Let K, L be 1-dimensional (finite) cell complexes. Prove that if $\varphi : \pi KK^0 \to \pi LL^0$ is any morphism, then there is a map $f : K \to L$ such that $\pi f = \varphi$. Prove also that if $f, g : K \to L$ are cellular maps such that $\pi f \simeq \pi g : \pi KK^0 \to \pi LL^0$, then f is homotopic to g.

6. Extend the results 9.1.5, 9.1.6, 9.1.7 to (infinite) CW-complexes. Prove that if G is any group, then there is a CW-complex K and a vertex x of K such that $\pi(K, x)$ is isomorphic to G. Deduce that if G is any groupoid then there is a CW-complex K such that πKK^0 is isomorphic to G. [You may assume that if G is any group then there is a free group F and a free subgroup R of F such that G is isomorphic to F/R.]
7. Prove that $\mathbb{R}^2$ and $\mathbb{R}^n$ for $n > 2$ are not homeomorphic.
8. Let $p : z \mapsto z^n + a_{n-1}z^{n-1} + \cdots + a_1 z + a_0$ and $q : z \mapsto z^n$ be polynomials with $a_i \in \mathbb{C}$. For $r > 0$ let $C_r = \{z \in \mathbb{R}^2 : |z| = r\}$. Prove that for r large enough, p and q restrict to homotopic maps $C_r \to \mathbb{R}^2 \setminus \{0\}$. Prove that for any $r > 0$ this restriction of q is essential and hence show that the polynomial p has a root. [This is known as the Fundamental Theorem of Algebra.]

9.2 The Jordan Curve Theorem

The Jordan Curve Theorem states that if C is a subset of the plane $\mathbb{R}^2$ such that C is homeomorphic to the circle $\mathbb{S}^1$, then $\mathbb{R}^2 \setminus C$ has exactly two components, one of them bounded and the other unbounded, and each with C as boundary. The set C is called a *simple closed curve in* $\mathbb{R}^2$. The bounded component of $\mathbb{R}^2 \setminus C$ is of course called the *inside* of the curve, and the unbounded component is called the *outside* of the curve.

This theorem is a classic instance of a result which at first sight seems intuitively obvious, but which needs some sophisticated machinery for its proof. In fact the theorem is not quite so obvious intuitively. Consider for example the computer generated simple closed curve in $\mathbb{R}^2$ shown in Fig. 9.10. How do you determine the inside and outside? (I am indebted to S. J. Abas for this diagram.)

Any method you choose for this curve might be defeated by a more complicated example. In any case, if you try and work out the inside and outside for this case, you begin to see the prospective complications of the problem.

In this section we shall use the final results of the last section to give a complete proof of the theorem; we also draw further consequences of the method. For this reason, we take what might seem a circuitous route to the theorem, by introducing a property which a space may or may not have.

Fig. 9.10

A topological space X is said to *have the Phragmen-Brouwer property* (here abbreviated to PBP) if X is connected and the following holds: *if* D *and* E *are disjoint, closed subsets of* X, *and if* a *and* b *are points in* $X \setminus (D \cup E)$ *which lie in the same component of* $X \setminus D$ *and in the same component of* $X \setminus E$, *then* a *and* b *lie in the same component of* $X \setminus (D \cup E)$. To express this more succinctly, we say a subset D of a space X *separates* the points a and b if a and b lie in distinct components of $X \setminus D$. Thus the PBP is that: if D and E are disjoint closed subsets of X and a, b are points of X not in $D \cup E$ such that neither D nor E separate a and b, then $D \cup E$ does not separate a and b.

A standard example of a space not having the PBP is the circle $\mathbb{S}^1$, since we can take $D = \{+1\}$, $E = \{-1\}$, $a = i$, $b = -i$. This example is typical, as the next result shows. But first we remark that our criterion for the PBP will involve fundamental groups, that is will involve paths, and so we need to work with path-components rather than components. However, if X is locally path-connected, then components and path-components of open sets of X coincide, and so for these spaces we can replace in the PBP 'component' by 'path-component'. This explains the assumption of locally path-connected in the results that follow.

9.2.1 *Let* X *be a path-connected and locally path-connected space whose fun-*

damental group (at any point) does not have the integers $\mathbb{Z}$ as a retract. Then X has the PBP.

Proof Suppose X does not have the PBP. Then there are disjoint, closed subsets D and E of X and points a and b of $X \setminus (D \cup E)$ such that $D \cup E$ separates a and b but neither D nor E separates a and b. Let $X_1 = X \setminus D$, $X_2 = X \setminus E$, $X_0 = X \setminus (D \cup E) = X_1 \cap X_2$. Let J be a subset of X_0 such that $a, b \in J$ and J meets each path-component of X_0 in exactly one point. Since D and E do not separate a and b, there are elements $\alpha \in \pi X_1(a, b)$ and $\beta \in \pi X_2(a, b)$. Since X is path-connected, the set J is representative in X_0, X_1 and X_2. By 6.7.2 the following diagram of morphisms induced by inclusions is a pushout of groupoids:

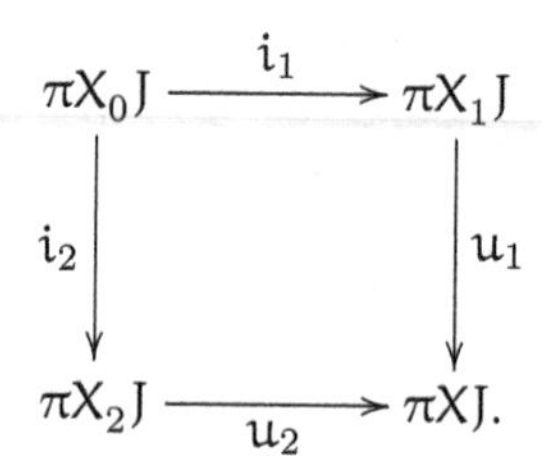

Since X_1 and X_2 are path-connected and J has more than one element, it follows from (9.1.9 (Corollary)) that πXJ has the integers $\mathbb{Z}$ as a retract. □

As an immediate application we obtain:

9.2.2 *The following spaces have the PBP: the sphere $\mathbb{S}^n$ for $n > 1$; $\mathbb{S}^2 \setminus \{a\}$ for $a \in \mathbb{S}^2$; $\mathbb{S}^n \setminus A$ if A is a finite set in $\mathbb{S}^n$ and $n > 2$.* □

In each of these cases the fundamental group is trivial.

An important step in our proof of the Jordan Curve Theorem is to show that if A is an *arc* in $\mathbb{S}^2$, that is a subspace of $\mathbb{S}^2$ homeomorphic to the unit interval $\mathbb{I}$, then the complement of A is path-connected. This follows from the following more general result.

9.2.3 *Let X be a path-connected and locally path-connected Hausdorff space such that for each x in X the space $X \setminus \{x\}$ has the PBP. Then any arc in X has path-connected complement.*

Proof Suppose A is an arc in X and $X \setminus A$ is not path-connected. Let a and b lie in distinct path-components of $X \setminus A$.

By choosing a homeomorphism $\mathbb{I} \to A$ we can speak unambiguously of the mid-point of A or of any subarc of A. Let x be the mid-point of A, so that A is the union of sub-arcs A' and A'' with intersection $\{x\}$. Since X is Hausdorff, the compact sets A' and A'' are closed in X. Hence $A' \setminus \{x\}$ and

$A'' \setminus \{x\}$ are disjoint and closed in $X \setminus \{x\}$. Also $A \setminus \{x\}$ separates a and b in $X \setminus \{x\}$ and so one at least of A', A'' separates a and b in $X \setminus \{x\}$. Write A_1 for one of A', A'' which does separate a and b. Then A_1 is also an arc in X.

In this way we can find by repeated bisection a sequence A_i, $i \geqslant 1$, of sub-arcs of A such that for all i the points a and b lie in distinct path-components of $X \setminus A_i$ and such that the intersection of the A_i for $i \geqslant 1$ is a single point, say y, of X.

Now $X \setminus \{y\}$ is path-connected, by definition of the PBP. Hence there is a path λ joining a to b in $X \setminus \{y\}$. But λ has compact image and hence lies in some $X \setminus A_i$. This is a contradiction. □

9.2.3 *(Corollary) The complement of any arc in $\mathbb{S}^n$ is path-connected.* □

In this theorem the case $n = 0$ is trivial, while the case $n = 1$ needs a special argument that the complement of any arc in $\mathbb{S}^1$ is an open arc. The case $n \geqslant 2$ follows from the above results.

We now prove one step along the way to the full Jordan Curve Theorem.

9.2.4 (The Jordan Separation Theorem) *The complement of a simple closed curve in $\mathbb{S}^2$ is not connected.*

Proof Let C be a simple closed curve in $\mathbb{S}^2$. Since C is compact and $\mathbb{S}^2$ is Hausdorff, C is closed, $\mathbb{S}^2 \setminus C$ is open, and so path-connectedness of $\mathbb{S}^2 \setminus C$ is equivalent to connectedness.

Write $C = A \cup B$ where A and B are arcs in C meeting only at a and b say. Let $U = \mathbb{S}^2 \setminus A$, $V = \mathbb{S}^2 \setminus B$, $W = U \cap V$, $X = U \cup V$. Then $W = \mathbb{S}^2 \setminus C$ and $X = \mathbb{S}^2 \setminus \{a, b\}$. Also X is path-connected, and, by 9.2.3 (Corollary), so also are U and V.

Let $x \in W$. Suppose that W is path-connected. By 6.7.2, the following diagram of morphisms induced by inclusion is a pushout of groups:

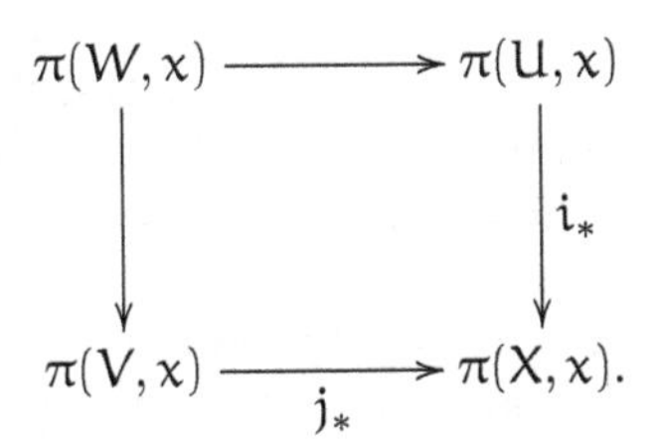

Now $\pi(X, x)$ is isomorphic to the group $\mathbb{Z}$ of integers. We derive a contradiction by proving that the morphisms i_* and j_* are trivial. We give the proof for i_*, as that for j_* is similar.

Let $f : \mathbb{S}^1 \to U$ be a map and let $g = if : \mathbb{S}^1 \to X$. Let γ be a parametrisation of A which sends 0 to b and 1 to a. Choose a homeomorphism

$h : \mathbb{S}^2 \setminus \{a\} \to \mathbb{R}^2$ which takes b to 0 and such that hg maps $\mathbb{S}^1$ into $\mathbb{R}^2 \setminus \{0\}$. Then $h\gamma(0) = 0$ and $\|h\gamma(t)\|$ tends to infinity as t tends to 1. Since the image of g is compact, there is an $r > 0$ such that $hg[\mathbb{S}^1]$ is contained in $B(0, r)$. Now there exists $0 < t_0 < 1$ such that the distance from 0 to $y = h\gamma(t_0)$ is $> r$. Define the path λ to be the part of $h\gamma$ reparametrised so that $\lambda(0) = 0$ and $\lambda(1) = y$.

Define $G : \mathbb{S}^1 \times \mathbb{I} \to \mathbb{R}^2$ by

$$G(z, t) = \begin{cases} hg(z) - \lambda(2t) & \text{if } 0 \leqslant t \leqslant \frac{1}{2}, \\ (2 - 2t)hg(z) - y & \text{if } \frac{1}{2} \leqslant t \leqslant 1. \end{cases}$$

Then G is well-defined. Also G never takes the value 0 (this explains the choices of λ and y). So G gives a homotopy in $\mathbb{R}^2 \setminus \{0\}$ from hg to the constant map at $-y$. So hg is inessential and hence g is inessential. This completes the proof that i_* is trivial. □

As we shall see, the Jordan Separation Theorem is used in the proof of the Jordan Curve Theorem.

9.2.5 (Jordan Curve Theorem) *If C is a simple closed curve in $\mathbb{S}^2$, then the complement of C has exactly two components, each with C as boundary.*

Proof As in the proof of 9.2.4, write C as the union of two arcs A and B meeting only at a and b say, and let $U = \mathbb{S}^2 \setminus A$, $V = \mathbb{S}^2 \setminus B$. Then U and V are path-connected and $X = U \cup V = \mathbb{S}^2 \setminus \{a, b\}$ has fundamental group isomorphic to $\mathbb{Z}$. Also $W = U \cap V = \mathbb{S}^2 \setminus C$ has at least two path-components, by 9.2.4.

If W has more than two path-components, then the fundamental group G of X contains a copy of the free group on two generators, by 9.1.9 (Corollary), and so G is non-abelian. This is a contradiction, since $G \cong \mathbb{Z}$. So W has exactly two path-components P and Q, say, and this proves the first part of 9.2.5.

Since C is closed in $\mathbb{S}^2$ and $\mathbb{S}^2$ is locally path-connected, the sets P and Q are open in $\mathbb{S}^2$. It follows that if $x \in \overline{P} \setminus P$ then $x \notin Q$, and hence $\overline{P} \setminus P$ is contained in C. So also is $\overline{Q} \setminus Q$, for similar reasons. We prove these sets are equal to C.

Let $x \in C$ and let N be a neighbourhood of x in $\mathbb{S}^2$. We prove N meets $\overline{P} \setminus P$. Since $\overline{P} \setminus P$ is closed and N is arbitrary, this proves that $x \in \overline{P} \setminus P$.

Write C in a possibly new way as a union of two arcs D and E intersecting in precisely two points and such that D is contained in $N \cap C$. Choose points p in P and q in Q. Since $\mathbb{S}^2 \setminus E$ is path-connected, there is a path λ joining p to q in $\mathbb{S}^2 \setminus E$. Then λ must meet D, since p and q lie in distinct path-components of $\mathbb{S}^2 \setminus E$. In fact if $s = \sup\{t \in \mathbb{I} : \lambda[0, t] \subseteq P\}$, then $\lambda(s) \in \overline{P} \setminus P$. It follows that N meets $\overline{P} \setminus P$.

So $\overline{P} \setminus P = C$ and similarly $\overline{Q} \setminus Q = C$. □

NOTES

There are many books containing a further discussion of this area. For more on the Phragmen-Brouwer property, see [Why42] and [Wil49]. Wilder lists five other properties which he shows for a connected and locally connected metric space are each equivalent to the PBP. The above proof of the Jordan Curve Theorem is adapted from [Mun75]. Because he does not have our Van Kampen theorem for non-connected spaces, he is forced into rather special covering space arguments to prove his replacements for 9.1.9 (Corollary) and for 9.2.1. As far as I am aware, 9.2.1 and 9.2.2 are not previously published.

A different kind of proof of the Jordan Curve Theorem is given in [Mae84]; this uses only the Brouwer Fixed Point Theorem and the Tietze Extension Theorem.

An important strengthening of the Jordan Curve Theorem is the Schoenflies Theorem: : *if* C *is a simple closed curve in* $\mathbb{S}^2$ *then each component of* $\mathbb{S}^2 \setminus C$ *is homeomorphic to* $\mathbb{R}^2$*; in fact, there is a homeomorphism* $h : \mathbb{R}^2 \to \mathbb{R}^2$ *which takes* C *to the standard circle* $\mathbb{S}^1$. For more information on this area, see for example [Moi77], [Bin83], [Sie05], and the web site `www.maths.ed.ac.uk/~aar/jordan/`.

For exercises in this area, see [Mun75].

An important part of algebraic topology deals with braids, links and mapping class groups [Bir75], but the results given in this book on groupoids have hardly been used in this area.

Chapter 10

Covering spaces, covering groupoids

The notion of a covering space is of interest for several reasons. First, the construction of a covering space $\widetilde{X}$ from a given space X (which we give in section 10.5) is useful since the structure of $\widetilde{X}$ is in a sense simpler than that of X ($\widetilde{X}$ has a 'smaller' fundamental group than X). Second, the method of studying a given space X by means of particular kinds of maps into X is of wide importance in topology—the covering maps onto X exemplify this, and indeed form a particular example of the fibre maps whose ramifications extend through topology, differential geometry, analysis, and even algebra. Third, and most important from the point of view of this book, the topology is here very nicely modelled in the notion of covering morphism of groupoids. With this account, we give a further aspect of combinatorial groupoid theory, and one which is important for applications to topology and to group theory. It may be that the results of the next chapter will also come to be seen in this light.

These results also imply that the fundamental group of the circle is infinite cyclic, and indeed this is the classical method of proof. However the groupoid viewpoint adds new insight to this and other results.

The main result is the equivalence of categories between the category of covering maps of a 'locally nice' space X and the category of covering morphisms of the fundamental groupoid πX. In this way, we really do show a tight analogy between topology and algebra. The method of proof gives tighter results than those usually given, and has also been applied to other situations, such as when X is a topological group.

We shall use additive notation for groupoids in this chapter.

10.1 Covering maps and covering homotopies

Throughout this chapter all spaces will be assumed *locally path-connected* (or proved to be so). This assumption is not essential for all the results (for an alternative approach see [HW60] or [Spa66]) but it does seem to lead to a smoother theory; also the study of covering maps for spaces which are not locally path-connected is something of a curiosity.

Definitions Let $p : \widetilde{X} \to X$ be a map of topological spaces. A subset U of X is called *canonical* (with respect to p) if U is open, path-connected, and each path-component of $p^{-1}[U]$ is open in $\widetilde{X}$ and is mapped by p homeomorphically onto U; each path-component of $p^{-1}[U]$ is also called *canonical*. The map p is called a *covering map* if each x in X has a canonical neighbourhood; and in such case $\widetilde{X}$ is called a *covering space* of X. The covering map p is called *connected* if both $\widetilde{X}$ and X are path-connected.

The local picture of a covering map is thus that of Fig. 10.1(a).

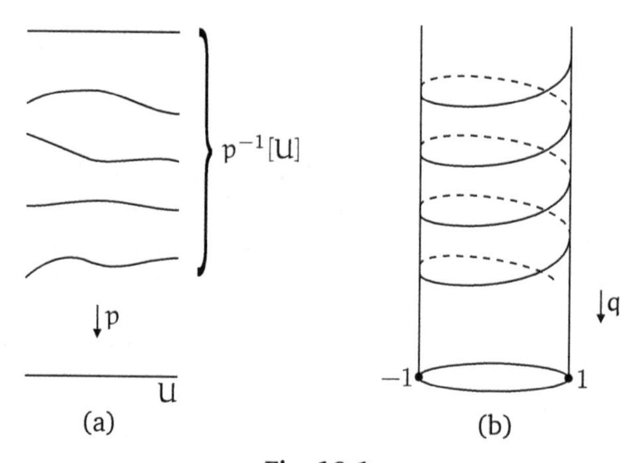

Fig. 10.1

EXAMPLES

1. Consider the map $p : \mathbb{R} \to \mathbb{S}^1$, $t \mapsto e^{2\pi i t}$. The sets

$$U_1 = \mathbb{S}^1 \setminus \{1\}, \qquad U_{-1} = \mathbb{S}^1 \setminus \{-1\}$$

form an open cover of $\mathbb{S}^1$. The path-components of $p^{-1}[U_1]$ (respectively $p^{-1}[U_{-1}]$) are open intervals $]n, n+1[$ (respectively $]n - \frac{1}{2}, n + \frac{1}{2}[$) for all $n \in \mathbb{Z}$. So p is a covering map. A good picture of p is obtained by writing $p = qr$ where $r : \mathbb{R} \to \mathbb{R}^2 \times \mathbb{R}$ is $t \mapsto (pt, t)$ and $q : \mathbb{R}^2 \times \mathbb{R} \to \mathbb{R}^2$ is the projection. Thus the image of r is the helix of Fig. 10.1(b).

2. Let $f : \mathbb{S}^1 \to \mathbb{S}^1$ be the map $z \mapsto z^n$, where n is a non-zero integer. It is easily proved by using the sets U_1, U_{-1} of Example 1 that f is a covering map.
3. If $p : \widetilde{X} \to X$ is a covering map, then for each x in X the subspace $p^{-1}[x]$ of $\widetilde{X}$ is a discrete space—this follows from the fact that if U is a canonical set then the path-components of $p^{-1}[U]$ are open in $\widetilde{X}$. It follows that the fundamental map $p : \mathbb{R}^{n+1}_* \to P^n(\mathbb{R})$ is not a covering map.
4. However, let $i : \mathbb{S}^n \to \mathbb{R}^{n+1}_*$ be the inclusion so that $h = pi$ is the Hopf map. Let U_i be the subset of $P^n(\mathbb{R})$ of points px such that the ith coordinate of x is non-zero. The sets U_i, $i = 1, \dots n+1$ form an open cover of $P^n(\mathbb{R})$. The path-components of $h^{-1}[U_i]$ are two open hemispheres of $\mathbb{S}^n$ each of which is mapped homeomorphically by h onto U_i. (For example, $h^{-1}[U_{n+1}] = \mathbb{S}^n \setminus \mathbb{S}^{n-1}$.) Therefore, h is a covering map.

Let $p : \widetilde{X} \to X$ be a covering map. If the diagram of maps

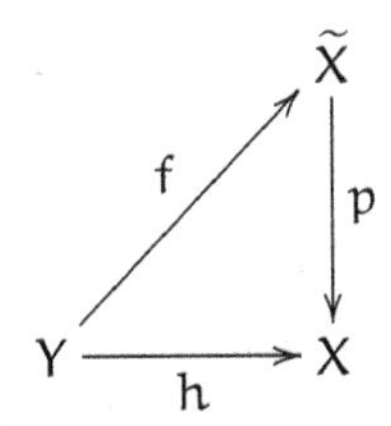

is commutative, then we say f is a *lifting* of h, or that f *covers* h. We shall see that the central properties of covering maps are concerned with liftings; for example, we have the following *covering homotopy property*.

10.1.1 *Let $p : \widetilde{X} \to X$ be a covering map and suppose given a commutative diagram of maps (in which i is $z \mapsto (z, 0)$)*

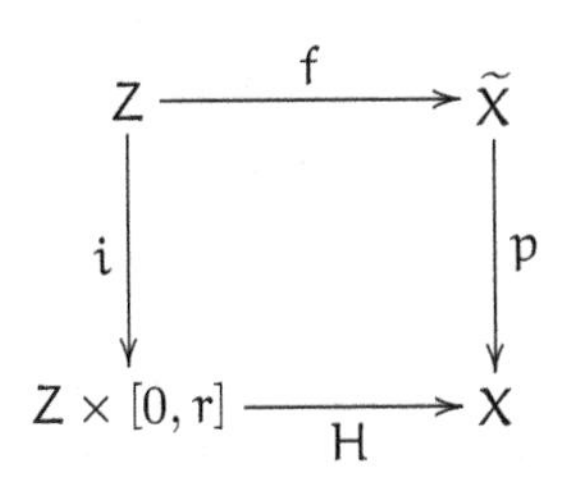

so that H is a homotopy of pf. Then H is covered by a unique homotopy $F : Z \times [0, r] \to \widetilde{X}$ of f.

Proof Since Z is locally path-connected, each path-component of Z is open in Z. Therefore, it is sufficient to work in each path-component of Z at a time, that is, to assume Z is path-connected.

Step 1. Suppose that $\operatorname{Im} H$ is contained in a canonical subset U of X. Since Z is path-connected, $\operatorname{Im} f$ is contained in a path-component V of $p^{-1}[U]$. Let p_V be the inverse of the homeomorphism $p \mid V, U$. Then the map

$$F : Z \times [0, r] \to \widetilde{X}$$
$$(z, t) \mapsto p_V H(z, t)$$

is a homotopy of f which covers H. Since $Z \times [0, r]$ is path-connected, it is also the only such homotopy of f.

Step 2. Suppose that the homotopy H is a sum

$$H = H_n + \cdots + H_1$$

such that the image of each H_i is contained in a canonical subset of X. By Step 1, we can define inductively unique homotopies $F_i : f_i \simeq f_{i+1}$, $i = 1, \ldots, n$, such that $f_1 = f$ and F_i covers H_i. Clearly,

$$F = F_n + \cdots + F_1$$

is a homotopy of f which covers H. If F' is any other homotopy of f which covers H, then the given subdivision of H determines a subdivision $F' = F'_n + \cdots + F'_1$. It follows by induction that $F'_i = F_i$, $i = 1, \ldots, n$ and so $F' = F$.

Step 3. Let $z \in Z$. We use Step 2 to show that the result is true with Z replaced by some neighbourhood M^z of z.

For each t in $[0, r]$ there are open neighbourhoods M_t, N_t of z, t respectively such that $H[M_t \times N_t]$ is contained in a canonical subset of X. The set $\{N_t : t \in [0, r]\}$ is an open cover of $[0, r]$ which, by compactness, has a finite subcover $\{N_t : t \in A\}$ say. Let

$$M = \bigcap_{t \in A} M_t$$

and let M^z be an open path-connected neighbourhood of z contained in M.

Let l be the Lebesgue number of the cover $\{N_t : t \in A\}$ and let n be an integer such that $0 < r/n < l/2$. For each $i = 1, \ldots, n$ the interval $L_i = [(i-1)r/n, i_r/n]$ is contained in some N_t, $t \in A$, and so $H[M^z \times L_i]$ is contained in a canonical subset of X.

Let H_i^z be the homotopy

$$M^z \times [0, r/n] \to X$$
$$(w, t) \mapsto H(w, (i-1)r/n + t)$$

and let

$$H^z = H \mid M^z \times [0, r] = H^z_n + \cdots + H^z_1.$$

By Step 2, H^z is covered by a unique homotopy F^z of $f \mid M^z$.

Step 4. Let $F : Z \times [0, r] \to \widetilde{X}$ be the function $(z, t) \mapsto F^z(z, t)$. The uniqueness part of the above argument shows that for each $z \in Z$

$$F \mid M^z \times [0, r] = F^z.$$

Since M^z is a neighbourhood of z, it follows that F is continuous. Clearly, F is the unique homotopy of f covering H. □

Our main use of the covering homotopy property is to prove the following *path lifting property*.

10.1.2 *Let $p : \widetilde{X} \to X$ be a covering map, let $\tilde{x} \in \widetilde{X}$ and let $x = p\tilde{x}$. Then paths a, b in X with initial point x lift uniquely to paths $\tilde{a}, \tilde{b}$ with initial point $\tilde{x}$; and a is equivalent to b if and only if $\tilde{a}$ is equivalent to $\tilde{b}$.*

Proof The existence and uniqueness of $\tilde{a}$ (and also $\tilde{b}$) is immediate from 10.1.1 by taking Z to consist of a single point z and defining

$$fz = \tilde{x}, \qquad H(z, t) = at, \qquad t \in [0, r], \qquad r = |a|.$$

Suppose next that a, b are paths from x to y of length r which are homotopic rel end points. Then, twisting the usual way of writing a homotopy, there is a map

$$H : \mathbb{I} \times [0, r] \to X$$

such that

$$\begin{aligned} H(s, 0) &= x, & H(s, r) &= y, & s &\in \mathbb{I} \\ H(0, t) &= at, & H(1, t) &= bt, & t &\in [0, r]. \end{aligned}$$

Let $f : \mathbb{I} \to \widetilde{X}$ be the constant map with value $\tilde{x}$. Then H is a homotopy of pf and so, by 10.1.1, H is covered by a unique homotopy F of f.

The paths in $\widetilde{X}$

$$t \mapsto F(0, t), \qquad t \mapsto F(1, t)$$

have initial point $\tilde{x}$ and cover a, b respectively; so these paths are $\tilde{a}, \tilde{b}$ respectively. The path $s \mapsto F(s, r)$ covers the constant path at y and so (by uniqueness) is itself constant; the path $s \mapsto F(s, 0)$ is constant by definition of F. Therefore, F (suitably twisted) is a homotopy rel end points $\tilde{a} \simeq \tilde{b}$.

Finally, if there are real numbers $s, s' \geqslant 0$ such that $s + a$, $s' + b$ are homotopic rel end points, then $s + \tilde{a}$, $s' + \tilde{b}$ cover $s + a$, $s' + b$ respectively, and hence $\tilde{a}$ is equivalent to $\tilde{b}$. □

Remark In fact, we shall use only the consequence 10.1.2 of 10.1.1 and not 10.1.1 itself. A simpler proof of 10.1.1 can be given in the case $Z = [0, q]$ by subdividing the rectangle $[0, q] \times [0, r]$ into rectangles so small that each is mapped by H into a canonical subset of X. But the interest of 10.1.1 is that, provided the word uniqueness is omitted, it applies to many other situations. For example, the fundamental map $\mathbb{K}_*^{n+1} \to P^n(\mathbb{K})$ satisfies this weaker covering homotopy property—this follows from Exercise 8 of Section 5.3 and results of [Dol63] (see also [Spa66, 2.7.12]).

EXERCISES

1. Which of the following maps are covering maps?

   (i) $\mathbb{R} \to \mathbb{R}, \quad x \mapsto x^2$.

   (ii) $\mathbb{R} \to \mathbb{R}, \quad x \mapsto x^3$.

   (iii) $\mathbb{R}_* \to \mathbb{R}_*, \quad x \mapsto x^2$.

   (iv) $\mathbb{S}^1 \to \mathbb{S}^1, \quad e^{i\theta} \mapsto e^{2i|\theta|} \quad (-\pi < \theta \leqslant \pi)$.

   (v) $]0, 3[\to \mathbb{S}^1, \quad t \mapsto e^{2\pi i t}$.

2. Let $f : Y' \to Y$, $g : X \to X'$ be homeomorphisms and let $p : Y \to X$ be a covering map. Prove that $gpf : Y' \to X'$ is a covering map.
3. Let $p : \widetilde{X} \to X$ be a covering map, let $f : Y \to \widetilde{X}$ be a map, let $i : A \to Y$ be an inclusion and let $G : A \times \mathbb{I} \to \widetilde{X}$ be a homotopy of $f \mid A$. Prove that, if $H : Y \times \mathbb{I} \to X$ is a homotopy of pf which agrees with pG on A, then H is covered by a homotopy F of f such that F extends G. Deduce that for any map $u : A \to \widetilde{X}$, p induces an injection $p_* : [(Y, i), (\widetilde{X}, u)] \to [(Y, i), (X, pu)]$.
4. Prove that 10.1.1 is true without the assumption that Z is locally path-connected.
5. Let $p : \widetilde{X} \to X$ be a covering map and let $f : Y \to X$ be a map. Let

$$\widetilde{X}_f = \{(y, \tilde{x}) \in Y \times \widetilde{X} : fy = p\tilde{x}\}$$

and let $p_f : \widetilde{X}_f \to Y$ be the restriction of $\widetilde{X}_f$ of the projection $Y \times \widetilde{X} \to Y$. Prove that $\widetilde{X}_f$ is locally path-connected. Prove also that p_f is a covering map. Prove also that f lifts to $\widetilde{X}$ if and only if there is a map $s : Y \to \widetilde{X}_f$ such that $p_f s = 1$.
6. Let $p : \widetilde{X} \to X$ be a covering map and let $i : A \to X$ be the inclusion of the subspace A of X. Prove that if $q = p \mid p^{-1}[A], A$ then there is a homeomorphism $g : \widetilde{X}_i \to p^{-1}[A]$ such that $qg = p_i$ (where $\widetilde{X}_i, p_i$ are as in the previous exercise).
7. Let G be the sheaf of germs of continuous functions from $\mathbb{R}$ to $\mathbb{R}$ [Exercise 3 of Section 2.10]. Prove that the projection $p : G \to \mathbb{R}$, $f^x \mapsto f(x)$, is not a covering map.

8. Give another proof of 10.1.1 on the following lines. First prove it is true when Z consists of a single point. In the general case, we then have that for each z in Z then map $H \mid \{z\} \times [0, r]$ lifts to a homotopy $F \mid \{z\} \times [0, r]$ of $f \mid \{z\}$. So we have a function $F : Z \times [0, r] \to \widetilde{X}$. Prove that F is continuous by supposing that F is non-continuous and obtaining a contradiction by considering the infinum of the set of t such that F is not continuous at some (z, t).

9. Let $p : \widetilde{X} \to X$ be a connected covering map which is non-trivial (i.e., it is not a homeomorphism). Let $Y = X \times X \times X \times \cdots$ be the countable product of X with itself, and let $\widetilde{X}^n$ be the n-fold product of $\widetilde{X}$ with itself. Let $\widetilde{Y}_n = \widetilde{X}^n \times Y$. Define $p_n : \widetilde{Y}_n \to Y$ by $(\tilde{x}_1, \ldots, \tilde{x}_n, x_1, x_2, \ldots) \mapsto (p\tilde{x}_1, \ldots, p\tilde{x}_n, x_1, x_2, \ldots)$. Prove that p_n is a covering map. Let

$$\widetilde{Z} = \sqcup_{n \geqslant 1} \widetilde{Y}_n$$

and let Z be the countably infinite topological sum of Y with itself. Let $q = \sqcup p_n : \widetilde{Z} \to Z$ and let $r : Z \to Y$ be the obvious projection. Prove that q and r are covering maps but that the composite rq is not a covering map.

10.2 Covering groupoids

In this section we shall show how covering spaces are modelled in the category of groupoids.

Let G be a groupoid. For each object x of G the *star* of x in G, denoted by $\mathrm{St}_G\, x$, is the union of the sets $G(x, y)$ for all objects y of G. Thus $\mathrm{St}_G\, x$ consists of all elements of G with initial point x. When no confusion will arise we abbreviate $\mathrm{St}_G\, x$ to $\mathrm{St}\, x$.

Definition Let $p : \widetilde{G} \to G$ be a morphism of groupoids. We say p is a *covering morphism* if for each object $\tilde{x}$ of $\widetilde{G}$ the restriction of p

$$\mathrm{St}_{\widetilde{G}}\, \tilde{x} \to \mathrm{St}_G\, p\tilde{x}$$

is bijective; in such case, we call $\widetilde{G}$ a *covering groupoid* of G. The covering morphism p is called *connected* if both $\widetilde{G}$ and G are connected.

For any morphism $p : \widetilde{G} \to G$ and object $\tilde{x}$ of $\widetilde{G}$ we call the subgroup $p[\widetilde{G}(\tilde{x})]$ of $G(p\tilde{x})$ the *characteristic group* of p at $\tilde{x}$—by an abuse of language, we also refer to this group as the characteristic group of $\widetilde{G}, \tilde{x}$. If p is a covering morphism, then p maps $\widetilde{G}(\tilde{x})$ isomorphically onto this characteristic group; also, if $a \in G(p\tilde{x})$, then there is a unique element $\tilde{a}$ of $\mathrm{St}\, \tilde{x}$ such that $p\tilde{a} = a$, but $\tilde{a}$ will be a loop, that is $\tilde{a}$ will belong to $\widetilde{G}(\tilde{x})$, if and only if a itself belongs to the characteristic group of $\widetilde{G}, \tilde{x}$.

The sense in which a characteristic group of a covering morphism p characterises p will be discussed later.

EXAMPLES

1. Recall that **I** is a simply-connected groupoid with two objects $0, 1$ and one element ι of $I(0,1)$. If 0 denotes ambiguously the zero of **I** at 0 or 1, we have

$$\mathrm{St}_{\mathbf{I}}\, 0 = \{0, \iota\}, \qquad \mathrm{St}_{\mathbf{I}}\, 1 = \{0, -\iota\}.$$

In the group $\mathbb{Z}_2$ (which has one object 0 say)

$$\mathrm{St}_{\mathbb{Z}_2}\, 0 = \{0, 1\}.$$

Hence if $p : \mathbf{I} \to \mathbb{Z}_2$ is the unique morphism such that $p\iota = 1$, then p is a covering morphism. Also each characteristic group of p is trivial.

2. In the group $\mathbb{Z}_3$ (with one object 0)

$$\mathrm{St}_{\mathbb{Z}_3}\, 0 = \{0, 1, -1\}.$$

Hence, although the morphism $p : \mathbf{I} \to \mathbb{Z}_3$ which sends ι to 1 and $-\iota$ to -1 is surjective on the elements, p is not a covering morphism.

3. For groups, the only covering morphisms are isomorphisms.

We now show the utility for topology of covering morphisms.

10.2.1 *Let $p : \widetilde{X} \to X$ be a covering map, let A be a subset of X and let $\widetilde{A} = p^{-1}[A]$. Then the induced morphism*

$$\pi p : \pi\widetilde{X}\widetilde{A} \to \pi X A$$

is a covering morphism.

Proof Let $\tilde{x} \in \widetilde{A}$ and let $p\tilde{x} = x$. For each path a in X with initial point x, let $\tilde{a}$ denote the unique covering path of $\widetilde{X}$ with initial point $\tilde{x}$. If the final point of a is in A, then the final point of $\tilde{a}$ is in $\widetilde{A}$. Also, the equivalence class of $\tilde{a}$ depends only on the equivalence class of a (by 10.1.2). So the function $\mathrm{cls}\, a \mapsto \mathrm{cls}\, \tilde{a}$ is inverse to the restriction of p which maps $\mathrm{St}\, \tilde{x} \mapsto \mathrm{St}\, x$. □

Once more we obtain:

10.2.1 *(Corollary 1) The circle $\mathbb{S}^1$ has fundamental group isomorphic to the integers $\mathbb{Z}$.*

Proof Let $p : \mathbb{R} \to \mathbb{S}^1$ be the covering map $t \mapsto e^{2\pi i t}$ so that

$$p' = \pi p : \pi\mathbb{R}\mathbb{Z} \to \pi(\mathbb{S}^1, 1)$$

is a covering morphism. Now $\pi\mathbb{R}\mathbb{Z}$ is 1-connected: so for each $n \in \mathbb{Z}$ there is a unique element τ_n of $\pi\mathbb{R}(0, n)$. Of course, τ_1 is represented by the path $\mathbb{I} \to \mathbb{R}$, $t \mapsto t$, while $\tau_{n+1} - \tau_n$—the unique element of $\pi\mathbb{R}(n, n+1)$—is represented by the path $\mathbb{I} \to \mathbb{R}$, $t \mapsto t+n$. It follows from the definition of p that

$$p'(\tau_{n+1} - \tau_n) = p'\tau_1.$$

Hence, if $\tau = p'\tau_1$, then

$$\begin{aligned} p'\tau_n &= p'(\tau_n - \tau_{n-1}) + \cdots + p'(\tau_2 - \tau_1) + p'\tau_1 \\ &= n\tau. \end{aligned}$$

Since p' is a covering morphism, $p'\tau_n \neq 0$, and so

$$n\tau \neq 0.$$

Also, if $a \in \pi(\mathbb{S}^1, 1)$, then $a = p'\tau_n$ for some n and hence

$$a = n\tau.$$

This shows that $\pi(\mathbb{S}^1, 1)$ is an infinite cyclic group with generator τ. □

10.2.1 *(Corollary 2) For $n > 1$, real projective n-space $P^n(\mathbb{R})$ has fundamental group isomorphic to $\mathbb{Z}_2$.*

Proof The Hopf map $h : \mathbb{S}^n \to P^n(\mathbb{R})$ is a covering map, and by the proof of 9.1.7 $\mathbb{S}^n$ is 1-connected for $n > 1$. Let $\tilde{x}, -\tilde{x}$ be antipodal points of $\mathbb{S}^n$ and let $x = h\tilde{x} = h(-\tilde{x})$. By 10.2.1, the morphism

$$\pi h : \pi\mathbb{S}^n\{x, -x\} \to \pi(P^n(\mathbb{R}), x)$$

is a covering morphism. However, the groupoid $\pi\mathbb{S}^n\{\tilde{x}, -\tilde{x}\}$ is isomorphic to **I**. It follows that the group $\pi(P^n(\mathbb{R}), x)$ has two elements, and so is isomorphic to $\mathbb{Z}_2$. □

We give in this section some simple results on covering groupoids.

10.2.2 *Let $p : \widetilde{G} \to G$ be a covering morphism such that G is connected. If a, b are any elements of G, then $p^{-1}[a]$, $p^{-1}[b]$ have the same cardinality.*

Proof Suppose $c \in G(x, y)$. It is an easy deduction from the definition of covering morphism that the functions

$$p^{-1}[c] \to p^{-1}[x], \qquad p^{-1}[c] \to p^{-1}[y]$$

which send an element of $p^{-1}[c]$ to its initial and its final point respectively, are both bijections.

Suppose now $a \in G(x, x_1)$, $b \in G(y, y_1)$. Since G is connected, there is an element c in $G(x, y)$ and we deduce

$$\divideontimes\, p^{-1}[a] = \divideontimes\, p^{-1}[x] = \divideontimes\, p^{-1}[y] = \divideontimes\, p^{-1}[b].$$

□

In particular, if $p^{-1}[a]$ has n elements for each a in G, then we call p an *n-fold covering morphism*. We shall see later that for connected covering morphisms, this number n is the index in $G(p\tilde{x})$ of the characteristic group of p at $\tilde{x}$ ($\tilde{x} \in \mathrm{Ob}(\widetilde{G})$).

10.2.3 *Let $r : K \to H$, $q : H \to G$ be morphisms of groupoids. If q and r are covering morphisms, so also is qr. If q and qr are covering morphisms, so also is r. If r and qr are covering morphisms and $\mathrm{Ob}(r)$ is surjective, then q is a covering morphism.*

Proof Let $x \in \mathrm{Ob}(K)$ and consider the composite $(qr)' = q'r'$

$$\mathrm{St}_K\, x \xrightarrow{r'} \mathrm{St}_H\, rx \xrightarrow{q'} \mathrm{St}_G\, qrx$$

where q', r' are induced by q, r. Clearly, if any two of q', r', and $q'r'$ are bijections, then so is the third. For the last part one needs that any y in $\mathrm{Ob}(H)$ is rx for some x. □

It is not hard to prove for spaces the result corresponding to the one part of 10.2.3. The other part is more tricky (it is false in general) and will be left till an exercise in section 10.5.

10.2.4 *If $r : Z \to Y$, $q : Y \to X$ are maps of spaces such that qr and q are covering maps, then r is a covering map.*

Proof Let $x \in X$, let U be a canonical neighbourhood of x for the map q, and let V be a canonical neighbourhood of x for the map qr. Let W be an open, path-connected neighbourhood of x contained in $U \cap V$. Clearly, W is canonical for both q and qr (since any open, path-connected subset of a canonical set is again canonical).

For each y in $q^{-1}[x]$, let W_y be the path-component of $q^{-1}[W]$ which contains y—these sets W_y are disjoint and open in Y. Clearly,

$$r^{-1}q^{-1}[W] = \bigcup\{r^{-1}[W_y] : y \in q^{-1}[W]\}$$

and the sets $r^{-1}[W_y]$ are disjoint and open in Z. Therefore, each path-component W' of $r^{-1}[W_y]$ is also a path-component of $r^{-1}q^{-1}[W]$. But $qr \mid W', W$ and $q \mid W_y, W$ are homeomorphisms. Therefore $r \mid W', W_y$ is a homeomorphism. □

EXERCISES

1. Let $f : \mathbb{S}^1 \to \mathbb{S}^1$ be the map $z \mapsto z^n$, where n is a non-zero integer. Prove that the induced map $\pi f : \pi(\mathbb{S}^1, 1) \to \pi(\mathbb{S}^1, 1)$ is multiplication by n.
2. Let $p : \widetilde{G} \to G$ be a covering morphism. Prove that if G is connected and $\widetilde{G}$ is non-empty then $\mathrm{Ob}(p)$ is surjective.
3. Prove that a 1-fold covering morphism is an isomorphism.
4. Let $p : \widetilde{X} \to X$ be a covering map of spaces such that X is path-connected. Prove that the sets $p^{-1}[x]$ have the same cardinality for all x in X, and that if $\widetilde{X}$ is non-empty then p is surjective.
5. Let $r : Z \to Y$, $q : Y \to X$ be covering maps such that $q^{-1}[x]$ is finite for each x in X. Prove that qr is a covering map.
6. Show that the notion of covering morphism for groupoids extends to functors of categories in such a way that if $p : \widetilde{X} \to X$ is a covering map of spaces, then $Pp : P\widetilde{X} \to PX$ is a covering functor of categories.

10.3 On lifting sums and morphisms

Let $p : \widetilde{G} \to G$ be a covering morphism. Because of the analogy with the fundamental groupoid of covering spaces, we say that an element $\tilde{a}$ of $\widetilde{G}$ *covers*, or is a *lifting*, of $p\tilde{a}$; similarly, we say a sum $\tilde{a}_n + \cdots + \tilde{a}_1$ in $\widetilde{G}$ covers, or is a lifting, of $p\tilde{a}_n + \cdots + p\tilde{a}_1$. The basic property of covering morphisms is that not only elements of G but also sums in G can be lifted into $\widetilde{G}$.

10.3.1 *Let $\tilde{x}$ be an object of $\widetilde{G}$ and let $p\tilde{x} = x$. If*

$$a = a_n + \cdots + a_1$$

belongs to $\mathrm{St}\, x$*, then there are unique elements $\tilde{a}_n, \ldots, \tilde{a}_1$ of $\widetilde{G}$ such that*
(a) $p\tilde{a}_i = a_i$, $i = 1, \ldots, n$
(b) $\tilde{a} = \tilde{a}_n + \cdots + \tilde{a}_1$ *is defined and belongs to* $\mathrm{St}\, \tilde{x}$.

Proof Since $p : \mathrm{St}\, \tilde{y} \to \mathrm{St}\, p\tilde{y}$ is bijective for each object $\tilde{y}$ of $\widetilde{G}$, the elements $\tilde{a}_i$ are uniquely defined by the inductive conditions (i) $p\tilde{a}_i = a_i$, (ii) $\tilde{a}_1 \in \mathrm{St}\, \tilde{x}$, (iii) $\tilde{a}_i \in \mathrm{St}\, \tilde{x}_i$ where $\tilde{x}_1 = \tilde{x}$ and $\tilde{x}_i$ is the final point of $\tilde{a}_{i-1}$ for $i > 1$. □

As mentioned in the last section, the characteristic group of $\widetilde{G}, \tilde{x}$ (that is, the subgroup $p[\widetilde{G}(\tilde{x})]$ of $\widetilde{G}(p\tilde{x})$) plays an important rule in the theory. The relationship of these groups for various $\tilde{x}$ is described in the next result.

First we need a definition: subgroups C of $G(x)$, D of $G(y)$ are called *conjugate* (in G) if there is an element c of $G(x, y)$ such that

$$c + C - c = D.$$

This relation implies easily that $-c + D + c = C$.

10.3.2 *Let C be the characteristic group of* $\widetilde{G}, \tilde{x}$.
(a) *If D is the characteristic group of* $\widetilde{G}, \tilde{y}$, *and* $\tilde{x}, \tilde{y}$ *lie in the same component of* $\widetilde{G}$, *then C and D are conjugate.*
(b) *If D is a subgroup of* $G(y)$ *and D is conjugate to C, then D is the characteristic group of* $\widetilde{G}, \tilde{y}$ *for some* $\tilde{y}$.

Proof (a) Let $\tilde{a}$ be an element of $\widetilde{G}(\tilde{x}, \tilde{y})$ and let $a = p\tilde{a}$. If $c \in C$, then c is covered by an element $\tilde{c}$ of $\widetilde{G}(\tilde{x})$. By 10.3.1, $a + c - a$ is covered by $\tilde{a} + \tilde{c} - \tilde{a}$, which is an element of $\widetilde{G}(\tilde{y})$. Therefore $a + c - a$ belongs to D. This proves (a).
(b) Suppose $a + C - a = D$ where $a \in G(x, y)$. Let $\tilde{a}$ be an element of $\mathrm{St}\,\tilde{x}$ covering a, let $\tilde{y}$ be the final point of $\tilde{a}$, and let D' be the characteristic group of $\widetilde{G}, \tilde{y}$. We prove that $D = D'$.

If $d \in D$, then $d = a + c - a$, where $c \in C$. It follows that d is covered by an element $\tilde{a} + \tilde{c} - \tilde{a}$ of $\widetilde{G}(\tilde{y})$, whence $d \in D'$. Conversely, if $d \in D'$, then d is covered by an element $\tilde{d}$ of $\widetilde{G}(\tilde{y})$, whence $d = a + c - a$ where $c = p(-\tilde{a} + \tilde{d} + \tilde{a}) \in C$. □

We shall now show the sense in which the characteristic group of $\widetilde{G}, \tilde{x}$ *is* characteristic. It is convenient to work in the category of pointed groupoids: a *pointed groupoid* G, x consists of a groupoid G and object x of G. A *pointed morphism* $G, x \to H, y$ of pointed groupoids consists of G, x and H, y and a morphism $f : G \to H$ such that $fx = y$; such a pointed morphism is usually called a morphism $G, x \to H, y$ and is often denoted simply by f. The *characteristic group* of a pointed morphism $f : G, x \to H, y$ is the characteristic group of f at x.

If $p : \widetilde{G}, \tilde{x} \to G, x$ is a pointed morphism, then we say p is a *covering morphism* if the morphism $p : \widetilde{G} \to G$ is a covering morphism. Similarly, if we are given a commutative diagram of pointed morphisms

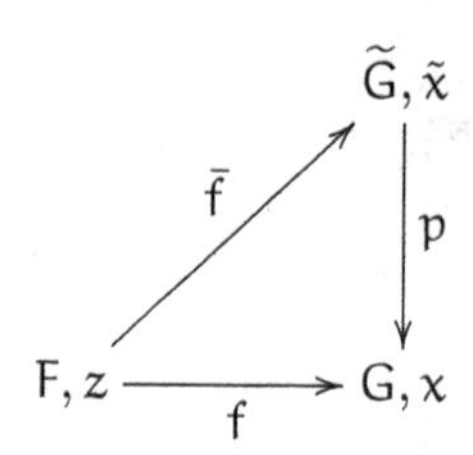

then we say $\tilde{f}$ is a *lifting* of f.

10.3.3 *Let* $p : \widetilde{G}, \tilde{x} \to G, x$ *be a covering morphism, and* $f : F, z \to G, x$ *a morphism such that* F *is connected. Then* f *lifts to a morphism* $\tilde{f} : F, z \to \widetilde{G}, \tilde{x}$ *if and only if the characteristic group of* f *is contained in that of* p*; and if this lifting exists, then it is unique.*

Proof Suppose first that $\tilde{f}$ exists; the relation $p\tilde{f} = f$ implies

$$f[F(z)] \subseteq p[\widetilde{G}(\tilde{x})]$$

and this proves the necessity of the condition.

Suppose, conversely, that the characteristic group of f is contained in the characteristic group C of p. Since p restricts to an isomorphism $\widetilde{G}(\tilde{x}) \to C$, the morphism

$$f : F(z) \to G(x)$$

lifts uniquely to a morphism

$$\tilde{f} : F(z) \to \widetilde{G}(\tilde{x}).$$

For each object v of F let τ_v be an element of $F(z, v)$ (with $\tau_z = 0$). If $a \in F(u, v)$ then a can be written uniquely as

$$a = \tau_v + a' - \tau_u$$

with $a' \in F(z)$; and if, further, $b \in F(v, w)$, then $(b + a)' = b' + a'$. Now each element $f\tau_v$ is covered by a unique element $\tilde{f}\tau_v$ of St $\tilde{x}$; so we define

$$\tilde{f}a = \tilde{f}\tau_v + \tilde{f}a' = \tilde{f}\tau_u$$

and it follows that

$$\begin{aligned} \tilde{f}b + \tilde{f}a &= \tilde{f}\tau_w + \tilde{f}b' + \tilde{f}a' - \tilde{f}\tau_u \\ &= \tilde{f}\tau_w + \tilde{f}(b + a)' = \tilde{f}\tau_u \\ &= \tilde{f}(b + a). \end{aligned}$$

Therefore $\tilde{f}$ is a morphism.

Clearly $\tilde{f}$ lifts f. Also any morphism which lifts f must agree with $\tilde{f}$ on $F(z)$ and on the elements τ_v; therefore, such a lifting must coincide with $\tilde{f}$. This proves the uniqueness of the lifting. □

10.3.3 *(Corollary 1) If* $p : \widetilde{G}, \tilde{x} \to G, x$ *and* $q : \widetilde{H}, \tilde{y} \to G, x$ *are connected covering morphisms with characteristic groups* C, D *respectively, and if* $C \subseteq$

D, then there is a unique covering morphism $r : \widetilde{G}, \tilde{x} \to \widetilde{H}, \tilde{y}$ *such that* $p = qr$. *If* $C = D$ *then* r *is an isomorphism.*

Proof By 10.3.3, p lifts uniquely to a morphism $r : \widetilde{G}, \tilde{x} \to \widetilde{H}, \tilde{y}$ such that $qr = p$. By 10.2.3, r is a covering morphism. Finally, if $C = D$ the usual universal argument shows that r is an isomorphism. □

10.3.3 *(Corollary 2) A 1-connected covering groupoid of* G *covers every covering groupoid of* G.

Proof If $p : \widetilde{G}, \tilde{x} \to G, x$ is a covering morphism and $\widetilde{G}$ is 1-connected, then the characteristic group of p is trivial and so contained in any subgroup of $G(x)$. □

Because of 10.3.3 (Corollary 2) a 1-connected covering groupoid of G is called a *universal covering groupoid* of G. The existence of such for connected G will be proved later; its uniqueness, as a pointed groupoid, is a consequence of the last part of 10.3.3 (Corollary 1).

Exercises

Throughout these exercises we suppose given a covering morphism $p : \widetilde{G}, \tilde{x} \to G, x$.

1. Let $a, b \in G(x, y)$ and let $c = -a + b$. Let $\tilde{a}, \tilde{b}, \tilde{c}$ be elements of $\operatorname{St} \tilde{x}$ covering a, b, c respectively. Prove that $\tilde{c}$ belongs to $\widetilde{G}(\tilde{x})$ if and only if $\tilde{a}, \tilde{b}$ have the same end point.
2. Prove that if p is connected then the following are equivalent: (i) p is an isomorphism; (ii) p is bijective on objects; (iii) p has characteristic group $G(x)$.
3. Prove that if Γ is a graph which generates G, then $\widetilde{\Gamma} = p^{-1}[\Gamma]$ is a graph which generates $\widetilde{G}$. Prove also that if Γ generates G freely, then $\widetilde{\Gamma}$ generates $\widetilde{G}$ freely.
4. The covering morphism p is called *regular* if p is connected and for all a in $G(x)$ the elements of $p^{-1}[a]$ are either all or none of them loops. Prove that p is regular if and only if p is connected and has characteristic group normal in $G(x)$.
5. A *cover transformation of* p is an isomorphism $h : \widetilde{G} \to \widetilde{G}$ such that $ph = p$. Prove that if p is regular, then the group of cover transformations is under composition a group anti-isomorphic to the quotient group $G(x)/p[\widetilde{G}(\tilde{x})]$.
6. Let $p : H \to G$ be a fibration of groupoids. Prove that if the diagram of morphisms

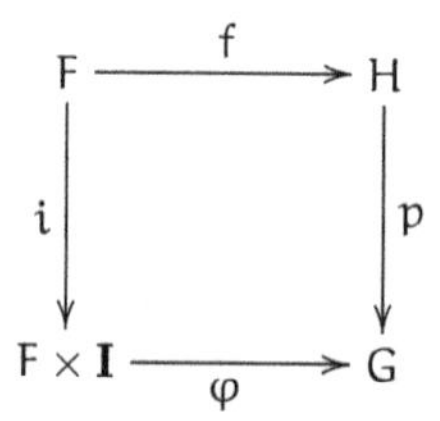

is commutative, where i is given by $a \mapsto (a, 0)$, then there is a morphism $\psi : F \times \mathbf{I} \to H$ such that $\psi i = f$, $p\psi = \varphi$. Prove further that if p is a covering morphism then ψ is unique. [This is the *covering homotopy property* for fibrations and coverings of groupoids; it would be possible, and indeed reasonable, to use this as the definition of these terms.]

7. Let H and K be groupoids. The groupoid $GPD(H, K)$ is defined to have objects the morphisms $H \to K$ and morphisms from f to g the pairs (h, f) where h is a homotopy $f \simeq g$. Make this construction explicit by defining the sum of homotopies and verifying that this does give a groupoid. Prove that if H, K are groups, then the object group of $GPD(H, K)$ at the morphism $f : H \to K$ is isomorphic to the centraliser of $f[H]$ in K.

8. Prove that if also G is a groupoid, then there is a bijection

$$\theta : \mathsf{Grpd}(G \times H, K) \to \mathsf{Grpd}(G, GPD(H, K)).$$

[This is a lengthy exercise.]

9. Let $p : K \to L$ be a morphism of groupoids. Show how p induces through composition a morphism $p_* : GPD(H, K) \to GPD(H, L)$ such that if p is a fibration, then so also is p_*. Interpret the exact sequence of this fibration for specific cases.

10.4 Existence of covering groupoids

Our main purpose in this section is to prove that if x is an object of the connected groupoid G, then any subgroup of $G(x)$ is the characteristic group of some covering morphism $\widetilde{G}, \tilde{x} \to G, x$. We shall deduce this from a more general construction of covering groupoids which is of considerable importance even for the case G is a group.

Let G be a groupoid. An *action of* G *on a set* consists of a set S, a function $w : S \to \mathrm{Ob}(G)$, and a partial function $G \times S \rightarrowtail S$, $(g, s) \mapsto g \cdot s$, which for each $x, y \in \mathrm{Ob}(G)$, assigns to an element (g, s), where $g \in G(x, y)$ and $s \in w^{-1}[x]$, an element $g \cdot s \in w^{-1}[y]$. The following rules are to be satisfied.

10.4.1 (i) If $x \in \mathrm{Ob}(G)$, $s \in w^{-1}[x]$, then $0_x \cdot s = s$.
(ii) If $g \in G(x, y)$, $h \in G(y, z)$, $s \in w^{-1}[x]$, then

$$(h + g) \cdot s = h \cdot (g \cdot s).$$

We also say G *acts on* S *via* w, and that S is a G-*set*.

EXAMPLE

1. Let $p : \widetilde{G} \to G$ be a covering morphism of groupoids, and let $S = \mathrm{Ob}(\widetilde{G})$, $w = \mathrm{Ob}(p)$. Then we obtain an action of G on S via w by assigning to $s \in S$

and $g \in \mathrm{St}_G\, ps$ the target of the unique lift of g with source s. This action is one of the essential features of 10.2.2.

It follows from the rules for an action that an element $g \in G(x, y)$ defines a bijection $g_{\#} : w^{-1}[x] \to w^{-1}[y]$, $s \mapsto g \cdot s$. The action is said to be *transitive* if for all x, y in $\mathrm{Ob}(G)$, $s \in w^{-1}[x]$, $t \in w^{-1}[y]$, there is a $g \in G(x, y)$ such that $g \cdot s = t$. In this case, all the sets $w^{-1}[x]$, $x \in \mathrm{Ob}(G)$, have the same cardinality.

If $s \in w^{-1}[x]$, the *group of stability of* s is the subgroup G_s of G of elements g such that $g \cdot s = s$. Such an element g is said to *stabilise*, or *fix*, s, and s is said to be a *fixed point* of g.

Given such an action, the *semidirect product groupoid* $S \rtimes G$ is defined to be the groupoid with object set S and elements of $(S \rtimes G)(s, t)$ the pairs (s, g) such that $g \in G(ps, pt)$ and $g \cdot s = t$. The addition in $S \rtimes G$ is defined to be

$$(t, h) + (s, g) = (s, h + g).$$

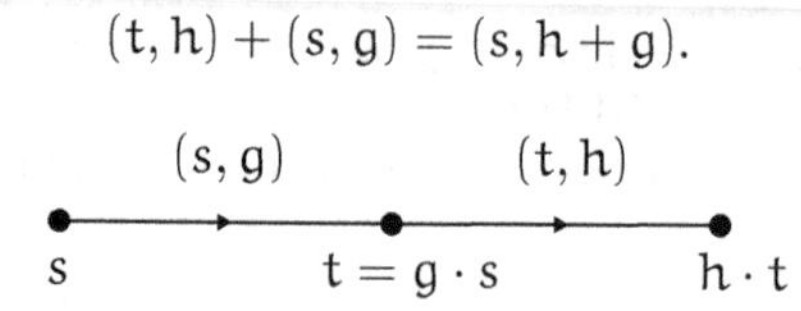

10.4.2 (a) *The above construction makes $S \rtimes G$ a groupoid.*
(b) *The projection $p : S \rtimes G \to G$, given on objects by $w : S \to \mathrm{Ob}(G)$ and on elements by $(s, g) \mapsto g$, is a covering morphism of groupoids.*
(c) *The groupoid $S \rtimes G$ is connected if and only if the action is transitive.*
(d) *If $s \in S$, then the object group $(S \rtimes G)(s)$ is mapped by p isomorphically to G_s, the group of stability of s.*
(e) *The action of G on S determined by the covering morphism p is the original action.*

Proof The proof of (a) is easy. Associativity is easily checked. The zero at s is $(s, 0_{ps})$, and the negative of (s, g) is $(g \cdot s, -g)$.

It is clear from the definition that p is a morphism of groupoids. Also p is a covering morphism, because if $g \in G(x, y)$, $s \in w^{-1}[x]$, then (s, g) is the unique element of $S \rtimes G$ which has source s and projects to g.

For (c), we need only note that $(S \rtimes G)(s, t)$ is non-empty if and only if there is a g in G such that $g \cdot s = t$.

For (d), $(S \rtimes G)(s)$ consists of those (s, g) such that $g \cdot s = s$, and so this object group is mapped by p to G_s.

Finally, (e) is clear. □

To use this result, we need a supply of actions of a groupoid. Here is a way of obtaining a transitive action.

10.4.3 *Let* x *be an object of the connected groupoid* G*, and let* C *be a subgroup of the object group* $G(x)$*. Let* S *be the set of (left) cosets*

$$a + C = \{a + c : c \in C\}$$

for a *in* $\mathrm{St}_G\, x$*. Let* $w : S \to \mathrm{Ob}(G)$ *send* $a + C$ *to the final point of* a*. Then* G *acts transitively on* S *by* $g \cdot (a + C) = g + a + C$*. The corresponding connected covering morphism* $p : S \rtimes G, \tilde{x} \to G, x$*, where* $\tilde{x}$ *is the coset* C*, has characteristic group* C*. Further,* $p^{-1}[x]$ *is the set* $G(x)/C$ *of left cosets of* C *in* $G(x)$*.*

Proof Suppose $a \in G(x, y)$, $g \in G(y, z)$. Then $w(a+C) = y$, $w(g+a+C) = z$. The rules 10.4.1(i) and (ii) are clearly satisfied. The action is transitive, because if $a + C$, $b + C$ are cosets, and $g = b - a$, then $g + a + C = b + C$. It follows that $S \rtimes G$ is connected.

The required characteristic group consists of the elements a of G such that the coset $a + C$ is defined and is C, and this holds if and only if $a \in C$. Hence the characteristic group is C.

The final statement is obvious. □

The pointed groupoid constructed in 10.4.3 is sometimes in the literature written $\mathrm{Tr}(G, C)$. It could also be written $(G/C) \rtimes G$. It is commonly called the *covering groupoid of* G *based on the subgroup* C *of* $G(x)$. The results of the previous section show that a pointed, connected covering morphism is determined up to isomorphism by its characteristic group, so the particular construction of $\mathrm{Tr}(G, C)$ is not always needed.

10.4.3 *(Corollary 1) Any connected groupoid has a universal covering groupoid.*

Proof If G is a connected groupoid, and $x \in \mathrm{Ob}(G)$, then a universal covering groupoid of G is $\mathrm{Tr}(G, 0_x)$, where 0_x is the trivial subgroup of $G(x)$. Notice that in this case the set on which G acts is simply $\mathrm{St}_G\, x$, and the action is via the target map $\tau : \mathrm{St}_G\, x \to \mathrm{Ob}(G)$. □

The universal covering groupoid of G constructed in this corollary is called the *universal covering groupoid of* G *based at* x. Note that there is no unique universal covering groupoid of G, but only a family $\mathrm{St}_G\, x \rtimes G$ of them indexed by the objects of G.

10.4.3 *(Corollary 2) A connected covering morphism* $\widetilde{G}, \tilde{x} \to G, x$ *is an* n*-fold covering if and only if its characteristic group has index* n *in* $G(x)$*.*

Proof By 10.3.3 (Corollary 1) $\widetilde{G}$ is isomorphic to $\mathrm{Tr}(G, C)$ where C is the characteristic group of $\widetilde{G}, \tilde{x}$. □

Here is a nice application of the idea of an action of a groupoid on a set. Let H be a groupoid, and let N be a totally disconnected normal subgroupoid of H. Then we can form the quotient groupoid H/N (section 8.3), which has the same objects as H, but the elements of $(H/N)(x,y)$ are the cosets $h+N(x)$, or, equally, $N(y)+h$, for all $h \in H(x,y)$. It is convenient to write such a coset as $h+N$, or $N+h$. There is a *quotient morphism* $p : H \to H/N$ sending an element h to the coset $h+N$. Note that the star of H/N at x consists of the cosets $h+N(x)$ for all h in $\mathrm{St}_H\, x$.

Now we give an action of the groupoid $H \times H$ on the set H/N via the function $w : H/N \to \mathrm{Ob}(H) \times \mathrm{Ob}(H)$ which sends an element $h+N \in (H/N)(x,y)$ to the pair (x,y). If $a \in H(x,u)$, $b \in H(y,v)$, and $h+N \in (H/N)(x,y)$, then we define

$$(a,b)\cdot(h+N) = b+h-a+N \in (H/N)(u,v).$$

This can be conveniently represented by the diagram

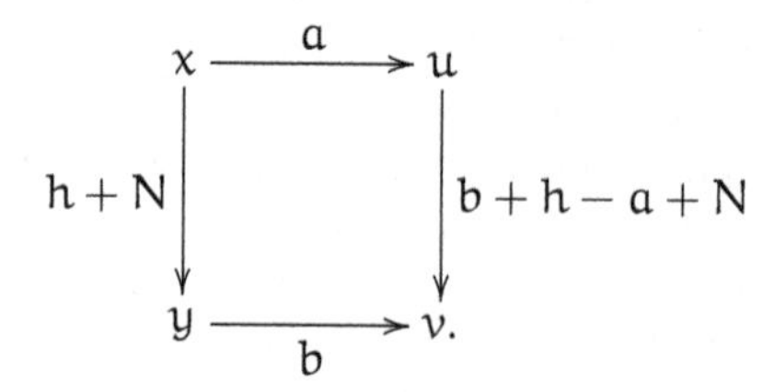

The rules for an operation are easy to verify, using the normality of N, which gives $h+N = N+h$ for all h in H. Hence if $h+N = k+N$ then

$$\begin{aligned} b+h-a+N &= b+h+N-a \\ &= b+k+N-a \\ &= b+k-a+N \end{aligned}$$

so that the operation is well defined. Clearly $(0_x, 0_y)\cdot(h+N) = h+N$. If further $c \in H(u,w)$, $d \in H(v,z)$, then

$$\begin{aligned} (c,d)\cdot(b+h-a+N) &= d+b+h-a-c+N \\ &= (c+a, d+b)\cdot(h+N), \end{aligned}$$

as required.

Let (x,y) be an object of $H \times H$. Then the group of stability of the action at (x,y) consists of pairs $(a,b) \in H(x)\times H(y)$ such that $b+h-a+N = h+N$, or, equivalently,

$$b+h+N = h+a+N.$$

Let $\Box_N H$ be the semidirect product groupoid $(H/N) \rtimes (H \times H)$. Then the object group of $\Box_N H$ at the object (x, x) consists of pairs (a, b) such that $a, b \in H(x)$ and $a + N(x) = b + N(x)$. Thus $(\Box_N H)(x, x)$ is isomorphic to the semidirect product group $N(x) \rtimes H(x)$, which consists of pairs $(n, a) \in N(x) \times H(x)$ with addition

$$(n, a) + (n', a') = (-a' + n + a' + n', a + a')$$

the isomorphism being given by

$$(a, b) \mapsto (-b + a, b).$$

A special case of this construction is when N is trivial, i.e. consists only of the zeros of H. Then we write 0 for N, and find that $(\Box_0 H)(x)$ is isomorphic to $H(x)$.

Going back to the general case, let ψ be the composite

$$\Box_N H \to H \times H \to H$$

of the covering morphism and the first projection. If x is an object of H, then the groupoid $\psi^{-1}[0_x]$ is precisely the covering groupoid $\mathrm{Tr}(H, N(x))$ of H determined by the subgroup $N(x)$ of $H(x)$. Thus the groupoid $\Box_N H$ gives a global description of the union of all the covering groupoids of H determined by the subgroups $N(x)$ for all objects x of H. This global description is, of course, restricted to the case of normal subgroups. We will study this case further under the term *regular covering groupoids* in section 10.6.

We will use these constructions in the next section to give a convenient topology on the groupoid $(\pi X)/N$ for a normal totally disconnected subgroupoid of the fundamental groupoid πX, for suitable X.

Remark In the theory of groups it is embarrassing that a subgroup C of a group G determines a quotient group G/C if C is normal in G, but in the general case it seems to determine nothing. In fact the action of G on the set of cosets G/C is important, and this action is nicely represented in the theory of groupoids by the covering morphism $(G/C) \rtimes G \to G$. Thus if we regard a group as a groupoid, then a subgroup of a group determines a structure of the same type. More generally, actions of groups on sets play an important role in the applications of groups, for example in enumeration questions such as determining the numbers of necklaces with given numbers of beads of various colours, and in chemistry determining the numbers of isomers of a given type. Some problems involving group actions are conveniently formulated in terms of the exact sequence of a covering discussed in the exercises.

EXERCISES

1. Let $p : \widetilde{G} \to G$ be a covering morphism of groupoids and let $\tilde{x} \in \mathrm{Ob}(\widetilde{G})$. Let S be the set $p^{-1}[p\tilde{x}]$. Show that the exact sequence of a fibration specialises to the exact sequence

$$0 \to \widetilde{G}(\tilde{x}) \to G(px) \to S \to \pi_0\widetilde{G} \to \pi_0 G,$$

where exactness is formulated in section 7.2 for a fibration of groupoids. Show how in the case G is a group, this sequence includes the information given by Lagrange's theorem on the index of a subgroup.
2. Let G be a group acting on a set S and let H be a group acting on a set T. Let $\theta : G \to H$ be a morphism of groups, and let $\varphi : S \to T$ be a function. Assume that for all $g \in G$ and $s \in S$, $\varphi(g \cdot s) = (\theta g) \cdot (\varphi s)$. Show that the function $\psi : S \rtimes G \to T \rtimes H$ given by $(s, g) \mapsto (\varphi s, \theta g)$ is a morphism of groupoids and that ψ is a fibration if and only if θ is surjective.
3. Continuing the notation of Exercise 2, let $K = \mathrm{Ker}\,\theta$, let $s \in S$ and let $F = \varphi^{-1}[\varphi s]$. Show that if $g \in K$ then $g \cdot F \subseteq F$, and that $\psi^{-1}[\varphi s] = F \rtimes K$. Assume now that θ is surjective. Show how the exact sequence of a fibration gives rise to an exact sequence of groups and pointed sets

$$1 \to K_s \to G_s \to H_{\varphi s} \to F/K \to S/G \to T/H$$

and describe the exactness properties in detail. Interpret this exact sequence in the case that T is a singleton.
4. Assume that $\theta : G \to H$ is an epimorphism of groups and that $K = \mathrm{Ker}\,\theta$. Let G, H and K act on themselves by conjugation. Prove that if $s \in G$ then there is an exact sequence

$$1 \to C_K(s) \to C_G(s) \to H \to [K] \to [G] \to [H]$$

where $C_G(s)$, the centraliser of s in G, is the set of $g \in G$ such that $g + s = s + g$, and $[G]$ is the set of conjugacy classes of G. [Exercises 2, 3, 4 are from [HK82].]
5. Let H be a groupoid with object set X. An *admissible section* of H is a function $u : X \to H$ such that ux has source x for all $x \in X$ and the function sending x to τux, the target of ux, is a bijection $X \to X$. Prove that the set $\Gamma(H)$ of admissible sections of H is a group under the product $vu : x \mapsto v\tau ux + ux$. Prove that $\Gamma(H)$ acts on X by $u \cdot x = \tau ux$. Let G be a group acting on X. Prove that there is a bijection between the set of morphisms $f : G \to \Gamma(H)$ such that $(fg) \cdot x = g \cdot x$ for all $x \in X$, and the set of morphisms $X \rtimes G \to H$ which are the identity on X. [The group of admissible sections has an old history in work of [Ehr80].]
6. Let $p : Y \to X$ be a function of sets X, Y and let H be a groupoid with object set Y. Define a groupoid $p_!(H)$ with object set X and an action of $p_!(H)$ on Y via p as follows. The elements from x to x' are the functions $u : p^{-1}[x] \to H$ such that $u(y)$ has source y and target in $p^{-1}[x']$ and $y \mapsto \tau u(y)$ is a bijection $p^{-1}[x] \to p^{-1}[x']$. The addition is given by

$$v + u : y \mapsto v(\tau uy) + u(y).$$

The action is $u \cdot y = \tau u(y)$. Generalise Exercise 5 to obtain, for G a groupoid on X operating on Y via p, a bijection of morphisms $f : G \to p_!(H)$ which are the identity on objects and satisfy $f(g) \cdot y = g \cdot y$, and morphisms $f^* : Y \rtimes G \to H$ which are the identity on objects. [Exercises 5 and 6 are in [Cha77].]

7. Prove that a subgroup C of a free group G is free, and that if G is of rank r and C is of index n in G, then C is of rank $nr - n + 1$. [Construct $\widetilde{G} = \mathrm{Tr}(G, C)$ and use Exercise 2 of Section 10.3 and 8.2.1(c).]

8. Let R be a ring and I a left ideal of R; then, considering I simply as a subgroup of the abelian group R, we can form the groupoid $\mathrm{Tr}(R, I)$. What additional structure can be put on this object by using the multiplication? [This is more of a research project than an exercise! Are the algebroids of [Mit85] relevant?]

9. Let G be a groupoid and let $\mathrm{Cov}(G)$ be the category whose objects are covering morphisms $p : \widetilde{G} \to G$ and whose morphisms from $p : \widetilde{G} \to G$ to $p' : \widetilde{G}' \to G$ are determined by p, p' and a morphism $r : \widetilde{G} \to \widetilde{G}'$ such that $p'r = p$. Prove that $\mathrm{Cov}(G)$ is equivalent to the category $\mathsf{Fun}(G, \mathrm{Set})$ [Exercise 6 of Section 6.6]. [If $p : \widetilde{G} \to G$ is a covering morphism, define a factor $Lp : G \to \mathrm{Set}$ which on objects sends $x \mapsto p^{-1}[x]$, and if $a \in G(x, y)$, $\tilde{x} \in p^{-1}[x]$, then $Lp(a)(\tilde{x})$ is to be the target of the element $\tilde{a}$ which starts at $\tilde{x}$ and covers a.]

10.5 Lifted topologies

In this section we apply the previous results to prove the existence of liftings of maps and the existence of covering spaces—that is, we prove topological theorems corresponding to 10.3.3 and 10.4.1. However this correspondence is not complete since the existence of a covering map $\widetilde{X} \to X$ implies a local condition on X. (We recall that all spaces are, in any case, assumed locally path-connected.)

We introduce some useful language. If $f : H \to G$ is a morphism of groupoids, and $x \in \mathrm{Ob}(G)$, then $\chi_f(x)$ denotes $G(x)$ if $f^{-1}[x]$ is empty, and otherwise denotes the intersection of the characteristic groups of f at y for all $y \in f^{-1}[x]$. Thus χ_f is a wide, totally disconnected subgroupoid of G.

The notion of characteristic group and of χ_f will be used for maps of spaces. More precisely, if $f : Y \to X$ is a map, then the *characteristic group* of f at y is simply the characteristic group of $\pi f : \pi Y \to \pi X$ at y; and this group is also called the characteristic group of $f : Y, y \to X, fy$ or even, by an abuse of language, the characteristic group of Y, y. Similarly, we write χ_f for $\chi_{\pi f}$, so that χ_f is a subgroupoid of πX.

Let X be a space and let χ be any wide subgroupoid of πX. A subset U of X is called *weakly χ-connected* if for all x in U the characteristic group at x of the inclusion $U \to X$ is contained in $\chi(x)$. Clearly, any subset of a weakly χ-connected set is again weakly χ-connected.

We say X itself is *semi-locally χ-connected* if each x in X has a weakly

χ-connected neighbourhood; if χ is simply connected, this is also expressed as: X is *semi-locally simply-connected.* Our first result shows the utility of these definitions.

10.5.1 *Let* $p : \widetilde{X} \to X$ *be a covering map. Then* X *is semi-locally* χ_p*-connected.*

Proof Let U be a canonical subset of X. If $x \in U$ and $\tilde{x} \in p^{-1}[x]$ then the inclusion $i : U, x \to X, x$ lifts to a map $U, x \to \widetilde{X}, \tilde{x}$. So the morphism $\pi U, x \to \pi X, x$ lifts, and it follows that the characteristic group of U, x is contained in the characteristic group of p at $\tilde{x}$. This is true for all $\tilde{x}$ in $p^{-1}[x]$ and so the characteristic group of U, x is contained in $\chi_p(x)$.

Thus any canonical subset of X is weakly χ_p-connected. By definition of covering map, any x in X has a canonical neighbourhood. □

The following corollary is immediate.

10.5.1 *(Corollary 1) If* X *is path-connected and has a 1-connected covering space, then* X *is semi-locally simply-connected.*

It may be proved using the methods of chapter 9 that the Hawaiian ear-ring [cf. Fig. 9.2, p. 342] is not semi-locally simply-connected (the origin being a 'bad' point) and so this space has no 1-connected covering space.

Suppose now that X is a space and

$$q : \widetilde{G} \to \pi X$$

is a covering morphism of groupoids. Let X be semi-locally χ_q-connected, and let

$$\widetilde{X} = \mathrm{Ob}(\widetilde{G}), \quad p = \mathrm{Ob}(q) : \widetilde{X} \to X.$$

We shall use q to 'lift' the topology of X to a topology on $\widetilde{X}$.

Let $\mathcal{U}$ be the set of all open, path-connected, weakly χ_q-connected subsets of X. If U is any element of $\mathcal{U}$ consider the diagram

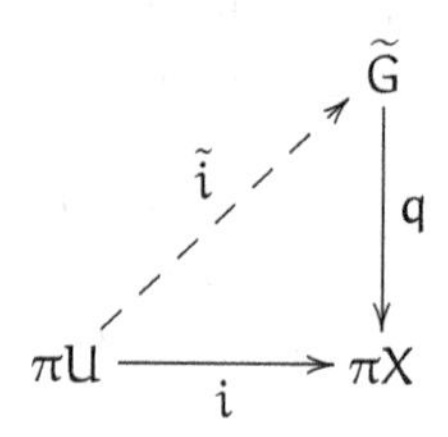

where i is induced by inclusion. By 10.3.3, i lifts to a morphism $\tilde{i} : \pi U \to G$; the set $\tilde{i}[U]$ is a subset $\widetilde{U}$ of $\widetilde{X}$ which we call a *lifting* of U. The set of all such liftings of U for all U in $\mathcal{U}$ is written $\widetilde{\mathcal{U}}$. Since X is semi-locally χ_q-connected, the set $\widetilde{\mathcal{U}}$ covers $\widetilde{X}$.

We need a lemma about these liftings.

Let V be path-connected and $g : V \to X$ a map such that $g[V]$ is contained in a set U of $\mathcal{U}$. Let $g' : \pi V \to \widetilde{G}$ be a lifting of $\pi g : \pi V \to \pi X$, and let $\widetilde{U}$ be a lifting of U.

10.5.2 *If* $\tilde{g} = \mathrm{Ob}(g')$, *and* $\tilde{g}[V]$ *has one point in common with* $\widetilde{U}$, *then* $\tilde{g}[V] \subseteq \widetilde{U}$.

Proof Let $\tilde{x} = \tilde{g}v$ be an element of $\tilde{g}[V] \cap \widetilde{U}$ and let $x = p\tilde{x} = gv$. Consider the diagram

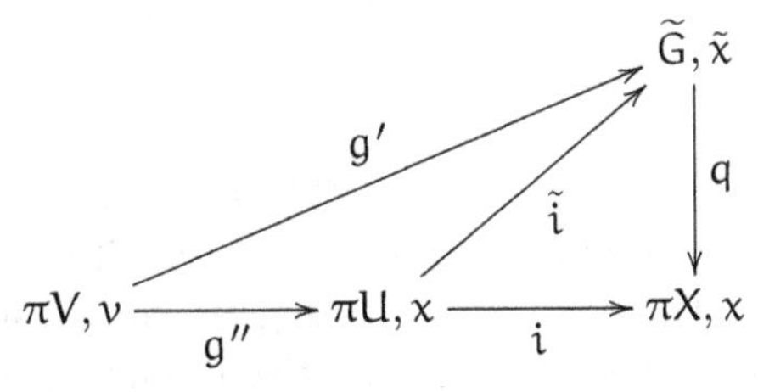

in which g'' is induced by the restriction of g and $\tilde{i}$ is a lifting of the morphism i induced by inclusion. Both g' and $\tilde{i}g''$ are liftings of $\pi g : \pi V, v \to \pi X, x$. But V is path-connected, and so $g' = \tilde{i}g''$ (by 10.3.3). It follows that $\tilde{g}[V] \subseteq \widetilde{U}$. □

A corollary of 10.5.2 is that if two liftings of an element U of $\mathcal{U}$ have a point in common, then they coincide—that is, two liftings of U either coincide or are disjoint.

We now prove that $\widetilde{\mathcal{U}}$ is a base for the open sets of a topology on $\widetilde{X}$ [cf. 5.6.3]. Let $\widetilde{U}, \widetilde{V}$ be respectively liftings of elements of U, V of $\mathcal{U}$, and suppose $\widetilde{w} \in \widetilde{U} \cap \widetilde{V}$. Let $w = p\widetilde{w}$ and let W be an open path-connected set such that $w \in W \subseteq U \cap V$; then W is weakly χ_q-connected and so has a lifting $\widetilde{W}$ such that $\widetilde{w} \in \widetilde{W}$. By 10.5.2, $\widetilde{W}$ is contained in both $\widetilde{U}$ and $\widetilde{V}$ and hence $\widetilde{W} \subseteq \widetilde{U} \cap \widetilde{V}$.

This completes the proof that $\widetilde{\mathcal{U}}$ is a base for the open sets of a topology on $\widetilde{X}$. This topology on $\widetilde{X}$ will be called the *topology of* X *lifted by* q or simply the *lifted topology*.

Suppose now that $\widetilde{X}$ has this topology.

10.5.3 *Let* $f : Z \to X$ *be a map. If* $\pi f : \pi Z \to \pi X$ *lifts to a morphism* $f' : \pi\mathbb{Z} \to \widetilde{G}$, *then*

$$\tilde{f} = \mathrm{Ob}(f') : Z \to \widetilde{X}$$

is continuous and is a lifting of $f : Z \to X$. *All liftings of* f *arise in this way.*

Proof Let $z \in Z$ and let $\widetilde{U}$ be a lifting of a set U of $\mathcal{U}$ such that $\tilde{f}z \in \widetilde{U}$. Then U is a neighbourhood of fz and so there is an open neighbourhood V of z, which may be assumed path-connected, such that $f[V] \subseteq U$. Let $j : V \to Z$ be the inclusion. Then $f'\pi(j)$ is a lifting of $\pi(fj)$. By 10.5.2, $\tilde{f}[V] \subseteq \widetilde{U}$. This proves continuity of $\tilde{f}$.

If $\tilde{f} : Z \to \widetilde{X}$ is any lifting of f, then $\pi\tilde{f}$ lifts πf, and so all liftings arise in the above way. □

On the other hand, the lifted topology is relevant to the case of a given covering map.

10.5.4 *Let* $p : \widetilde{X} \to X$ *be a covering map. The topology of* $\widetilde{X}$ *is that of* X *lifted by* $\pi p : \pi\widetilde{X} \to \pi X$.

Proof Let U be a canonical subset of X. If $\widetilde{U}$ is a path-component of $p^{-1}[U]$, then the inclusion $U \to X$ lifts to a map $U \to \widetilde{X}$ with image $\widetilde{U}$. It follows that $\widetilde{U}$ is a lifting of U. Clearly, these liftings $\widetilde{U}$ for all canonical subsets U of X form a base for the open sets of the given topology on $\widetilde{X}$.

On the other hand, let $\mathcal{U}$ be the set of open, path-connected, weakly χ_p-connected subsets of X. We know that each canonical subset of X belongs to $\mathcal{U}$. Let $U \in \mathcal{U}$, and let $\widetilde{U}$ be a lifting of U. Because each point of U has a canonical neighbourhood contained in U, it is easy to prove that $\widetilde{U}$ is open in $\widetilde{X}$ and is mapped by p homeomorphically on to U. Therefore U is canonical. Hence the lifted topology coincides with the given topology. □

By combining 10.5.4 with 10.5.3 and the corollaries to 10.3.3 we obtain the following results.

10.5.4 *(Corollary 1) If* $p : \widetilde{X}, \tilde{x} \to X, x$ *and* $q : \widetilde{Y}, \tilde{y} \to X, x$ *are connected covering maps with characteristic groups* C, D *respectively, and if* $C \subseteq D$, *then there is a unique map* $r : \widetilde{X}, \tilde{x} \to \widetilde{Y}, \tilde{y}$ *such that* $p = qr$. *Further,* r *is a covering map, and is a homeomorphism if* $C = D$.

10.5.4 *(Corollary 2) A 1-connected covering space of* X *covers every covering space of* X.

Because of 10.5.4 (Corollary 2) a 1-connected covering space of X is called a *universal covering space* of X. It is a special case of the last part of 10.5.4 (Corollary 1) that any two universal covering spaces of a connected space X are homeomorphic.

We now show that the lifted topology does give rise to a covering space. Let $q : \widetilde{G} \to \pi X$ be a covering morphism, let $\widetilde{X} = \mathrm{Ob}(\widetilde{G})$, $p = \mathrm{Ob}(q)$. Suppose also that X is semi-locally χ_q-connected and that $\mathcal{U}$ is the set of open, path-connected, weakly χ_q-connected subsets of X.

10.5.5 *The lifted topology is the only topology on $\widetilde{X}$ such that*
(a) $p : \widetilde{X} \to X$ *is a covering map;*
(b) *there is an isomorphism* $r : \widetilde{G} \to \pi\widetilde{X}$ *which is the identity on objects and such that the following diagram commutes.*

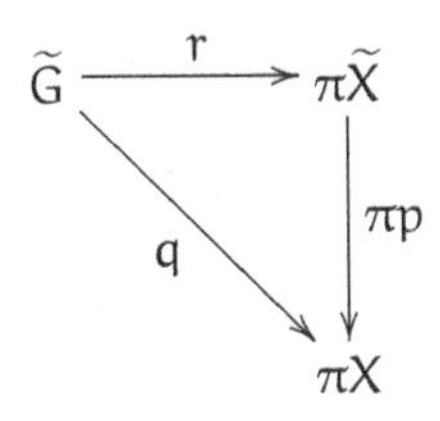

Proof We first prove that if $\widetilde{X}$ has the lifted topology then p is a covering map.

Let $\widetilde{\mathcal{U}}$ be the set of liftings of elements of $\mathcal{U}$, so that $\widetilde{\mathcal{U}}$ is a base for the open sets of the lifted topology on X. If $U \in \mathcal{U}$, then $p^{-1}[U]$ is a union of elements of $\widetilde{\mathcal{U}}$; therefore p is continuous. Also, if $\widetilde{U} \in \widetilde{\mathcal{U}}$, then $p[\widetilde{U}] \in \mathcal{U}$; therefore p is open. If $\widetilde{U}$ is a lifting of a set U of $\mathcal{U}$ then $p \mid \widetilde{U}, U$ is a bijection and hence a homeomorphism; since also $\widetilde{U}$ is open in $\widetilde{X}$, it follows that $\widetilde{U}$ is canonical in $\widetilde{X}$ and U is canonical in X. Therefore p is a covering map.

We now define a morphism $r : \widetilde{G} \to \pi\widetilde{X}$. On objects, r is to be the identity. Let $\alpha \in \widetilde{G}(\tilde{x}, \tilde{y})$, and suppose $q\alpha \in \pi X(x, y)$. Let $a : \mathbb{I} \to X$ be a representative of $q\alpha$; then a induces a morphism $\pi a : \pi\mathbb{I} \to \pi X$ such that

$$(\pi a)(\iota) = q\alpha$$

where ι is the unique element of $\pi\mathbb{I}(0, 1)$.

Since $\mathbb{I}$ is 1-connected, πa lifts uniquely to a morphism

$$a' : \pi\mathbb{I}, 0 \to \widetilde{G}, \tilde{x}.$$

Notice that $a'(\iota)$ is a lifting of $q\alpha$, so that

$$a'(\iota) = \alpha.$$

By 10.5.3 $\tilde{a} = \mathrm{Ob}(a') : \mathbb{I} \to \widetilde{X}$ is continuous, and we define

$$r\alpha = \mathrm{cls}\, \tilde{a};$$

clearly, $r\alpha \in \pi\widetilde{X}(\tilde{x}, \tilde{y})$. Also, $r\alpha$ is independent of the choice of representative a in $q\alpha$, since different representatives a_1, a_2 are equivalent and so have equivalent liftings $\tilde{a_1}, \tilde{a_2}$.

Suppose that $\beta \in \widetilde{G}(\tilde{y}, \tilde{z})$. Then $r(\beta + \alpha)$ and $r\beta + r\alpha$ both lift $q(\beta + \alpha)$. Hence $r(\beta + \alpha) = r\beta + r\alpha$. This proves that r is a morphism and, by definition of r, $(\pi p)r = q$. By 10.2.3, r itself is a covering morphism. This implies that for each $\tilde{x}, \tilde{y}$ in $\widetilde{X}$

$$r : \widetilde{G}(\tilde{x}, \tilde{y}) \to \pi\widetilde{X}(\tilde{x}, \tilde{y})$$

is injective. It is also surjective, since any γ in $\pi\widetilde{X}(\tilde{x}, \tilde{y})$ is covered by an element $\tilde{\gamma}$ of $\mathrm{St}_{\widetilde{G}}\, \tilde{x}$; if $\tilde{\gamma} \in \widetilde{G}(\tilde{x}, \tilde{y}')$ then $r\tilde{y} = r\tilde{y}'$ and so $\tilde{y} = \tilde{y}'$. Hence r is an isomorphism. This proves that (a) and (b) are satisfied by the lifted topology.

The uniqueness of a topology satisfying (a) and (b) follows from 10.5.4: since r is the identity on objects, the topology of X lifts by q and by πp to the same topology on $\widetilde{X}$. □

We now discuss covering spaces of cell complexes.

10.5.6 *Let $p : \widetilde{X} \to X$ be a finite covering map and let X be a cell complex. Then $\widetilde{X}$ can be given a cell structure for which p is a cellular map.*

Proof Let $h_\alpha : \mathbb{E}^{n_\alpha} \to X$ be a characteristic map for an n_α-cell of X. By 10.5.3, 10.5.4, and since $\mathbb{E}^{n_\alpha}$ is 1-connected, h_α lifts to a map $\tilde{h}_\alpha : \mathbb{E}^{n_\alpha} \to \widetilde{X}$. Since $\mathbb{E}^{n_\alpha}$ is compact and $\widetilde{X}$ is Hausdorff, $\tilde{h}_\alpha$ is a closed map; it is also injective on $\mathbb{B}^{n_\alpha}$ (since h_α is) and so $\tilde{h}_\alpha$ maps $\mathbb{B}^{n_\alpha}$ homeomorphically to a subset $\tilde{e}^{n_\alpha}$ of $\widetilde{X}$. That is, the open cells of $\widetilde{X}$ are liftings of the open cells of X. By the uniqueness of liftings of maps, these cells of $\widetilde{X}$ are disjoint; clearly they cover $\widetilde{X}$.

It is clear that $\widetilde{X}^n = p^{-1}[X^n]$. It follows that for each lifting $\tilde{h}_\alpha$ of h_α, we have $\tilde{h}_\alpha[\mathbb{S}^{n-1}] \subseteq \widetilde{X}^{n-1}$. The number of liftings of h_α is finite. So the maps $\tilde{h}_\alpha$ are characteristic maps of a cell structure on $\widetilde{X}$. The cellularity of p is obvious. □

A similar result is true for any covering space of a cell complex. But if the covering is infinite then the number of cells of $\widetilde{X}$ is infinite, and so it is necessary for this to discuss the topology of infinite cell complexes. The general topology useful for proving a covering space of a CW-complex is a CW-complex is given in [Mas67, p. 183–184].

We now give an entertaining application of these results on covering spaces.

10.5.7 *If $n \geqslant 2$ there is no map $f : \mathbb{S}^n \to \mathbb{S}^1$ such that $f(-x) = -fx$ for all x in $\mathbb{S}^n$.*

Proof Suppose there is such an f and consider the diagram

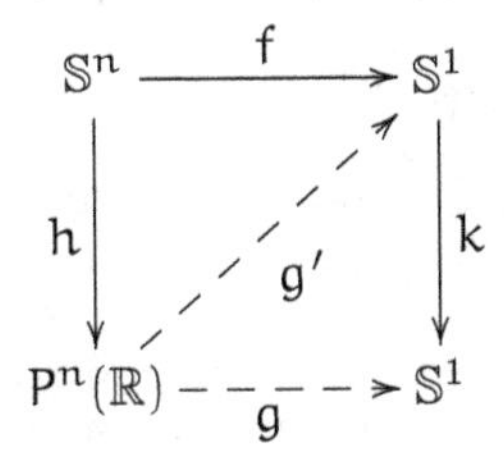

in which h is the Hopf map and k is the map $z \mapsto z^2$. Both h, k are covering maps which identify antipodal points of the spheres they are defined on.

By the given condition on f, $kf(-x) = kfx$ for all x in $\mathbb{S}^n$. Therefore f defines a map $g : P^n(\mathbb{R}) \to \mathbb{S}^1$ such that $gh = kf$. Since $P^n(\mathbb{R})$, $\mathbb{S}^1$ have fundamental groups $\mathbb{Z}_2$, $\mathbb{Z}$ respectively, any characteristic group of g is 0. So g lifts to a map g' such that $kg' = g$. Notice that

$$kg'h = gh = kf$$

so that $g'h$ and f are liftings of the same map. Hence $g'h$ and f agree at all points if they agree at one.

Let $x \in \mathbb{S}^n$. Then the points $g'hx, fx$ are either the same or antipodal points. In the latter case, $g'h(-x) = g'hx = -fx = f(-x)$. So $g'h, f$ agree at either x or $-x$. Hence they agree everywhere, i.e., $g'h = f$. But this implies that $f(-x) = fx$, and so we have a contradiction. □

10.5.7 *(Corollary 1) Let* $n \geqslant 2$ *and let* $g : \mathbb{S}^n \to \mathbb{R}^2$ *be a map. Then there exists a point* x_0 *in* $\mathbb{S}^n$ *such that* $g(x_0) = g(-x_0)$.

Proof Let $h : \mathbb{S}^n \to \mathbb{R}^2$ be the map $x \mapsto g(x) - g(-x)$. Then $h(-x) = -h(x)$ for all x in $\mathbb{S}^n$.

Suppose there is no x_0 in $\mathbb{S}^n$ such that $g(x_0) = g(-x_0)$. Then $h(x_0)$ is never zero, and so the map $f : \mathbb{S}^n \to \mathbb{S}^1$, $x \mapsto h(x)/\|h(x)\|$, is well-defined. But then $f(-x) = -f(x)$ for all x in $\mathbb{S}^n$; this is impossible by 10.5.7. □

The above result with $\mathbb{R}^2$ replaced by $\mathbb{R}^n$ is known as the Borsuk-Ulam Theorem. However, this more general theorem cannot be proved with the methods treated in this book.

Here is a nice application of the lifted topology.

10.5.8 *Let* X *be a semi-locally simply connected space, and let* N *be a totally disconnected normal subgroupoid of* πX. *Then the set of elements of the quotient groupoid* $(\pi X)/N$ *may be given a topology such that the projection* $q = (\sigma, \tau) : (\pi X)/N \to X \times X$ *is a covering map and for* $x \in X$ *the target* *map* $\tau : \mathrm{St}_{\pi X}\, x \to X$ *is the covering map determined by the subgroup* $N(x)$ *of* $\pi(X, x)$. *Further there is a canonical isomorphism of groupoids*

$$\pi((\pi X)/N) \cong \square_N \pi X.$$

Hence the fundamental group of the space $(\pi X)/N$ *at* x *is isomorphic to the semidirect product group* $N(x) \rtimes \pi(X, x)$.

Proof Let $H = \pi X$. In section 10.4 we constructed a covering groupoid $\square_N H$ of the product groupoid $H \times H$ such that $\square_N H$ has object set H/N. We now use 10.5.5 to lift the topology of X to give a topology on the object set H/N of $\square_N H$. The formulae for the fundamental groupoid and fundamental group follow.

In order to identify the subspace $\mathrm{St}_{(\pi X)/N}\, x$ of $(\pi X)/N$ with the covering space $\widetilde{X}_{N,x}$ of X we have to examine in more detail the lifted topology on $(\pi X)/N$. This topology has a basis of sets $(\pi V) + g + (\pi U)$ for $g : x \to y$ in $(\pi X)/N$ and U and V canonical neighbourhoods of x and y (where πU and πV also denote here their images in $(\pi X)/N$). The restriction of this topology to $\mathrm{St}_{(\pi X)/N}\, x$ has a basis of sets $(\pi V) + g$, for $g \in \mathrm{St}_{(\pi X)/N}\, x$. So the subspace topology is the lifted topology, as required. $\square$

It is also true that $(\pi X)/N$ is what we call a *topological groupoid*—that is, all the structure maps of this groupoid are continuous for the given topology. Thus all the covering spaces $\widetilde{X}_{N,x}$ of X are nicely tied together in the one topological groupoid structure $(\pi X)/N$. However, we do not prove this here, but refer the reader to [BD75]. For more information on topological groupoids, we refer the reader to [Mac05] and the references in [Bro87].

Finally for this section, we should mention an important fact about universal covers of surfaces. Many readers will be familiar with the classification of compact, connected surfaces without boundary, as given in many texts. The universal cover $\widetilde{S}$ of such a surface S is known to be either the sphere $\mathbb{S}^2$ or the plane $\mathbb{R}^2$. For a proof, see for example [Sti80]. The key point of the proof is to show that the universal cover of the orientable surface of genus two is $\mathbb{R}^2$.

EXERCISES

1. Prove that if $q : \widetilde{X} \to X$, $r : \widetilde{Y} \to Y$ are covering maps, then so also is $q \times r : \widetilde{X} \times \widetilde{Y} \to X \times Y$.
2. Let $q : \widetilde{X} \to \pi X$ be a covering morphism and let X be semi-locally χ_q-connected. Let $\mathcal{U}$ be a set of open, path-connected weakly χ_q-connected subsets of X such that $\mathcal{U}$ is a base for the open sets of X. Let $\widetilde{\mathcal{U}}$ be the set of liftings of elements of $\mathcal{U}$. Prove that $\mathcal{U}$ is a base for the open sets of the lifted topology on $\widetilde{X} = \mathrm{Ob}(\widetilde{G})$.

3. Let $r : Z \to Y$, $q : Y \to X$ be covering maps. Prove that $qr : Z \to X$ is a covering map if and only if X is semi-locally $\mathcal{X}_{qr}$-connected.
4. Prove that the Borsuk-Ulam Theorem does imply 10.5.7 (Corollary 1).
5. Prove that no subspace of $\mathbb{R}^2$ is homeomorphic to $\mathbb{S}^2$.
6. Let $n \geqslant 2$ and $g : \mathbb{S}^n \to \mathbb{R}^2$ be a map such that $g(-x) = -g(x)$ for all x in $\mathbb{S}^n$. Prove that there is a point x_0 in $\mathbb{S}^n$ such that $g(x_0) = 0$.

10.6 The equivalence of categories

In this section we give a more sophisticated and comprehensive view of the previous material by showing that there is a complete 'translation system' from covering spaces to covering groupoids. The 'translation system' or dictionary is given by an equivalence of categories. This itself is illustrative of a modern trend in mathematics. Very often an equivalence of categories can hold a lot of information, giving two different views of what is essentially the same kinds of objects. Each view can be more appropriate and useful in some circumstances than the other. The interplay can then be quite powerful. In our particular case, the equivalence between covering spaces of X and covering groupoids of πX gives a complete translation between topology and algebra, and in an elegant form. This translation is important because of the wide occurrence of covering spaces in mathematics. For a survey of this, see [Mag76] and [DD77, DD79].

We now define the categories we wish to prove equivalent. First we give the topological example.

Let X be a Hausdorff, locally connected space. The category $\mathsf{CovTop}(X)$ of covering spaces of X has objects the covering maps $p : Y \to X$ and has as arrows or morphisms the commutative diagrams of maps, where p and q are covering maps,

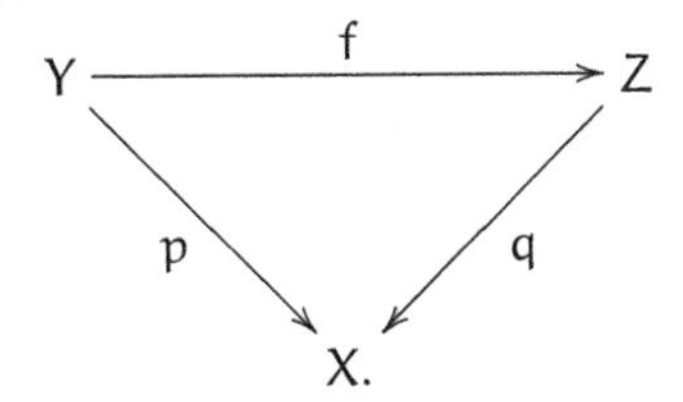

It is convenient to write such a diagram as a triple (f, p, q). The composition in $\mathsf{CovTop}(X)$ is then given by $(g, q, r)(f, p, q) = (gf, p, r)$. Note that we make no assumption that X, Y or Z are connected. By 10.2.4, f also is a covering map.

Let G be a groupoid. The category $\mathsf{CovGrp}(G)$ of covering morphisms of G has objects the covering morphisms $p : H \to G$ and has as arrows or morphisms the commutative diagrams of morphisms, where p and q are covering morphisms,

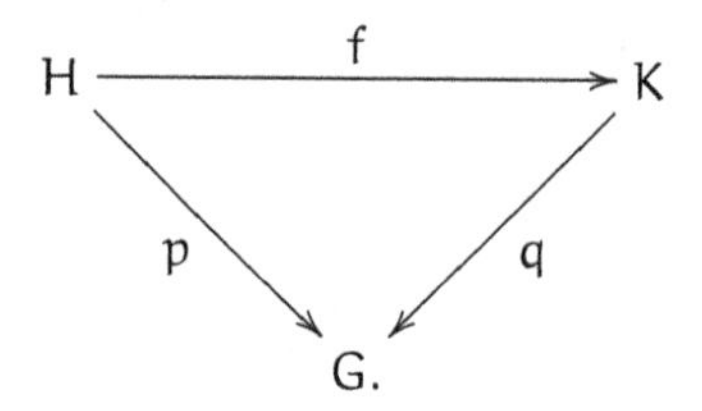

Again it is convenient to write such a diagram as a triple (f, p, q), and the composition is as in the topological case. By 10.2.3, f also is a covering morphism.

Now we come to our main result.

10.6.1 *If X is a Hausdorff, locally connected and semi-locally 1-connected space, then the fundamental groupoid functor π induces an equivalence of categories*

$$\pi_! : \mathsf{CovTop}(X) \to \mathsf{CovGrp}(\pi X).$$

Proof If $p : Y \to X$ is a covering map of spaces, then $\pi(p) : \pi Y \to \pi X$ is a covering morphism of groupoids, by 10.2.1. Since π is a functor, we also obtain the functor $\pi_!$.

To prove $\pi_!$ an equivalence of categories, we construct a functor $\rho : \mathsf{CovGrp}(\pi X) \to \mathsf{CovTop}(X)$ and prove that there are homotopies of functors $1 \simeq \rho\pi_!$, $1 \simeq \pi_!\rho$. The technical work for this has already been done, and it is simply a matter of assembling the facts.

Let $q : \widetilde{G} \to \pi X$ be a covering morphism of groupoids. We suppose the space X is given. Let $\widetilde{X} = \mathrm{Ob}(\widetilde{G})$, and let $p = \mathrm{Ob}(q) : \widetilde{X} \to X$. By 10.5.5, there is a topology on $\widetilde{X}$ making p a covering map, and there is an isomorphism $r : \widetilde{G} \to \pi\widetilde{X}$. We have to prove that this topology on $\widetilde{X}$ is natural with respect to morphisms in the category $\mathsf{CovGrp}(\pi X)$, and that r also is natural.

Let (f, q, s) be a morphism in $\mathsf{CovGrp}(\pi X)$, where $f : \widetilde{G} \to \widetilde{H}$. Let $\mathcal{U}$ be the set of all open, path-connected, subsets U of X such that the inclusion $U \to X$ maps the fundamental group of U trivially. In section 10.5 we showed how to lift the cover $\mathcal{U}$ to give covers $\widetilde{\mathcal{U}}_q$, $\widetilde{\mathcal{U}}_s$ of $\widetilde{X} = \mathrm{Ob}(\widetilde{G})$ and $\widetilde{Y} = \mathrm{Ob}(\widetilde{H})$ respectively. These covers are bases for the topologies of $\widetilde{X}$ and $\widetilde{Y}$ respectively. Let $\tilde{x} \in \widetilde{X}$, and suppose $\widetilde{U}_s$ is an element of $\widetilde{\mathcal{U}}_s$ containing $f\tilde{x}$. Then $U = s[\widetilde{U}_s]$ belongs to $\mathcal{U}$ and U lifts to an element $\widetilde{U}_q$ of $\widetilde{\mathcal{U}}_q$ containing

$\tilde{x}$. Further $f[\widetilde{U}_q] = \widetilde{U}_s$. So $f : \widetilde{X} \to \widetilde{Y}$ is continuous. This defines ρ on morphisms, and it is clear that ρ is a functor.

The homotopy of functors $r : 1 \simeq \pi_! \rho$ is defined in essence in 10.5.5. The only extra fact which has to be proved is that r is natural. So let (f, q, s) be a morphism in $\mathsf{CovGrp}(\pi X)$ as above. We have to prove that the following diagram is commutative:

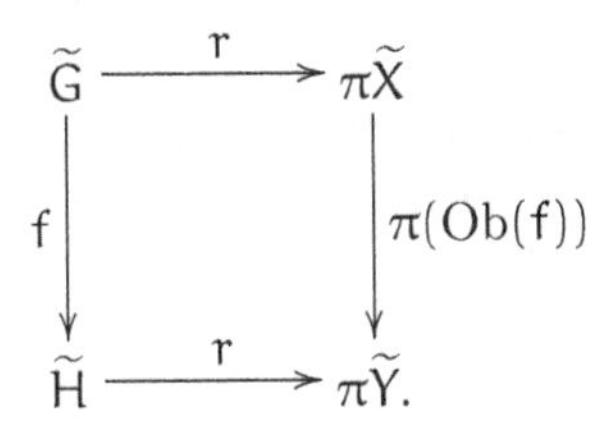

$$\begin{array}{ccc} \widetilde{G} & \xrightarrow{r} & \pi\widetilde{X} \\ {\scriptstyle f}\downarrow & & \downarrow{\scriptstyle \pi(\mathrm{Ob}(f))} \\ \widetilde{H} & \xrightarrow{r} & \pi\widetilde{Y}. \end{array}$$

This requires examining the definition of r. Let α be an element of $\widetilde{G}$ which starts at $\tilde{x}$. Let $a : \mathbb{I} \to X$ be a representative of $q\alpha \in \pi X$. Then a induces a morphism $\pi a : \pi\mathbb{I} \to \pi X$ such that $(\pi a)(\iota) = q\alpha$. Further, πa lifts uniquely to a morphism $a' : \pi\mathbb{I}, 0 \to \widetilde{G}, \tilde{x}$. Then $r\alpha$ is the class of the path $\mathrm{Ob}(a') : \mathbb{I} \to \widetilde{X}$. Let $\beta = f\alpha$, and apply the same process to give $b' : \pi\mathbb{I}, 0 \to \widetilde{H}, f\tilde{x}$, where $b = \mathrm{Ob}(f)a$. Since b' is uniquely determined by b, it follows that $rf(\alpha) = \pi(\mathrm{Ob}(f))r(\alpha)$.

Finally we have to construct a homotopy $\theta : 1 \simeq \rho\pi_!$. But $\widetilde{X} = \mathrm{Ob}(\pi\widetilde{X})$ and the topology of $\widetilde{X}$ is precisely the lifted topology. So in fact $1 = \rho\pi_!$. So the proof is complete. □

The significance of this result is that it shows that the functor $\pi_!$ allows us to translate a topological problem on covering spaces to an algebraic problem on covering groupoids. This is one of those rare examples of a complete and faithful translation. Usually a functor 'forgets' some of the original formulation. Such forgetting has the advantage of translating a problem into a simpler one, and one possibly with more opportunities for calculation, but it does mean that some of the original geometry is lost.

As one example of the use of this translation, we will study the automorphisms of an object of the category $\mathsf{CovGrp}(G)$. This gives the *group of covering isomorphisms* of a covering morphism, and our results will then, by virtue of the equivalence of categories, yield results on the automorphisms of an object of $\mathsf{CovTop}(X)$. But first we solve a more general problem for which we require some notation.

Let A be a group, and let F, C be subgroups of A. We define

$$M_A(F, C) = \{a \in A : F \subseteq a + C - a\}.$$

Let us temporarily write this set as M. In general it is not true that $a, b \in M$ implies $a+b \in M$. However, this is true if $F = C$, in which case the addition gives M the structure of a monoid (i.e. it satisfies the associativity and unit axioms, but inverses need not exist); in general, M is not in this case a group. To obtain a group in this case we consider the usual *normaliser* of F in A, namely

$$N_A(F) = \{a \in A : F = a + F - a\}.$$

Then $N_A(F)$ is the largest subgroup of A in which F is a normal subgroup.

10.6.2 *Let* Z, G, H *be connected groupoids, and let* $p : H \to G$ *be a covering morphism. Let* $f : Z \to G$ *be a morphism of groupoids, and let* $L(f, p)$ *be the set of liftings of* f *to morphisms* $Z \to H$. *Let* $z \in Ob(Z)$, $x \in Ob(H)$ *and suppose* $px = fz$. *Let*

$$A = G(px), \quad F = f[Z(z)], \quad C = p[H(x)].$$

Then there is a surjection φ *from* $M_A(F, C)$ *to* $L(f, p)$ *with the property that* $\varphi a = \varphi b$ *if and only if there is an element* $c \in C$ *such that* $b = a + c$.

Proof Let $M = M_A(F, C)$ and let $a \in M$. Then a lifts uniquely to an element $\tilde{a}$ in $St_G\, x$. Let y be the end point of $\tilde{a}$. By 10.3.2, the characteristic group of p at y is $a + C - a$. Since $F \subseteq a + C - a$, it follows from 10.3.3 that f lifts uniquely to a morphism $\tilde{f}$ such that $\tilde{f}(z) = y$. We set $\varphi(a) = \tilde{f}$.

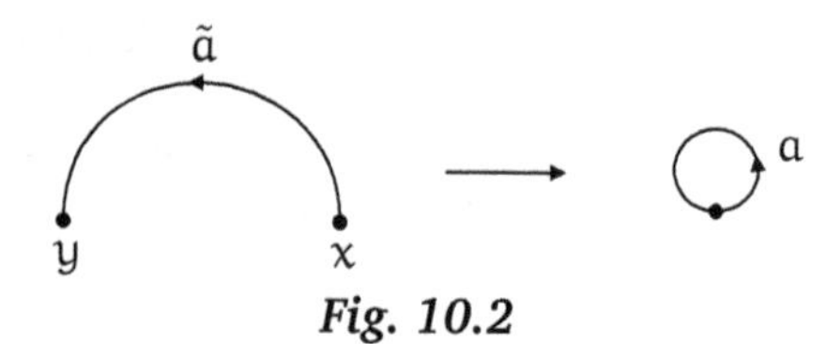

Fig. 10.2

If $\tilde{f} : Z \to H$ is any lifting of f, then since H is connected, there is an element $\tilde{a}$ in $H(x, \tilde{f}(z))$. If $a = p\tilde{a}$ then $\varphi(a) = \tilde{f}$. This proves that φ is surjective.

Suppose now that $a, b \in M$, and that $\tilde{a}, \tilde{b}$ are lifts of a, b starting at x. If $b = a + c$ where $c \in C$, then c has a lift $\tilde{c} \in H(x)$ and then $\tilde{a} + \tilde{c}$ also lifts b, and has the same target as $\tilde{a}$. Hence $\varphi(a) = \varphi(b)$. On the other hand, if $\varphi(a) = \varphi(b)$ then $\tilde{a}$ and $\tilde{b}$ have the same target and so $\tilde{c} = -\tilde{a} + \tilde{b}$ is well defined. Let $c = p(\tilde{c})$. Then $b = a + c$. This proves the result. □

In the case $f = p$ we can make the result a bit stronger. Recall that a *covering isomorphism* $h : H \to H$ of a covering morphism $p : H \to G$ is an isomorphism of groupoids such that $ph = p$. Clearly the set of these covering isomorphisms forms under composition a group, and we write this group as $Aut(p)$.

10.6.3 *Let* $p : H \to G$ *be a covering morphism of connected groupoids, let* $x \in \mathrm{Ob}(H)$ *and let* $C = p[H(x)]$ *be the characteristic group of* p *at* x. *Then* $\mathrm{Aut}(p)$ *is anti-isomorphic to the quotient of the group* $N_{G(px)}(C)$ *by its normal subgroup* C.

Proof Let $A = G(px)$. Then $N = N_A(C)$ is the normaliser of C in A and C is normal in N so that the quotient group N/C is well defined. We know that if $a, b \in N$ then $\varphi(a), \varphi(b)$ are morphisms $h, k : H \to H$, say, such that $ph = h$, $pk = k$. We now prove that $\varphi(b + a) = hk$.

Let $\tilde{a}, \tilde{b} \in \mathrm{St}_H\, x$ have final points y and u, and lift a, b respectively. Then $h(x) = y$, $k(x) = u$, and so $h(u) = hk(x)$. Hence $h(\tilde{b}) \in H(y, hk(x))$. But $h(\tilde{b}) + \tilde{a}$ is defined, belongs to $\mathrm{St}_H\, x$, and lifts $b + a$. Since $phk = p$, it follows that $\varphi(b + a) = hk$.

Since $\varphi(0)$ is the identity $H \to H$, and φ is an anti-morphism, it follows that $h = \varphi(a)$ is an isomorphism. The result follows. □

The next result is useful when combined with the previous one. First, we say that an element g of a groupoid G is a *loop* if it belongs to some vertex group of G, i.e. if g has the same source and target. As motivation for the result, consider the following picture of two graphs whose labellings show how they cover the space $X = \mathbb{S}^1 \vee \mathbb{S}^1$ consisting of a wedge of two circles labelled a and b.

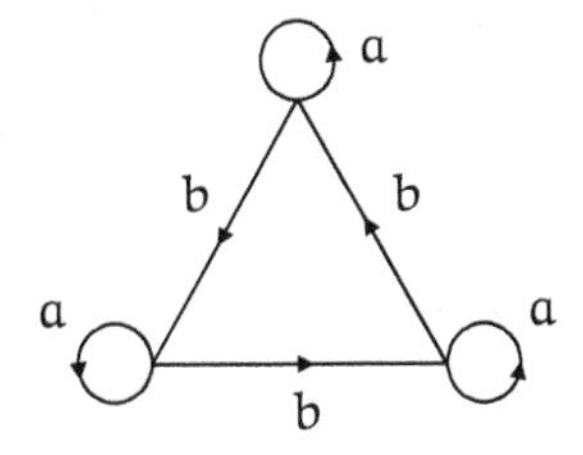

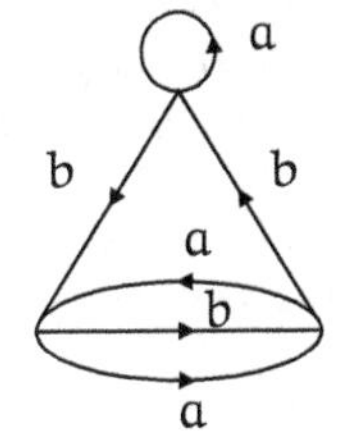

Fig. 10.3

In the first example, the edges with a given label are either all or none of them loops. In the second example, the edges labelled a consist of one loop and two non loops.

10.6.4 *Let* $p : H \to G$ *be a covering morphism of groupoids. Consider the following conditions:*
(a) *for all loops* a *in* G, *either all or no lifts of* a *are loops;*
(b) *for all objects* x *of* H, *the characteristic group* $p[H(x)]$ *is normal in* $G(px)$.
Then (a) $\Rightarrow$ (b), *and if* H *is connected,* (b) $\Rightarrow$ (a).

Proof (a) $\Rightarrow$ (b) Let x be an object of H and let $C = p[H(x)]$. Let $c \in C$, $a \in G(px)$. We need to prove that $-a + c + a \in C$. Let $b \in H(x, y)$ lift a.

By (a), c lifts to an element d of $H(y)$. Then $-b + d + b$ belongs to $H(x)$ and is mapped by p to $-a + c + a$. Hence $-a + c + a \in C$.

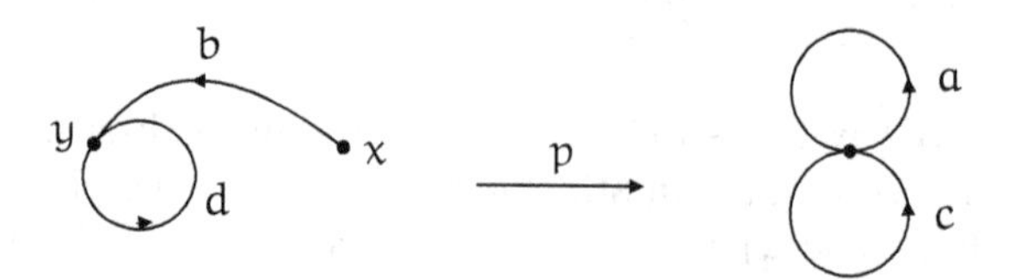

Fig. 10.4

(b) $\Rightarrow$ (a) Let x, y be objects of H such that $p(x) = p(y)$, and let $c \in C = p[H(x)]$. Since H is connected, there is an element b in $H(x, y)$. Let $a = p(b)$. Now $-a + c + a \in C$, since C is normal in $G(px)$, and so $-a + c + a = p(e)$ where $e \in H(x)$. Then

$$p(b + e - b) = a - a + c + a - a = c.$$

But $b + e - b$ belongs to $H(y)$ and is a loop. So c lifts also to a loop at y. □

A covering morphism $p : H \to G$ of groupoids is called *regular* if H and G are connected and one of the conditions (a) or (b) of the previous result is satisfied.

10.6.4 *(Corollary 1) If* $p : H \to G$ *is a regular covering morphism of groupoids, and* C *is a characteristic group of* p *at an object* x *of* H, *then the group of covering isomorphisms of* p *is anti-isomorphic to the quotient group* $H(x)/C$.

Proof This is immediate from 10.6.3 and 10.6.4. □

We can now use the equivalence of categories to translate these ideas and results to the topological case. Let $p : Y \to X$ be a covering map. If a space Z, and also Y and X, are path connected, then the set of lifts of a map $f : Z \to X$ to a map $Z \to Y$ can be entirely determined from 10.6.2, by considering the corresponding problem for the fundamental groupoids. Also, p is called *regular* if the corresponding covering morphism of groupoids $\pi p : \pi Y \to \pi X$ is a regular covering morphism. So we immediately obtain:

10.6.4 *(Corollary 2) Let* $p : Y \to X$ *be a regular covering map of spaces. Let* $y \in Y$. *Then the group* $\mathrm{Aut}(p)$, *of homeomorphisms* $h : Y \to Y$ *such that* $ph = p$, *is anti-isomorphic to the quotient group of the fundamental group* $\pi(X, py)$ *by the normal subgroup* $p_*[\pi(Y, y)]$. *In particular, if* X *is connected and* Y *is a universal cover of* X, *then the group of homeomorphisms* $h : Y \to Y$ *such that* $ph = p$ *is anti-isomorphic to the fundamental group* $\pi(X, x)$ *at any point* x *in* X. □

The importance of this last result is that the fundamental group is shown to occur as a group of automorphisms. This links two basic motivating examples for group theory, namely symmetry groups and fundamental groups. This link has proved useful in a number of areas of mathematics. There is even a proof of the Van Kampen theorem for the fundamental group entirely in terms of covering spaces, see for example [DD77, DD79]. However a proof of theorem 6.7.2 in this spirit is not available in the literature.

There is also a notion of *symmetry groupoid*, whose applications to differential geometry and topology were initiated by [Ehr80]. For further references on this, see [Mac05] and the survey article [Bro87].

EXERCISES

1. Prove that a connected 2-fold covering morphism is regular.
2. We say two covering spaces Y, Z of $X = \mathbb{S}^1 \vee \mathbb{S}^1$ are equivalent if Y is homeomorphic to Z. Find representatives of all equivalence classes of 3-fold covering spaces of X. Which of these are regular? For each connected 3-fold covering, use the method of proof of the Nielsen–Schreier theorem (see 10.8.2) to give a subgroup H of the fundamental group F of X at the base point, such that H determines the covering. In the regular cases, write down the quotient group F/H.
3. Carry out a similar exercise to the previous one but for 4-fold covering spaces of $\mathbb{S}^1 \vee \mathbb{S}^1$.
4. Illustrate the application of 10.6.2 and 10.6.3 by explicit examples for the case G and Z are groups.
5. With reference to the paragraph before 10.6.2, give an example where $a, b \in M$ but $a + b \notin M$. Also, give an example of a group G, subgroup H of G and element $g \in G$ such that $g + H - g$ is a proper subset of H. [An example of this type is given in [Ros78, p. 62].]
6. Construct regular and non-regular but connected 3-fold coverings of the graphs shown in Fig. 10.5:

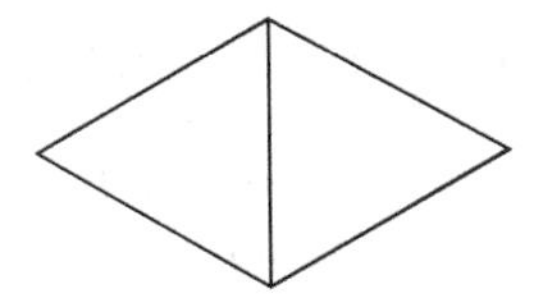
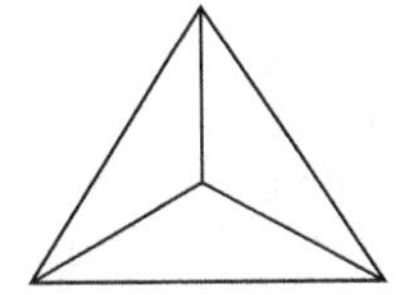

Fig. 10.5

10.7 Induced coverings and pullbacks

In this section we deal with the important topics of restricting coverings to subspaces, and pulling back coverings.

We have already defined *pushouts* in any category. The dual notion of *pullback* arose earlier than pushout, in the theory of fibre spaces, and in fact suggested the term pushout for the dual term. We have defined pullbacks in a general category in an exercise in chapter 6. Here we take a more concrete approach.

Suppose given two maps $f : A \to X$ and $p : Y \to X$. The *pullback* of f and p, also called the *fibre product*, is the subspace P of $A \times Y$ given by

$$P = A\, {}_f\!\times_p\, Y = \{(a, y) \in A \times Y : fa = py\}.$$

It is common to write also $P = A \times_X Y$, since we can think of A and Y as spaces over X.

There is a diagram

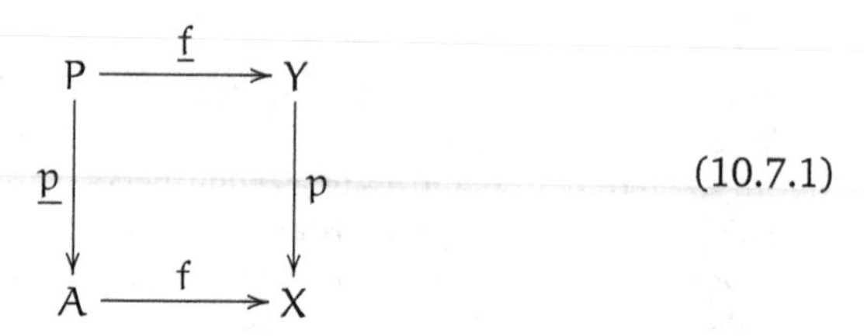

in which $\underline{p} : (a, y) \mapsto a, \underline{f} : (a, y) \mapsto y$. Then the diagram (10.7.1) is commutative: $p\underline{f} = f\underline{p}$. It is also a pullback in the following sense: given any commutative diagram

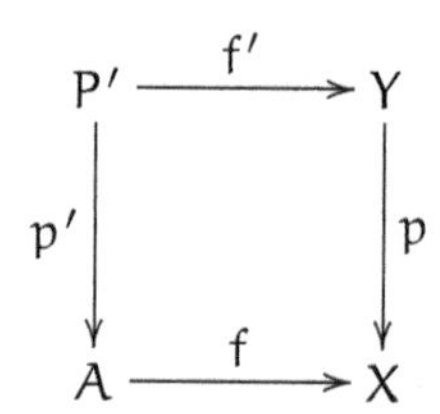

there is a unique map $r : P' \to P$ such that $\underline{f}r = f'$, $\underline{p}r = p'$. Here r is simply the restriction of the map (p', f').

If f is injective then the point (a, y) of P is determined entirely by the element $y \in p^{-1}f[A]$. If further f is an embedding, then the map $P \to p^{-1}f[A]$, $(a, y) \mapsto y$, is a homeomorphism with inverse $y \mapsto (f^{-1}p(y), y)$. This makes the important point that if f is the inclusion of a subspace A of X, then we may identify $A \times_X Y$ with $p^{-1}[A]$.

10.7.2 *If $p : Y \to X$ is a covering map, then for any map $f : A \to X$, the induced map $\underline{p} : A \times_X Y \to A$ is also a covering map. Further, if p is an n-fold covering map, so also is $\underline{p}$.*

Proof We prove that if U is a canonical open set in X, then $f^{-1}[U]$ is a canonical open set in A. Suppose then that $\varphi : p^{-1}[U] \to U \times F$ is a homeomorphism of the form $y \mapsto (py, \varphi_2 y)$, with inverse ψ, where F is a discrete space. Then we may define maps

$$\varphi' : \underline{p}^{-1}f^{-1}[U] \to f^{-1}[U] \times F, \qquad \psi' : f^{-1}[U] \times F \to \underline{p}^{-1}f^{-1}[U],$$
$$(a, y) \mapsto (a, \varphi_2 y), \qquad (a, z) \mapsto (a, \psi(fa, z))$$

and check that

$$\begin{aligned} \psi'\varphi'(a, y) &= \psi'(a, \varphi_2 y) = (a, \psi(fa, \varphi_2 y)) \\ &= (a, \psi(py, \varphi_2 y)) = (a, \psi\varphi y) = (a, y), \\ \varphi'\psi'(a, z) &= \varphi'(a, \psi(fa, z)) = (a, \varphi_2\psi(fa, z)) = (a, z). \end{aligned}$$

This proves that $\underline{p}$ is a covering map and also the statement about n-coverings. □

There are similar notions of pullback for groupoids. Let $f : L \to G$ and $p : H \to G$ be groupoid morphisms. The *pullback square* defined by f and p is the diagram

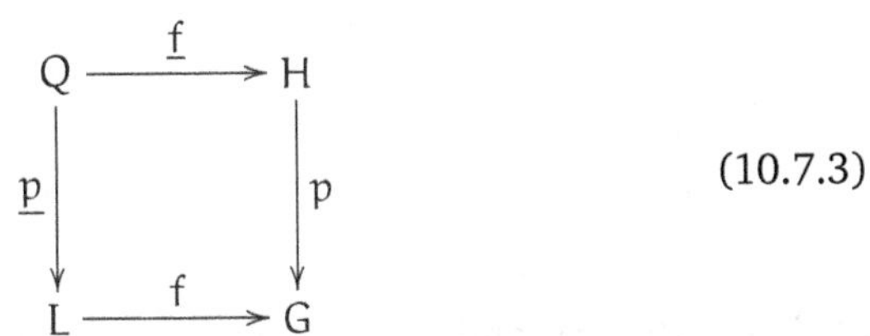

(10.7.3)

in which Q is the subgroupoid of the product groupoid $L \times H$ of pairs (l, h) (of elements or objects) such that $fl = ph$. The groupoid Q is also called the *fibre product* of L and H over G, and is written $L \times_G H$. The morphisms $\underline{f}$ and $\underline{p}$ are given respectively by $(l, h) \mapsto l$, $(l, h) \mapsto h$. We shall also need the fact that:

10.7.4 *If* $f : L \to G$ *and* $p : H \to G$ *are morphisms of groupoids, and* p *is a covering morphism, then* $\underline{p} : L \times_G H \to L$ *is also a covering morphism.*

Proof This is a straightforward check. Let $Q = L \times_G H$ and let $(x, y) \in \mathrm{Ob}(Q)$. Then $fx = py$. We have to prove that $\underline{p} : \mathrm{St}_Q(x, y) \to \mathrm{St}_L\, x$ is bijective. Let $l \in \mathrm{St}_L\, x$. Since p is a covering morphism, and $fx = py$, there is an element $h \in \mathrm{St}_H\, y$ such that $fl = ph$. Clearly $\underline{p}(l, h) = l$, and $(l, h) \in \mathrm{St}_Q(x, y)$. Also (l, h) is the only such element, since p is a covering morphism. □

One special case of the pullback construction is important. Let $p : H \to G$ be a morphism and let $f : L \to G$ be the inclusion of the subgroupoid L of G. Then the element (l, h) of $L \times_G H$ can be identified with the element

h, subject to the sole condition that $ph \in L$. Hence we can identify $L \times_G H$ and $p^{-1}[L]$.

As may be expected, it is of interest to compare pullbacks of spaces and pullbacks of the corresponding fundamental groupoids.

10.7.5 *Suppose given a pullback* (10.7.1) *of spaces. Then there is an induced morphism of groupoids*

$$\theta : \pi(A \times_X Y) \to \pi A \times_{\pi X} \pi Y$$

which is the identity on objects. Further, if p *is a covering map of spaces, then* θ *is an isomorphism.*

Proof The morphism θ is determined by the morphisms $\pi(\underline{p})$, $\pi(\underline{f})$, and the properties of the pullback of groupoids. Clearly θ is the identity on objects.

Consider the diagram

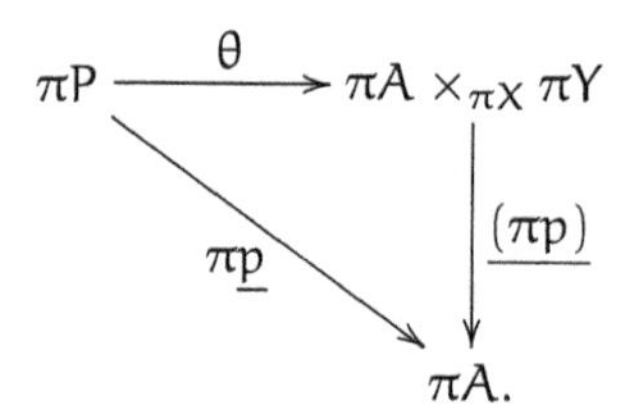

We know that $\pi\underline{p}$ and $\underline{(\pi p)}$ are covering morphisms. It follows that θ is a covering morphism. But θ is the identity on objects. Therefore θ is an isomorphism. □

Suppose then that (10.7.1) is a pullback of spaces and that p is a covering morphism. Our aim is to find information on πP. By the last result, it is sufficient to prove a result about pullbacks of covering morphisms of groupoids. The result we want is conveniently expressed in terms of an exact sequence.

10.7.6 *Suppose that* (10.7.3) *is a pullback of groupoids and that* p *is a covering morphism. Let* $(l, y) \in \mathrm{Ob}(Q)$, *so that* $fl = py = x$, *say. Then there is a sequence*

$$Q(l,y) \xrightarrow{i} L(l) \times H(y) \xrightarrow{\delta} G(x) \xrightarrow{\Delta} \pi_0 Q \xrightarrow{q} \pi_0 L \times_{\pi_0 G} \pi_0 H$$

in which

(1) i *is the inclusion;*

(2) $L(l) \times H(y)$ *acts on the set* $G(x)$ *by* $(\lambda, \eta) \cdot \gamma = f\lambda + \gamma - p\eta$, *and* ∂ *is the function* $(\lambda, \eta) \mapsto f\lambda - p\eta$;

(3) $\Delta\gamma = \mathrm{cls}(l, z)$ *where* z *is the end point of the lift* $\widetilde{\gamma}$ *of* γ *with initial point* y*;*
(4) $q : \mathrm{cls}(b, z) \mapsto (\mathrm{cls}\, b, \mathrm{cls}\, z)$.
This sequence is exact in the sense that:
(a) i *is a monomorphism and* $\partial(\lambda, \eta) = \partial(\mu, \xi)$ *if and only if* $(-\lambda + \mu, -\eta + \xi) \in Q$*;*
(b) $\Delta\gamma = \Delta\delta$ *if and only if there are elements* $\lambda \in L(l)$, $\eta \in H(y)$ *such that* $\delta = f\lambda + \gamma - p\eta$*;*
(c) $\mathrm{Im}\,\Delta = q^{-1}(\mathrm{cls}\, l, \mathrm{cls}\, y)$*;*
(d) q *is surjective.*

Proof The functions in the sequence are well defined so that it only remains to verify exactness.
(a) The morphism i is an inclusion and so a monomorphism. By definition of ∂, we have $\partial(\lambda, \eta) = \partial(\mu, \xi)$ if and only if $f\lambda - p\eta = f\mu - p\xi$. This is equivalent to $f(-\lambda + \mu) = p(-\eta + \xi)$, which is itself equivalent to $(-\lambda + \mu, -\eta + \xi) \in Q$.
(b) Suppose $\delta = f\lambda + \gamma - p\eta$ as given. If $\widetilde{\gamma} \in H(y, z)$ covers γ, then $\widetilde{\gamma} - \eta$ covers $\gamma - p\eta$ and has initial point y, so that $z = \Delta\gamma = \Delta(\gamma - p\eta)$. If $\widetilde{\lambda} \in H(z, w)$ covers $f\lambda$, then $(\lambda, \widetilde{\lambda})$ joins (l, z) to (l, w) in Q, and so $\Delta(\gamma - p\eta) = \Delta(f\lambda + \gamma - p\eta)$.

Suppose conversely that $\Delta\gamma = \Delta\delta$. Let $\widetilde{\gamma} \in H(y, z)$ cover γ and let $\widetilde{\delta} \in H(y, w)$ cover δ. Since (l, z), (l, w) lie in the same component of Q, there are elements $\lambda \in L(l, l)$, $\kappa \in H(z, w)$ such that $f\lambda = p\kappa$. Let $\eta = -\widetilde{\delta} + \kappa + \widetilde{\gamma} \in H(y)$. Then

$$f\lambda = p\kappa = p\widetilde{\delta} + p\eta - p\widetilde{\gamma} = \delta + p\eta - \gamma$$

which gives δ as required.
(c) Note that if $\widetilde{\gamma} \in H(y, z)$ covers $\gamma \in G(x)$, then

$$q\Delta(\gamma) = (\mathrm{cls}\, l, \mathrm{cls}\, z) = (\mathrm{cls}\, l, \mathrm{cls}\, y).$$

Suppose conversely that $q(\mathrm{cls}(b, w)) = (\mathrm{cls}\, l, \mathrm{cls}\, y)$. Then $fb = pw$ and there are elements $\lambda \in L(l, b)$, $\kappa \in H(y, w)$. Let $\eta \in H(z, w)$ lift $f\lambda$, and let $\gamma = -f\lambda + p\eta$. Then $\gamma \in G(x)$ and $-\eta + \kappa$ belongs to $H(y, z)$ and lifts γ. Hence $\Delta(\gamma) = \mathrm{cls}(l, z)$. But $\mathrm{cls}(l, z) = \mathrm{cls}(b, w)$ since $(\lambda, \eta) : (l, z) \to (b, w)$ in Q.
(d) Finally we prove that q is surjective. Let $\mathrm{cls}\, b \in \pi_0 L$, $\mathrm{cls}\, w \in \pi_0 H$ be such that $\mathrm{cls}\, fb = \mathrm{cls}\, pw$ in $\pi_0 G$. Then there is an element $\kappa : pw \to fb$ in G. Let $\eta : w \to z$ lift κ. Then (b, z) is an object of Q and $q(\mathrm{cls}(b, z)) = (\mathrm{cls}\, b, \mathrm{cls}\, z) = (\mathrm{cls}\, b, \mathrm{cls}\, w)$. □

10.7.6 *(Corollary 1) If* L *and* H *are connected groupoids, then there is a* *bijection from* $\pi_0 Q$ *to the set of distinct double cosets*

$$f[L(l)]\gamma p[H(y)]$$

for $\gamma \in G(x)$.

Proof This just says that the orbits of the action of $L(l) \times H(y)$ on $G(x)$ are these double cosets. □

10.7.6 *(Corollary 2) Suppose that* L *and* H *are connected groupoids and that* H *is simply connected, so that* $p : H \to G$ *is a universal covering morphism. Then the pullback* $Q = L \times_G H$ *is connected if and only if* $f : L(l) \to G(x)$ *is surjective; and* Q *is simply connected if and only if* $f : L(l) \to G(x)$ *is injective.*

Proof Since H is 1-connected and L is connected, the exact sequence of 10.7.6 becomes

$$1 \to Q(l, y) \to L(l) \to G(x) \to \pi_0 Q \to 1,$$

from which the statements follow immediately. □

These results on pullbacks of covering morphisms of groupoids have, by 10.7.5, immediate applications to pullbacks of covering maps of spaces. We leave the reader to make the obvious conclusions.

We refer the reader to the exercises for a different treatment of the exact sequence in 10.7.6; it is deduced from the sequence for a covering of groupoids by using the notion of *homotopy pullback*. This treatment is more elegant but would have taken us too far afield. The advantage of the use of homotopy pullbacks is that one obtains exact sequences for *any* pair of morphisms of groupoids $L \to G$, $H \to G$, and in particular there are exact sequences for the case L, G, H are groups. It is shown in [BHK83] how the exact sequence of 10.7.6 can be used in conjunction with non-abelian cohomology with coefficients in a groupoid to deduce the Van Kampen theorem for the fundamental groupoid.

EXERCISES

1. [Exercises 1–5 are taken from [BHK84].] Let $f : L \to G$ and $p : H \to G$ be morphisms of groupoids. The *homotopy pullback* $Z = Z(f, p)$ is defined as follows. The objects of Z are to consist of all triples (l, γ, y) where $l \in Ob(L)$, $y \in Ob(H)$, and $\gamma : fl \to py$ in G. An element of $Z((l, \gamma, y), (l', \gamma', y'))$ is a triple (γ, λ, η) such that $\lambda \in L(l, l')$, $\eta \in H(y, y')$ and $p\eta + \gamma = \gamma' + f\lambda$. Define $r : Z \to L \times H$ by $(l, \gamma, y) \mapsto (l, y)$ on objects, $(\gamma, \lambda, \eta) \mapsto (\lambda, \eta)$ on elements. Prove that r is a covering morphism and that the induced function $\pi_0 Z \to \pi_0 L \times_{\pi_0 G} \pi_0 H$ is surjective. Explain how Z derives from an action of the groupoid $L \times H$.

2. Continuing the notation of Exercise 1, let $Q = L \times_G H$ be the pullback of f and p. Define $\varphi : Q \to Z$ by $(l, y) \mapsto (l, 1_{fl}, y)$ on objects, $(\lambda, \eta) \mapsto (1, \lambda, \eta)$ on elements. Prove that φ is full and faithful and that the induced morphism $\pi_0\varphi : \pi_0 Q \to \pi_0 Z$ is injective. Prove that $\pi_0\varphi$ is surjective, and hence that φ is a homotopy equivalence, if and only if for each object (l, γ, y) of Z there are elements $\lambda : l \to l'$ in L, $\eta : y' \to y$ in H such that $fl' = py'$ and $\gamma = p\eta + f\lambda$. Use this result, Exercise 1, and the exact sequence of a fibration to deduce 10.7.6.
3. Let G be a group and let L and H be subgroups of G. For $\gamma \in G$ let $\kappa_\gamma : H \cap \gamma^{-1}L\gamma \to L \times H$ send $\eta \mapsto (\gamma\eta\gamma^{-1}, \eta)$. By considering the exact sequence at the base point $(1, \gamma, 1)$ of the homotopy pullback of the inclusions $L \to G$, $H \to G$, prove that G is bijective with the disjoint union of the cosets $(L \times H)/\operatorname{Im}\kappa_\gamma$ for γ in a set of representatives of the double cosets $L\gamma H$. Hence show that if G is finite and $|G|$ denotes the order of G then

$$|G| = |L||H|(\Sigma_\gamma |H \cap \gamma^{-1}L\gamma|^{-1}).$$

4. Let L and H be subgroups of the group G such that $G = LH$. Let $\gamma \in L \cap H$. Prove that there is an exact sequence

$$1 \to C_{L\cap H}(\gamma) \to C_L(\gamma) \times C_H(\gamma) \to C_G(\gamma) \to [L \cap H] \to [L] \times_{[G]} [H] \to 1$$

where $C_G(\gamma)$ is the centraliser of γ in G and $[G]$ is the set of conjugacy classes of G. [Let the given groups act on their underlying sets by conjugation and consider the induced morphisms of groupoids $L \rtimes L \to G \rtimes G \leftarrow H \rtimes H$.]
5. Let $f : L \to G$ be a morphism of groupoids and let $P_f = Z(f, 1_G)$. Define $i : L \to P_f$ by $l \mapsto (l, 1_{fl}, fl)$ on objects and $\lambda \mapsto (\lambda, f\lambda)$ on elements. Prove that i is a homotopy equivalence. Define $p : P_f \to G$ by $(l, \gamma, y) \mapsto y$ on objects and $(\lambda, \eta) \mapsto \eta$ on elements. Prove that $pi = f$ and that p is a fibration of groupoids. Interpret the exact sequence of the fibration p in the case f is the inclusion of a subgroup of the group G.
6. Let $\mathsf{Set}_\bullet$ be the category of pointed sets: thus an object of $\mathsf{Set}_\bullet$ is a pair (X, x) where $x \in X$, and a morphism $(X, x) \to (Y, y)$ is a pair (f, x) where $f : X \to Y$ is a function and $f(x) = y$. Let $p : \mathsf{Set}_\bullet \to \mathsf{Set}$ be the 'forgetful' functor which on objects sends $(X, x) \mapsto X$ and on morphisms sends $(f, x) \mapsto f$. Prove that p is a covering morphism. Let $q : \widetilde{G} \to G$ be any covering morphism of groupoids. Prove that there is a functor $q' : G \to \mathsf{Set}$ and an isomorphism $r : \widetilde{G} \to G \times_{\mathsf{Set}} \mathsf{Set}_\bullet$ such that $p\underline{r} = q$.

10.8 Applications to subgroup theorems in group theory

We give some other applications of covering spaces to what are called *subgroup theorems*. The general problem is the following. *Let* H *be a subgroup of the group* G *and suppose that information on* G *is given: deduce useful information on* H. The classical result of this type is that a subgroup of a free group is also a free group.

Covering morphisms give a method for dealing with this kind of problem. For if H is a subgroup of the group G, then H is a vertex group of the covering morphism $p : \widetilde{G} \to G$ determined by H. If the information on G can be lifted to information on $\widetilde{G}$, then the problem may be changed into one of obtaining information on a vertex group of a groupoid from information on the groupoid itself. Before going ahead with the theory, we work out a specific example.

For the rest of this section we use multiplicative notation for the composition in groups and in groupoids.

EXAMPLE

1. Let F be the free group on the two elements x and y. Let H be the normal subgroup of F generated by the elements x^3, y^2, and $xyxy$. The quotient group F/H is well known to be the permutation group S_3 on the set $\{1, 2, 3\}$, where x represents the cyclic permutation (123) and y represents the transposition (23). Thus H is the kernel of the quotient morphism $q : F \to S_3$. The directed graph $X = \{x, y\}$ which generates F can be pictured as the wedge of two circles. Fig. 10.6(i) (do not pay attention to the dots for the moment) is a picture of what is called the *Cayley graph* of the group S_3 for the generators x, y. What it shows is how the generators act on the elements of the group. So an arrow labelled x is an arrow $w \to xw$, and an arrow labelled y is an arrow $w \to yw$, for $w \in S_3$. We call this graph $\widetilde{X}$. Then $\widetilde{X} = p^{-1}[X]$ where $p : \widetilde{F} \to F$ is the covering morphism determined by the subgroup H of F, so that $\widetilde{F} = S_3 \rtimes F$, where S_3 is here taken as a set on which F acts.

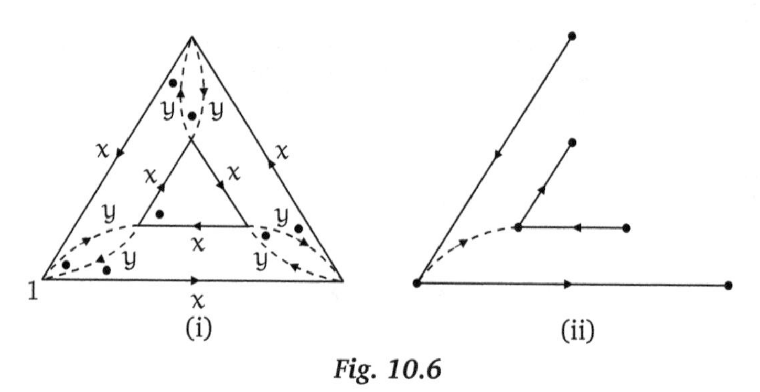

Fig. 10.6

We shall prove below that $\widetilde{F}$ is the free groupoid on the graph $\widetilde{X}$. Hence we can use the methods of section 8.2 to obtain a basis for the vertex group $\widetilde{F}(1)$ of $\widetilde{F}$, a vertex group which is isomorphic to H. We choose a maximal tree T in $\widetilde{X}$, for example that given by Fig. 10.6(ii). This tree determines a retraction $r : \widetilde{F} \to \widetilde{F}(1)$, where we may identify $\widetilde{F}(1)$ and H, and the basis for H consists of the elements of $r[\Delta]$ where Δ consists of the edges of $\widetilde{X}$ not in T. In this instance the basis for H is then given by the seven elements

$$\mathbf{y}y,\ y^{-1}x\mathbf{y}x,\ x^{-1}\mathbf{y}x^{-1}y,\ x\mathbf{x}x,\ y^{-1}x\mathbf{x}xy,\ x\mathbf{y}xy,\ y^{-1}x^{-1}\mathbf{y}x^{-1}$$

where in the above the elements which come from Δ are written in bold type.

Another interesting feature which can be drawn from Fig. 10.6(i) is that of an *identity among relations*. Write $r = x^3$, $s = y^2$, $t = xyxy$. These are elements of the free group F. The following word w in F, in which v^u means $u^{-1}vu$,

$$w = t(s^{-1})^{xy}t^{y^{-1}xy}(r^{-1})^{y}s^{-1}(s^{-1})^{x^{-1}y^{-1}}t^{x}$$

represents a path in $\widetilde{X}$ which starts from the base point 1, then goes out to a base point, as represented by a dot, of a 2-cell of our planar diagram of $\widetilde{X}$, goes once round the 2-cell, and then returns to the base point 1, before traversing another such path, till each 2-cell has been traversed once. A direct check shows that $w = r$, so that $wr^{-1} = 1$ in F. This word wr^{-1} in r, s, t and their inverses and conjugates is called an *identity among the relations* for the presentation $\langle x, y : r, s, t\rangle$ of the group S_3. These ideas are the start of a 2-dimensional theory corresponding to that of free groups and fundamental groupoids. For more information, see [BH82] and [Bro70].

Now we return to subgroup theorems. The key result for our purposes is the following, which links the methods of universal groupoids of chapter 8 with those of covering morphisms. We use the notion of pullback from the last section.

10.8.1 *Let $f : L \to G$ be a universal groupoid morphism, and let $p : \widetilde{G} \to G$ be a covering morphism. Consider the pullback diagram*

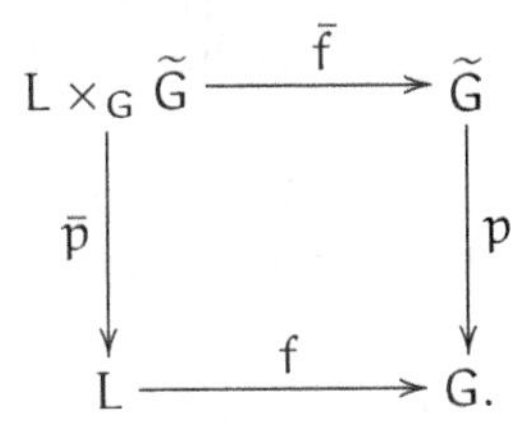

Then $\bar{f}$ is universal.

Proof We use the description of universal groupoid morphisms given in section 8.1. Let h be a non-identity element of $\widetilde{G}$. We prove that h has a unique representation as a product $h = (\bar{f}q_n)\cdots(\bar{f}q_1)$ where the q_i are non-identities in $Q = L \times_G \widetilde{G}$ and no product $q_{i+1}q_i$ exists in Q. Let $g = ph$. Since f is universal, and ph is not an identity of G, there is a unique representation $ph = (fl_n)\cdots(fl_1)$ where the l_i are non-identities in L and no product $l_{i+1}l_i$ exists in L. Hence we can uniquely write $h = h_n \cdots h_1$ where $ph_i = fl_i$. Setting $q_i = (l_i, h_i)$ gives the required representation.

We still have to prove this representation is unique. Suppose also $h = (\bar{f}q'_m)\cdots(\bar{f}q'_1)$, where the q'_i satisfy conditions similar to those for the q_i. Since $\bar{p}$ is a covering morphism, the $\bar{p}q'_i$ are not identities in L. Also the product $(\bar{f}q'_{i+1})(\bar{f}q'_i)$ exists in H, and therefore the product $(\bar{p}q'_{i+1})(\bar{p}q'_i)$ does not exist in L. Therefore $m = n$ and $\bar{p}q'_i = l_i$, since f is universal. It now follows that $q'_i = q_i$ for all i. Hence $\bar{f}$ is universal. □

10.8.1 *(Corollary 1) Let* $p : H \to G$ *be a covering morphism and suppose that* G *is the free groupoid on a graph* X*. Then* H *is the free groupoid on the graph* $p^{-1}[X]$.

Proof Let $\varphi : D(X) \to X$ be the dispersion of the graph X (see section 8.2). According to 8.2.1(a), the induced morphism $\bar{\varphi} : F(D(X)) \to G$ is strictly universal. Let Q be the pullback $F(D(X)) \times_G H$. It is easy to check that Q may be identified with $F(D(\widetilde{X}))$ where $\widetilde{X}$ has one element $\tilde{x}$ for each element of $p^{-1}[X]$. By 10.8.1, $\bar{f} : Q \to H$ is universal. Hence H is free on $p^{-1}[X]$. □

10.8.1 *(Corollary 2) Let the groupoid* G *be given as the free product* $G_1 * \cdots * G_n$ *of groupoids* $G_1, \ldots, G_n$*. Let* $p : H \to G$ *be a covering morphism. Then the groupoid* H *is the free product of the groupoids* $p^{-1}[G_1], \ldots, p^{-1}[G_n]$.

Proof This again is a simple consequence of 10.8.1, since a free product may be described in terms of strictly universal morphisms, as in section 8.1. □

Now we can give our first subgroup theorem. Recall that the index $[G : L]$ of a subgroup L of a group G is the number of cosets of L in G.

10.8.2 (Nielsen-Schreier theorem) *A subgroup* L *of a free group* G *is itself a free group. Further, if* G *is of rank* r *and* L *is of index* i *in* G*, then* L *is of rank* $l = ri - i + 1$*. Hence*

$$[G : L] = \frac{l-1}{r-1}.$$

Proof Let X be a basis of G. Let $p : H \to G$ be the covering morphism determined by the subgroup L of G. By 10.8.1 (Corollary 1), H is the free groupoid on $p^{-1}[X]$. By 8.2.3, the vertex groups of H are also free groups.

This proves the first part of the theorem. The second part follows also from 8.2.3, since if X has r elements, then $p^{-1}[X]$ has ri elements and i vertices. □

Note also that the method of proof of 8.2.3, which involves choosing a maximal tree in the generating graph $p^{-1}[X]$, also gives a method of writing down generators of the characteristic group L. This is essentially the procedure we followed in our example which involved the symmetric group S_3.

A more elaborate analysis gives another famous theorem of group theory.

10.8.3 (Kurosch theorem) *Let the group* G *be given as a free product* $*_{\lambda\in\Lambda}G^{\lambda}$, *where the* G^{λ} *are subgroups of* G. *Let* L *be a subgroup of* G. *Then* L *may be written as a free product*

$$(*_{\lambda,\mu}L^{\lambda,\mu}) * F$$

with the following properties:

(a) *each* $L^{\lambda,\mu}$ $(\lambda \in \Lambda, \mu \in M^{\lambda})$ *is of the form* $L \cap x_{\lambda\mu}^{-1}G^{\lambda}x_{\lambda\mu}$, *where, as* μ *varies in* M^{λ}, $x_{\lambda\mu}$ *runs through a (suitably chosen) set of representatives of the double cosets* $G^{\lambda}xL$ *of* L *and* G^{λ} *in* G;

(b) F *is a free group; if* Λ *is finite with* l *elements, and* L *has finite index* i *in* G, *then* F *has rank* $li-i-m+1$ *where* m *is the total number of double cosets* $G^{\lambda}xL$ $(\lambda \in \Lambda)$.

Proof Let $p : H \to G$ be the covering morphism determined by the subgroup L of G. If 1 denotes the coset L of G/L, then p maps the object group $H(1)$ isomorphically to L. According to 10.8.1 (Corollary 2), we can write H as a free product $*_{\lambda}H^{\lambda}$ where $H^{\lambda} = p^{-1}[G^{\lambda}]$. The remainder of the proof is similar to that of 9.1.8; in that result, we were amalgamating two connected groupoids, and identifying the vertex group of the result. Here we are amalgamating a number of possibly non-connected groupoids, but the amalgamation is only by identifying objects.

For each λ let $H^{\lambda\mu}$ $(\mu \in M^{\lambda})$ be the components of H^{λ}. Then $H^{\lambda} = *_{\mu}H^{\lambda\mu}$ and therefore $H = *_{\lambda,\mu}H^{\lambda\mu}$. Now $H^{\lambda\mu}$ may be written $K^{\lambda\mu} * T^{\lambda\mu}$ where $K^{\lambda\mu}$ is an arbitrary vertex group of $H^{\lambda\mu}$ and $T^{\lambda\mu}$ is a wide tree groupoid in $H^{\lambda\mu}$. Hence $H = K * T$ where $K = *_{\lambda\mu}K^{\lambda\mu}$ is totally disconnected and $T = *_{\lambda\mu}T^{\lambda\mu}$ is free. We can write $T = T' * F'$ where T' is a tree groupoid generated by elements from various $T^{\lambda\mu}$ and F' is a free group at the object 1.

We now argue as in the proof of 8.2.3. Consider the diagram

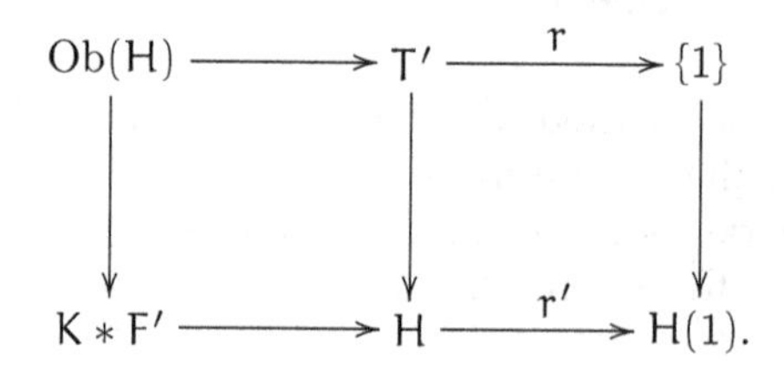

The left-hand square of inclusions is a pushout, and the right-hand square, where r and r' are retractions determined by T', is also a pushout by 6.7.3. Hence the composite $K * F' \to H(1)$ is universal. Let $L^{\lambda\mu} = pr'[K^{\lambda\mu}]$, $F = p[F']$. Then

$$L = p[H(1)] = (*_{\lambda\mu} L^{\lambda\mu}) * F.$$

We now have to verify the stated description of $L^{\lambda\mu}$. By 10.7.6 (Corollary 1) the components of H^{λ} correspond to the distinct double cosets $G^{\lambda}xL$. There is a unique element $h_{\lambda\mu}$ of T' from the object 1 of H to the unique object of $K^{\lambda\mu}$, and $x_{\lambda\mu} = ph_{\lambda\mu}$ lies in the double coset $G^{\lambda}xL$ corresponding to the component $H^{\lambda\mu}$. Hence

$$L^{\lambda\mu} = pr'[K^{\lambda\mu}] = x_{\lambda\mu}^{-1} p[K^{\lambda\mu}] x_{\lambda\mu}.$$

But $p[K^{\lambda\mu}]$ is the stabiliser in G^{λ} of the coset $x_{\lambda\mu}L$, i.e.

$$p[K^{\lambda\mu}] = G^{\lambda} \cap x_{\lambda\mu} L x_{\lambda\mu}^{-1},$$

which gives immediately the stated formula (a) for $L^{\lambda\mu}$.

In the finite case H^{λ} has i objects and a finite number m_{λ} of components. So $*_{\mu} T^{\lambda\mu}$ is freely generated by $i - m_{\lambda}$ elements, and hence T is freely generated by $li - m$ elements, where m is the sum of the m_{λ}. Hence F', and so also F, is freely generated by $li - m - (i-1)$ elements. □

The above proofs are taken from [Hig05]. It should be emphasised that the proofs follow some traditional lines (compare with [Mas67], [SW79]) but the reformulation in terms of covering groupoids allows for a more geometric view of the algebra, while at the same time avoiding using topological information, for example that a covering space of a CW-complex is also a CW-complex. Once again we can see that groupoids model the geometry more closely than do groups. A further advantage of the groupoid proofs is that they have analogues for topological groups: see [BH75] and [Nic81].

The groupoid methods have also led to more general results even in the abstract case. A notable example of this is the following result for whose proof we refer the reader to [Hig66] and [Hig05].

10.8.4 (Higgins' theorem) *Let* G, B *be groups with free decompositions* $G = *_\lambda G^\lambda$, $B = *_\lambda B^\lambda$ $(\lambda \in \Lambda)$, *and let* $\psi : G \to B$ *be a group morphism such that* $\psi[G^\lambda] = B^\lambda$ *for all* $\lambda \in \Lambda$. *Let* H *be a subgroup of* G *such that* $\psi[H] = B$. *Then* H *has a decomposition* $H = *_\lambda H^\lambda$ *such that* $\psi[H^\lambda] = B^\lambda$ *for all* $\lambda \in \Lambda$.

The following corollary was first given a topological proof by Stallings (for an account, see [SW79]).

10.8.4 (Corollary: Grushko's theorem) *Let* B *be a group with a free decomposition* $B = *_\lambda B^\lambda$, *and let* F *be a free group. Let* $\varphi : F \to B$ *be an epimorphism. Then* F *has a decomposition* $F = *_\lambda F^\lambda$ *such that* $\varphi[F^\lambda] = B^\lambda$ *for all* λ.

An interesting consequence of Grushko's theorem is that if B is a free group of rank n and X is a subset of B with n elements which generates B, then X generates B freely. We leave the proof to the reader. Another proof of this result uses Nielsen transformations (see [LS77]).

Braun in [Bra04] has proved a conjecture of Higgins which combines the Kurosch Theorem and the Higgins Theorem as follows:

10.8.5 (Higgins' conjecture) *Let* $\Theta\colon G = \prod^*_{\lambda \in \Lambda} G_\lambda \to B = \prod^*_{\lambda \in \Lambda} B_\lambda$ *be a group homomorphism such that* $G_\lambda \Theta = B_\lambda$ *for all* $\lambda \in \Lambda$. *Let* $H \subseteq G$ *be a subgroup such that* $H\Theta = B$. *Then* $H = \prod^*_{\lambda \in \Lambda} H_\lambda$ *such that* $H_\lambda \Theta = B_\lambda$ *where* $H_\lambda = \prod^*_{x_\lambda} (H \cap G_\lambda^{x_\lambda}) * F_\lambda$ *such that* $x_\lambda \Theta = 1$ *for all* x_λ, *and for each* λ *the* x_λ *runs through a suitable set of representatives of double cosets* $G_\lambda x H$ *such that* $G_\lambda H$ *is represented by* 1. *Furthermore, the* F_λ *are free.*

There is another proof of 10.8.1 which also yields a result for the case that p is a fibration of groupoids. This proof uses the sophisticated statement that the pullback functor

$$p^* : \mathsf{Grpd}/G \to \mathsf{Grpd}/\widetilde{G}$$

has a right adjoint p_* if and only if p is a fibration of groupoids. An application of this to the case that p is an epimorphism of groups is given in [BH87] (see also [HK88]). However the study of adjoint functors, and the important result that if Γ and Δ are functors such that Γ is a left adjoint of Δ then Γ preserves colimits and Δ preserves limits, would take us too far afield.

EXERCISES

1. Let F be the free group on two generators x and y, and let G be the dihedral group with generators x, y and relations $x^4 = y^2 = xyxy = 1$. Let $q : F \to G$ be the

quotient morphism given by the identity on the generators, and let H be the kernel of q. Find a free basis for H. Can you formulate and prove a generalisation of this for the general dihedral group?
2. Carry out a similar exercise to the previous one but for the group $\mathbb{Z}_2 \times \mathbb{Z}_2$ with generators x, y and relations $x^2 = y^2 = xyx^{-1}y^{-1} = 1$. For this presentation, find an identity among the relations, in an analogous manner to that found for the group S_3.
3. Compare the method of proof of the Nielsen-Schreier theorem given here in terms of groupoids with proofs found in books on group theory using Schreier transversals. Indeed, show that a choice of Schreier transversal is equivalent to the choice of maximal tree as explained here.
4. Let the group G be the free product $\mathbb{Z}_3 * \mathbb{Z}_2$ of cyclic groups of order 3 and 2 respectively, with generators x, y say. Let $q : G \to S_3$ be the morphism which is the identity on the generators, where S_3 has the generators given earlier. Let H be the kernel of q. Find a free decomposition of H in accordance with the Kurosch theorem.
5. Let B be a free group of rank n and let X be a generating subset of B such that X has n elements. Prove that X generates B freely.
6. Read from a category theory text the theory of adjoint functors as far as the result on preservation of limits and colimits referred to above. Discuss the existence of some left and right adjoints to some standard functors, such as (i) the functor $\mathsf{Top} \to \mathsf{Set}$ which sends a topological space to its underlying set, (ii) the functor $\mathsf{Grpd} \to \mathsf{Set}$ which sends a groupoid to its object set.

NOTES

A number of books on homotopy theory or algebraic topology show the link between covering spaces and fibrations. A different approach to covering spaces which does not use paths is given in [Che46]. The relation of covering spaces to other areas of mathematics is shown there and in [DD77], [Mag76], [BG82].

Covering spaces form one of the older parts of algebraic topology, partly because of the link with Riemann surfaces. By contrast, the development of covering groupoids is in the latter half of the 20th century. Higgins in [Hig64] introduces the groupoid $\mathrm{Tr}(G, C)$ for C a subgroup of the group G, and mentions the generalisation to groupoids. In fact, these methods were suggested to him by a reading of [HW60] on covering spaces (private communication). Higgins defined covering groupoids, but assuming connectedness, in lecture notes and in [Hig05]. An equivalent definition was independently given in [GZ67] which also shows the connection between covering groupoids of a group G and functors $G \to \mathsf{Set}$. However the similar construction of a groupoid $\widetilde{G}$ for the case of a group operating on a set appears in [Rei32], (without any applications) and in [Ehr65, p. 49-50].

Indeed the collected works of Ehresmann ([Ehr80]) show his long interest in the idea that mathematical structure can be seen as a category operating on a set, or, equivalently, as what he calls a *hypermorphism* of categories.

An earlier groupoid proof of the subgroup theorems is due to [Has60].

Another interesting relation is between graphs, groups and surfaces—see [Big74], [Big84] and [Whi84]. In particular, the Cayley graph of a group G with a set X of generators is seen as the 1-skeleton of the universal cover $p : \widetilde{K} \to K$ of a CW-complex such that K^0 is a single point, K^1 has 1-cells in one-one correspondence with the elements of X, and K has fundamental group G. Thus the Cayley graph $\widetilde{K}^1$ is a directed graph without loops whose edges are labelled via the map with the elements of X.

Covering spaces have an important generalisation to *branched covering spaces*, or *covering spaces with singularities*. These occur in the theory of Riemann surfaces, in knot theory, and in low-dimensional topology. For more information, see [vE86], [Fox57], [Hun82], [Lin80], [Whi84].

It is not hard to prove that a connected covering space of a connected topological group has the structure of a topological group. The non-connected case is more tricky. This was studied by [Tay54], and was re-examined by [BM94] using the equivalence of categories of section 10.6 in an essential way.

Chapter 11

Orbit spaces, orbit groupoids

In this chapter we give an introduction to the notion of *orbit space by the action of a discrete group*, and the corresponding notion of *orbit groupoid* of the action of a group on a groupoid. This gives another example of the utility of the groupoid viewpoint: under reasonable conditions, the fundamental groupoid of an orbit space X/G is isomorphic to the orbit groupoid $(\pi X)//G$ of the induced action on the fundamental groupoid. This seems the best possible result! The proof, as for the Van Kampen theorem in previous chapters, is by verifying the appropriate universal property. We then give general methods of calculation of orbit groupoids to determine some topological examples.

11.1 Groups acting on spaces

In this section we show some of the theory of a *group acting on a space* and the associated notion of *orbit space*. It is difficult to exaggerate the importance of these ideas. The idea of a group action is related to basic intuitions on the notion of symmetry, and specific groups occur naturally as acting on spaces. For example, in chapter 5 we discussed various groups of isometries on the spaces $\mathbb{K}^n$ for $\mathbb{K}$ the reals, complex numbers, or quaternions. Let G be such a group of isometries. Then each element g of G is a function $g : \mathbb{K}^n \to \mathbb{K}^n$ and so if $x \in \mathbb{K}^n$ then $g(x)$ is defined. It is convenient to write $g \cdot x$ instead of $g(x)$ as the basic notation for a group action. The reason for the changed notation is that for a general action of a group G on $\mathbb{K}^n$ we may have distinct elements g, h of G such that for all x in $\mathbb{K}^n$

$g \cdot x = h \cdot x$. Thus in general the elements of G *act* as functions on $\mathbb{K}^n$ but are not themselves functions on $\mathbb{K}^n$. This distinction is important.

Associated with an action of a group G on a set X will be an *orbit set* written X/G, whose elements are the *orbits* of the action, namely the classes of X under the equivalence relation $x \sim y$ if and only if there is a g in G such that $y = g \cdot x$. Thus the orbits are the sets in X swept out by the action of G. The very word *orbit* calls to mind the origin of the word in celestial mechanics, where the group acting is the additive group $\mathbb{R}$ of real numbers, so that an element $t \in \mathbb{R}$ acts on the position of a planet at time s to give the position at time $t + s$. The *orbit* of the planet is then the totality of all its positions throughout time. Thus the theory of group actions on a space is also part of the study of *dynamical systems*.

In this section we give an introduction to these ideas. We show some of the theory of a group G acting on a topological space X, and describe the *orbit topological space*, which is written X/G.

There arises the problem of relating topological invariants of the orbit space X/G to those of X and the group action. In particular, it is a complicated and interesting question to find, if at all possible, relations between the fundamental groups and groupoids of X and X/G. This we shall do for a particular family of actions which arise commonly, namely the *discontinuous actions.* The resulting theory generalises that of regular covering spaces, and has a number of important applications. A useful special case of a discontinuous action is the action of a finite group on a Hausdorff space (see below); there are in the literature many interesting cases of discontinuous actions of infinite groups (see [Bea83]).

We now come to formal definitions.

Let G be a group, with its group operation written as multiplication, and let X be a set. An *action* of G on X is a function $G \times X \to X$, written $(g, x) \mapsto g \cdot x$, satisfying the following properties for all x in X and g, h in G:

11.1.1 (i) $1 \cdot x = x$,
(ii) $g \cdot (h \cdot x) = (gh) \cdot x$.

Thus the first rule says that the identity of G acts as identity, and the second rule says that two elements of G, acting successively, act as the product of the two elements.

There are some standard notions associated with such an action. First, an equivalence relation is defined on X by $x \sim y$ if and only if there is an element g of G such that $y = g \cdot x$. This is an equivalence relation. Reflexivity follows since G has an identity. Symmetry follows from the existence of inverses in G, using 11.1.1(i), (ii). Transitivity follows from the product of two elements in G being in G. The equivalence classes under this relation are the *orbits* of the action. The set of these orbits is written X/G.

Suppose given an action of the group G on the set X. If $x \in X$, then the *group of stability of* x is the subgroup of G

$$G_x = \{g \in G : g \cdot x = x\}.$$

The elements of G_x are said to *stabilise* x, that is, they leave x fixed by their action. If G_x is the whole of G, then x is said to be a *fixed point* of the action. The set of fixed points of the action is often written X^G. The action is said to be *free* if all groups of stability are trivial.

Another useful condition is the notion of *effective* action of a group. This requires that if $g, h \in G$ and for all x in X, $g \cdot x = h \cdot x$, then $g = h$. In this case the elements of G are entirely determined by their action on X.

We now turn to the topological situation. Let X be a topological space, and let G be a group. An *action* of G on X is again a function $G \times X \to X$ with the same properties as given in 11.1.1, but with the additional condition that when G is given the discrete topology then the function $(g, x) \mapsto g \cdot x$ is continuous. This amounts to the same as saying that for all $g \in G$, the function $g_\# : x \mapsto g \cdot x$ is continuous. Note that $g_\#$ is a bijection with inverse $(g^{-1})_\#$, and since these two functions are continuous, each is a homeomorphism.

Let X/G be the set of orbits of the action and let $p : X \to X/G$ be the *quotient map*, which assigns to each x in X its orbit. For convenience we will write the orbit of x under the action as $\bar{x}$. So the defining property is that $\bar{x} = \bar{y}$ if and only if there is a g in G such that $y = g \cdot x$. Now a topology has been given for X. We therefore give the orbit space X/G the identification topology with respect to the map p. This topology will always be assumed in what follows. The first result on this topology, and one which is used a lot, is as follows.

11.1.2 *The quotient map* $p : X \to X/G$ *is an open map.*

Proof Let U be an open set of X. For each $g \in G$ the set $g \cdot U$, by which is meant the set of $g \cdot x$ for all x in U, is also an open set of X, since $g_\#$ is a homeomorphism, and $g \cdot U = g_\#[U]$. But

$$p^{-1}p[U] = \bigcup_{g \in G} g \cdot U.$$

Since the union of open sets is open, it follows that $p^{-1}p[U]$ is open, and hence $p[U]$ is open. □

Definition An action of the group G on the space X is called *discontinuous* if the stabiliser of each point of X is finite, and each point x in X has a neighbourhood V_x such that any element g of G not in the stabiliser of x satisfies $V_x \cap g \cdot V_x = \varnothing$.

Suppose G acts discontinuously on the space X. For each x in X choose such an open neighbourhood V_x of x. Since the stabiliser G_x of x is finite, the set

$$U_x = \bigcap\{g \cdot V_x : g \in G_x\}$$

is open; it contains x since the elements of G_x stabilise x. Also if $g \in G_x$ then $g \cdot U_x = U_x$. We say U_x is *invariant* under the action of the group G_x. On the other hand, if $h \notin G_x$ then

$$(h \cdot U_x) \cap U_x \subseteq (h \cdot V_x) \cap V_x = \varnothing.$$

An open neighbourhood U of x which satisfies $(h \cdot U) \cap U = \varnothing$ for $h \notin G_x$ and is invariant under the action of G_x is called a *canonical neighbourhood* of x. Note that any neighbourhood N of x contains a canonical neighbourhood: the proof is obtained by replacing V_x in the above by $N \cap V_x$. The image in X/G of a canonical neighbourhood U of x is written $\overline{U}$ and called a *canonical neighbourhood* of $\bar{x}$.

In order to have available our main example of a discontinuous action, we prove:

11.1.3 *An action of a finite group on a Hausdorff space is discontinuous.*

Proof Let G be a finite group acting on the Hausdorff space X. Then the stabiliser of each point of X is a subgroup of G and so is finite.

Let $x \in X$. Let $x_0, x_1, \dots, x_n$ be the distinct points of the orbit of x, with $x_0 = x$. Suppose $x_i = g_i \cdot x$, $g_i \in G$, $i = 1, \dots, n$, and set $g_0 = 1$. Since X is Hausdorff, we can find pairwise disjoint open neighbourhoods N_i of x_i, $i = 0, \dots, n$. Let

$$N = \bigcap\{g_i^{-1} \cdot N_i : i = 0, 1, \dots n\}.$$

Then N is an open neighbourhood of x. Also, if $g \in G$ does not belong to the stabiliser G_x of x, then for some $j = 1, \dots, n$, $g \cdot x = x_j$, whence $g \cdot N \subseteq N_j$. Hence $N \cap g \cdot N = \varnothing$, and the action is discontinuous. □

Our main result in general topology on discontinuous actions is the following.

11.1.4 *If the group G acts discontinuously on the Hausdorff space X, then the quotient map $p : X \to X/G$ has the path lifting property: that is, if $\bar{a} : \mathbb{I} \to X/G$ is a path in X/G and x_0 is a point of X such that $p(x_0) = \bar{a}(0)$, then there is a path $a : \mathbb{I} \to X$ such that $pa = \bar{a}$ and $a(0) = x_0$.*

Proof If there is a lift a of $\bar{a}$ then there is an element g of G such that $g \cdot a(0) = x_0$ and so $g \cdot a$ is a lift of $\bar{a}$ starting at x_0. So we may ignore x_0 in what follows.

Since the action is discontinuous, each point $\bar{x}$ of X/G has a canonical neighbourhood. By the Lebesgue covering lemma [3.6.4], there is a subdivision

$$\bar{a} = \bar{a}_n + \cdots + \bar{a}_1$$

of $\bar{a}$ such that the image of each $\bar{a}_i$ is contained in a canonical neighbourhood. So if the path lifting property holds for each canonical neighbourhood in X/G, then it holds for X/G.

Now $p : X \to X/G$ is an open map. Hence for all x in X the restriction $p_x : U_x \to \overline{U}_x$ of p to a canonical neighbourhood U_x is also open, and hence is an identification map. So we can identify $\overline{U}_x$ with the orbit space $(U_x)/G_x$. The key point in this case is that the group G_x is finite.

Thus it is sufficient to prove the path lifting property for the case of the action of a *finite* group G, and this we do by induction on the order of G. That is, we assume that path lifting holds for any action of any proper subgroup of G on a Hausdorff space, and we prove that path lifting holds for the action of G. The case $|G| = 1$ is trivial.

Again let $\bar{a}$ be as in the proposition, and we are assuming G is finite. Let F be the set of fixed points of the action. Then F is the intersection for all $g \in G$ of the sets $X^g = \{x \in X : g \cdot x = x\}$. Since X is Hausdorff, the set X^g is closed in X, and hence F is closed in X. So $p[F]$ is closed, since

$$p^{-1}p[F] = \bigcup_{g \in G} g \cdot F = F.$$

Let A be the subspace of $\mathbb{I}$ of points t such that $\bar{a}(t)$ belongs to $p[F]$; that is, $A = \bar{a}^{-1}p[F]$. Then A is closed.

The restriction of the quotient map p to $p' : F \to p[F]$ is a homeomorphism. So $\bar{a} \mid A$ has a unique lift to a map $a \mid A : A \to X$. So we have to show how to lift $\bar{a} \mid (\mathbb{I} \setminus A)$ to give a map $a \mid \mathbb{I} \setminus A$ and then show that the function $a : \mathbb{I} \to X$ defined by these two parts is continuous.

In order to construct $a \mid \mathbb{I} \setminus A$, we first assume $A = \{1\}$.

Let S be the set of $s \in \mathbb{I}$ such that $\bar{a} \mid [0, s]$ has a lift to a map a_s starting at x. Then S is non-empty, since $0 \in S$. Also S is an interval. Let $u = \sup S$. Suppose $u < 1$. Then there is a $y \in X \setminus F$ such that $p(y) = \bar{a}(u)$. Choose a canonical neighbourhood U of y. If $u > 0$, there is a $\delta > 0$ such that $\bar{a}[u - \delta, u + \delta] \subseteq \overline{U}$. Then $u - \delta \in S$ and so there is a lift $a_{u-\delta}$ on $[0, u - \delta]$. Also the stabiliser of y is a proper subgroup of G, since $y \notin F$, and so by the inductive assumption there is a lift of $\bar{a} \mid [u - \delta, u + \delta]$ to a path starting at $a_{u-\delta}(u - \delta)$. Hence we obtain a lift $a_{u+\delta}$, contradicting the definition of u.

We get a similar contradiction to the case $u = 0$ by replacing in the above $u - \delta$ by 0. It follows that $u = 1$. We are not quite finished because all we have thus ensured is that there is a lift on $[0, s]$ for each $0 \leqslant s < 1$, but this is not the same as saying that there is a lift on $[0, 1[$. We now prove that such a lift exists.

By the definition of u, and since $u = 1$, for each integer $n \geqslant 1$ there is a lift a^n of $\bar{a} \mid [0, 1 - n^{-1}]$. Also there is an element g_n of G such that

$$g_n \cdot a^{n+1}(1 - n^{-1}) = a^n(1 - n^{-1}).$$

Hence a^n and $g_n \cdot (a^{n+1} \mid [1 - n^{-1}, 1 - (n+1)^{-1}])$ define a lift of $\bar{a} \mid [0, 1 - (n+1)^{-1}]$. Starting with $n = 1$, and continuing in this way, gives a lift of $\bar{a} \mid [0, 1[$. This completes the construction of the lift on $\mathbb{I} \setminus A$ in the case $A = \{1\}$.

We now construct a lift of $\bar{a} \mid \mathbb{I} \setminus A$ in the general case. Since A is closed, $\mathbb{I} \setminus A$ is a union of disjoint open intervals each with end points in $\{0, 1\} \cup A$. So the construction of the lift is obtained by starting at the mid-point of any such interval and working backwards and forwards, using the case $A = \{1\}$, which we have already proved.

The given lift of $\bar{a} \mid A$ and the choice of lift of $\bar{a} \mid \mathbb{I} \setminus A$ together define a lift $a : \mathbb{I} \to X$ of $\bar{a}$ and it remains to prove that a is continuous.

Let $t \in \mathbb{I}$. If $t \notin A$, then a is continuous at t by construction. Suppose then $t \in A$ so that $y = a(t) \in F$. Let N be any neighbourhood of y. Then N contains a canonical neighbourhood U of y. If $g \in G$ then $g \cdot y = y$, and so U is invariant under the action of G. Hence $p^{-1}p[U] = U$. Since $\bar{a}$ is continuous, there is a neighbourhood M of t such that $\bar{a}[M] \subseteq \overline{U}$. Since $pa = \bar{a}$ it follows that $a[M] \subseteq p^{-1}[\overline{U}] = U$. This proves continuity of a, and the proof of the proposition is complete. □

In our subsequent results we shall use the path lifting property rather than the condition of the action being discontinuous.

EXERCISES

1. Let $\lambda \in \mathbb{R}$ and let the additive group $\mathbb{R}$ of real numbers act on the torus $T = \mathbb{S}^1 \times \mathbb{S}^1$ by

$$t \cdot (e^{2\pi i\theta}, e^{2\pi i\varphi}) = (e^{2\pi i(\theta + t)}, e^{2\pi i(\varphi + \lambda t)})$$

for $t, \theta, \varphi \in \mathbb{R}$. Prove that the orbit space has the indiscrete topology if and only if λ is irrational. [You may assume that the group generated by 1 and λ is dense in $\mathbb{R}$ if and only if λ is irrational.]

2. Let G be a group and let X be a G-space. Prove that the quotient map $p : X \to X/G$ has the following universal property: if Y is a space and $f : X \to Y$ is a map such that $f(g \cdot x) = f(x)$ for all $x \in X$ and $g \in G$, then there is a unique map $f^* : X/G \to Y$ such that $f^*p = f$.

3. Let X, Y, Z be G-spaces and let $f : X \to Z$, $h : Y \to Z$ be G-maps (i.e. $f(g \cdot x) = g \cdot f(x)$ for all $g \in G$ and $x \in X$, and similarly for h). Let $W = X \times_Z Y$ be the pullback. Prove that W becomes a G-space by the action $g \cdot (x, y) = (g \cdot x, g \cdot y)$. Prove also that if $Z = X/G$ and f is the quotient map, then W/G is homeomorphic to Y.

11.2 Groups acting on groupoids

Our aim now is to determine the fundamental groupoid of the orbit space X/G. In general it is difficult to say much. However we can give reasonable and useful conditions for which the question can be completely answered. From our point of view the result is also of interest in that our statement and proof use groupoids in a crucial way. This use could be overcome, but at the cost of complicating both the statement of the theorem and its proof.

In order to make the transition from the topology to the algebra, it is necessary to introduce the notion of a group acting on a groupoid.

Let G be a group and let Γ be a groupoid. We will write the group structure on G as multiplication, and the groupoid structure on Γ as addition. An *action* of G on Γ assigns to each $g \in G$ a morphism of groupoids $g_\# : \Gamma \to \Gamma$ with the properties that $1_\# = 1 : \Gamma \to \Gamma$, and if $g, h \in G$ then $(hg)_\# = h_\# g_\#$. If $g \in G$, $x \in \mathrm{Ob}(\Gamma)$, $a \in \Gamma$, then we write $g \cdot x$ for $g_\#(x)$, $g \cdot a$ for $g_\#(a)$. Thus the rules 11.1.1 apply also to this situation, as well as the laws **11.1.1** (iii) $g \cdot (a + b) = g \cdot a + g \cdot b$, and (iv) $g \cdot 0_x = 0_{g \cdot x}$ for all $g \in G$, $x \in \mathrm{Ob}(\Gamma)$, $a, b \in \Gamma$ such that $a + b$ is defined.

The action of G on Γ is *trivial* if $g_\# = 1$ for all g in G.

11.2.1 *Let G be a group acting on a groupoid Γ. An* orbit groupoid *of the action is a groupoid $\Gamma//G$ together with a morphism $p : \Gamma \to \Gamma//G$ such that:*
(a) *If $g \in G$, $\gamma \in \Gamma$, then $p(g \cdot \gamma) = p(\gamma)$.*
(b) *The morphism p is universal for* (a), *i.e. if $\varphi : \Gamma \to \Phi$ is a morphism of groupoids such that $\varphi(g \cdot \gamma) = \varphi(\gamma)$ for all $g \in G$, $\gamma \in \Gamma$, then there is a unique morphism $\varphi^* : \Gamma//G \to \Phi$ of groupoids such that $\varphi^* p = \varphi$.*
The morphism $p : \Gamma \to \Gamma//G$ is then called an orbit morphism.

The universal property (b) implies that $\Gamma//G$, if it exists, is unique up to a canonical isomorphism. At the moment we are not greatly concerned with proving any general statement about the *existence* of the orbit groupoid. One can argue that $\Gamma//G$ is obtained from Γ by imposing the relations $g \cdot \gamma = \gamma$ for all $g \in G$ and all $\gamma \in \Gamma$; however we have not yet explained quotients in this generality. We will later prove existence by giving a *construction* of $\Gamma//G$ which will be useful in interpreting our main theorem. But our next result will give conditions which ensure that the induced morphism $\pi X \to \pi(X/G)$ is an orbit morphism, and our proof will not *assume* general results on the existence of the orbit groupoid. The reason we can do this is that our proof directly verifies a universal property.

First we must point out that if the group G acts on the space X, then G acts on the fundamental groupoid πX, since each g in G acts as a homeomorphism of X and $g_{\#} : \pi X \to \pi X$ may be defined to be the induced morphism. This is one important advantage of groupoids over groups: by contrast, the group G acts on the fundamental group $\pi(X, x)$ only if x is a fixed point of the action.

Suppose now that G acts on the space X. Our purpose is to give conditions on the action which enable us to prove that

$$p_* : \pi X \to \pi(X/G)$$

determines an isomorphism $(\pi X)//G \to \pi(X/G)$, by verifying the universal property for p_*. We require the following conditions:

11.2.2 (a) *The projection* $p : X \to X/G$ *has the path lifting property: i.e. if* $\bar{a} : \mathbb{I} \to X/G$ *is a path, then there is a path* $a : \mathbb{I} \to X$ *such that* $pa = \bar{a}$.
(b) *If* $x \in X$, *then* x *has an open neighbourhood* U_x *such that*
(i) *if* $g \in G$ *does not belong to the stabiliser* G_x *of* x, *then* $U_x \cap (g \cdot U_x) = \varnothing$;
(ii) *if* a *and* b *are paths in* U_x *beginning at* x *and such that* pa *and* pb *are homotopic rel end points in* X/G, *then there is an element* $g \in G_x$ *such that* $g \cdot a$ *and* b *are homotopic in* X *rel end points.*

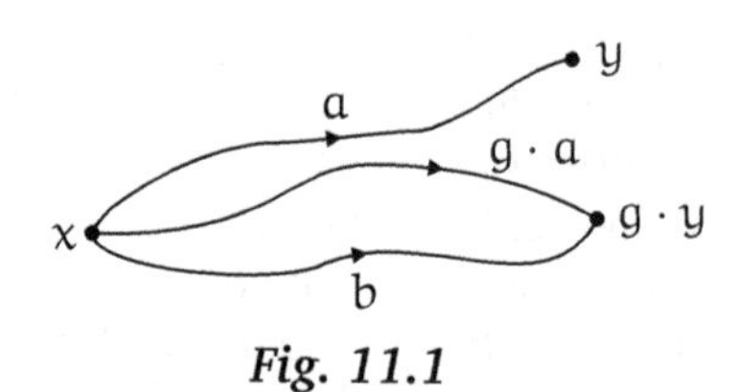

Fig. 11.1

For a discontinuous action, 11.2.2(b)(i) trivially holds, while 11.2.2(a) holds by virtue of 11.1.4. However, 11.2.2(b)(ii) is an extra condition. It does hold if X is semi-locally simply-connected, since then for sufficiently small U and x, $g \cdot y \in U$, any two paths in U from x to $g \cdot y$ are homotopic in X rel end points; so 11.2.2(b)(ii) is a reasonable condition to use in connection with covering space theory.

A neighbourhood U_x of $x \in X$ satisfying 11.2.2(a) and (b) will be called a *strong canonical neighbourhood* of x. The image $p[U_x]$ of U_x in X/G will be called a *strong canonical neighbourhood* of px.

11.2.3 *If the action of* G *on* X *satisfies* 11.2.2(a) *and* (b) *above, then the induced morphism* $p_* : \pi X \to \pi(X/G)$ *makes* $\pi(X/G)$ *the orbit groupoid of* πX *by the action of* G.

Proof Let $\varphi : \pi X \to \Phi$ be a morphism to a groupoid Φ such that $\varphi(g \cdot \gamma) = \varphi(\gamma)$ for all $\gamma \in \pi X$ and $g \in G$. We wish to construct a morphism $\varphi^* : \pi(X/G) \to \Phi$ such that $\varphi^* p = \varphi$.

Let $\bar{a}$ be a path in X/G. Then $\bar{a}$ lifts to a path a in X. Let $[b]$ denote the homotopy class rel end points of a path b. We prove that $\varphi[a]$ in Φ is independent of the choice of $\bar{a}$ in its homotopy class and of the choice of lift a; hence we can define $\varphi^*[\bar{a}]$ to be $\varphi[a]$.

Suppose given two homotopic paths $\bar{a}$ and $\bar{b}$ in X/G, with lifts a and b which without loss of generality we may assume start at the same point x in X. (If they do not start at the same point, then one of them may be translated by the action of G to start at the same point as the other.) Let $h : \mathbb{I} \times \mathbb{I} \to X/G$ be a homotopy rel end points $\bar{a} \simeq \bar{b}$. The method now is not to lift the homotopy h itself, but to lift pieces of a subdivision of h; it is here that the method differs from that used in the theory of covering spaces given in section 10.1.

Subdivide $\mathbb{I} \times \mathbb{I}$, by lines parallel to the axes, into small squares each of which is mapped by h into a strong canonical neighbourhood in X/G. This subdivision determines a sequence of homotopies $h_i : \bar{a}_{i-1} \simeq \bar{a}_i$, $i = 1, 2, \ldots, n$, say, where $\bar{a}_0 = \bar{a}$, $\bar{a}_n = \bar{b}$. Keep i fixed for the present. Each h_i is further expressed by the subdivision as a composite of homotopies h_{ij} $(j = 1, 2, \ldots, m)$ as shown in the following picture in which for convenience the boundaries of the h_{ij} are labelled:

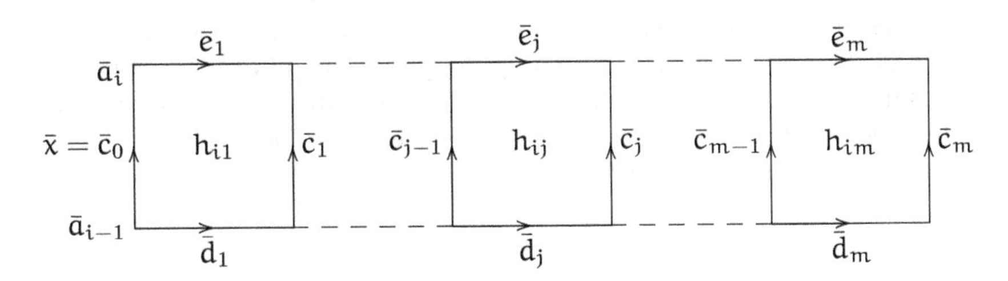

Choose lifts a_{i-1}, a_i of $\bar{a}_{i-1}$, $\bar{a}_i$ respectively; express a_{i-1} as a sum $a_{i-1} = d_m + \cdots + d_1$ and a_i as a sum $a_i = e_m + \cdots + e_1$ where d_j lifts $\bar{d}_j$ and e_j lifts $\bar{e}_j$. Choose for each j a lift c_j of $\bar{c}_j$ (with c_0 the constant path at x). For fixed j choose $f, g, h \in G$ such that $g \cdot d_j$ has the same initial point as c_{j-1} and the sums

$$f \cdot c_j + g \cdot d_j, \qquad h \cdot e_j + c_{j-1}$$

are defined. This is possible because of the boundary relations between the projections in X/G of the various paths.

Now our assumption 11.2.2(b)(ii) implies that there is an element $k \in G$ such that the following paths in X

$$k \cdot (f \cdot c_j + g \cdot d_j), \qquad h \cdot e_j + c_{j-1}$$

are homotopic rel end points in X. On applying φ to homotopy classes of paths in X and using equations such as $\varphi(g \cdot \gamma) = \varphi(\gamma)$ we find that

$$\begin{aligned}\varphi[e_j] + \varphi[c_{j-1}] &= \varphi[h \cdot e_j] + \varphi[c_{j-1}] \\ &= \varphi[h \cdot e_j + c_{j-1}] \\ &= \varphi[k \cdot (f \cdot c_j + g \cdot d_j)] \\ &= \varphi[k \cdot f \cdot c_j] + \varphi[g \cdot d_j] \\ &= \varphi[c_j] + \varphi[d_j].\end{aligned}$$

This proves that

$$\varphi[e_j] = \varphi[c_j] + \varphi[d_j] - \varphi[c_{j-1}].$$

It follows easily that

$$\varphi[a_{i-1}] = \varphi[a_i],$$

and hence by induction on i that $\varphi[a] = \varphi[b]$.

From this it follows that $\varphi^* : \pi(X/G) \to \Phi$ is a well defined function such that $\varphi^* p = \varphi$. The uniqueness of φ^* is clear since p_* is surjective on elements, by the path lifting property of 11.2.2(a). The proof that φ^* is a morphism is simple. This completes the proof of 11.2.3. □

In the next sections, we introduce some further constructions in the theory of groupoids and groups acting on groupoids, in order to interpret 11.2.3 in a manner suitable for calculations. Once again, we will find that an apparently abstract result involving a universal property can, when appropriately interpreted, lead to specific calculations.

Exercises

1. Let G and Γ be groupoids and let $w : \Gamma \to \mathrm{Ob}(G)$ be a morphism where $\mathrm{Ob}(G)$ is considered as a groupoid with identities only. An *action of* G *on* Γ *via* w is an assignment to each $g \in G(x, y)$ and $\gamma \in w^{-1}[a]$ an element $g \cdot \gamma \in w^{-1}[b]$ and with the usual rules: $h \cdot (g \cdot \gamma) = (hg) \cdot \gamma$; $1 \cdot \gamma = \gamma$; $g \cdot (\gamma + \delta) = g \cdot \gamma + g \cdot \delta$. In this case Γ is called a G-groupoid. Show how to define a category of G-groupoids so that this category is equivalent to the functor category $\mathsf{Fun}(G, \mathsf{Set})$.
2. Prove that if Γ is a G-groupoid via w, then $\pi_0\Gamma$ becomes a G-set via $\pi_0(w)$.

3. If Γ is a G-groupoid via w, then the action is *trivial* if for all $a, b \in \mathrm{Ob}(G)$, $g, h \in G(a, b)$ and $\gamma \in w^{-1}[a]$, we have $g \cdot \gamma = h \cdot \gamma$. Prove that the action is trivial if for all $a \in \mathrm{Ob}(G)$, the action of the group $G(a)$ on the groupoid $w^{-1}[a]$ is trivial. Prove also that Γ contains a unique maximal subgroupoid Γ^G on which G acts trivially. Give examples to show that Γ^G may be empty.
4. Continuing the previous exercise, define a G-*section* of w to be a morphism $s : \mathrm{Ob}(G) \to \Gamma$ of groupoids such that $ws = 1$ and s commutes with the action of G, where G acts on $\mathrm{Ob}(G)$ via the source map by $g \cdot a = b$ for $g \in G(a, b)$. Prove that Γ^G is non-empty if w has a G-section, and that the converse holds if G is connected. Given a G-section s, let $\Gamma^G(s)$ be the set of functions $u : \mathrm{Ob}(G) \to \Gamma$ such that $wu = 1$ and u commutes with the action of G (but we do not assume u is a morphism). Show that $\Gamma^G(s)$ forms a group under addition of values, and that if G is connected and $a \in \mathrm{Ob}(G)$, then $\Gamma^G(s)$ is isomorphic to the group $\Gamma(sa)^{G(a)}$ of fixed points of $\Gamma(sa)$ under the action of $G(a)$.

11.3 General normal subgroupoids and quotient groupoids

The theory of quotient groupoids is modelled on that of quotient groups, but differs from it in important respects. In particular, the First Isomorphism Theorem of group theory (that every surjective morphism of groups is obtained essentially by factoring out its kernel) is no longer true for groupoids, so we need to characterise those groupoid morphisms (called quotient morphisms) for which this isomorphism theorem holds. The next two propositions achieve this; they were first proved in [Hig63].

Let $f : K \to H$ be a morphism of groupoids. Then f is said to be a *quotient morphism* if $\mathrm{Ob}(f) : \mathrm{Ob}(K) \to \mathrm{Ob}(H)$ is surjective and for all x, y in $\mathrm{Ob}(K)$, $f : K(x, y) \to H(fx, fy)$ is also surjective. Briefly, we say f is *object surjective* and *full*.

11.3.1 *Let* $f : K \to H$ *be a quotient morphism of groupoids. Let* $N = \mathrm{Ker}\, f$. *The following hold:*
(a) *If* $k, k' \in K$, *then* $f(k) = f(k')$ *if and only if there are elements* $m, n \in N$ *such that* $k' = m + k + n$.
(b) *If* x *is an object of* K, *then* $H(fx)$ *is isomorphic to the quotient group* $K(x)/N(x)$.

Proof (a) If k, k' satisfy $k' = m + k + n$ where $m, n \in N$, then clearly $f(k) = f(k')$.

Suppose conversely that $f(k) = f(k')$, where $k \in K(x, y)$, $k' \in K(x', y')$.

Then $f(x) = f(x')$, $f(y) = f(y')$.

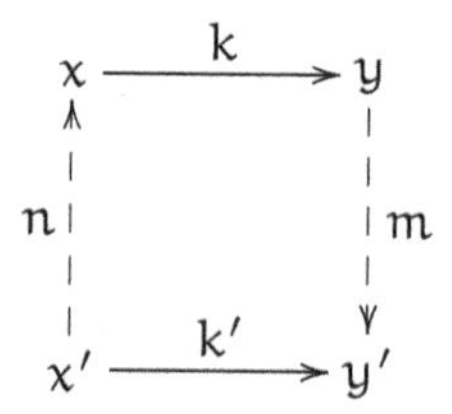

Since $f : K(y, y') \to H(fy, fy')$ is surjective, there is an element $m \in K(y, y')$ such that $f(m) = 0_{f(y)}$. Similarly, there is an element $n \in K(x', x)$ such that $f(n) = 0_{f(x)}$. It follows that if

$$n' = -k' + m + k + n \in K(x'),$$

then $f(n') = 0_{f(x')}$, and so $n' \in N$. Hence $k' = m + k + n - n'$, where $m, n - n' \in N$. This proves (a).

(b) By definition of quotient morphism, the restriction $f' : K(x) \to H(fx)$ is surjective. Also by (a), if $f'(k) = f'(k')$ for $k, k' \in K(x)$, then there are $m, n \in N(x)$ such that $k' = m + k + n$. Since $N(x)$ is normal in $K(x)$, there is an m' in $N(x)$ such that $m + k = k + m'$. Hence $k' + N(x) = k + N(x)$. Conversely, if $k' + N(x) = k + N(x)$ then $f'(k') = f'(k)$. So f' determines an isomorphism $K(x)/N(x) \to H(fx)$. □

We recall the definition of normal subgroupoid.

Let G be a groupoid. A subgroupoid N of G is called *normal* if N is wide in G (i.e. $\text{Ob}(N) = \text{Ob}(G)$) and, for any objects x, y of G and $a \in G(x, y)$, $aN(x)a^{-1} \subseteq N(y)$, from which it easily follows that

$$aN(x)a^{-1} = N(y).$$

We now prove a converse of the previous result. That is, we suppose given a normal subgroupoid N of a groupoid K and use 11.3.1(a) as a model for constructing a quotient morphism $p : K \to K/N$.

The object set of K/N is to be $\pi_0 N$, the set of components of N. Recall that a normal subgroupoid is, by definition, wide in K, so that $\pi_0 N$ is also a quotient set of $X = \text{Ob}(K)$. Define a relation on the elements of K by $k' \sim k$ if and only if there are elements m, n in N such that $m + k + n$ is defined and equal to k'. It is easily checked, using the fact that N is a subgroupoid of K, that $\sim$ is an equivalence relation on the elements of K. The set of equivalence classes is written K/N. If $\text{cls}\, k$ is such an equivalence class, and $k \in K(x, y)$, then the elements $\text{cls}\, x$, $\text{cls}\, y$ in $\pi_0 N$ are independent of the choice of k in its equivalence class. So we can write $\text{cls}\, k \in K/N(\text{cls}\, x, \text{cls}\, y)$. Let $p : K \to K/N$ be the quotient function. So far, we have not used normality of N. Not surprisingly, normality is used to give K/N an addition which makes it into a groupoid.

Suppose

$$\operatorname{cls} k_1 \in (K/N)(\operatorname{cls} x, \operatorname{cls} y), \qquad \operatorname{cls} k_2 \in (K/N)(\operatorname{cls} y, \operatorname{cls} z).$$

Then we may assume $k_1 \in K(x, y)$, $k_2 \in K(y', z)$, where $y \sim y'$ in $\pi_0 N$. So there is an element $l \in N(y, y')$, and we define

$$\operatorname{cls} k_2 + \operatorname{cls} k_1 = \operatorname{cls}(k_2 + l + k_1).$$

We have to show that this addition is well defined. Suppose then

$$k_1' = m_1 + k_1 + n_1,$$
$$k_2' = m_2 + k_2 + n_2,$$

where $m_1, n_1, m_2, n_2 \in N$. Choose any l' such that $k_2' + l' + k_1'$ is defined. Then we have the following diagram, in which a, a' are to be defined:

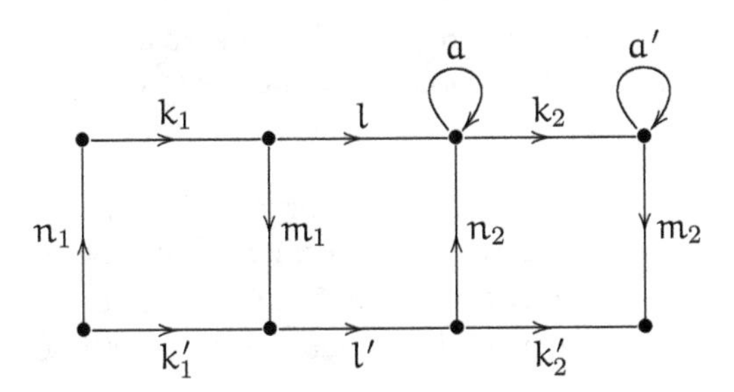

Let $a = n_2 + l' + m_1 - l$. Then $a \in N$, and $l = -a + n_2 + l' + m_1$. Since N is normal there is an element $a' \in N$ such that $a' + k_2 = k_2 - a$. Hence

$$\begin{aligned} k_2 + l + k_1 &= k_2 - a + n_2 + l' + m_1 + k_1 \\ &= -a' + k_2 + n_2 + l' + m_1 + k_1 \\ &= -a' - m_2 + k_2' + l' + k_1' - n_1. \end{aligned}$$

Since $a', m_2, n_1 \in N$, we obtain $\operatorname{cls}(k_2 + l + k_1) = \operatorname{cls}(k_2' + l' + k_1')$ as was required.

Now we know that the addition on K/N is well defined, it is easy to prove that the addition is associative, has identities, and has inverses. We leave the details to the reader. So we know that K/N becomes a groupoid.

11.3.2 *Let N be a normal subgroupoid of the groupoid K, and let K/N be the groupoid just defined. Then*

(a) *the quotient function* $p : k \mapsto \operatorname{cls} k$ *is a quotient morphism* $K \to K/N$ *of groupoids;*
(b) *if* $f : K \to H$ *is any morphism of groupoids such that* $\operatorname{Ker} f$ *contains* N, *then there is a unique morphism* $f^* : K/N \to H$ *such that* $f^*p = f$.

Proof The proof of (a) is clear. Suppose f is given as in (b). If $m + k + n$ is defined in K and $m, n \in N$, then $f(m + k + n) = f(k)$. Hence f^* is well defined on K/N by $f^*(\operatorname{cls} k) = f(k)$. Clearly $f^*p = f$. Since p is surjective on objects and elements, f^* is the only such morphism. □

In order to apply these results, we need generalisations of some facts on normal closures which were given in section 8.3 for the case of a family $R(x)$ of subsets of the object groups $K(x)$, $x \in \operatorname{Ob}(K)$, of a groupoid K. The argument here is based on [Hig05, Exercise 4, Chapter 12].

Suppose that R is *any* set of elements of the groupoid K. The *normal closure* of R in K is the smallest normal subgroupoid $N(R)$ of K containing R. Clearly $N(R)$ is the intersection of all normal subgroupoids of K containing R, but it is also convenient to have an explicit description of $N(R)$.

11.3.3 *Let* $\langle R \rangle$ *be the wide subgroupoid of* K *generated by* R. *Then the normal closure* $N(R)$ *of* R *is the subgroupoid of* K *generated by* $\langle R \rangle$ *and all conjugates* khk^{-1} *for* $k \in K$, $h \in \langle R \rangle$.

Proof Let $\widehat{R}$ be the subgroupoid of K generated by $\langle R \rangle$ and all conjugates khk^{-1} for $k \in K$, $h \in \langle R \rangle$. Clearly any normal subgroupoid of K containing R contains $\widehat{R}$, so it is sufficient to prove that $\widehat{R}$ is normal.

Suppose then that $k + a - k$ is defined where $k \in K$ and $a \in \widehat{R}$ so that

$$a = r_1 + c_1 + r_2 + c_2 + \cdots + r_l + c_l + r_{l+1}$$

where each $r_i \in \langle R \rangle$ and each $c_i = k_i + h_i - k_i$ is a conjugate of a loop h_i in $\langle R \rangle$ by an element $k_i \in K$.

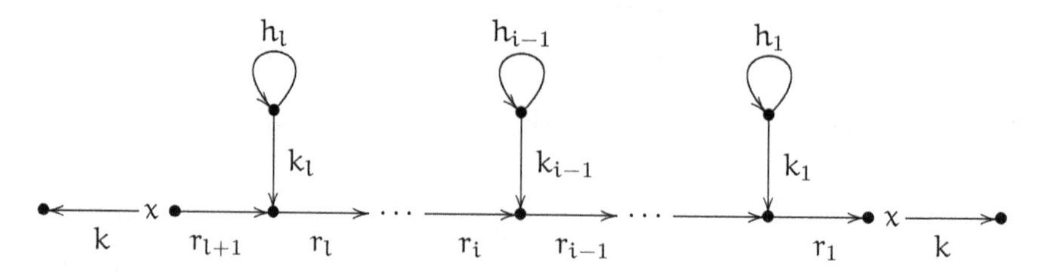

Then a is a loop, since $k + a - k$ is defined, and so also is

$$b = r_1 + r_2 + \cdots + r_{l+1}.$$

Let

$$d_i = k + r_1 + \cdots + r_i + c_i - r_i - \cdots - r_1 - k$$

so that d_i is a conjugate of a loop in $\langle R \rangle$ for $i = 1, \ldots, l$. Then it is easily checked that

$$k + a - k = d_1 + \cdots + d_l + k + b - k$$

and hence $k + a - k \in \widehat{R}$. □

Notice that the loop b in the proof belongs to $\langle R \rangle$ rather than to R, and this shows why it is not enough just to take $N(R)$ to be the subgroupoid generated by R and conjugates of loops in R.

The elements of $N(R)$ as constructed above may be called the *consequences* of R.

Exercises

1. Let $f : G \to H$ be a groupoid morphism with kernel N. Prove that the following are equivalent: (i) f is a quotient morphism; (ii) f is surjective and any two vertices of G having the same image in H lie in the same component of N.
2. Prove that a composite of quotient morphisms is a quotient morphism.
3. Let H be a subgroupoid of the groupoid G with inclusion morphism $i : H \to G$. Let $f : G \to H$ be a morphism with kernel N. Prove that the following are equivalent: (i) f is a deformation retraction; (ii) f is piecewise bijective and $fi = 1_H$; (iii) f is a quotient morphism, N is simply connected, and $fi = 1_H$.
4. Suppose the following diagram of groupoid morphisms is a pushout

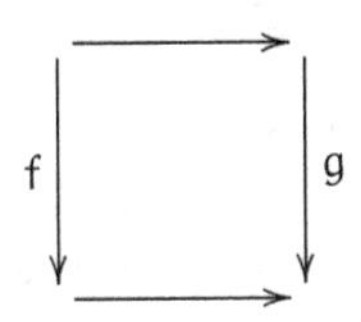

and f is a quotient morphism. Prove that g is a quotient morphism.
5. Let $f, g : H \to G$ be two groupoid morphisms. Show how to construct the coequaliser $c : G \to C$ of f, g as defined in Exercise 4 of Section 6.4. Show how this gives a construction of the orbit groupoid. [Hint: First construct the coequaliser $\sigma : \mathrm{Ob}(G) \to Y$ of the functions $\mathrm{Ob}(f), \mathrm{Ob}(g)$, then construct the groupoid $U_\sigma(G)$, and finally construct C as a quotient of $U_\sigma(G)$.]

11.4 The semidirect product groupoid

We next give the definition of the *semidirect product* of a group with a groupoid on which it acts. Let G be a group and let Γ be a groupoid with G acting on the left. The *semidirect product groupoid* $\Gamma \rtimes G$ has object set $\mathrm{Ob}(\Gamma)$ and arrows $x \to y$ the set of pairs (γ, g) such that $g \in G$ and $\gamma \in \Gamma(g \cdot x, y)$. The sum of $(\gamma, g) : x \to y$ and $(\delta, h) : y \to z$ in $\Gamma \rtimes G$ is defined to be

$$(\delta, h) + (\gamma, g) = (\delta + h \cdot \gamma, hg).$$

This is easily remembered from the following picture.

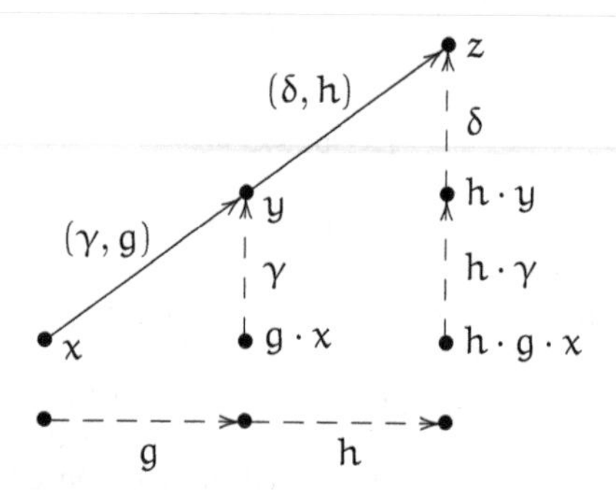

11.4.1 *The above addition makes* $\Gamma \rtimes G$ *into a groupoid and the projection*

$$q : \Gamma \rtimes G \to G, \quad (\gamma, g) \mapsto g,$$

is a fibration of groupoids. Further:
(a) q *is a quotient morphism if and only if* Γ *is connected;*
(b) q *is a covering morphism if and only if* Γ *is discrete;*
(c) q *maps* $(\Gamma \rtimes G)(x)$ *isomorphically to* G *for all* $x \in \mathrm{Ob}(\Gamma)$ *if and only if* Γ *has trivial object groups and* G *acts trivially on* $\pi_0\Gamma$.

Proof The proof of the axioms for a groupoid is easy: we have

$$\begin{aligned}(\gamma, g) &= (\gamma, g) + (0_x, 1)\\ &= (0_{g\cdot x}) + (\gamma, g),\\ -(\gamma, g) &= (g^{-1} \cdot (-\gamma), g^{-1}).\end{aligned}$$

We leave the reader to check associativity.

To prove that q is a fibration, let $g \in G$ and $x \in \mathrm{Ob}(\Gamma)$. Then $(0_{g\cdot x}, g)$ has source x and maps by q to g.

We now prove (a). Let x, y be objects of Γ. Suppose q is a quotient morphism. Then q maps $(\Gamma \rtimes G)(x, y)$ surjectively to G and so there is an element (γ, g) such that $q(\gamma, g) = 1$. So $g = 1$ and $\gamma \in \Gamma(x, y)$. This proves Γ is connected.

Suppose Γ is connected. Let $g \in G$. Then there is a $\gamma \in \Gamma(g \cdot x, y)$, and so $q(\gamma, g) = g$. Hence q is a quotient morphism.

We now prove (b). Suppose Γ is discrete, so that Γ may be thought of as a set on which G acts. Then $\Gamma \rtimes G$ is simply the covering groupoid of the action as constructed in section 10.4. So q is a covering morphism.

Let x be an object of Γ. If γ is an element of Γ with source x then $(\gamma, 1)$ is an element of $\Gamma \rtimes G$ with source x and which lifts 1. So if q is a covering morphism then the star of Γ at any x is a singleton, and so Γ is discrete.

The proof of (c) is best handled by considering the exact sequence based at $x \in \mathrm{Ob}(\Gamma)$ of the fibration q. This exact sequence is (by 7.2.9)

$$1 \to \Gamma(x) \to (\Gamma \rtimes G)(x) \xrightarrow{q'} G \to \pi_0\Gamma \to \pi_0(\Gamma \rtimes G) \to 1.$$

It follows that q' is injective if and only if $\Gamma(x)$ is trivial. Exactness also shows that q' is surjective for all x if and only if the action of G on $\pi_0\Gamma$ is trivial. □

Here is a simple application of the definition of semidirect product which will be used later.

11.4.2 *Let G be a group and let Γ be a G-groupoid. Then the formula*

$$(\gamma, g) \cdot \delta = \gamma + g \cdot \delta$$

for $\gamma, \delta \in \Gamma$, $g \in G$, defines an action of $\Gamma \rtimes G$ on the set Γ via the target map $\tau : \Gamma \to \mathrm{Ob}(\Gamma)$.

Proof This says in the first place that if $(\gamma, g) \in (\Gamma \rtimes G)(y, z)$ and δ has target y, then $\gamma + g \cdot \delta$ has target z, as is easily verified. The axioms for an action are easily verified. The formula for the action also makes sense if one notes that

$$(\gamma, g)(\delta, 1) = (\gamma + g \cdot \delta, g).$$

□

If X is a G-space, and $x \in X$, let $\sigma(X, x, G)$ be the object group of the semidirect product groupoid $\pi X \rtimes G$ at the object x. This group is called by Rhodes in [Rho66] and [Rho68] the *fundamental group of the transformation group* (although he defines it directly in terms of paths). The following result from [Rho68] gives one of the reasons for its introduction.

11.4.2 *(Corollary) If X is a G-space, $x \in X$, and the universal cover $\widetilde{X}_x$ exists, then the group $\sigma(X, x, G)$ has a canonical action on $\widetilde{X}_x$.*

Proof By 10.5.8, we may identify the universal cover $\widetilde{X}_x$ of X at x with $\mathrm{St}_{\pi X}\, x$. The function $\pi X \to \pi X$, $\delta \mapsto -\delta$, transports the action of $\pi X \rtimes G$

on πX via the target map τ to an action of the same groupoid on πX via the source map σ. Hence the object group $(\pi X \rtimes G)(x)$ acts on $\mathrm{St}_{\pi X}\, x$ by

$$(\gamma, g) * \delta = -((\gamma, g) \cdot (-\delta)) = g \cdot \gamma - \gamma.$$

The continuity of the action follows easily from the detailed description of the lifted topology (see also the remarks on topological groupoids after 10.5.8). □

EXERCISES

1. Suppose the groupoid G acts on the groupoid Γ via $w : \Gamma \to \mathrm{Ob}(G)$ as in Exercise 1 of Section 11.2. Define the *semidirect product groupoid* $\Gamma \rtimes G$ to have object set $\mathrm{Ob}(\Gamma)$ and elements the pairs $(\gamma, g) : x \to y$ where $g \in G(wx, wy)$ and $\gamma \in \Gamma(g \cdot x, y)$. The sum in $\Gamma \rtimes G$ is given by $(\delta, h) + (\gamma, g) = (\delta + h \cdot \gamma, hg)$. Prove that this does define a groupoid, and that the projection $p : \Gamma \rtimes G \to G$, $(\gamma, g) \mapsto g$, is a fibration of groupoids. Prove that the quotient groupoid $(\Gamma \rtimes G)/\mathrm{Ker}\, p$ is isomorphic to $(\pi_0\Gamma \rtimes G)$.
2. Let G and Γ be as in Exercise 1, and let the groupoid H act on the groupoid Δ via $\nu : \Delta \to \mathrm{Ob}(H)$. Let $f : G \to H$ and $\theta : \Gamma \to \Delta$ be morphisms of groupoids such that $\nu\theta = \mathrm{Ob}(f)w$ and $\theta(g \cdot \gamma) = f(g) \cdot \theta(y)$ whenever the left-hand side is defined. Prove that a morphism of groupoids $(\theta, f) : \Gamma \rtimes G \to \Delta \rtimes H$ is defined by $(\gamma, g) \mapsto (\theta(\gamma), f(g))$. Investigate conditions on f and θ for (θ, f) to have the following properties: (i) injective, (ii) connected fibres, (iii) quotient morphism, (iv) discrete kernel, (v) covering morphism. In the case that (θ, f) is a fibration, investigate the exact sequences of the fibration.
3. Generalise the corollary to 11.4.2 from the case of the universal cover to the case of a regular covering space of X determined by a subgroup N of $\pi(X, x)$.
4. Let $\mathbf{E} : 1 \to A \to E \xrightarrow{p} G \to 1$ be an exact sequence of groups. Prove that there is an action of G on a connected groupoid Γ and an object x of Γ such that the above exact sequence $\mathbf{E}$ is isomorphic to the exact sequence of the fibration $\Gamma \rtimes G \to G$ at the object x. [Hint: the groupoid Γ is the action groupoid of the right action of E on the set G via p.] [cf. [BD75].]

11.5 Semidirect product and orbit groupoids

Now we start using the semidirect product to compute orbit groupoids. The next two results may be found in [HT82], [Tay82] and [Tay88].

11.5.1 *Let* N *be the normal closure in* $\Gamma \rtimes G$ *of the set of elements of the form* $(0_x, g)$ *for all* $x \in \mathrm{Ob}(\Gamma)$ *and* $g \in G$. *Let* p *be the composite*

$$\Gamma \xrightarrow{i} \Gamma \rtimes G \xrightarrow{\nu} (\Gamma \rtimes G)/N,$$

in which the first morphism is $\gamma \mapsto (\gamma, 1)$ *and the second morphism* ν *is the quotient morphism. Then*
(a) p *is a surjective fibration;*
(b) p *is an orbit morphism and so determines an isomorphism*

$$\Gamma/\!/G \cong (\Gamma \rtimes G)/N;$$

(c) *the function* $\mathrm{Ob}(\Gamma) \to \mathrm{Ob}(\Gamma/\!/G)$ *is an orbit map, so that* $\mathrm{Ob}(\Gamma/\!/G)$ *may be identified with the orbit set* $\mathrm{Ob}(\Gamma)/G$.

Proof Let $\Delta = (\Gamma \rtimes G)/N$. We first derive some simple consequences of the definition of Δ. Let $\gamma \in \Gamma(x, y)$, $g, h \in G$. Then

$$(0_{g\cdot y}, g) + (\gamma, 1) = (g \cdot \gamma, g), \tag{1}$$

$$(\gamma, 1) + (0_x, h) = (\gamma, h). \tag{2}$$

It follows that in Δ we have

$$\nu(h \cdot \gamma, 1) = \nu(\gamma, g). \tag{3}$$

Note also that the set R of elements of $\Gamma \rtimes G$ of the form $(0_{g\cdot x}, g)$ is a subgroupoid of G, since

$$(0_{hg\cdot x}, h) + (0_{g\cdot x}, g) = (0_{hg\cdot x}, hg),$$

and $-(0_{g\cdot x}, g) = (0_x, g^{-1})$. It follows that $\pi_0 N = \pi_0 R = \mathrm{Ob}(\Gamma)/G$, the set of orbits of the action of G on $\mathrm{Ob}(\Gamma)$. Hence $\mathrm{Ob}(p)$ is surjective. This proves (c), once we have proved (b).

We now prove easily that $p : \Gamma \to \Delta$ is a fibration. Let $(\gamma, g) : x \to y$ in $\Gamma \rtimes G$ be a representative of an element of Δ, and suppose $\nu z = \nu x$, where $z \in \mathrm{Ob}(\Gamma)$. Then z and x belong to the same orbit and so there is an element h in G such that $h \cdot x = z$. Clearly $h \cdot \gamma$ has source z and by (3), $p(h \cdot \gamma) = \nu(\gamma, g)$.

Suppose now $g \in G$ and $\gamma : x \to y$ in Γ. Then by (3) $p(g \cdot \gamma) = p(\gamma)$. This verifies 11.2.1(a).

To prove the other condition for an orbit morphism, namely 11.2.1(b), suppose $\varphi : \Gamma \to \Phi$ is a morphism of groupoids such that Φ has a trivial action of the group G and $\varphi(g \cdot \gamma) = \varphi(\gamma)$ for all $\gamma \in \Gamma$ and $g \in G$. Define $\varphi' : \Gamma \rtimes G \to \Phi$ on objects by $\mathrm{Ob}(\varphi)$ and on elements by $(\gamma, g) \mapsto \varphi(\gamma)$. That φ is a morphism follows from the trivial action of G on Φ, since

$$\begin{aligned}
\varphi'((\delta, h) + (\gamma, g)) &= \varphi(\gamma + h.\delta) \\
&= \varphi(\gamma) + \varphi(h.\delta) \\
&= \varphi(\gamma) + \varphi(\delta) \\
&= \varphi'(\delta, h) + \varphi'(\gamma, g).
\end{aligned}$$

Also $\varphi'(0_x, g) = \varphi(0_x) = 0_{\varphi x}$, and so $N \subseteq \operatorname{Ker} \varphi'$. By 11.3.2(b), there is a unique morphism $\varphi^* : (\Gamma \rtimes G)/N \to \Phi$ such that $\varphi^* \nu = \varphi'$. It follows that $\varphi^* p = \varphi^* \nu i = \varphi' i = \varphi$. The uniqueness of φ^* follows from the fact that p is surjective on objects and on elements.

Finally, the isomorphism $\Gamma//G \cong \Delta$ follows from the universal property. □

In order to use the last result, we analyse the morphism $p : \Gamma \to \Gamma//G$ in some special cases. The construction of the orbit groupoid given in 11.5.1 is what makes this possible.

11.5.2 *The orbit morphism $p : \Gamma \to \Gamma//G$ is a fibration whose kernel is generated as a subgroupoid of Γ by all elements of the form $\gamma - g \cdot \gamma$ where g stabilises the initial point of γ. Furthermore,*
(a) if G acts freely on Γ, by which we mean no non-identity element of G fixes an object of Γ, then p is a covering morphism;
(b) if Γ is connected and G is generated by those of its elements which fix some object of Γ, then p is a quotient morphism; in particular, p is a quotient morphism if the action of G on $\operatorname{Ob}(\Gamma)$ has a fixed point;
(c) if Γ is a tree groupoid, then each object group of $\Gamma//G$ is isomorphic to the factor group of G by the (normal) subgroup of G generated by elements which have fixed points.

Proof We use the description of p given in 11.5.1, which already implies that p is a fibration.

Let R be the subgroupoid of $\Gamma \rtimes G$ consisting of elements $(0_{g \cdot x}, g)$, $g \in G$. Let N be the normal closure of R. By the construction of the normal closure in 11.3.3, the elements of N are sums of elements of R and conjugates of loops in R by elements of $\Gamma \rtimes G$. So let $(0_{g \cdot x}, g)$ be a loop in R. Then $g \cdot x = x$. Let $(\gamma, h) : x \to y$ in $\Gamma \rtimes G$, so that $\gamma : h \cdot x \to y$. Then we check that

$$(\gamma, h) + (0_x, g) - (\gamma, h) = (\gamma - hgh^{-1} \cdot \gamma, hgh^{-1}).$$

Writing $k = hgh^{-1}$, we see that $(\gamma - k \cdot \gamma, k) \in N$ if k stabilises the initial point of γ.

Now $\gamma \in \operatorname{Ker} p$ if and only if $(\gamma, 1) \in N$. Further, if $(\gamma, 1) \in N$ then $(\gamma, 1)$ is a consequence of R and so $(\gamma, 1)$ is equal to

$$(\gamma_1 - k_1 \cdot \gamma_1, k_1) + (\gamma_2 - k_2 \cdot \gamma_2, k_2) + \cdots + (\gamma_r - k_r \cdot \gamma_r, k_r)$$

for some γ_i, k_i where k_i stabilises the initial point of γ_i, $i = 1, \ldots, r$. Let $h_1 = 1$, $h_i = k_1 \ldots k_{i-1}$ $(i \geqslant 2)$, $\delta_i = h_i \cdot \gamma$, $g_i = h_i k_i h_i^{-1}$, $i \geqslant 1$. Then

$$(\gamma, 1) = (\delta_1 - g_1 \cdot \delta_1 + \delta_2 - g_2 \cdot \delta_2 + \cdots + \delta_r - g_r \cdot \delta_r, 1)$$

and so γ is a sum of elements of the form $\delta - g \cdot \delta$ where g stabilises the initial point of δ. This proves our first assertion.

The proof of (a) is simple. We know already that $p : \Gamma \to \Gamma//G$ is a fibration. If G acts freely, then by the result just proved, p has discrete kernel. It follows that if $x \in \mathrm{Ob}(\Gamma)$, then $p : \mathrm{St}_\Gamma x \to \mathrm{St}_{\Gamma//G}\, px$ is injective. Hence p is a covering morphism.

Now suppose Γ is connected and G is generated by those of its elements which fix some object of Γ. To prove p a quotient morphism we have to show that for $x, y \in \mathrm{Ob}(\Gamma)$, the restriction $p' : \Gamma(x,y) \to (\Gamma//G)(px, py)$ is surjective.

Let (γ, g) be an element $x \to y$ in $\Gamma \rtimes G$, so that $\gamma : g \cdot x \to y$ in Γ. Using the notation of 11.5.1, we have to find $\delta \in \Gamma(x,y)$ such that $p\delta = \nu(\gamma, g)$. As shown in 11.5.1, $\nu(\gamma, g) = \nu(\gamma, 1)$. By assumption, $g = g_n g_{n-1} \cdots g_1$ where g_i stabilises an object x_i, say. Since Γ is connected, there are elements

$$\delta_1 \in \Gamma(x_1, x), \quad \delta_i \in \Gamma(x_i, (g_{i-1} \cdots g_1) \cdot x), \quad i \geqslant 1.$$

The situation is illustrated below for $n = 2$.

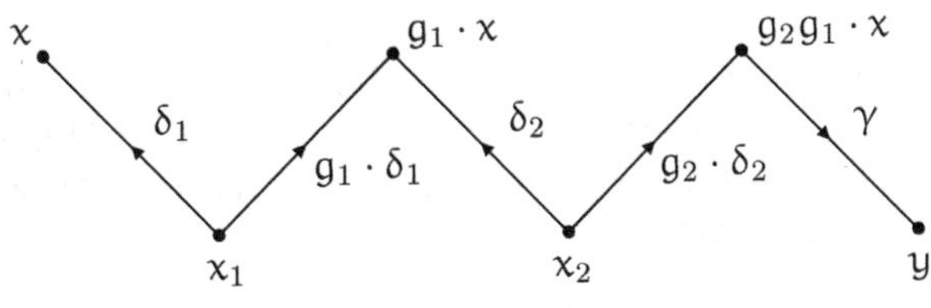

Fig. 11.2

Let

$$\delta = \gamma + (g_n \cdot \delta_n - \delta_n) + \cdots + (g_1 \cdot \delta_1 - \delta_1) : x \to y.$$

Then $p(\delta) = \nu(\gamma, 1)$. This proves (b).

For the proof of (c), let x be an object of Γ. Since Γ is a tree groupoid, the projection $(\Gamma \rtimes G)(x) \to G$ is an isomorphism which sends the element $(0_x, g)$, where $g \cdot x = x$, to the element g. Also if g fixes x and $h \in G$ then hgh^{-1} fixes $h \cdot x$. Thus the image of $N(x)$ is the subgroup K of G generated by elements of G with a fixed point, and K is normal in G. Let $\bar{x}$ denote the orbit of x. By 11.3.1(b), and 11.5.1, the group $(\Gamma//G)(\bar{x})$ is isomorphic to the quotient of $(\Gamma \rtimes G)(x)$ by $N(x)$. Hence $(\Gamma//G)(\bar{x})$ is also isomorphic to G/K. □

In 11.5.2, the result (a) relates the work on orbit groupoids to work on covering morphisms. The result (b) will be used below. The result (c) is particularly useful in work on discontinuous actions on Euclidean or hyperbolic space.

In the case of a discontinuous action of a group G on a space X which satisfies the additional condition 11.2.2(b), and which has a fixed point x, we obtain from 11.5.2 a convenient description of $\pi(X/G, \bar{x})$ as a quotient of $\pi(X, x)$. In the more general case, $\pi(X/G, \bar{x})$ has to be computed as a quotient of $\sigma(X, x, G) = (\pi X \rtimes G)(x)$, as given in 11.5.1. Actually the case corresponding to 11.5.2(c) was the first to be discovered, for the case of simplicial actions [Arm65]. The general case followed from the fact that if X has a universal cover $\widetilde{X}_x$ at x, then $\sigma(X, x, G)$ acts on $\widetilde{X}_x$ with orbit space homeomorphic to X/G.

We give next two computations for actions with fixed points.

EXAMPLE

1. Let the group $\mathbb{Z}_2 = \{1, g\}$ act on the circle $\mathbb{S}^1$ in which g acts by reflection in the x-axis. The orbit space of the action can be identified with E^1_+, which is contractible, and so has trivial fundamental group.

To see how this agrees with the previous results, let $\Gamma = \pi\mathbb{S}^1$, $G = \mathbb{Z}_2$. It follows from 11.2.3 that the induced morphism $\pi\mathbb{S}^1 \to \pi(\mathbb{S}^1/G)$ is an orbit morphism, and so $\pi(\mathbb{S}^1/G) \cong \Gamma/\!/G$. By 11.5.1, $p : \Gamma \to \Gamma/\!/G$ is a quotient morphism, since the action has a fixed point 1 (and also -1). Let the two elements $a_\pm \in \Gamma(1, -1)$ be represented by the paths $[0,1] \to \mathbb{S}^1$, $t \mapsto e^{\pm i\pi t}$ respectively. The non trivial element g of G satisfies $g \cdot a_+ = a_-$. Hence the kernel of the quotient morphism $p : \Gamma(1) \to (\Gamma/\!/G)(p1)$ contains the element $-a_- + a_+$. But this element generates $\Gamma(1) \cong \mathbb{Z}$. So we confirm the fact that $(\Gamma/\!/G)(p1)$ is the trivial group. □

Before our next result we state and prove a simple group theoretic result. First let H be a group. It is convenient to write the group structure on H as multiplication. The *abelianisation* H^{ab} of H is formed from H by imposing the relations $hk = kh$ for all $h, k \in H$. Equivalently, it is the quotient of H by the (normal) subgroup generated by all commutators $hkh^{-1}k^{-1}$, for all $h, k \in H$.

11.5.3 *The quotient* $(H \times H)/K$ *of* $H \times H$ *by the normal subgroup* K *of* $H \times H$ *generated by the elements* (h, h^{-1}) *is isomorphic to* H^{ab}.

Proof We can regard $H \times H$ as the group with generators $[h]$, $\langle k \rangle$ for all $h, k \in H$ and relations $[hk] = [h][k]$, $\langle hk \rangle = \langle h \rangle \langle k \rangle$, $[h]\langle k \rangle = \langle k \rangle [h]$ for all $h, k \in H$, where we may identify $[h] = (h, 1)$, $\langle k \rangle = (1, k)$, $[h]\langle k \rangle = (h, k)$. Factoring out by K imposes the additional relations $[h]\langle h^{-1} \rangle = 1$, or equivalently $[h] = \langle h \rangle$, for all $h \in H$. It follows that $(H \times H)/K$ is obtained from H by imposing the additional relations $hk = kh$ for all $h, k \in H$. □

Definition (The symmetric square of a space) Let $G = \mathbb{Z}_2$ be the cyclic group of order 2, with non-trivial element g. For a space X, let G act on the product space $X \times X$ by interchanging the factors, so that $g \cdot (x, y) = (y, x)$. The fixed point set of the action is the diagonal of $X \times X$. The orbit space is called the *symmetric square* of X, and is written Q^2X.

11.5.4 *Let X be a connected, Hausdorff, semi-locally 1-connected space, and let $x \in X$. Let $\langle x \rangle$ denote the class in Q^2X of (x, x). Then the fundamental group $\pi(Q^2X, \langle x \rangle)$ is isomorphic to $\pi(X, x)^{ab}$, the fundamental group of X at x made abelian.*

Proof Since $G = \mathbb{Z}_2$ is finite, the action is discontinuous. Because of the assumptions on X, we can apply 11.2.3, and hence also the results of this section. We deduce that $p_* : \pi(X) \times \pi(X) \to \pi(Q^2X)$ is a quotient morphism and that if $x \in X$, $z = p(x, x)$, then the kernel of the quotient morphism

$$p' : \pi(X, x) \times \pi(X, x) \to \pi(Q^2X, z)$$

is the normal subgroup K generated by elements $(a, b) - g \cdot (a, b) = (a - b, b - a)$, $a, b \in \pi X(y, x)$, for some $y \in X$. Equivalently, K is the normal closure of the elements $(c, -c)$, $c \in \pi(X, x)$. The result follows. □

11.6 Full subgroupoids of orbit groupoids

To prepare for our next application to orbit groupoids, we give the following result, due to Taylor [Tay88]. First recall that a morphism $\theta : \Sigma \to \Gamma$ of groupoids is *full* if for all $x, y \in \mathrm{Ob}\,\Sigma$, θ is surjective as a function $\Sigma(x, y) \to \Gamma(\theta x, \theta y)$.

11.6.1 *Let $\theta : \Sigma \to \Gamma$ be a morphism of G-groupoids, and consider the following diagram induced by θ:*

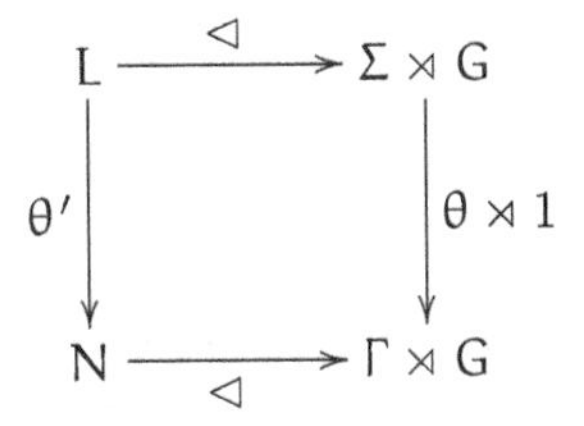

where: L is the normal subgroupoid of $\Sigma \rtimes G$ generated by the elements $(0_x, g)$ for $x \in \mathrm{Ob}(\Sigma)$, $g \in G$, and similarly for N and Γ; and θ' is the restriction of

$\theta \rtimes 1$. *Suppose*

(*) *the image of* $\mathrm{Ob}(\theta)$ *meets each component of the fixed point groupoid of each* $g \in G$,

and θ *is full, and injective on objects. Then* θ' *is full.*

Proof Let $x, y \in \mathrm{Ob}(\Sigma)$ and let $\nu \in N(\theta x, \theta y)$. We have to prove $\nu = \theta\lambda$ for some $\lambda \in L(x, y)$.

Write $\nu = \nu_r + \cdots + \nu_1$ where each ν_i is of the form $(0_u, g)$ or a conjugate of such an element. We may assume this sum is minimal in the sense that no intermediate vertex of the ν_i is in the image of $\mathrm{Ob}\,\theta$, since otherwise we may restrict to a part of the sum. Note that $\nu_1 : z \to \theta y$ for some $z \in \mathrm{Ob}\,\Gamma$.

If $\nu_1 = (0_u, g)$ for some u, g, then $0_u : g \cdot z \to \theta y$ and hence $u = g \cdot z = \theta y$. So $z = g^{-1} \cdot \theta y = \theta(g^{-1} \cdot y)$. By minimality, $\nu = \nu_1$, and so $z = \theta x$. Hence $\theta y = \theta(g \cdot x)$ so that, as θ is injective on objects, $y = g \cdot x$ and so $\nu = \theta'(0_x, g)$ where $(0_x, g) \in L(x, y)$.

If $\nu_1 = (\gamma, h) + (0_u, g) - (\gamma, h)$ for some object u of Γ, then ν_1 is a loop, and so $\nu_1 : \theta y \to \theta y$. Hence again, $\nu = \nu_1$, $\theta x = \theta y$, and so $x = y$. Since ν is a loop, so also is $(0_u, g)$, and so the object u must be a fixed point of g. By hypothesis (*), there is an element $\beta : u \to v$ of Γ such that $g \cdot \beta = \beta$ and $v = \theta z$ for some z. Now

$$\begin{aligned}\nu_1 &= (\gamma, h) + (\beta, 1) - (\beta, 1) + (0_u, g) + (\beta, 1) - (\beta, 1) - (\gamma, h)\\ &= (\delta, h) + (0_v, g) - (\delta, h)\end{aligned}$$

where $\delta = \gamma + h \cdot \beta : h \cdot \theta v \to v$.

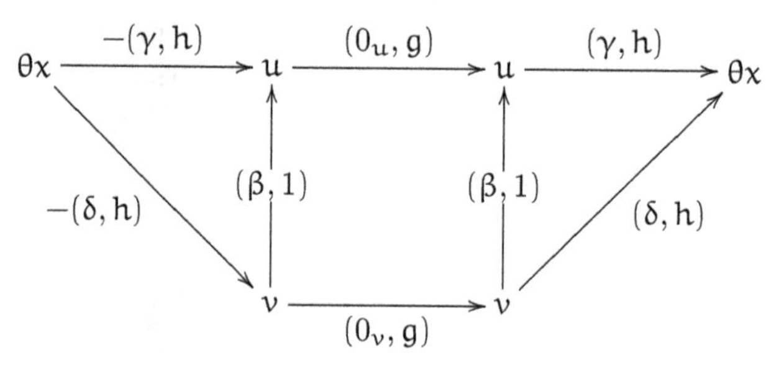

Since θ is full, there exists $\tau \in \Sigma(h \cdot z, x)$ such that $\theta\tau = \delta$. The element $\lambda = (\tau, h) + (0_z, g) - (\tau, h)$ then lies in $L(x, x)$ and $\theta'\lambda = \nu$, as required. □

11.6.2 *Let* $\theta : \Sigma \to \Gamma$ *be an injective, full* G*-morphism, and suppose that the image of* $\mathrm{Ob}(\Sigma)$ *meets each component of the fixed point groupoid of each element of* G*. Then the induced morphism*

$$\theta_G : \Sigma//G \to \Gamma//G$$

is injective.

Proof We use the isomorphisms $\Sigma//G \to (\Sigma \rtimes G)/L$, $\Gamma//G \to (\Gamma \rtimes G)/N$ given by 11.5.1. In the following diagram:

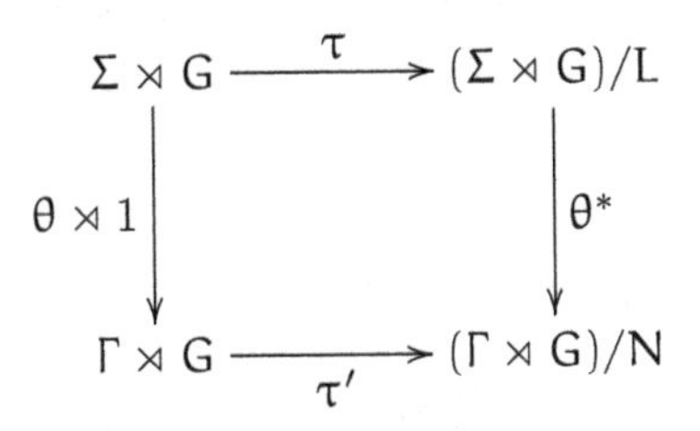

τ, τ' are the quotient morphisms, and θ^* is the induced morphism. Then θ injective implies $\theta \rtimes 1$ is injective, as is easily checked. Let $\tau(\gamma, g), \tau(\delta, h) \in \Sigma \rtimes G$ be such that $\theta^*\tau(\gamma, g) = \theta^*\tau(\delta, h)$. Then $\tau'(\theta\gamma, g) = \tau'(\theta\delta, h)$. So there exists $\nu_1, \nu_2 \in N$ such that

$$(\theta\gamma, g) = \nu_1 + (\theta\delta, h) + \nu_2.$$

Now ν_1, ν_2 have vertices in the image of $\mathrm{Ob}\,\Sigma$. So, by the previous proposition, there exists $\lambda_1, \lambda_2 \in L$ which map to ν_1, ν_2 respectively, and $\alpha = \lambda_1 + (\delta, h) + \lambda_2$ is defined in $\Sigma \rtimes G$, has the same source and target as (γ, g), and maps by $\theta \rtimes 1$ to the same as does (γ, g). Since $\theta \rtimes 1$ is injective, $(\gamma, g) = \alpha$, so $\tau(\gamma, g) = \tau(\delta, h)$, and so θ^* is injective. □

If A is a subset of $\mathrm{Ob}\,\Gamma$, then ΓA will denote as usual the full subgroupoid of Γ on the object set A.

11.6.2 *(Corollary 1) Let* A *be a* G*-invariant subset of* $\mathrm{Ob}\,\Gamma$ *which contains at least one vertex from each component of the fixed point groupoid of each group element. Then* $(\Gamma A)//G$ *is a full subgroupoid of* $\Gamma//G$ *so that the restriction of the orbit morphism* $\Gamma \to \Gamma//G$ *to* $\Gamma A \to (\Gamma//G)(A/G)$ *is itself an orbit morphism.*

Proof The inclusion $i : \Gamma A \to \Gamma$ is injective and full, and hence so too is the induced morphism $i_A^* : \Gamma A \to \Gamma//G$, by 11.6.2. Thus i_A^* embeds $(\Gamma A)//G$ in $\Gamma//G$ as a full subgroupoid. □

11.6.2 *(Corollary 2) If the action of* G *on* X *satisfies the conditions of* 11.2.2, *and* A *is a* G*-stable subset of* X *meeting each path component of the fixed point set of each element of* G, *then* $\pi(X/G)(A/G)$, *the fundamental groupoid of* X/G *on the set* A/G, *is canonically isomorphic to the orbit groupoid*

$(\pi XA)//G$ of πXA.

The results of this section answer some cases of the following question:

Question *Suppose $p : K \to H$ is a connected covering morphism, and $x \in \mathrm{Ob}(K)$. Then p maps $K(x)$ isomorphically to a subgroup of the group $H(px)$. What information in addition to the value of $K(x)$ is needed to reconstruct the group $H(px)$?*

There is an exact sequence

$$0 \to K(x) \to H(px) \to H(px)/p[K(x)] \to 0$$

in which $H(px)$ and $K(x)$ are groups while $H(px)/p[K(x)]$ is a pointed set with base point the coset $p[K(x)]$. Suppose now that p is a regular covering morphism. Then the group G of covering transformations of p is anti-isomorphic to $H(px)/p[K(x)]$, by 10.6.4 (Corollary 1). Also G acts freely on K.

11.6.3 *If $p : K \to H$ is a regular covering morphism, then p is an orbit morphism with respect to the action on K of the group G of covering transformations of p. Hence if $x \in \mathrm{Ob}(K)$, then $H(px)$ is isomorphic to the object group $(K \rtimes G)(x)$.*

Proof Let G be the group of covering transformations of p. Then G acts on K. Let $q : K \to K//G$ be the orbit morphism. If $g \in G$ then $pg = p$, and so G may be considered as acting trivially on H. By the condition 11.2.1(a), there is a unique morphism $\varphi : K//G \to H$ of groupoids such that $\varphi q = p$. By 11.5.2(a), q is a covering morphism. By 10.2.3, φ is a covering morphism. But φ is bijective on objects, because p is regular. Hence φ is an isomorphism.

Since G acts freely on K, the group $N(x)$ of 11.5.1 is trivial. So the description of $H(px)$ follows from 11.5.1(b). □

The interest of the above results extends beyond the case where G is finite, since general discontinuous actions occur in important applications in complex function theory, concerned with Fuchsian groups and Kleinian groups. Unfortunately, a description of these applications would take us too far beyond our allotted space, and we simply refer the reader to [Bea83] and [IK87].

Another important type of result in the area of Fuchsian groups is the theorem of Macbeath and Swan (for a recent account see [RT85]) which says that if G acts on the connected space X and V is a connected open neighbourhood of $x \in X$ such that $GV = X$, then there is an exact sequence of the form

$$1 \to N \to \pi(X, x) \to \Gamma \to G \to 1$$

where Γ is a group given by a presentation in terms of the intersections of V with its transforms by elements of G. This result yields presentations of G for example for special linear groups over number fields. Related results are also obtained by means of groups acting on graphs ([Ser80]).

However the utility of groupoid methods in these areas has only recently been found.

Finally, we should mention the book [Mon87], which relates in a geometric spirit the notions of covering space, group actions, tessellations and the modern theory of *orbifolds*.

EXERCISES

1. Let X^n be the n-fold product of X with itself, and let the symmetric group S_n act on X^n by permuting the factors. The orbit space is called the *n-fold symmetric product of* X and is written Q^nX. Prove that for $n \geqslant 2$ the fundamental group of Q^nX at an image of a diagonal point $(x, \ldots, x)$ is isomorphic to the fundamental group of X at x made abelian.
2. Investigate the fundamental groups of quotients of X^n by the actions of various proper subgroups of the symmetric group for various n and various subgroups. [Try out first the simplest cases which have not already been done in order to build up your confidence. Try and decide whether or not it is reasonable to expect a general formula.]

NOTES

The results in this chapter arose in the following circuitous way. In studying group actions, [Rho66, Rho68] introduced $\sigma(X, x, G)$, which he called the *fundamental group of the transformation group*. This construction was then used by Armstrong in a series of papers from 1966 to 1984 (see the references in [Arm65]), to describe the fundamental group of an orbit space of a discontinuous action. [BD75] pointed out that $\sigma(X, x, G)$ is isomorphic to the object group at x of the semidirect product $\pi X \rtimes G$. This idea was developed by [Tay82], [Tay88] and [HT82] in relation to Armstrong's results.

The path lifting property 11.1.4 for the quotient map by the action of a discontinuous group is taken in essence from [Bre72, Theorem 6.2], but with more detail given. The proof of 11.2.3 is taken from [BH86] and is essentially a groupoid version of a result of [Arm82], which is there expressed in terms of Rhodes' fundamental group of a transformation group.

Another important topic is the relation between an action of a group G on a space X, the fundamental group of X, and a so-called *fundamental region*, by which is meant a connected open set whose translates by the action of G cover X. The paper [RT85] gives a proof of results which Macbeath and Swan have applied to give presentations of special linear groups of rings of algebraic integers. For more information, see the references in the cited paper.

For work on fibre bundles, see [Ste51], [Hus66], [Mac05] and [IK87].

Chapter 12

Conclusion

The initial aim for this book in the 1960s was to give an account of some known basic ideas in algebraic topology, particularly homotopy type and the fundamental group. It then became an exploration of the use of groupoids in this area, for reasons explained in the Prefaces. The present retitling and reorganisation emphasise this direction. The length of time to make such changes does suggest the difficulty in recognising a change of paradigm, even for a person involved in the change!

This use of the word 'paradigm' comes from the famous book of Thomas Kuhn on 'The structure of scientific revolutions' [Kuh62]. This is discussed in relation to mathematics in Corfield's book [Cor03]. You are invited to investigate the topic of paradigms and revolutions in mathematics!

It may seem strange to discuss matters of this type in a text—professional mathematicians may argue that this is a distraction from getting on with stating and proving theorems. However this book is also addressed to the beginner, and matters of methodology are important at various stages of any activity. Human beings rarely perform an activity well without some mode of analysis and some knowledge at a higher level.

Einstein wrote in 1916 [Ein90]:

> When I think of the ablest students whom I have encountered in teaching—i.e., those who have distinguished themselves by their independence and judgement and not only mere agility—I find that they have a concern for the theory of knowledge. They like to start discussions concerning the aims and methods of the sciences, and showed unequivocally by the obstinacy with which they defend their views that this subject seemed important to them.

> This is not really astonishing. For when I turn to science not for some superficial reason such as money-making or ambition, and also not (or at least exclusively) for the pleasure of the sport, the delights of brain-athletics, then the following questions must burningly interest me as a disciple of science: What goal will be reached by the science to which I am dedicating myself? To what extent are its general results 'true'? What is essential and what is based only on the accidents of development?... Concepts which have proved useful for ordering things easily assume so great an authority over us, that we forget their terrestrial origin and accept them as unalterable facts. They then become labelled as 'conceptual necessities', 'a priori situations', etc. The road of scientific progress is frequently blocked for long periods by such errors. It is therefore not just an idle game to exercise our ability to analyse familiar concepts, and to demonstrate the conditions on which their justification and usefulness depend, and the way in which these developed, little by little...

For some, the main fascination in mathematics is the challenge of problems. Certainly, the solution of a famous problem will give the solver fame, in at least the mathematical world. Yet problems can be stated only at a given level of conceptualisation; so others see the progress of mathematics as strongly involving the development of a rigorous language for:
description; verification; deduction; and calculation.
It can be argued that it is the development of such a language which has been the main contribution of mathematics to culture, science and technology over the centuries.

The elaboration of concepts is a key to this development. The problems of doing this are indicated by Alexander Grothendieck, who in correspondence with me in the 1980s, refers to 'the difficulty of bringing new concepts out of the dark', and again to the fact that 'mathematics was held back for centuries for lack of the trivial concept of zero'. Elsewhere he suggests that solving a problem is not like cracking a nut with a hammer, but more like gradually softening the shell with water till it can be peeled away. The immediate fame of Grothendieck was also that he developed new language to invent and describe structures which could then be used to solve well known problems. This made it easier for many to judge his achievement against known targets. His enduring fame will be in the new worlds and methods he opened up.

Shakespeare's Theseus in 'A Midsummer Night's Dream' describes the rôle of the poet as:

> The Poet's eye in a fine frenzy rolling
> Doth glance from heaven to earth, from earth to heaven,
> And as imagination bodies forth the forms of things unknown
> The Poet's pen turns them to shapes, and gives to airy nothing
> A local habitation and a name.

The last line could also be seen as relating to the mathematicians theories and concepts.

In the preface to this edition we argued the case for the importance of analogy in mathematics, and that we obtain analogies in mathematics not between objects but between relations between objects. In so doing we try to find the landscape, the structures, in which these objects lie. The slogan of Ehresmann [Ehr66] was 'to find the structure of everything'! This search for new structures to express intuitions is for some of us a main part of the fascination of mathematics; the solution of already formulated problems is a useful test of the new concepts, more exciting is the view of new problems and questions.

Many mathematical structures have been found as a result of trying to understand, to make things clear, or even through laziness, to get one exposition instead of several. In obtaining such an exposition, one is again exploiting or finding analogies. This is the main point of abstract ideas in mathematics.

The fact that the Van Kampen theorem for the fundamental groupoid gave complete information on the fundamental group even in the non-connected case was a surprise to me, as it seemed to conflict with current methods in algebraic topology and homological algebra. These latter methods tend to give information in the form of exact sequences, and so do not yield complete information. For example, they seemed unlikely to be able to yield 8.4.1.

On the other hand the groupoid results seem to obtain their power because a groupoid has structure in the two dimensions 0 and 1, and this is what is required in order to model the geometric identifications. This suggested that higher dimensional results could be obtained by using algebraic gadgets which had structure in a wider range of dimensions, so replacing exact sequences by computations of colimits. Such a replacement would probably not be complete, but it is of interest to see to what extent this philosophy can lead to the investigation of interesting new algebraic gadgets for modelling geometry.

Another input was the proof of the cellular approximation theorem in section 7.6. This proof seemed to have elements of a higher dimensional analogue of the proof of 6.7.2; that proof produced an algebraic result because it had an algebraic structure to match the geometry. An analogous algebraic structure was lacking in higher dimensions.

Thus we have an anomaly: the examination and unpacking of this anomaly has suggested a variety of new ideas.

The earliest move towards higher dimensional algebraic structures in homotopy theory was the definition of the homotopy groups $\pi_n(X, x)$ of a space X with base point x in [Č32]. However it was quickly proved that these groups are abelian for $n \geqslant 2$, and this was thought to make them not a sensible model of higher dimensional phenomena. For example, it seemed unlikely that they could be handled by techniques of generators and relations in a manner similar to techniques available for the fundamental group. [H. Hopf told E. Dyer in 1966 of his embarrassment that he had urged Čech to withdraw his full paper on higher homotopy groups for the 1932 International Congress, so that all that was published was a brief paragraph.]

The striking early results in this area were the proof by Hopf in 1930 that the 'Hopf map' $\mathbb{S}^3 \to \mathbb{S}^2$ is essential, and the results of Hurewicz in 1935 linking homotopy and homology (see for example [Whi78, Hat02] for both these results). Such results inspired important work on higher homotopy groups, so that the initial disappointment with their abelian nature came to be looked on as a quirk of history.

The generalised Van Kampen theorems as given in [BH78] and [BL87] bring to homotopy theory a novel style of result, involving 'higher homotopy groupoids', in which the basic tools are non-abelian. As sample results, the first paper gives the relative Hurewicz theorem (a tough nut in basic homotopy theory) in the form of a relation between the relative homotopy group $\pi_n(X, A, x)$ and the absolute homotopy group $\pi_n(X \cup CA, x)$ which is deduced from such a Van Kampen theorem, while the second paper introduces a non-abelian tensor product of groups which yields new computations of some third homotopy groups including new proofs that $\pi_3(\mathbb{S}^2, x) \cong \mathbb{Z}$ and $\pi_4(\mathbb{S}^3, x) \cong \mathbb{Z}_2$. These methods suggest the existence of a truly non-abelian homotopy theory whose extent is yet to be determined.

The key to these developments was the observation that groupoids have a composition ab which is defined under geometric conditions: the final point of b must be the initial point of a. From this starting point, it is easy to see how to consider partially defined compositions of squares in two directions. It is this partial composition which allows non abelian structures in higher dimensions.

One interest in the development of new methods in homotopy theory is the wide occurrence of the notion of 'deformation' as a tool for the classification of structures. Indeed it was for the classification of solutions of dynamical problems such as the motion of three bodies under gravitational forces, these solutions being regarded as paths in a suitable phase space, that was one of the influences that led Poincaré to work out the notion of fundamental group in the first place. Another was the notion of monodromy in complex variable theory. There is now a body of work on groupoids, monodromy and holonomy.

There seems to be a lot of mileage in the naive approach that, since 1-dimensional phenomena in homotopy theory are well modelled by the fundamental groupoid, then n-dimensional phenomena should be modelled by n-dimensional groupoids. These latter algebraic structures are generalisations of groups. In view of the wide occurrence of groups in mathematics, it will be interesting to see what impact such an approach to generalised groups will have on other areas of mathematics.

This book has tried to show the influence of groupoids on one area of mathematics. The paper 'From groups to groupoids', [Bro87], gives more history and explains wider impacts. For the area of Lie groupoids, see for example [Mac05]. For the impact on physics, and functional analysis, see [Con94] and later work on 'Non commutative geometry', as well as [Pat99]. The book [BJ01] shows the use of groupoids in the important area of Galois theory. The book [MLM94] gives a wider framework for 'internal groupoid'.

Our chapter 8 is titled 'Combinatorial groupoid theory', a novel term: it represents at this stage, and as far it is distinct from combinatorial group theory, a small subject. Two books which apply such ideas in directions other than Higgins, [Hig05], and this one, are [Coh89] and [DV96]. There is also a large subject of 'computational group theory'. There is clearly work to be done to test an extension of this!

The diagram of the relation of groupoids to other areas of mathematics given on the back cover suggests further areas of work.

A major work is to return to the roots of algebraic topology in geometry and analysis, and see if new outlooks can provide methods to answer questions of the pioneers. What may be the applications of these new non-abelian methods to higher dimensional local-to-global problems? This concluding chapter may be the end of the beginning.

Appendix A

Functions, cardinality, universal properties

A.1 Functions

Let A, B be sets. A subset F of $A \times B$ is called *functional* if for each $a \in A$, there is exactly one $b \in B$ such that $(a, b) \in F$. A *function* f *from* A *to* B will consist of the set A, the set B and a functional subset F of $A \times B$; more precisely, f is the triple (A, B, F). The set A is a *domain* of f, the set B is the *codomain* of f, and the functional subset F of $A \times B$ is the *graph* of f. (The codomain is sometimes called the *range* of f, but the word range is also used for what we shall call the *image* of f.) If $a \in A$ then the element b of B such that $(a, b) \in F$ is called the *value* of f on a, and is written $f(a)$, or fa, or even f_a; we also say f *sends, or maps,* a to fa. In order to emphasise the fact that f depends on both A and B we often write $f : A \to B$, or $A \xrightarrow{f} B$, for f. When a particular name is not required, a function from A to B is written simply $A \to B$.

Our definition ensures that two functions $f : A \to B$, $f' : A' \to B'$ are equal if and only if they have the same domain (i.e., $A = A'$), the same codomain (i.e., $B = B'$) and the same graph.

Functions are often defined by formulae. Thus the formulae $2x^2 + \sin x$ defines a function $\mathbb{R} \to \mathbb{R}$ whose value on a real number x is $2x^2 + \sin x$; this function we denote by

$$\begin{aligned} \mathbb{R} &\to \mathbb{R} \\ x &\mapsto 2x^2 + \sin x \end{aligned}$$

and when the domain and codomain can be understood from the context we shall often denote such a function simply by $x \mapsto 2x^2 + \sin x$. We also allow that the domain of a function specified in this way shall be the maximum domain on which the formula gives a unique answer; thus $x \mapsto \log x + (x^2 - 1)^{-1}$ denotes a function with domain $\{x \in \mathbb{R} : x > 0$ and $x \neq 1\}$.

A.1.1 Let $f : A \to B$, $g : B \to C$ be functions. The *composite* of these two functions is the function

$$\begin{aligned} gf : A &\to C \\ x &\mapsto gfx. \end{aligned}$$

Sometimes, when extra clarity is essential, we write $g \circ f$ for gf. If $h : C \to D$ is another function, then we have the associative law

$$h(gf) = (hg)f.$$

The *identity function* on A is the function

$$\begin{aligned} 1, \quad \text{or} \quad 1_A : A &\to A \\ x &\mapsto x. \end{aligned}$$

If $f : A \to B$, then $f1_A = f$, $1_B f = f$.

A.1.2 A function $f : A \to B$ is *surjective* (and is a *surjection*) if for each b in B, there is an a in A such that $fa = b$; f is *injective* (and is an *injection*) if for all a, a' in A, $fa = fa'$ implies $a = a'$. Finally, f is *bijective* (and is a *bijection*) if f is injective and surjective. For example, the identity $1 : A \to A$ is a bijection.

A.1.3 Let $f : A \to B$, $g : B \to A$ be functions such that $gf = 1_A$. Then we call g a *left-inverse* of f, and f a *right inverse* of g. If $a \in A$, then $a = gfa$, and so g is surjective. If $a, a' \in A$ and $fa = fa'$, then $a = gfa = gfa' = a'$; so f is injective.

Suppose further that $fg = 1_B$. Then f is surjective, g is injective. Thus the two relations $gf = 1_A$, $fg = 1_B$ imply that both f and g are bijective. If these two relations hold, we say g is an *inverse* of f. This inverse is unique because if, further, $g'f = 1_A$ then

$$g' = g'1_B = g'fg = 1_A g = g.$$

If g is *the* inverse of f, then f is the inverse of g, and we write

$$g = f^{-1}, \qquad f = g^{-1}.$$

Suppose now $f : A \to B$ is a bijection. Then the subset $\{(fa, a) \in B \times A : a \in A\}$ is functional and so defines a function $g : B \to A$ which sends each b in B to the unique a in A such that $b = fa$. Thus g is the inverse of f.

A.1.4 Let $f : A \to B$ be a function, let $A' \subseteq A$, $B' \subseteq B$ and suppose that $a \in A'$ implies $fa \in B'$. The function

$$\begin{aligned} A' &\to B' \\ a &\mapsto fa \end{aligned}$$

is called the *restriction* (or *cut down*) of f to A, B and is written

$$f \mid A', B'.$$

We also write $f \mid A', B$ (i.e., in case $B' = B$) simply as $f \mid A'$.

In particular, the restriction

$$1_A \mid A' : A' \to A$$

of the identity function is called the *inclusion function* of A' into A. It is clearly injective. We emphasise that if $A' \neq A$, then the inclusion function $A' \to A$ is not the same as the identity function $A' \to A'$, since these functions have different codomains.

A.1.5 Let $f : A \to B$ be a function and let X, Y be any sets. The *image* of X by f is the subset $f[X]$ of B consisting of the elements $f(a)$ for all a in $A \cap X$. That is,

$$f[X] = \{b \in B : \exists\, a \in A \cap X \quad \text{such that} \quad f(a) = b\}.$$

We note that $f[X] = f[X \cap A]$; but it is convenient to allow $f[X]$ to be defined for any set X, rather than restrict X to be a subset of A. The set $f[A]$ is called the *image of* f and is written $\operatorname{Im} f$.

The *inverse image* of Y by f is the set $f^{-1}[Y]$ of elements a of A such that $f(a) \in Y$. That is,

$$f^{-1}[Y] = \{a \in A : f(a) \in Y\}.$$

In this case, $f^{-1}[Y] = f^{-1}[Y \cap \operatorname{Im} f]$, but again it is convenient to allow $f^{-1}[Y]$ to be defined for any set Y.

The purpose of the square bracket notation $f[X]$ is to avoid ambiguity between the image of a set and the value of the function. For example, $f[\varnothing]$ is always the empty set, but if $A = \{\varnothing\}$, then $f(\varnothing)$ will be an element of B, not necessarily empty.

We shall make use of some abbreviations. For any y, we write $f^{-1}[y]$ for $f^{-1}[\{y\}]$. In some circumstances we omit the square brackets. For example, $f^{-1}f[X]$ means $f^{-1}[f[X]]$, and if $[a, b[$ is an interval of $\mathbb{R}$, then $f[a, b[$ means $f[[a, b[]$, and $f^{-1}[a, b[$ means $f^{-1}[[a, b[]$.

EXAMPLES In the following examples we consider only functions whose domains are subsets of $\mathbb{R}$.

1. Let f be the function $x \mapsto x^2$. Then

$$f[-2,1] = [0,4], \qquad f^{-1}[-2,4] = [0,2].$$

2. Let f be the function $x \mapsto -\log x$. Then

$$f[0,3] = [-\log 3, \rightarrow[, \qquad f^{-1}[0,1[\ = \ ]e^{-1},1].$$

A.1.6 We use in this book a number of relations between images and inverse images, and union and intersection. We state these here and leave to the reader their proof and illustration with examples. Let $f : A \to B$, $g : B \to C$ be functions, X, Y sets and $(X_j)_{j \in J}$ a family of sets.

A.1.6(1) $f^{-1}f[X] \supseteq X \cap A$.

A.1.6(2) $ff^{-1}[Y] = Y \cap f[A]$.

A.1.6(3) $f[\bigcup_{j \in J} X_j] = \bigcup_{j \in J} f[X_j]$.

A.1.6(4) $f[\bigcap_{j \in J} X_j] \subseteq \bigcap_{j \in J} f[X_j]$.

A.1.6(5) $f^{-1}[\bigcup_{j \in J} X_j] = \bigcup_{j \in J} f^{-1}[X_j]$.

A.1.6(6) $f^{-1}[\bigcap_{j \in J} X_j] = \bigcap_{j \in J} f^{-1}[X_j]$.

A.1.6(7) $(gf)[X] = gf[X]$.

A.1.6(8) $(gf)^{-1}[Y] = f^{-1}g^{-1}[Y]$.

A.1.6(9) $f^{-1}[B \setminus Y] = A \setminus f^{-1}[Y]$.

Remark There are two other definitions of function than the one adopted here. The first, and usual, definition is that a function is a set F of ordered pairs with the property that $(x, y), (x, y') \in F \Rightarrow y = y'$. That is, a function is identified with its graph. From F we can recover the domain of F, namely the set $\{x : \exists y \text{ such that } (x, y) \in F\}$, and the image of F, namely the set $\{y : \exists x \text{ such that } (x, y) \in F\}$. There are definite advantages to this definition, particularly in analysis; for example, the inverse of an injective function is easy to define. However, this definition is not sufficient for our purposes. We really do require that two functions $A \to B$, $A' \to B'$ are the same if and only if they have the same graphs, and $A = A'$, $B = B'$. In fact, when A, B are sets with structures, the functions will be indexed also with these structures.

A compromise between the two definitions, with the advantages of both, is to define a function to be a triple $f = (A, B, F)$ where F is a subset of $A \times B$ such that $(x, y), (x, y') \in F \Rightarrow y = y'$. The domain of f is then a subset of A. A function $f : A \to B$ is then called a *mapping* if its domain is all of A.

This definition is in fact required in one part of chapter 5, and a function of this type will then be written $f : A \rightarrowtail B$ and called a partial function from A to B. For the rest of the book we require the domain of a function $A \to B$ to be all of A and so we keep to the first definition given.

Exercises

1. Let $f : A \to B$, $g : B \to C$ be functions. Prove that

(a) If f and g are injective, so is gf.
(b) If f and g are surjective, so is gf.
(c) If f and g are bijective, so is gf.
(d) If gf is surjective, so also is g.
(e) If gf is injective, so also is f.

2. Let $f : A \to B$ be a function.

(a) Prove that f is surjective if and only if for all C and all functions $g, g' : B \to C$, the relation $gf = g'f$ implies $g = g'$.
(b) Prove that f is injective if and only if for all C and all functions $g, g' : C \to A$, the relation $fg = fg'$ implies $g = g'$.

3. Prove the relations given in A.1.6.
4. Let $f : X \to X'$, $g : Y \to Y'$ be functions. We define the *cartesian product* of f and g to be the function

$$\begin{aligned} f \times g : X \times Y &\to X' \times Y' \\ (x, y) &\mapsto (fx, gy). \end{aligned}$$

Prove that

(a) $f \times g$ is injective $\Leftrightarrow f, g$ are injective,
(b) $f \times g$ is surjective $\Leftrightarrow f, g$ are surjective.

Prove also that if $f' : X' \to X''$, $g' : Y' \to Y''$ then

$$(g' \times f')(g \times f) = g'g \times f'f,$$

and that $1_X \times 1_Y = 1_{X \times Y}$.
5. Let $f : X \to Y$ be a function, let $A \subseteq X$ and let $g = f \mid A, f[A]$. Prove that if $U \subseteq f[A]$, $V \subseteq Y$, then

$$U = f[A] \cap V \Leftrightarrow g^{-1}[U] = A \cap f^{-1}[V].$$

A.2 Finite, countable and uncountable sets

We shall not attempt an account of all the properties we need of the basic structures of mathematics (that is, such structures as $\mathbb{N}$, $\mathbb{Z}$, $\mathbb{Q}$, $\mathbb{R}$). In this section, we begin by starting without proof some results on counting which are particularly relevant. The details are in many books.

For each natural number $n > 0$, $\mathbb{N}_n$ denotes the set of natural number less than n; $\mathbb{N}_0$ denotes the empty set.

A.2.1 *If there is a bijection $\mathbb{N}_m \to \mathbb{N}_n$, then $m = n$.*

This result implies that the following definition makes sense. Let $n \in \mathbb{N}$. A set X *has n elements* if there is a bijection $\mathbb{N}_n \to X$. Such a bijection $f : \mathbb{N}_n \to X$ labels the elements of X as $f(0), \dots, f(n-1)$ and so, in effect, counts them. By A.2.1, such a counting process leads to a unique answer.

A set X is *finite* if X has n elements for some natural number n.

A.2.2 *A subset X of $\mathbb{N}$ is finite if and only if X is bounded above.*

Here by X is *bounded above* is meant that there is a natural number greater than every element of X. In such case, the set X has a greatest element. But $\mathbb{N}$ itself has no greatest element. Therefore, $\mathbb{N}$ if *infinite*, that is, $\mathbb{N}$ is not finite.

Another consequence of A.2.2 and of its proof is that if X is a finite subset of $\mathbb{N}$ with n elements, then every subset Y of X is finite with at most n elements. Further, if Y has n elements, then $Y = X$. An immediate consequence is:

A.2.3 *If X is a finite set with n elements, then every subset Y of X is finite with at most n elements. Further, if Y has n elements, then $Y = X$.*

The process of comparing a set X with $\mathbb{N}_n$ generalises to arbitrary sets. Two sets X and Y have the *same cardinality*, written $\divideontimes X = \divideontimes Y$, if there is a bijection $X \to Y$. This relation between sets is an equivalence relation [cf. Glossary].

A set X is *countably infinite* if $\divideontimes X = \divideontimes \mathbb{N}$; X is *countable* if X is finite or countably infinite; and otherwise, X is *uncountable*. We shall prove later that uncountable sets exist.

A.2.4 *Any subset of* $\mathbb{N}$ *is countable.*

Consequently, any subset of a countable set is again countable. But there are non-finite proper subsets of $\mathbb{N}$, for example $\mathbb{N}\setminus\{0\}$, the set of even numbers, the set of prime numbers, and so on; all these must, by A.2.4, be countably infinite. Thus the last part of A.2.3 does not generalise to infinite sets. In fact, though we do not prove this, a set X is infinite if and only if there is a proper subset Y of X such that $\divideontimes X = \divideontimes Y$.

When X is a finite set, we write $\divideontimes X$ for the number of elements of X. It is not hard to prove, and we leave it as an exercise to the reader, that if X and Y are finite then

$$\divideontimes(X \times Y) = \divideontimes X \cdot \divideontimes Y$$

and that if also X and Y are disjoint, then

$$\divideontimes(X \cup Y) = \divideontimes X + \divideontimes Y.$$

It might be thought that if X, Y are countably infinite, then $X \times Y$ is not countable. Surprisingly, this is false.

A.2.5 $\mathbb{N} \times \mathbb{N}$ *and* $\mathbb{N}$ *have the same cardinality.*

Proof Consider the function

$$\begin{aligned} f : \mathbb{N} \times \mathbb{N} &\to \mathbb{N} \\ (x, y) &\mapsto 2^x 3^y. \end{aligned}$$

Clearly, f is an injection, and so $\mathbb{N} \times \mathbb{N}$ is countable. But $\mathbb{N} \times \mathbb{N}$ is not finite since it contains the infinite set $\mathbb{N}\times\{0\}$. Therefore $\mathbb{N}\times\mathbb{N}$ is countably infinite. □

A more elementary proof of A.2.5 is suggested in Exercise 14 of Section A.2. We use A.2.5 to prove two important results (A.2.7 and A.2.8). First we need:

A.2.6 *Let* X *be countable and* $f : X \to Y$ *a surjection. Then* Y *is countable, and is finite if* X *is finite.*

Proof If X is empty, then so also is Y and the result follows. We suppose then that X is non-empty.

We first note that there is a surjection $g : \mathbb{N} \to X$. In fact, if X is infinite, then there is a bijection $g : \mathbb{N} \to X$; if X is finite, there is a bijection $e : \mathbb{N}_n \to X$ for some $n > 0$, and the function

$$\begin{aligned} g : \mathbb{N} &\to X \\ m &\mapsto \begin{cases} e(m), & m < n \\ e(n-1), & m \geqslant n \end{cases} \end{aligned}$$

is a surjection.

The function $h = fg : \mathbb{N} \to Y$ is a surjection. Hence for each y in Y, the set $h^{-1}[y]$ is non-empty, and so has a smallest element $k(y)$. Then $hk(y) = y$, whence the function $k : Y \to \mathbb{N}$ is injective [A.1.6]. So Y is countable by A.2.4.

If X is finite, then $k[Y] \subseteq \mathbb{N}_n$, and so Y is finite. □

We say a family $(X_\lambda)_{\lambda \in L}$ *is countable* if L, the set of indices, is countable; and $(X_\lambda)_{\lambda \in L}$ is *finite* if L is finite.

A.2.7 *The union of a countable family of countable sets is countable.*

Proof Let $(X_\lambda)_{\lambda \in L}$ be a countable family such that each X_λ is countable. Let $X = \bigcup_{\lambda \in L} X_\lambda$, $M = \{\lambda \in L : X_\lambda \neq \varnothing\}$. Then M is countable and $X = \bigcup_{\lambda \in M} X_\lambda$. If $M = \varnothing$, then $X = \varnothing$ and the result follows. So suppose $M \neq \varnothing$.

As shown in the proof of A.2.6, there is a surjection $f : \mathbb{N} \to M$ and for each λ in M there is a surjection $g_\lambda : \mathbb{N} \to X_\lambda$. Consider the function

$$\begin{aligned} g : \mathbb{N} \times \mathbb{N} &\to X \\ (m, n) &\mapsto g_{f(m)}(n). \end{aligned}$$

We prove that g is a surjection; this implies by A.2.6 and A.2.5 that X is countable.

Let $x \in X$. Then $x \in X_\lambda$ for some $\lambda \in M$; further, $\lambda = f(m)$ for some $m \in \mathbb{N}$ and $x = g_\lambda(n)$ for some $n \in \mathbb{N}$. Thus $x = g_{f(m)}(n)$. □

A.2.8 *The set $\mathbb{Q}$ of rational numbers is countable.*

Proof Any rational number is of the form m/n or $-m/n$ for m, n in $\mathbb{N}$ and $n \neq 0$. Let $X = \{(m, n) \in \mathbb{N} \times \mathbb{N} : n \neq 0\}$. Then X is a subset of $\mathbb{N} \times \mathbb{N}$ and so is countable.

Let $\mathbb{Q}^{\geqslant 0}$ ($\mathbb{Q}^{\leqslant 0}$) denote the set of non-negative (non-positive) rational numbers. The function $(m, n) \mapsto m/n$ is a surjection $X \to \mathbb{Q}^{\geqslant 0}$. Therefore $\mathbb{Q}^{\geqslant 0}$ is countable. Similarly, $\mathbb{Q}^{\leqslant 0}$ is countable and so $\mathbb{Q} = \mathbb{Q}^{\geqslant 0} \cup \mathbb{Q}^{\leqslant 0}$ is countable. □

In order to construct uncountable sets from countable ones we need the operation $\mathcal{P}(X)$, the *power* of a set X. By definition, $\mathcal{P}(X)$ is the set of all subsets of X. For example, if X is empty, then $\mathcal{P}(X)$ consists of one element, the empty set. If $X = \{0, 1\}$, then $\mathcal{P}(X)$ consists of $\varnothing, \{0\}, \{1\}, X$. In the exercises we shall suggest a proof that when X is finite

$$\divideontimes\, \mathcal{P}(X) = 2^{\times X}.$$

Here we prove:

A.2.9 *For any set X there is an injection* $X \to \mathcal{P}(X)$, *but no bijection* $X \to \mathcal{P}(X)$.

Proof The function $x \mapsto \{x\}$ is an injection $X \to \mathcal{P}(X)$.

Suppose $f : X \to \mathcal{P}(X)$ is a bijection with inverse $g : \mathcal{P}(X) \to X$. We derive a contradiction by a marvellous argument due to G. Cantor.

For each x in X, $f(x)$ is a subset of X. So it makes sense to ask whether or not $x \in f(x)$. Let

$$A = \{x \in X : x \notin f(x)\} \qquad (*)$$

and let $a = g(A)$ so that $f(a) = A$. We now ask: does $a \in A$? If $a \in A$, then by (*) $a \notin f(a)$. Since $f(a) = A$, we have a contradiction. On the other hand, if $a \notin A$, then, since $A = f(a)$, we have $a \notin f(a)$. So $a \in A$ by (*), and we still have a contradiction. This shows that a bijection $X \to \mathcal{P}(X)$ cannot exist. □

This result implies that $\mathcal{P}(\mathbb{N})$ is uncountable.

A.2.10 *The set* $\mathbb{R}$ *of real numbers is uncountable.*

Proof It is sufficient to prove that $\mathbb{R}$ has an uncountable subset. For this we construct an injection

$$\chi : \mathcal{P}(\mathbb{N}) \to \mathbb{R}.$$

For any subset X of $\mathbb{N}$ we define the *characteristic function* of X

$$\chi_X : \mathbb{N} \to \{0, 1\}$$
$$n \mapsto \begin{cases} 0 & \text{if } n \notin X \\ 1 & \text{if } n \in X. \end{cases}$$

Clearly, $\chi_X = \chi_{X'}$ if and only if $X = X'$. We now define

$$\chi : \mathcal{P}(\mathbb{N}) \to \mathbb{R}$$
$$X \mapsto \sum_{n=0}^{\infty} \chi_X(n)/2^{2n}.$$

Certainly the series for $\chi(X)$ is convergent, and so $\chi(X)$ is a well-defined real number. In fact, $\chi(X)$ is a binary decimal of the form

$$n_0.0n_1 0n_2 0n_3 0 \cdots . \qquad (**)$$

It follows that if $\chi(X) = \chi(X')$, then $\chi_X(n) = \chi_{X'}(n)$ for all n in $\mathbb{N}$, whence $X = X'$ (the presence of the 0's in (**) ensures that there is no difficulty with repeated 1's). □

The argument of A.2.10 can be refined to show that $※\mathbb{R} = ※\mathcal{P}(\mathbb{N})$, but we do not need this fact.

The reader should keep well aware of the intuitive meaning of A.2.10. The rational numbers are 'dense' in the real line in the sense that there are rational numbers arbitrarily close to any real number; also between any two rational numbers there is another. Thus the rational numbers appear to fill up the real line. But this is illusory: by A.2.10 there are many more real numbers than rational numbers.

A further point is that very little is known about the irrational numbers, and particularly about the non-algebraic numbers (for whose definition see Exercise 14). It can be argued that by the process of mathematical reasoning we can acquire at most a countable number of facts about $\mathbb{R}$; yet $\mathbb{R}$ is uncountable and so its properties must remain largely unexplored.

A.2.11 *The sets $\mathbb{R} \times \mathbb{R}$ and $\mathbb{R}$ have the same cardinality.*

Proof Let $A =]0, 1]$. The function

$$\begin{aligned} f : \mathbb{Z} \times A &\to \mathbb{R} \\ (n, a) &\mapsto n + a \end{aligned}$$

is a bijection. Also $\mathbb{Z}$ is countably infinite (the function $n \mapsto 2n$ $(n \geqslant 0)$, $n \mapsto -2n - 1$ $(n < 0)$ is a bijection $\mathbb{Z} \to \mathbb{N}$) and so there is a bijection $g : \mathbb{Z} \times \mathbb{Z} \to \mathbb{Z}$. We shall prove below that there is a bijection $h : A \times A \to A$; the composite

$$\mathbb{R} \times \mathbb{R} \xrightarrow{f^{-1} \times f^{-1}} \mathbb{Z} \times A \times \mathbb{Z} \times A \xrightarrow{1 \times T \times 1} \mathbb{Z} \times \mathbb{Z} \times A \times A \xrightarrow{g \times h} \mathbb{Z} \times A \xrightarrow{f} \mathbb{R}$$

in which T is the bijection $(a, n) \mapsto (n, a)$, is the required bijection $\mathbb{R} \times \mathbb{R} \to \mathbb{R}$.

Let B be the set of all sequences $(m_r)_{r \geqslant 0}$ of natural numbers. We construct a bijection $i : A \to B$ as follows. Each a in A can be written uniquely as a binary decimal,

$$.a_1 a_2 a_3 \cdots$$

where $a_t = 0$ or 1 and the expression does not end in repeated 0's (that is, it is false that $a_r = 0$ for r sufficiently large). We define $i(a)$ to be the sequence $(m_r)_{r \geqslant 0}$ such that m_0 is the number of 0's between the decimal point and the first 1, and m_r $(r > 0)$ is the number of 0's between the rth and the $(r + 1)$st 1. For example, if $a = .01100101 \cdots$ then $i(a)$ is initially the sequence $1, 0, 2, 1, \cdots$.

There is an obvious bijection $j : B \times B \to B$, where $(l_r) = j((m_r), (n_r))$ is the sequence $m_0, n_0, m_1, n_1, \cdots$, that is, $l_{2r} = m_r$, $l_{2r+1} = n_r$. The composite

$$A \times A \xrightarrow{i \times i} B \times B \xrightarrow{j} B \xrightarrow{i^{-1}} A$$

is the required bijection. □

EXERCISES

1. Prove that the function $x \mapsto x/(1+|x|)$ is a bijection $\mathbb{R} \to]-1, 1[$.
2. Let a, b be real numbers such that $a < b$. Construct a bijection $[0,1] \to [a,b]$.
3. Construct bijections $[0,1] \to [0,1[\to]0,1[$.
4. Let X, Y be disjoint sets such that $※X = ※Y = ※\mathbb{R}$. Prove that $※(X \cup Y) = ※\mathbb{R}$.
5. Let X, Y be finite sets with m, n elements respectively. Prove that Y^X, the set of all functions $X \to Y$, has n^m elements.
6. Let A be a set. For each subset X of A define the *characteristic function* of X

$$\chi_X : A \to \{0,1\}$$
$$a \mapsto \begin{cases} 0 & \text{if } a \notin X \\ 1 & \text{if } a \in X. \end{cases}$$

Prove that the function $X \mapsto \chi_X$ is a bijection $\mathcal{P}(A) \to \{0,1\}^A$. Deduce that if $※A = n$, then $※\mathcal{P}(A) = 2^n$.
7. Continuing the notion of Exercise 6, prove that if X and $X_1, \cdots, X_n$ are subsets of A then

(a) $\chi(A \setminus X) = 1 - \chi(X)$,

(b) $\chi(X_1 \cap \cdots \cap X_n) = \chi(X_1)\chi(X_2)\cdots\chi(X_n)$,

(c) $\chi(X_1 \cup \cdots \cup X_n) = 1 - (1-\chi(X_1))(1-\chi(X_2))\cdots(1-\chi(X_n))$.

Use the characteristic function to verify that

$$X \cap (X_1 \cup \cdots \cup X_n) = (X \cap X_1) \cup \cdots \cup (X \cap X_n).$$

8. We define the *projections*

$$p_1 : X \times Y \to X \qquad p_2 : X \times Y \to Y$$
$$(x,y) \mapsto x \qquad (x,y) \mapsto y.$$

Let $\Delta : Z \to Z \times Z$ be the *diagonal map* $z \mapsto (z,z)$. Prove that the functions

$$\rho : (X \times Y)^Z \to X^Z \times Y^Z \qquad \sigma : X^Z \times Y^Z \to (X \times Y)^Z$$
$$f \mapsto (p_1 f, p_2 f) \qquad (f,g) \mapsto (f \times g)\Delta$$

satisfy $\rho\sigma = 1$, $\sigma\rho = 1$. Deduce that if l, m, n are natural numbers, then

$$(lm)^n = l^n m^n.$$

9. Let X, Y be disjoint sets. Prove that there is a bijection

$$Z^{X \cup Y} \to Z^X \times Z^Y.$$

Deduce that if l, m, n are natural numbers, then

$$l^{m+n} = l^m l^n.$$

10. Let X, Y, Z be sets. Prove that the exponential map

$$\begin{aligned} e : X^{Z \times Y} &\to (X^Y)^Z \\ f &\mapsto (z \mapsto (y \mapsto f(z, y))) \end{aligned}$$

is a bijection. (The notation means that $e(f)$ is the function such that $e(f)(z)(y) = f(z, y)$.) Deduce that if l, m, n are natural numbers, then

$$l^{nm} = (l^m)^n.$$

11. Let X, Y be finite sets with m, n elements respectively where $m \leqslant n$. Determine the number of bijections $X \to X$ and the number of injections $X \to Y$.
12. Read the proof of A.2.10 given in [Die60, 2.2.17].
13. Prove that the function

$$\begin{aligned} f : \mathbb{N} \times \mathbb{N} &\to \mathbb{N} \\ (x, y) &\mapsto \tfrac{1}{2}(x + y)(x + y + 1) + y \end{aligned}$$

is a bijection.
14. A real number α is called *algebraic* if α satisfies an equation

$$\alpha^n + a_1 \alpha^{n-1} + \cdots + a_{n-1} \alpha + a_n = 0$$

where $a_i \in \mathbb{Z}$. Prove that the set of algebraic numbers is countable.
15. Prove that the plane $\mathbb{R} \times \mathbb{R}$ is not the union of countably many lines.

A.3 Products and the axiom of choice

Let $(X_\lambda)_{\lambda \in L}$ be a family of sets and let X be the set of all families $x = (x_\lambda)_{\lambda \in L}$ such that $x_\lambda \in X_\lambda$. Then X is called a *product* of the family $(X_\lambda)_{\lambda \in L}$ and is denoted by

$$\prod_{\lambda \in L} X_\lambda .$$

The elements x_λ of X_λ is called the *λth coordinate* of x. Thus we have functions

$$\begin{aligned} p_\lambda : X &\to X_\lambda \\ x &\mapsto x_\lambda ; \end{aligned}$$

the function p_λ is called the *λth projection*. (Note that a family $(x_\lambda)_{\lambda \in L}$ such that $x_\lambda \in X_\lambda$ is a function $x : L \to X'$ where $X' = \bigcup_{\lambda \in L} X$, and that x_λ is the same as $x(\lambda)$, the value of x at λ.)

Suppose, in particular, that $L = \{0, 1\}$. The function

$$\prod_{\lambda \in L} X_\lambda \to X_0 \times X_1$$
$$x \mapsto (x_0, x_1)$$

is a bijection, and this shows that the product $\prod_{\lambda \in L} X_\lambda$ is a reasonable generalisation of the cartesian product $X_0 \times X_1$.

If one of the sets X_λ is empty, then so also is $\prod_{\lambda \in L} X_\lambda$. The converse of this statement is the Axiom of Choice.

Axiom of Choice. If $\prod_{\lambda \in L} X_\lambda$ is empty, then one of the sets X_λ is empty. Or, alternatively, if each set X_λ is non-empty, then $\prod_{\lambda \in L} X_\lambda$ is non-empty.

We shall discuss later why this is an axiom rather than a theorem. For the moment, we illustrate the axiom by a simple consequence.

Let X, Y be sets. We say $※ Y \leqslant ※ X$ if there is an injection $Y \to X$.

A.3.1 *Let X, Y be sets and $f : X \to Y$ a surjection. Then $※ Y \leqslant ※ X$.*

Proof We suppose X is non-empty, so that Y is non-empty. Then for each y in Y the set $f^{-1}[y]$ is non-empty. By the Axiom of Choice there is an element k of $\prod_{y \in Y} f^{-1}[y]$.

Now k is a function $Y \to X$. Since $k(y) \in f^{-1}[y]$ for each $y \in Y$, it follows that $fk = 1_Y$. Therefore, k is injective and so $※ Y \leqslant ※ X$. □

The reader should compare this proof carefully with that of A.2.6. In A.2.6 an injection $k : Y \to \mathbb{N}$ was constructed explicitly using properties of the natural numbers; no such method is available here and we must rely instead on the Axiom of Choice.

The following theorem was proved by Cantor using the Axiom of Choice. Later, proofs were found not using this axiom, and for this reason the theorem is known as the Schröder-Berstein theorem.

A.3.2 *Let X, Y be sets. If $※ X \leqslant ※ Y$ and $※ Y \leqslant ※ X$, then $※ X = ※ Y$.*

On the other hand, the proof of the following theorem does involve the Axiom of Choice.

A.3.3 *If X and Y are sets, then either $※ X \leqslant ※ Y$ or $※ Y \leqslant ※ X$.*

Neither of these theorems is essential for this book, and so we omit the proofs.

For any set X there is a set Y such that $※ Y > ※ X$ (for example, $Y = \mathcal{P}(X)$). An obvious problem is to determine all infinite cardinalities. With out present notation we can state only the following result.

A.3.4 *There are sets* $A_{\mathbb{N}}$*, and* A_n *for each* n *in* $\mathbb{N}$*, such that* $A_0 = \mathbb{N}$ *and*

$$※ A_0 < ※ A_1 < \cdots < ※ A_n < \cdots < ※ A_{\mathbb{N}}.$$

Further, if X *is not finite and has cardinality less than that of* $A_{\mathbb{N}}$*, then* $※ X = ※ A_n$ *for some* n *in* $\mathbb{N}$.

There are also greater cardinalities than that of $A_{\mathbb{N}}$, but to describe these, and also to prove A.3.2, the theory of ordinal numbers is required [cf., for example, the Appendix to [Kel55]].

We can now ask: where, if anywhere, in the list given in A.3.2 does the cardinality of $\mathbb{R}$ fall? We have already proved that $※ \mathbb{R} > \mathbb{N}$; Cantor spent a large portion of his life trying to prove the

Continuum Axiom: $※ \mathbb{R} = ※ A_1$.

It is now known that his attempts were bound to fail, although we can describe only roughly why this is so.

In order to decide whether or not this axiom, and the Axiom of Choice, can be proved, it is first necessary to describe precisely what is meant by a proof. This involves setting up carefully the system of logic used in our normal arguments. Second, within this system it is necessary to construct a Set Theory; such a theory is an axiomatic system, with undefined terms such as set, membership, and so on, together with axioms guaranteeing the existence or non-existence of certain sets and governing the use of the undefined terms.

Once Set Theory is set up in this way, we can then ask: do the Axiom of Choice and the Continuum Axiom follow from the other axioms of Set Theory? It was proved by P. Cohen in 1962 that these two axioms are independent of each other and of the other axioms of Set Theory. This means that for each of these axioms we obtain three theories, in which the axiom is either asserted, or denied, or simply left out (and so there are nine theories from these two axioms!).

Few of these theories have been investigated, and there are other varieties of Continuum Axiom which are the subject of current research. The Axiom of Choice plays an important role in many branches of mathematics, and we shall use it without further mention. The Continuum Axiom has some (possibly undesirable) consequences in measure theory. There are also applications in general topology of other Axioms. For a further discussion of these topics we refer the reader to [Göd47], [Coh66], [Joh02], and [IK87].

A.4 Universal properties

One purpose of this section is to give an abstract characterisation of the product of sets and to introduce the 'dual' notion of sum of sets (the word dual is explained in chapter 6). The kind of argument used is of importance in discussing adjunction spaces and computations of the fundamental group. I suggest you read through this section quickly and return to master it when you find the methods are needed.

Let $(X_\lambda)_{\lambda \in L}$ be a family of sets. We have seen that associated with the product $\prod_{\lambda \in L} X_\lambda$ is a family of projections

$$p_\lambda : \prod_{\lambda \in L} X_\lambda \to X_\lambda.$$

The crucial property of these projections is that they determine completely functions *into* the product.

A.4.1 *Let* Y *be a set and let there be given a family of functions*

$$f_\lambda : Y \to X_\lambda, \quad \lambda \in L.$$

Then there is a unique function $f : Y \to \prod_{\lambda \in L} X_\lambda$ *such that*

$$p_\lambda f = f_\lambda, \quad \lambda \in L.$$

Proof An element x of $\prod X_\lambda$ is determined completely by its family of 'coordinates' $x_\lambda = p_\lambda x$. Therefore, we can define f by saying that for each y in Y, $f(y)$ is to have coordinates $f_\lambda(y)$ for each $\lambda \in L$. This is equivalent to $p_\lambda f(y) = f_\lambda(y)$. □

We now show that this property characterises the product in a certain sense.

A.4.2 *Let* $p'_\lambda : X' \to X_\lambda$, $\lambda \in L$ *be a family of functions. Then the following conditions are equivalent.*

(a) *There is a bijection* $p' : X' \to \prod X_\lambda$ *such that*

$$p_\lambda p' = p'_\lambda, \quad \lambda \in L.$$

(b) *For any* Y *and any family of functions* $f_\lambda : Y \to X_\lambda$, $\lambda \in L$, *there is a unique function* $f : Y \to X'$ *such that*

$$p'_\lambda f = f_\lambda, \quad \lambda \in L.$$

Proof In order to bring out the structure of the argument we make some definitions. All families in what follows will be indexed by L; the family (X_λ) is given.

By an *object* we mean a pair $(Y, (f_\lambda))$ consisting of a set Y and a family $f_\lambda : Y \to X_\lambda$ of functions. Such an object is abbreviated to (Y, f_λ). By a *map* $f : (Y, f_\lambda) \to (Z, g_\lambda)$ of objects we mean a function $f : Y \to Z$ such that

$$g_\lambda f = f_\lambda, \quad \text{all } \lambda \in L.$$

We then have the following properties.
(i) The identity $1 : (Y, f_\lambda) \to (Y, f_\lambda)$ is a map.
(ii) If $f : (Y, f_\lambda) \to (Z, g_\lambda)$ and $g : (Z, g_\lambda) \to (W, h_\lambda)$ are maps, then so also is $gf : (Y, f_\lambda) \to (W, h_\lambda)$. (Since $h_\lambda gf = g_\lambda f = f_\lambda$.)
(iii) If $f : (Y, f_\lambda) \to (Z, g_\lambda)$ is a map, and $f : Y \to Z$ is a bijection, then $f^{-1} : (Z, g_\lambda) \to (Y, f_\lambda)$ is a map. (The relation $g_\lambda f = f_\lambda$ implies that $f_\lambda f^{-1} = g_\lambda f f^{-1} = g_\lambda$.)

Let $X = \prod X_\lambda$. The property A.4.1 can now be stated as: *for any object* (Y, f_λ) *there is exactly one map* $f : (Y, f_\lambda) \to (X, p_\lambda)$. Similarly, the property (b) above can be stated: *for any object* (Y, f_λ) *there is exactly one map* $f : (Y, f_\lambda) \to (X', p'_\lambda)$.

We now show that (b) $\Rightarrow$ (a). By A.4.1 there is exactly one map $p' : (X', p'_\lambda) \to (X, p_\lambda)$. By (b) there is exactly one map $p : (X, p_\lambda) \to (X', p'_\lambda)$. By (ii), $p'p : (X, p_\lambda) \to (X, p_\lambda)$ is a map; but there is only one map $(X, p_\lambda) \to (X, p_\lambda)$ and by (i) this is the identity. Hence $p'p = 1$. Similarly, $pp' : (X', p'_\lambda) \to (X', p'_\lambda)$ is a map and so $pp' = 1$. These relations show that p' is a bijection, while the relations given in (a) simply state that p' is a map.

We now show that (a) $\Rightarrow$ (b). Since p' is a bijection it has an inverse, which we write $p : X \to X'$. By (iii), $p : (X, p_\lambda) \to (X', p'_\lambda)$ is a map. Let (Y, f_λ) be an object. By A.4.1 there is a unique map $f : (Y, f_\lambda) \to (X, p_\lambda)$ and by (ii) $pf : (Y, f_\lambda) \to (X', p'_\lambda)$ is a map. So we have constructed a map as required, but we must show that this is unique.

Suppose then $g : (Y, f_\lambda) \to (X', p'_\lambda)$ is a map. Then $p'g : (Y, f_\lambda) \to (X, p_\lambda)$ is a map and hence $p'g = f$. It follows that $g = pp'g = pf$. □

The property A.4.1 of the product is called a *universal property*, and the method of proof of A.4.2 will be called the *usual universal argument*.

The important thing about a product is that it has projections which satisfy the universal property. There is in many cases no canonical choice for the actual set we take as the product: for example, if L consists of three elements 1,2,3, then we can take for the product either $\prod_{\lambda \in L} X_\lambda$ (as defined above) or

$$(X_1 \times X_2) \times X_3, \quad \text{or} \quad X_1 \times (X_2 \times X_3)$$

and there seems no reason to prefer one to another. The fact that in each case we have projections satisfying the universal property ensures that there are bijections between any two of these sets.

We therefore now *define* a product of the family $(X_\lambda)_{\lambda \in L}$ of sets to be a set X with projections $p_\lambda : X \to X_\lambda$, $\lambda \in L$, such that the universal property of A.4.1 holds. The set X is then also written $\prod_{\lambda \in L} X_\lambda$.

In the case L is finite, say $L = \{1, \ldots, n\}$, then a product of $X_1, \ldots, X_n$ is written $X_1 \times \cdots \times X_n$. If $f_\lambda : Y \to X_\lambda$, $\lambda = 1, \ldots, n$ is a family of functions, then the unique function $f : Y \to X_1 \times \cdots \times X_n$ such that $p_\lambda f = f_\lambda$ is written $(f_1, \ldots, f_n)$, and f_λ is called the λth component of f.

Sum of sets

We shall also need the less familiar construction of the *sum* (also called the *disjoint union*, or *coproduct*) of sets. We define this *ab initio* by a universal property.

Let $(X_\lambda)_{\lambda \in L}$ be a family of sets. A *sum* of the family is a set X together with functions $i_\lambda : X_\lambda \to X$, called *injections*, with the following property: for any Y and any family of functions $f_\lambda : X_\lambda \to Y$, $\lambda \in L$, there is a unique function $f : X \to Y$ such that $f i_\lambda = f_\lambda$, $\lambda \in L$.

The formal difference between the definitions of product and sum are that we are given maps *from* the product and *into* the *sum*. So we distinguish between the two universal properties by saying that the product is ι-universal (ι for initial) and the sum is φ-universal (φ for final).

A.4.3 *Let $i_\lambda : X_\lambda \to X$ ($\lambda \in L$) be a sum of the family $(X_\lambda)_{\lambda \in L}$, and let $i'_\lambda : X_\lambda \to X'$ ($\lambda \in L$) be a family of functions. Then the following conditions are equivalent.*
(a) *There is a bijection $i' : X \to X'$ such that*

$$i' i_\lambda = i'_\lambda, \quad \lambda \in L.$$

(b) *X' and the family $(i'_\lambda)_{\lambda \in L}$ is a sum of the family $(X_\lambda)_{\lambda \in L}$.*

Proof The method of proof is similar to that of A.4.2 and we only outline it.

An *object* (Y, f_λ) is defined to be a pair consisting of a set Y and a family $f_\lambda : X_\lambda \to Y$ of functions. A *map* $f : (Y, f_\lambda) \to (Z, g_\lambda)$ is defined to be a function $f : Y \to Z$ such that

$$f f_\lambda = g_\lambda, \quad \lambda \in L.$$

The properties analogous to (i), (ii), (iii) in the proof of A.4.2 are easily verified.

The object (X, i_λ) is thus a sum if and only if there is exactly one map $(X, i_\lambda) \to (Y, f_\lambda)$ for any object (Y, f_λ). The remainder of the proof is analogous to that of A.4.2 and is left to the reader. □

If $i_\lambda : X_\lambda \to X$ is a sum then we write

$$X = \bigsqcup_{\lambda \in L} X_\lambda.$$

In particular, if $L = \{1, \ldots, n\}$ then we write

$$X = X_1 \sqcup \cdots \sqcup X_n.$$

We must now show that sums exist. Suppose first of all that the family (X_λ) consists of disjoint sets. Let $X = \bigcup_{\lambda \in L} X_\lambda$ and let $i_\lambda : X_\lambda \to X$ be the inclusion. Then, clearly, we have a sum of the family, since first a function $f : X \to Y$ is completely determined by the functions $f_\lambda = f \mid X_\lambda = f i_\lambda$, and second, any such family defines a function f. Thus for families of disjoint sets, we can always take $\bigsqcup_{\lambda \in L} X_\lambda$ to be the union $\bigcup_{\lambda \in L} X_\lambda$.

The situation is different if the sets are not disjoint. For example, two functions $f_1 : X_1 \to Y$, $f_2 : X_2 \to Y$ define a function $f : X_1 \cup X_2 \to Y$ such that $f \mid X_1 = f_1$, $f \mid X_2 = f_2$ if and only if f_1, f_2 agree on $X_1 \cap X_2$, a condition which is vacuous if $X_1 \cap X_2 = \varnothing$ but not otherwise.

A.4.4 *Any family* $(X_\lambda)_{\lambda \in L}$ *of sets has a sum.*

Proof The idea of the proof is to replace the given family by a family of disjoint sets and then take the union. In fact we replace X_λ by $X'_\lambda = X_\lambda \times \{\lambda\}$; then X'_λ meets X'_μ if and only if $\lambda = \mu$.

Let $X = \bigcup_{\lambda \in L} X'_\lambda$ and let

$$\begin{aligned} i_\lambda : X_\lambda &\to X \\ x &\mapsto (x, \lambda). \end{aligned}$$

Let $f_\lambda : X_\lambda \to Y$ be a family of functions. Then the function

$$\begin{aligned} f : X &\to Y \\ (x, \lambda) &\mapsto f_\lambda(x) \end{aligned}$$

is well-defined and is the only function $X \to Y$ whose composite with each i_λ is f_λ. □

A.4.4 *(Corollary 1) If* $i_\lambda : X_\lambda \to X$ *is a sum, then each* i_λ *is injective.*

Proof If $i_\lambda : X_\lambda \to X$ is the sum constructed in A.4.4, then it is obvious that each i_λ is injective. If $i'_\lambda : X_\lambda \to X'$ is any sum, then there is a bijection $i' : X \to X'$ such that $i' i_\lambda = i'_\lambda$, and it follows that each i'_λ is injective. □

In the case of a finite family $X_1, \ldots, X_n$, the function $f : X_1 \sqcup \cdots \sqcup X_n \to Y$ such that $f i_\lambda = f_\lambda$, $\lambda = 1, \ldots, n$, is sometimes written

$$(f_1, \ldots, f_n)^t.$$

Equivalence relations

We need in chapter 4 some facts on equivalence relations in addition to those given in the Glossary.

Let R be a relation on a set A. We define a new relation E on A by $aEb \Leftrightarrow$ there is a sequence $a_1, \cdots, a_n$ of elements of A such that
(a) $a_1 = a, \quad a_n = b$,
(b) for each $i = 1, \ldots, n-1$,

$$a_i R a_{i+1} \quad \text{or} \quad a_{i+1} R a_i \quad \text{or} \quad a_i = a_{i+1}.$$

A.4.5 *The relation E is the smallest equivalence relation on A containing R.*

Proof We first prove that E is an equivalence relation. Let $a, b, c \in A$. Clearly aEa, since the sequence a, a satisfies (a) and (b). Suppose aEb, and that $a_1, \ldots, a_n$ satisfies (a) and (b). Then the sequence $a_n, \ldots, a_1$ ensures that bEa.

Finally, if aEb and bEc, then by splicing together two sequences we obtain that aEc.

Suppose E' is an equivalence relation containing R. We have to prove that E is contained in E', i.e., that $aEb \Rightarrow aE'b$.

Let aEb, and let $a_1, \ldots, a_n$ be a sequence satisfying (a) and (b). Now E' contains R. So $a_i E' a_{i+1}$ for each $i = 1, \ldots, n-1$ (by (b)). Hence $a_1 E' a_n$, i.e., $aE'b$. □

Because of A.4.5, we call E the *equivalence relation generated by* R. Let $p : A \to A/E$ be the projection $a \mapsto \text{cls}\, a$. Then p can be characterised by a φ-universal property.

A.4.6 (a) *For all a, b in A, $aRb \Rightarrow pa = pb$.*
(b) *If $f : A \to B$ is any function such that*

$$aRb \Rightarrow fa = fb \quad \textit{all } a, b \textit{ in } A.$$

then there is a unique function $f^* : A/E \to B$ such that

$$f^* p = f.$$

Proof The proof of (a) is obvious, since $R \subseteq E$ and

$$pa = pb \quad \Leftrightarrow \quad \text{cls}\, a = \text{cls}\, b \quad \Leftrightarrow \quad aEb.$$

Let $f : A \to B$ be a function, and let E_f be the relation on A given by

$$aE_f b \quad \Leftrightarrow \quad fa = fb.$$

Then E_f is an equivalence relation. Suppose also $aRb \Rightarrow fa = fb$. Then E_f contains R and so E_f contains E. Hence if $\text{cls}\, a = \text{cls}\, b$, then $fa = fb$. Therefore the function

$$\begin{aligned} f^* : A/E &\to B \\ \text{cls}\, a &\mapsto fa \end{aligned}$$

is well-defined, and clearly $f^*p = f$. The uniqueness of f^* follows from the fact that p is surjective. □

The usual universal argument shows that if $p' : A \to A'$ is any function satisfying A.4.6(a) and (b) (with p replaced by p') then there is a unique bijection $p^* : A/E \to A'$ such that $p^*p = p'$. Also, given a bijection $p^* : A/E \to A'$ for some A', then $p' = p^*p$ satisfies A.4.6(a) and (b) (with p replaced by p');

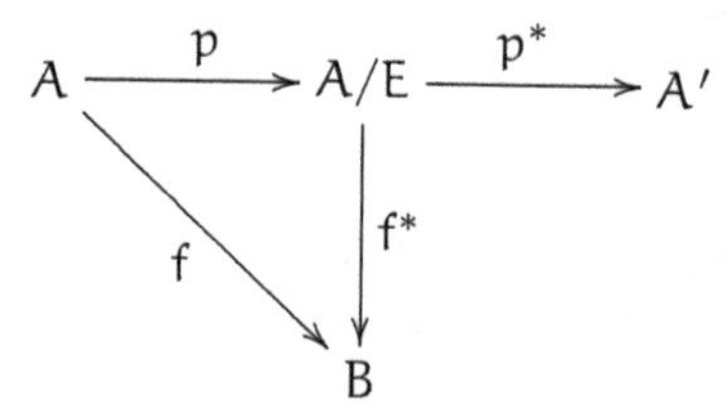

(for if $f : A \to B$ satisfies $aRb \Rightarrow fa = fb$, then we can construct $f^* : A/E \to B$ such that $f^*p = f$, and then take $f^*(p^*)^{-1} : A' \to B$).

EXERCISES

1. Prove that for any sets X_1, X_2, X_3, there are bijections

$$\begin{aligned} X_1 \sqcup X_2 &\to X_2 \sqcup X_1, \\ X_1 \sqcup (X_2 \sqcup X_3) &\to (X_1 \sqcup X_2) \sqcup X_3, \\ X_1 \times (X_2 \sqcup X_3) &\to (X_1 \times X_2) \sqcup (X_1 \times X_3). \end{aligned}$$

2. Let $f_1 : X_1 \to Y_1$, $f_2 : X_2 \to Y_2$ be functions. The function

$$f_1 \sqcup f_2 : X_1 \sqcup X_2 \to Y_1 \sqcup Y_2$$

is by definition $(i_1 f_1, i_2 f_2)^t$. Prove that
(a) $f_1 \sqcup f_2$ is injective $\Leftrightarrow f_1, f_2$ are injective.
(b) $f_1 \sqcup f_2$ is surjective $\Leftrightarrow f_1, f_2$ are surjective.
Prove also that if $g_1 : Y_1 \to Z_1$, $g_2 : Y_2 \to Z_2$, then

$$(g_1 \sqcup g_2)(f_1 \sqcup f_2) = g_1 f_1 \sqcup g_2 f_2.$$

3. An n-diagram is defined to be a pair of functions

$$\{x\} \xrightarrow{j} X \xrightarrow{f} X$$

such that $x \in X$ and j is the inclusion. A *map* of n-diagrams is a commutative diagram,

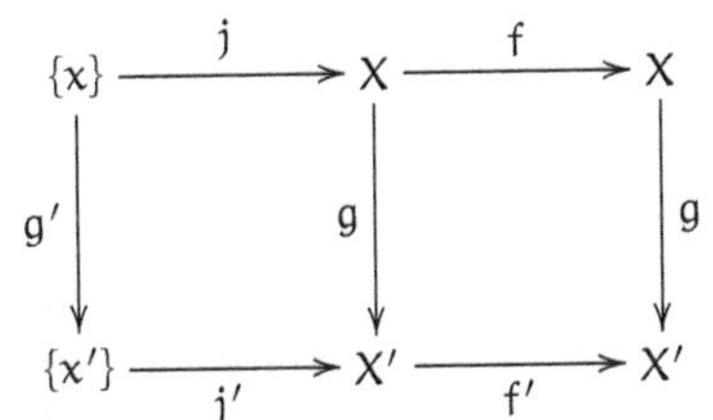

Let $\mathcal{N}$ be the n-diagram $\{0\} \to \mathbb{N} \xrightarrow{s} \mathbb{N}$ where s is the function $n \mapsto n+1$. Prove that $\mathcal{N}$ is φ-universal: that is, if $\mathcal{X}$ is any n-diagram, then there is a unique map $\mathcal{N} \to \mathcal{X}$ of n-diagrams. Show that this property expresses the principle of inductive (or recursive) definition of a function. Show that if also $\mathcal{N}'$ is φ-universal, then there are unique maps $\varphi : \mathcal{N} \to \mathcal{N}'$, $\psi : \mathcal{N}' \to \mathcal{N}$ such that $\varphi\psi$, $\psi\varphi$ are the respective identity maps.

Deduce from the φ-universal property the following: if $h : \mathbb{N} \times Y \to Y$ is a function, and $y_0 \in Y$, then there is a unique function $k : \mathbb{N} \to Y$ such that

$$k(0) = y_0, \qquad k(n+1) = h(n, k(n)), \quad n \in \mathbb{N}.$$

Show that the φ-universal property implies the following form of the principle of induction: if $A \subseteq \mathbb{N}$ and (i) $0 \in A$, (ii) for all n, $n \in A \Rightarrow n+1 \in A$, then $A = \mathbb{N}$.

NOTES

The most important use of the principle of induction is to define functions. It seems reasonable therefore to axiomatise the natural numbers in a way which immediately allows such a definition. The above universal property does exactly this. For more information, seek information on *natural number objects*. But note that we have shown in Chapter 6 that the integers $\mathbb{Z}$ can be defined by a pushout from the finite groupoid **I**. It is easy to define the natural numbers by an analogous pushout in the category of small categories.

As explained in the Preface to this edition, the notion of universal argument has proved very fruitful for obtaining analogies. This suggests the principle, which is quite well established, that in general definitions should if possible be given in terms of universal properties; afterwards, the specific construction should, even must, be given, but this may seem to have a certain *ad hoc* nature, and the construction should not be the definition. Of course, in this book we have offended against this principle in a number of places, sometimes for pedagogic reasons, and the reader is invited to rewrite those parts in a different order, and to evaluate where the payoff for using universal properties is clear!

Glossary of terms from set theory

Agree Two functions $f, g : X \to Y$ agree at a point x in X if $f(x) = g(x)$; and f, g agree on a subset A of X if $f \mid A = g \mid A$.

Anti-symmetric cf. *Relation*.

Belongs to Synonyms for a belongs to A are: a is a member of A, a is an element of A, a is in A, A contains a, $a \in A$, $A \ni a$.

Bounded Let X be a set and $\leqslant$ a partial order relation on X. Let A be a subset of X. An *upper bound* for A is an element u of X such that $a \in A \Rightarrow a \leqslant u$. A *supremum*, $\sup A$, for A is an upper bound u for A such that if v is any upper bound for A then $u \leqslant v$. The terms *lower bound, infinum* ($\inf A$) are defined as for upper bound and supremum but with $\leqslant$ replaced by $\geqslant$. The set A is *bounded above* if it has an upper bound, *bounded below* if it has a lower bound, and *bounded* if it is bounded above and below. The order relation is *complete* if every non-empty subset of X which is bounded above has a supremum (and this implies that every non-empty subset of X which is bounded below has an infinum).

Cartesian product The Cartesian product of two sets A, B is the set $A \times B$ of all ordered pairs (a, b) for a in A, b in B.

Class There is a necessity to rescue set theory from some rather simple contradictions. One way of doing this is by distinguishing classes and sets. Here class is the general notion and a set is a class which is a member of some other class [cf. [Kel55]]. A more useful method for category theory is to assume axioms postulating the existence of 'universes' [cf. [Son62]]. Another approach is to axiomatise category theory instead of set theory [cf. [Law66]]. See also [Bla84].

Classifier notation This is the notation $\{x : P(x)\}$; it denotes the set of all

x having property $P(x)$ (if such exists; otherwise it has no denotation). We write $\{x \in A : P(x)\}$ for $\{x : P(x)$ and $x \in A\}$. If A can be understood from the context it is common to abbreviate $\{x \in A : P(x)\}$ to $\{x : P(x)\}$. If $a \in A$, then the set $\{x \in A : x \neq a\}$ is written $A^{\neq a}$.

Commutative The following diagrams of functions

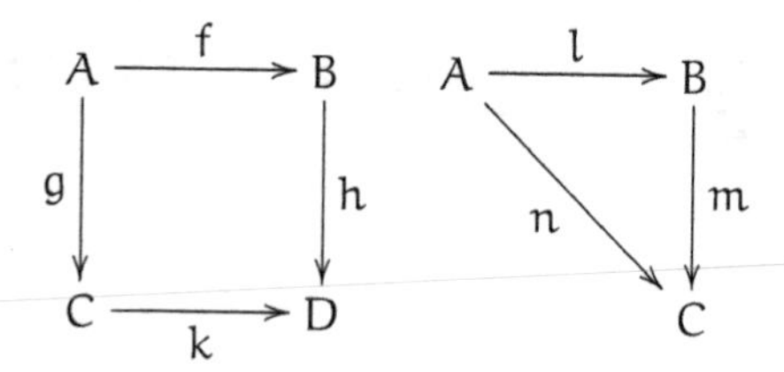

are *commutative* if $hf = kg$ (in the first case) and $n = ml$ (in the second case). This definition is extended to more complicated diagrams in the obvious way.

Complement If A is a subset of B, the complement of A in B is the difference $B \setminus A$. When B can be understood from the context, we refer to $B \setminus A$ simply as the complement of A.

Complete See *Bounded*.

De Morgan laws Let A, B, X be sets, and let $(A_i)_{i \in I}$ be a family of subsets of X. The De Morgan laws are:

$$X \setminus (A \cup B) = (X \setminus A) \cap (X \setminus B), \qquad X \setminus (A \cap B) = (X \setminus A) \cup (X \setminus B)$$
$$X \setminus \bigcup_{i \in I} X_i = \bigcap_{i \in I} (X \setminus X_i), \qquad X \setminus \bigcap_{i \in I} X_i = \bigcup_{i \in I} (X \setminus X_i).$$

Difference If X, A are sets, the difference of X and A is the set $X \setminus A = \{x \in X : x \notin A\}$. The difference satisfies

$$X \setminus (X \setminus A) = X \cap A.$$

Disjoint Two sets A, B are disjoint if $A \cap B = \emptyset$.

Distributivity law If A, B, X are sets, and $(X_i)_{i \in I}$ is a family of sets, then

$$X \cap (A \cup B) = (X \cap A) \cup (X \cap B)$$
$$X \cap \bigcup_{i \in I} X_i = \bigcup_{i \in I} (X \cap X_i)$$

Element An object or member of a set.

Empty set The set $\emptyset$ with no elements.

Equality Two sets are equal if and only if they have the same elements.

Equivalence relation An equivalence relation on a set A is a relation which is symmetric, transitive, and reflexive on A. If $\sim$ is an equivalence relation on A, then for each $a \in A$ the set $\text{cls}\, a = \{b \in A : b \sim a\}$ is called the equivalence class of a. Two sets $\text{cls}\, a, \text{cls}\, b$ either coincide, or are disjoint. The set of all such equivalence classes is written $A/\sim$. The function $A \to A/\sim$, $a \mapsto \text{cls}\, a$ is called the *projection*. We also abbreviate $a \in \text{cls}\, a \in A/\sim$ to $a \in\in A/\sim$.

Factors Let $n : A \to C$ be a function and B a set. It is possible that there are functions $I : A \to B$, $m : B \to C$ such that $n = ml$. The existence of such l, m is expressed by saying n factors through B. Given m, the existence of such an l is expressed by saying n factors through l. Given l, the existence of such an m is expressed by saying n factors through l. This definition is extended to other categories [cf. chapter 6] than Set in the obvious way.

Family A family $(X_\lambda)_{\lambda \in L}$ with indexing set L is a function $\lambda \mapsto X_\lambda$ with domain L. We allow the empty family, in which $L = \emptyset$. The restriction of $\lambda \mapsto X_\lambda$ to a subset of L is called a subfamily of $(X_\lambda)_{\lambda \in L}$.

Greatest element See *Maximal element*.

Inclusion See *Subset*.

Integers The set $\mathbb{Z}$ of integers has elements $0, \pm 1, \pm 2, \ldots$; $\mathbb{Z}$ may be constructed from the set $\mathbb{N}$ of natural numbers.

Intersection The intersection of two sets A, B is the set $A \cap B$ of elements belonging to both A and B; thus

$$A \cap B = \{x : x \in A \text{ and } x \in B\}.$$

The intersection of a non-empty family of sets $(X_\lambda)_{\lambda \in L}$ is the set

$$\bigcap_{\lambda \in L} X_\lambda = \{x : x \in X_\lambda \text{ for all } \lambda \in L\}.$$

The intersection of an empty family is defined only when we are dealing with families of subsets of a fixed set X, and then the intersection is X itself.

Interval Let X be a set and $\leqslant$ an order relation on X. A subset A of X is an interval if the conditions $x \leqslant z \leqslant y$ and $x, y \in A$ imply $z \in A$. Particular intervals of X are X itself, the empty set and the subsets of X given by

$$\begin{aligned} [x,y] &= \{z : x \leqslant z \leqslant y\}, & [x,\rightarrow[&= \{z : x \leqslant z\} \\ ]\leftarrow,x] &= \{z : z \leqslant x\}, & [x,y[&= \{z : x \leqslant z < y\} \\ ]x,y] &= \{z : x < z \leqslant y\}, &]x,\rightarrow[&= \{z : x < z < y\} \\ ]x,\rightarrow[&= \{z : x < z\}, &]\leftarrow,x[&= \{z : z < x\} \end{aligned}$$

for x, y in X such that $x \leqslant y$. The first three intervals are *closed*, the next two *half-open* and the last three, as well as X and $\varnothing$, are *open*. The *completeness* of the order is equivalent to the fact that any interval of X is of one of the forms given above. An interval $[x, \rightarrow[$ is also written $X^{\geqslant x}$; similar notations are used for other intervals involving $\rightarrow$ or $\leftarrow$. The *end points* of a non-empty interval $]x, y]$ are x and y; then *end point* of $]\leftarrow, x]$ is x; similar terms are applied to the other intervals.

Least element See *Maximal element.*

Maximal element Let X be a set, $\leqslant$ a partial order relation on X, and A a subset of X. An element a of A is maximal if no element of A is larger than a, i.e., if $a' \in A$ and $a \leqslant a'$ implies $a = a'$. An element a of A is a greatest element if a is larger than every element of A, i.e., if $a' \in A$ implies $a' \leqslant a$. A greatest element is maximal, and is unique. A maximal element need not be a greatest element, nor need it be unique; however it will have both of these properties if $\leqslant$ is an order relation, and in this case the maximal element of A (when it exists) is written $\max A$. The terms minimal, least, $\min A$ are defined as for maximal, greatest, $\max A$, but with $\leqslant$ replaced by $\geqslant$, and larger replaced by smaller.

Meet Two sets A, B meet if $A \cap B \neq \varnothing$; in such case we also say A meets B, B meets A.

Natural number The set $\mathbb{N}$ of natural numbers has elements $0, 1, 2, \ldots$; $\mathbb{N}$ may be described axiomatically or constructed explicitly.

Order relation A partial order relation on X is a relation which is transitive, anti-symmetric, and reflexive on X. A partial order is an *order relation* if it is a total relation. A partial order is often written $\leqslant$, in which case $x < y$ means $x \leqslant y$ and $x \neq y$, while $y \geqslant x$ means $x \leqslant y$.

Ordered pair Intuitively, the ordered pair (a, b) consists of the elements a, b taken in order. The crucial property is that $(a, b) = (a', b')$ if and only if $a = a'$, $b = b'$. This property may be derived from the definition

$$(a, b) = \{\{a\}, \{a, b\}\}.$$

Point The word point is a geometric synonym for element, object.

Positive Greater than 0.

Proper See *Subset*.

Rational number A rational number is a ratio m/n for m, n integers and $n \neq 0$. The set $\mathbb{Q}$ of all rational numbers may be constructed from $\mathbb{Z}$.

Real number The set $\mathbb{R}$ of real numbers satisfies axioms given, in many books; $\mathbb{R}$ may be constructed from $\mathbb{Q}$ by Dedekind sections or Cantor sequences.

Reflexive See *Relation*.

Relation A relation on A is a subset R of $A \times A$. If $(x, y) \in R$ we write also xRy. The relation R is *reflexive* if xRx for all $x \in A$; *symmetric* if $xRy \Rightarrow yRx$ for all $x, y \in A$; *anti-symmetric* if xRy and $yRx \Rightarrow x = y$ for all $x, y \in A$; *transitive* if xRy and $yRz \Rightarrow xRz$ for all $x, y, z \in A$; *total on* A if xRy or yRx for all $x, y \in A$.

Sequence A family with indexing set a subset of $\mathbb{N}$.

Set A set is a 'collection of objects viewed as a whole'. A rigorous treatment of sets requires an axiomatic theory [cf. [IK87]]. The set whose elements are exactly $x_1, x_2, \ldots, x_n$ is written $\{x_1, x_2, \ldots, x_n\}$.

Subset The set A is a subset of B if every element of A is an element of B, that is, if $\forall x \; x \in A \Rightarrow x \in B$. Synonyms for A is a subset of B are: $A \subseteq B$, $B \supseteq A$, A is contained in B, A is included in B, B contains A. The set A is a proper subset of B if $A \subseteq B$ and $A \neq B$.

Symmetric See *Relation*.

Total See *Relation*.

Transitive See *Relation*.

Union The union of two sets A, B is the set $A \cup B$ whose elements are those of A and those of B. That is,

$$A \cup B = \{x : x \in A \text{ or } x \in B\}.$$

The union of a family $(X_\lambda)_{\lambda \in L}$ of sets is the set

$$\bigcup_{\lambda \in L} X_\lambda = \{x : x \in X_\lambda \text{ for some } \lambda \in L\};$$

in particular, the union of the empty family is the empty set.

Zorn's lemma This lemma, which is equivalent to the Axiom of Choice, states that if $\mathcal{T}$ is a partially ordered set such that any ordered subset has an upper bound in $\mathcal{T}$, then $\mathcal{T}$ contains a maximal element.

Bibliography

General reference:
For a general background and history of many of the topics in this book, we refer to [IK87].

[AB80] Abd-Allah, A. and Brown, R. 'A compact-open topology on partial maps with open domain'. *J. London Math. Soc.* **21** (2) (1980) 480–486.

[ABS64] Atiyah, M. F., Bott, R. and Schapiro, A. 'Clifford modules'. *Topology* **3** Suppl. (1964) 3–38.

[Alb63] Albert, A. (ed.). *Studies in modern algebra*. Mathematical Association of America, Englewood Cliffs (1963).

[Ark62] Arkowitz, M. 'The generalized Whitehead product'. *Pacific J. Math.* **12** (1962) 7–24.

[Arm65] Armstrong, M. A. 'On the fundamental group of an orbit space'. *Proc. Camb. Phil. Soc.* **61** (1965) 639–646.

[Arm82] Armstrong, M. A. 'Lifting homotopies through fixed points'. *Proc. Roy. Soc. Edinburgh* **A93** (1982) 123–128. (Also **96** (1984) 201–205).

[Bar55a] Barratt, M. G. 'Track groups I'. *Proc. London Math. Soc.* **5** (3) (1955) 71–106.

[Bar55b] Barratt, M. G. 'Track groups II'. *Proc. London Math. Soc.* **5** (3) (1955) 285–329.

[Bau88] Baues, H. J. *Algebraic homotopy*. Cambridge University Press, Cambridge (1988).

[BB78a] Booth, P. I. and Brown, R. 'On the application of fibred mapping spaces to exponential laws for bundles, ex-spaces and other categories of maps'. *Gen. Top. Appl.* **8** (1978) 165–179.

[BB78b] Booth, P. I. and Brown, R. 'Spaces of partial maps, fibred mapping spaces and the compact-open topology'. *Gen. Top. Appl.* **8** (1978) 181–195.

[BD75] Brown, R. and Danesh-Naruie, G. 'The fundamental groupoid as a topological groupoid'. *Proc. Edinburgh Math. Soc.* **19** (1975) 237–244.

[Bea83] Beardon, A. F. *The geometry of discrete groups, Graduate Texts in Mathematics*, Volume 91. Springer-Verlag, Berlin-Heidelberg-New York (1983).

[BG82] Bongart, K. and Gabriel, P. 'Covering spaces in representation theory'. *Inv. Math.* **65** (1982) 331–378.

[BH70] Brown, R. and Heath, P. R. 'Coglueing homotopy equivalences'. *Math. Z.* **113** (1970) 313–362.

[BH75] Brown, R. and Hardy, J. P. L. 'Subgroups of free topological groups and free products of topological groups'. *J. London Math. Soc.* **10** (2) (1975) 431–440.

[BH78] Brown, R. and Higgins, P. J. 'On the connection between the second relative homotopy groups of some related spaces'. *Proc. London Math. Soc.* **36** (3) (1978) 193–212.

[BH81] Brown, R. and Higgins, P. J. 'Colimit theorems for relative homotopy groups'. *J. Pure Appl. Alg.* **22** (1981) 11–41.

[BH82] Brown, R. and Huebschman, J. 'Identities among relations'. In R. Brown and T. L. Thickstun (eds.), 'Low-Dimensional Topology', *London Mathematical Society Lecture Note Series*, Volume 46. Cambridge University Press (1982), 153–202.

[BH86] Brown, R. and Higgins, P. J. 'On the fundamental groupoid of an orbit space'. Preprint 86.13, U.C.N.W. Pure Math. (1986).

[BH87] Brown, R. and Heath, P. R. 'Lifting amalgamated sums and other colimits of groups and topological groups'. *Math. Proc. Camb. Phil. Soc.* **101** (1987) 273–280.

[BHK83] Brown, R., Heath, P. R. and Kamps, H. K. 'Groupoids and the Mayer-Vietoris sequence'. *J. Pure Appl. Alg.* **30** (1983) 109–129.

[BHK84] Brown, R., Heath, P. R. and Kamps, H. K. 'Coverings of groupoids and Mayer-Vietoris type sequences'. In H. L. Bentley *et al.* (eds.), 'Categorical Topology: Proc. Conf. Toledo, Ohio, 1983', Heldermann-Verlag, Berlin (1984), 147–162.

[BHRW84] Bentley, H. L., Herrlich, H., Rajagopalan, M. and Wolff, H. *Categorical topology: Proceedings conference Toledo, Ohio, 1983*. Heldermann-Verlag, Berlin (1984).

[Big74] Biggs, N. *Algebraic graph theory, Cambridge tracts in mathematics*, Volume 67. Cambridge University Press, London (1974).

[Big84] Biggs, N. 'Homological coverings of graphs'. *J. London Math. Soc.* **30** (2) (1984) 1–14.

[Bin83] Bing, R. H. *The geometric topology of 3-manifolds, Amer. Math. Soc. Coll. Publ.*, Volume 40. American Mathematical Society, Providence, R.I. (1983).

[Bir75] Birman, J. S. *Braids, links and mapping class groups*. Annals of Mathematics Studies. Princeton Univ. Press, Princeton (1975).

[BJ01] Borceux, F. and Janelidze, G. *Galois theories, Cambridge Studies in Advanced Mathematics*, Volume 72. Cambridge University Press, Cambridge (2001).

[BK72] Bousfield, A. K. and Kan, D. *Homotopy limits: completions and localisations, Lecture Notes in Math.*, Volume 304. Springer-Verlag, Berlin-Heidelberg-New York (1972).

[BL87] Brown, R. and Loday, J.-L. 'Van Kampen theorems for diagrams of spaces'. *Topology* **26** (1987) 311–335.

[Bla84] Blass, A. 'The interaction between category theory and set theory'. In J. W. Gray (ed.), 'Mathematical Applications of Category Theory', *Contemporary Mathematics*, Volume 30. AMS Annual Meeting (89th: January 5–9, 1983: Denver, Colo.), American Mathematical Society (1984), 5–29.

[BM94] Brown, R. and Mucuk, O. 'Covering groups of non-connected topological groups revisited'. *Math. Proc. Camb. Phil. Soc.* **115** (1994) 97–110.

[BN79] Brown, R. and Nickolas, P. 'Exponential laws for topological categories, groupoids and groups, and mapping spaces of colimits'. *Cah. Top. Géom. Diff.* **20** (1979) 1–20.

[Boa60] Boas, R. P. *A primer of real functions, Carus Mathematical Monographs*, Volume 13. Mathematical Association of America (1960). Distributed by John Wiley & Sons, Inc.

[Bor94] Borceux, F. *Handbook of categorical algebra. 2, Encyclopedia of Mathematics and its Applications*, Volume 51. Cambridge University Press, Cambridge (1994). Categories and structures.

[Bou66] Bourbaki, N. *General topology*. Addison-Wesley, Reading, Mass. (1966).

[BR84] Brown, R. and Razak Salleh, A. 'A van Kampen theorem for unions of non-connected spaces'. *Arch. Math.* **42** (1984) 85–88.

[Bra26] Brandt, H. 'Über eine Verallgemeinerung des Gruppenbegriffes'. *Math. Ann.* **96** (1926) 360–366.

[Bra04] Braun, G. 'A proof of Higgins's conjecture'. *Bull. Austral. Math. Soc.* **70** (2) (2004) 207–212.

[Bre67] Bredon, G. E. *Sheaf theory*. McGraw-Hill, New York (1967).

[Bre72] Bredon, G. E. *Introduction to compact transformation groups*. Academic Press, New York-London (1972).

[Bro61] Brown, R. *Function spaces and FD-complexes*. D. Phil. Thesis, Oxford University (1961).

[Bro63] Brown, R. 'Ten topologies for $X \times Y$'. *Quart J. Math. Oxford* **14** (2) (1963) 303–319.

[Bro64] Brown, R. 'Function spaces and product topologies'. *Quart J. Math. Oxford* **15** (2) (1964) 238–250.

[Bro65] Brown, R. 'On a method of P. Olum'. *J. London Math. Soc.* **40** (1965) 303–304.

[Bro67] Brown, R. 'Groupoids and van Kampen's theorem'. *Proc. London Math. Soc.* **17** (3) (1967) 385–340.

[Bro70] Brown, R. 'Fibrations of groupoids'. *J. Alg.* **15** (1970) 103–132.

[Bro73] Brown, R. 'Sequentially proper maps and a sequential compactification'. *J. London Math. Soc.* **7** (2) (1973) 515–522.

[Bro82] Brown, R. 'Higher dimensional group theory'. In R. Brown and T. L. Thickstun (eds.), 'Low-Dimensional Topology', *London Mathematical Society Lecture Note Series*, Volume 46. Cambridge University Press (1982), 215–238.

[Bro84] Brown, R. 'Some non-abelian methods in homotopy theory and homological algebra'. In H. L. Bentley *et al.* (eds.), 'Categorical Topology: Proceedings conference Toledo, Ohio, 1983', Heldermann-Verlag, Berlin (1984), 108–146.

[Bro86] Brown, R. 'Convenient categories of topological spaces: historical note'. Preprint 86.4, U.C.N.W. Pure Maths. (1986).

[Bro87] Brown, R. 'From groups to groupoids: A brief survey'. *Bull. London Math. Soc.* **19** (1987) 113–134.

[Bro88] Brown, R. 'How mathematics gets into knots'. London Mathematical Society (1988). Video.

[BT80] Booth, P. I. and Tillotson, J. 'Monoidal closed categories and convenient categories of topological spaces'. *Pacific J. Math.* **88** (1980) 33–53.

[BW85] Barr, M. and Wells, C. *Toposes, triples and theories*. Springer-Verlag, Berlin-Heidelberg-New York (1985).

[Cai61] Cairns, S. S. *Introductory topology*. Ronald Press, New York (1961).

[CF63] Crowell, R. H. and Fox, R. H. *Knot theory*. Ginn & Co., Boston (1963). Reprint Springer-Verlag, Berlin-Heidelberg-New York (1982).

[Cha77] Chase, S. 'On representation of small categories and some constructions in algebra and combinatorics'. Preprint, Cornell University (1977).

[Che46] Chevalley, C. *Theory of Lie groups*. Princeton Univ. Press, Princeton (1946).

[CJ64] Cockroft, W. H. and Jarvis, T. M. 'An introduction to homotopy theory and duality I'. *Bull. Soc. Math. Belgique* **16** (1964) 407–428. (Also **17** (1965) 3–26).

[Cla68] Clark, A. 'Quasitopology and compactly generated spaces'. Preprint, Brown University (1968).

[CM82] Chandler, B. and Magnus, W. *The history of combinatorial group theory: a case study in the history of ideas, Studies in the History of Mathematics and the Natural Sciences*, Volume 9. Springer-Verlag, Berlin-Heidelberg-New York (1982).

[Coh66] Cohen, P. *Set theory and the continuum hypothesis*. W. A. Benjamin, New York (1966).

[Coh89] Cohen, D. E. *Combinatorial group theory: a topological approach, London Mathematical Society Student Texts*, Volume 14. Cambridge University Press, Cambridge (1989).

[Con94] Connes, A. *Noncommutative geometry*. Academic Press Inc., San Diego, CA (1994).

[Cor03] Corfield, D. *Towards a philosophy of real mathematics*. Cambridge University Press, Cambridge (2003).

[CP86] Cordier, J.-M. and Porter, T. 'Vogt's theorem on categories of homotopy coherent diagrams'. *Math. Proc. Cambridge Phil. Soc.* **100** (1986) 65–90.

[Cro59] Crowell, R. H. 'On the van Kampen theorem'. *Pacific J. Math.* **9** (1959) 43–50.

[Csa63] Csazar, A. *Foundations of general topology*. Pergamon Press, Oxford (1963).

[Day68] Day, B. J. *Relationship of Spanier's quasi-topological spaces to k-spaces*. M. Sc. dissertation, Sydney (1968).

[Day72] Day, B. J. 'A reflection theorem for closed categories'. *J. Pure Appl. Algebra* **2** (1972) 1–11.

[DD77] Douady, A. and Douady, R. *Algebres et théories Galoisiennes*, Volume 1. CEDIC, Paris (1977).

[DD79] Douady, A. and Douady, R. *Algebres et théories Galoisiennes*, Volume 2. CEDIC, Paris (1979).

[Die60] Dieudonné, J. *Foundations of modern analysis*. Academic Press, New York (1960).

[Die71] Dieck, T. tom. 'Partitions of unity in the theory of fibrations'. *Comp. Math.* **23** (1971) 159–167.

[DK70] Day, B. J. and Kelly, G. M. 'On topological quotient maps preserved by pullback and products'. *Proc. Camb. Phil. Soc.* **67** (1970) 553–558.

[Dol63] Dold, A. 'Partitions of unity in the theory of fibrations'. *Ann. Math.* **78** (1963) 223–255.

[Dol66] Dold, A. *Halbexacte homotopiefunktoren, Lecture Notes in Math.*, Volume 12. Springer-Verlag, Heidelberg (1966).

[Dol72] Dold, A. *Lectures on algebraic topology*. Springer-Verlag, Heidelberg (1972).

[Dow52] Dowker, C. H. 'The topology of metric complexes'. *Amer. J. Math.* **74** (1952) 555–577.

[DT78] Dostal, M. and Tindell, R. 'The Jordan curve theorem revisited'. *Jber. d. Dt. Math.-Verein.* **80** (1978) 111–128.

[Dug68] Dugundji, J. *Topology*. Allyn and Bacon, New York (1968).

[Dun85] Dunwoody, M. J. 'The accessibility of finitely presented groups'. *Invent. Math.* **81** (1985) 449–457.

[DV96] Dicks, W. and Ventura, E. *The group fixed by a family of injective endomorphisms of a free group, Contemporary Mathematics*, Volume 195. American Mathematical Society, Providence, RI (1996).

[Dyc72] Dychoff, R. 'Factorisation theorems and projective spaces in topology'. *Math. Z.* **127** (1972) 256–264.

[Dyc76] Dychoff, R. 'Categorical methods in dimension theory'. In E. Binz and H. Herrlich (eds.), 'Categorical Topology, Proceedings 1975', *Lecture Notes in Math.*, Volume 540. Springer-Verlag (1976), 115–119.

[Dyc84] Dychoff, R. 'Total reflections, partial products, and hereditary factorisations'. *J. Pure Appl. Alg* **17** (1984) 101–113.

[Eck62] Eckmann, B. 'Homotopy and cohomology theory'. In 'Proc. Inter. Congress of Mathematicians Stockholm', (1962), 59–73.

[Ehr65] Ehresmann, C. *Catégories et structures*. Dunod, Paris (1965).

[Ehr66] Ehresmann, C. 'Catégories topologiques'. *Proc. Kon. Akad. v. Wet.* **69** (1966) 133–175.

[Ehr80] Ehresmann, C. 'Oeuvres completes et commentées / Charles Ehresmann'. Issued as supplement to Cahiers de Topologie et Géometrie Differentielle in 4 parts from 1980–1984. (1980). With commentary by Andrée Charles Ehresmann.

[Ein90] Einstein, A. (1916) Quoted in *Math. Intelligencer* **12** (2) (1990) 31.

[EJM01] Emma J. Moore. *Graphs of groups: word computations and free crossed resolutons*. Ph.D. Thesis, University of Wales, Bangor (2001).

[EM45] Eilenberg, S. and Mac Lane, S. 'The general theory of natural equivalences'. *Trans. Amer. Math. Soc.* **58** (1945) 231–294.

[Eng68] Engelking, R. *Outline of general topology*. Translated from the Polish by K. Sieklucki. North-Holland Publishing Co., Amsterdam (1968).

[Eps62] Epstein, D. B. A. 'The theory of ends'. In M. K. Fort, Jr. (ed.), 'Topology of 3-manifolds and Related Topics', Prentice-Hall, Englewood Cliffs (1962), 110–117.

[ES52] Eilenberg, S. and Steenrod, N. E. *Foundations of algebraic topology*. Princeton Univ. Press, Princeton (1952).

[FA49] Fox, R. H. and Artin, E. 'Some wild cells and spheres in three-dimensional space'. *Ann. Math.* **50** (1949) 264–265.

[Fir74] Firby, P. A. 'On compactification of mappings'. *Proc. Edinburgh Math. Soc.* **19** (1974) 105–108.

[Fox57] Fox, R. H. 'Covering spaces with singularities'. In 'Algebraic Geometry and Topology, a Symposium in Honour of S. Lefschetz', Princeton Univ. Press (1957), 243–257.

[Fox61] Fox, R. H. 'A quick trip through knot theory'. In 'The Topology of 3-Manifolds and Related Topics', Prentice-Hall, Englewood Cliffs (1961).

[FP90] Fritsch, R. and Piccinini, R. A. *Cellular structures in topology, Cambridge Studies in Advanced Mathematics*, Volume 19. Cambridge University Press, Cambridge (1990).

[Fra65] Franklin, S. P. 'Spaces in which sequences suffice'. *Fundamenta Mathematicae* **57** (1965) 107–115.

[Fre64] Freyd, P. *Abelian categories*. Harper and Row, New York (1964).

[Fuc83] Fuchs, M. 'Extending local homotopy equivalences'. *Rend. Circ. Mat. Palermo* **32** (1983) 217–223.

[Göd47] Gödel, K. 'What is Cantor's continuum problem'. *Amer. Math. Month.* **54** (1947) 515–525.

[God58] Godement, R. *Topologie algébrique et théorie des faisceaux*. Actualités Sci. Ind. No. 1252. Publ. Math. Univ. Strasbourg. No. 13. Hermann, Paris (1958).

[Gri54] Griffiths, H. B. 'The fundamental group of two spaces with a common point'. *Quart. J. Math. Oxford* **5** (2) (1954) 175–190. (Correction **6** (2) (1955) 154–155).

[Gri56] Griffiths, H. B. 'Infinite products of semi-groups and local connectivity'. *Proc. London Math. Soc.* **6** (3) (1956) 455–480.

[GZ67] Gabriel, P. and Zisman, M. *Categories of fractions and homotopy theory*. Springer-Verlag, Heidelberg (1967).

[Hae84] Haefliger, A. 'Groupoid d'holonomie et classifiants'. *Astérisque* **116** (1984) 70–97.

[Hal58] Halmos, P. R. *Finite dimensional vector spaces*. van Nostrand, Princeton (1958).

[Hal60] Halmos, P. R. *Naive set theory*. van Nostrand, Princeton (1960).

[Har86] Harasani, H. A. *Topos theoretic methods in general topology*. Ph.D. thesis, University of Wales, Bangor (1986).

[Has60] Hasse, M. 'Einige bemerkungen über graphen, kategorien und gruppoide'. *Math. Nachr.* **22** (1960) 255–270.

[Hat02] Hatcher, A. *Algebraic topology*. Cambridge University Press, Cambridge (2002).

[Hau14] Hausdorff, F. *Grundzuge der Mengenlehre*. Veit & Co., Leipzig (1914). Reprint Chelsea, New York (1949).

[Hea70] Heath, P. R. 'Induced homotopy equivalences on mapping spaces and duality'. *Canadian J. Math.* **22** (1970) 697–700.

[Hea73] Heath, P. R. 'Groupoid operations and fibre homotopy equivalence. I'. *Math. Z.* **130** (1973) 207–233.

[Hea78] Heath, P. R. *An introduction to homotopy theory via groupoids and universal constructions, Queen's papers in pure and applied mathematics*, Volume 49. Queen's University, Kingston, Ont. (1978).

[Her66] Hermann, R. *Lie groups for physicists*. W. A. Benjamin, Inc., New York-Amsterdam (1966).

[Her71] Herrlich, H. 'Categorical topology'. *Gen. Top. Appl.* **1** (1971) 1–15.

[Her72] Herrlich, H. 'A generalisation of perfect maps'. In 'Proc. Third Prague Symp. General Topology', Prague (1972), 187–191.

[Her83] Herrlich, H. 'Are there convenient subcategories of Top?' *Gen. Top. Appl.* **15** (1983) 263–271.

[Her87] Herrlich, H. 'Topological improvements of structured sets'. *Topology Appl.* **27** (1987) 145–155.

[Hig63] Higgins, P. J. 'Algebras with a scheme of operators'. *Math. Nach.* **27** (1963) 115–132.

[Hig64] Higgins, P. J. 'Presentations of groupoids, with applications to groups'. *Proc. Camb. Phil. Soc.* **60** (1964) 7–20.

[Hig66] Higgins, P. J. 'Grushko's theorem'. *J. Algebra* **4** (1966) 365–372.

[Hig74] Higgins, P. J. *Introduction to topological groups, London Mathematical Society Lecture Note Series*, Volume 15. Cambridge University Press, Cambridge (1974).

[Hig76] Higgins, P. J. 'The fundamental groupoid of a graph of groups'. *J. London Math. Soc.* **13** (2) (1976) 145–149.

[Hig05] Higgins, P. J. 'Categories and groupoids'. *Reprints in Theory and Applications of Categories* **7** (2005) 1–195.

[Hil53] Hilton, P. J. *An introduction to homotopy theory*. Cambridge University Press, London (1953).

[Hil65] Hilton, P. J. *Homotopy theory and duality*. Gordon and Breach, New York (1965).

[Hir66] Hirzebruch, F. *Topological methods in algebraic geometry*. Springer-Verlag, Heidelberg (1966).

[HK82] Heath, P. R. and Kamps, K. H. 'On exact orbit sequences'. *Ill. J. Math.* **26** (1982) 149–154.

[HK88] Heath, P. R. and Kamps, K. H. 'Lifting colimits of (topological) groupoids and (topological) categories'. Preprint, Fern Universität (1988).

[HNN49] Higman, G., Neumann, B. H. and Neumann, H. 'Embedding theorems for groups'. *J. London Math. Soc.* **24** (1949) 247–254.

[Hoe86] Hoehnke, H.-J. '66 Jahre brandtches gruppoid, Heinrich Brandt 1886–1986'. Wissenschaftliche Beiträge, Martin-Luther-Universität, Halle/S. (1986).

[Hou72] Houghton, C. H. 'Ends of groups and the associated first cohomology groups'. *J. London Math. Soc.* **6** (2) (1972) 81–92.

[Hou74] Houghton, C. H. 'Ends of locally compact groups and their coset spaces'. *J. Australian Math. Soc.* **17** (1974) 274–284.

[Hou75] Houghton, C. H. 'Wreath products of groupoids'. *J. London Math. Soc.* **10** (2) (1975) 179–188.

[Hou81] Houghton, C. H. 'Homotopy classes of maps to an aspherical complex'. *J. London Math. Soc.* (2) (1981) 332–348.

[HQ84] Haefliger, A. and Quach Ngcoc Du. 'Appendice: une présentation du groupe fondamental d'une orbifold'. *Astérisque* **116** (1984) 98–107.

[HS79] Herrlich, H. and Strecker, G. E. *Category theory, Sigma Series in Pure Mathematics,* Volume 1. Heldermann Verlag, Berlin, second edition (1979).

[HT82] Higgins, P. J. and Taylor, J. 'The fundamental groupoid and homotopy crossed complex of an orbit space'. In K. H. Kamps *et al.* (eds.), 'Category Theory: Proceedings Gummersbach 1981', *Lecture Notes in Math.*, Volume 962. Springer-Verlag (1982), 115–122.

[Hu59] Hu, S. T. *Homotopy theory*. Academic Press, New York (1959).

[Hu64] Hu, S. T. *Elements of general topology*. Holden-Day, San Francisco (1964).

[Hu65] Hu, S. T. *Theory of retracts*. Wayne State University Press, Detroit (1965).

[Hum94] Humphries, S. P. 'Quotients of Coxeter complexes, fundamental groupoids and regular graphs'. *Math. Z.* **217** (2) (1994) 247–273.

[Hun82] Hunt, J. H. V. 'Branched coverings as uniform completions of unbranched coverings (resumé)'. *Contemp. Math* (1982) 141–155.

[Hus66] Husemoller, D. *Fibre bundles*. McGraw-Hill, New York (1966).

[HW48] Hurewicz, W. and Wallman, H. *Dimension theory*. Princeton Univ. Press (1948).

[HW60] Hilton, P. J. and Wylie, S. *Homology theory, an introduction to algebraic topology*. Cambridge University Press, London (1960).

[HY61] Hocking, J. G. and Young, G. S. *Topology*. Addison-Wesley, Reading Mass. (1961).

[IK87] Iyanaga, S. and Kawada, Y. (eds.). *Encyclopaedic dictionary of mathematics*. MIT Press, Cambridge, Mass. and London, England, third edition (1987). Produced by Mathematical Society of Japan; reviewed by K. O. May.

[Jam58] James, I. M. 'The intrinsic join'. *Proc. London Math. Soc.* **8** (3) (1958) 507–535.

[Jam84] James, I. M. *General topology and homotopy theory*. Springer-Verlag, Berlin-Heidelberg-Tokyo (1984).

[Joh79] Johnstone, P. T. 'A topological topos'. *Proc. London Math. Soc.* **38** (3) (1979) 237–271.

[Joh82] Johnstone, P. T. *Stone spaces, Cambridge Studies in Advanced Mathematics*, Volume 3. Cambridge University Press, Cambridge (1982).

[Joh83] Johnstone, P. T. 'The point of pointless topology'. *Bull. American Math. Soc. (New Series)* **8** (1983) 409–419.

[Joh02] Johnstone, P. T. *Sketches of an elephant: a topos theory compendium Volume 1, Oxford Logic Guides*, Volume 43. Oxford Science Publications, Oxford (2002).

[JT66] James, I. M. and Thomas, E. 'Note on the classification of cross sections'. *Topology* **4** (1966) 351–360.

[JTTW63] James, I. M., Thomas, E., Toda, H. and Whitehead, G. W. 'On the symmetric square of a sphere'. *J. Math. Mech.* **12** (1963) 771–776.

[Kam33] Kampen, E. H. Van. 'On the connection between the fundamental groups of some related spaces'. *Amer. J. Math.* **55** (1933) 261–267.

[Kam72a] Kamps, K. H. 'Kan-Bedingungen und abstrakte homotopietheorie'. *Math. Z.* **124** (1972) 215–236.

[Kam72b] Kamps, K. H. 'Zur homotopietheorie von gruppoiden'. *Arch. Math.* **23** (1972) 610–618.

[Kam87] Kamps, K. H. 'Note on the gluing theorem for homotopy equivalences and van Kampen's theorem'. *Cah. Top. Géom. Diff. Cat.* **28** (1987) 303–306.

[Kau87] Kauffman, L. H. *On knots, Annals of Mathematical Studies,* Volume 115. Princeton Univ. Press, Princeton, NJ (1987).

[Kel55] Kelley, J. L. *General topology*. van Nostrand, Princeton (1955).

[Kel69] Kelly, G. M. 'Monomorphisms, epimorphisms and pullbacks'. *J. Austr. Math. Soc.* **9** (1969) 124–142.

[Kie77] Kieboom, R. W. 'On the rather weak and very weak homotopy-extension properties'. *Arch. Math.* **28** (1977) 308–311.

[KP97] Kamps, K. H. and Porter, T. *Abstract homotopy and simple homotopy theory*. World Scientific Publishing Co. Inc., River Edge, NJ (1997).

[Kuh62] Kuhn, T. *The structure of scientific revolutions*. University of Chicago Press, Chicago (1962).

[Kur63] Kurosch, A. G. *General algebra*. Chelsea Publishing Co., New York (1963). Trans. from the Russian by K. Hirsch.

[Lam77] Lamartin, W. F. 'On the foundations of k-group theory'. Diss. Math. 146, Warsaw (1977).

[Lam80] Lam, S. P. 'A note on ex-homotopy equivalence'. *Proc. Ned. Akad. v. Wetensch.* **42** (1980) 33–37.

[Lan65] Lang, S. *Algebra*. Addison-Wesley, Reading Mass. (1965).

[Law66] Lawvere, F. W. 'The category of categories as a foundation for mathematics'. In 'Proceedings of the Conference on Categorical Algebra, La Jolla, 1965', Springer-Verlag, Heidelberg (1966), 1–20.

[Law86] Lawvere, F. W. 'Introduction'. In F. W. Lawvere and S. H. Schanuel (eds.), 'Categories in Continuum Physics: Buffalo 1982', *Lecture Notes in Math.*, Volume 1174. Springer-Verlag, Berlin-Heidelberg-New York (1986).

[Lef49] Lefschetz, S. *Introduction to topology*. Princeton Univ. Press, Princeton (1949).

[Lil73] Lillig, J. 'A union theorem for cofibrations'. *Arch. Math.* **24** (1973) 410–415.

[Lin80] Lines, D. 'Revêtements ramifiés'. *Ens. Math.* **26** (1980) 173–182.

[Lod82] Loday, J.-L. 'Spaces with finitely many nontrivial homotopy groups'. *J. Pure Appl. Algebra* **24** (2) (1982) 179–202.

[LS77] Lyndon, R. C. and Schupp, P. E. *Combinatorial group theory, Erg. d. Mat. u.i. Grendz.*, Volume 89. Springer-Verlag, Berlin-Heidelberg-New York (1977).

[LS86] Lambek, J. and Scott, P. J. *Introduction to higher order categorical logic, Cambridge Studies in Adv. Math.*, Volume 7. Cambridge University Press, Cambridge (1986).

[Mac65] Mac Lane, S. 'Categorical algebra'. *Bull. Amer. Math. Soc.* **71** (1965) 40–106.

[Mac66] Mackey, G. W. 'Ergodic theory and virtual groups'. *Math. Ann.* **166** (1966) 187–207.

[Mac71] MacLane, S. *Categories for the working mathematician*. Springer-Verlag, New York (1971). Graduate Texts in Mathematics, Vol. 5.

[Mac87] Mackenzie, K. *Lie groupoids and Lie algebroids in differential geometry, London Mathematical Society Lecture Note Series*, Volume 124. Cambridge University Press, Cambridge (1987).

[Mac05] Mackenzie, K. *General Theory of Lie Groupoids and Lie Algebroids, London Mathematical Society Lecture Note Series*, Volume 213. Cambridge University Press, Cambridge (2005).

[Mae84] Maehara, R. 'The Jordan Curve Theorem via the Brouwer fixed point theorem'. *Amer. Math. Month.* (1984) 641–643.

[Mag76] Magid, A. 'Covering spaces of algebraic curves'. *Amer. Math. Month.* **83** (1976) 614–621.

[Man64] Manheim, J. H. *The genesis of point set topology*. Pergamon Press, Oxford (1964).

[Man77] Mandelbrot, B. B. *Fractals: form, chance and dimension*. W. H. Freeman, San Francisco (1977).

[Mas67] Massey, W. M. *Algebraic topology: an introduction*. Harcourt, Brace & World, Inc., New York (1967).

[May99] May, J. P. *A concise course in algebraic topology:*. Chicago Lectures in Mathematics. University of Chicago Press, Chicago,ILL (1999).

[Mey84] Meyer, J.-P. 'Bar and cobar constructions I'. *J. Pure Appl. Alg.* **33** (1984) 163–207.

[Mic66] Michael, E. A. 'c-spaces'. *J. Math. Mech.* **15** (1966) 983–1102.

[Mil58] Milnor, J. W. 'Some consequences of a theorem of Bott'. *Ann. Math.* **68** (1958) 444–449.

[Mil63] Milnor, J. W. *Morse theory, Annals of Mathematics Studies*, Volume 51. Princeton Univ. Press, Princeton (1963).

[Mil65] Milnor, J. W. *Topology from the differentiable viewpoint*. University Press of Virginia, Charlottesville (1965).

[Mit65] Mitchell, B. *Theory of categories*. Academic Press, New York (1965).

[Mit85] Mitchell, B. *On the Galois theory of separable algebroids*. Mem. Amer. Math. Soc. **333**, Providence, R.I. (1985).

[MLM94] Mac Lane, S. and Moerdijk, I. *Sheaves in geometry and logic*. Universitext. Springer-Verlag, New York (1994). A first introduction to topos theory, Corrected reprint of the 1992 edition.

[MM86] Morgan, J. W. and Morrison, I. 'A van Kampen theorem for weak joins'. *Proc. London Math. Soc.* **53** (3) (1986) 562–576.

[Moi77] Moise, E. E. *Geometric topology in dimensions 2 and 3*. Springer-Verlag, New York (1977).

[Mok64] Mokobodzki, G. 'Nouvelle méthode pour demontrer la paracompacité des espaces metrisables'. *Ann. Inst. Fourier Grenoble* **14** (2) (1964) 539–542.

[Mon87] Montesinos, J. M. *Classical tessellations and three-manifolds*. Springer-Verlag, Berlin-Heidelberg-New York (1987).

[Mun75] Munkres, J. R. *Topology: a first course*. Prentice-Hall, Englewood Cliffs (1975).

[MZ55] Montgomery, D. and Zippin, L. *Topological transformation groups*. Interscience, New York (1955).

[Nag65] Nagata, J.-I. *Modern dimension theory*. North-Holland, Amsterdam (1965).

[Neu65] Neuwirth, L. P. *Knot groups, Annals of Mathematics Studies*, Volume 56. Princeton Univ. Press, Princeton (1965).

[Nic81] Nickolas, P. 'A Kurosch theorem for topological groups'. *Proc. London Math. Soc.* **42** (3) (1981) 461–477.

[Olu58] Olum, P. 'Non-abelian cohomology and van Kampen's theorem'. *Ann. Math.* **68** (1958) 658–667.

[PAPR86] Pitt, D., Abramsky, S., Poigné, A. and Rydeheard, D. (eds.). *Category theory and computer programming: Tutorial and workshop, Guildford, U.K., September 1985, Lecture Notes in Computer Science*, Volume 240. Springer-Verlag, Berlin (1986).

[Pat99] Paterson, A. L. T. *Groupoids, inverse semigroups, and their operator algebras, Progress in Mathematics*, Volume 170. Birkhäuser Boston Inc., Boston, MA (1999).

[Pea75] Pears, A. R. *The dimension theory of general spaces*. Cambridge University Press, Cambridge (1975).

[Pic86] Pickering, A. *Constructing quarks: a socialogical history of particle physics*. U. Chicago Press (1986).

[Pon46] Pontrjagin, L. *Topological groups*. Princeton Univ. Press, Princeton (1946).

[Por69] Porteous, I. *Topological geometry*. van Nostrand, New York (1969).

[PP79] Popescu, N. and Popescu, L. *The theory of categories*. Edituri Acad. Bucharest (1979).

[Pup58] Puppe, D. 'Homotopie-Mengen und ihre induzierten Abildungen I'. *Math. Zeit.* **69** (1958) 299–344.

[Pup67] Puppe, D. 'Bemerkung über die Erweiterung von Homotopien'. *Archiv d. Math.* **18** (1967) 81–88.

[Qui67] Quillen, D. *Homotopical algebra, Lecture Notes in Math.*, Volume 43. Springer-Verlag, Berlin-Heidelberg-New York (1967).

[Raz76] Razak Salleh, A. *Union theorems for groupoids and double groupoids*. Ph.D. thesis, University of Wales, Bangor (1976).

[Rei32] Reidemeister, K. *Einführung die kombinatorische topologie*. F. Vieweg & Sohn, Braunschweig, Berlin (1932). Reprint Chelsea, New York (1950).

[Rei83] Reinhart, B. L. *Differential geometry of foliations*. Erg. d. Mat. u.i. Grendz. Springer-Verlag, Berlin-Heidelberg-New York (1983).

[Rho66] Rhodes, F. 'On the fundamental group of a transformation group'. *Proc. London Math. Soc.* **16** (3) (1966) 635–650.

[Rho68] Rhodes, F. 'On lifting transformation groups'. *Proc. Amer. Math. Soc.* **19** (1968) 905–908.

[Ros78] Rose, J. S. *A course in group theory*. Cambridge University Press, Cambridge (1978).

[RT85] Razak Salleh, A. and Taylor, J. 'On the relation between the fundamental groupoids of the classifying space and the nerve of an open cover'. *J. Pure Appl. Alg.* **37** (1985) 81–93.

[Rut72] Rutter, J. 'Fibred joins of fibrations and maps I'. *Bull. London Math. Soc.* **4** (1972) 187–190.

[Rut74] Rutter, J. 'Fibred joins of fibrations and maps II'. *J. London Math. Soc.* **8** (1974) 453–459.

[Sch72] Schubert, H. *Categories*. Springer-Verlag, Berlin-Heidelberg-New York (1972).

[Sch79] Schupp, P. E. 'Groups acting on trees, ends and cancellation diagrams'. *Math. Intell.* **1** (1979) 205–214.

[SE62] Steenrod, N. E. and Epstein, D. B. *Cohomology operations, Annals of Mathematics Studies,* Volume 50. Princeton Univ. Press, Princeton (1962).

[Sei31] Seifert, H. 'Konstruction drei dimensionaler geschlossener Raume'. *Berichte Sächs. Akad. Leipzig, Math.-Phys. Kl.* **83** (1931) 26–66.

[Ser80] Serre, J.-P. *Trees.* Springer-Verlag, Berlin-Heidelberg-New York (1980). Translated from the French by J. Stillwell.

[Shi84] Shitanda, Y. 'Sur la théorie d'homotopie abstrait'. *Mem. Fac. Sci. Kyushu Univ.* **A38** (1984) 183–198.

[Sie05] Siebenmann, L. 'The Osgood-Schoenflies theorem revisited'. *Hopf Archive* (2005) 21 pages.

[Sim63] Simmons, G. F. *Introduction to topology and modern analysis.* McGraw-Hill, New York (1963).

[Sma63] Smale, S. 'A survey of some recent developments in differential topology'. *Bull. Amer. Math. Soc.* **69** (1963) 131–145.

[Son62] Sonner, J. 'On the formal definition of categories'. *Math. Z.* **80** (1962) 163–176.

[Spa62] Spanier, E. H. 'Secondary operations on mappings and cohomology'. *Ann. Math.* **75** (1962) 260–282.

[Spa63] Spanier, E. H. 'Quasi-topologies'. *Duke Math J.* **30** (1963) 1–14.

[Spa66] Spanier, E. H. *Algebraic topology.* McGraw-Hill, New York (1966).

[SS78] Steen, L. A. and Seebach Jr., J. A. *Counterexamples in topology.* Springer-Verlag, Berlin-Heidelberg-New York, second edition (1978).

[ST34] Seifert, H. and Threllfall, W. *Lehrbuch der topologie.* Teubner Verlagsgesellschaft, Stuttgart (1934).

[ST80] Seifert, H. and Threllfall, W. *Textbook of topology.* Academic Press, New York-London (1980). Translated from the German by M. A. Goldman.

[Ste51] Steenrod, N. E. *The topology of fibre bundles.* Princeton Univ. Press, Princeton (1951).

[Ste67] Steenrod, N. E. 'A convenient category of topological spaces'. *Michigan Math. J.* **14** (1967) 133–152.

[Sti80] Stillwell, J. *Classical topology and combinatorial group theory, Graduate Texts in Mathematics*, Volume 72. Springer-Verlag, Berlin-Heidelberg-New York (1980).

[Str66] Strøm, A. 'Note on cofibrations'. *Math. Scand.* **19** (1966) 11–14.

[Str69] Strøm, A. 'Note on cofibrations II'. *Math. Scand.* **22** (1969) 130–142.

[Str72] Strøm, A. 'The homotopy category is a homotopy category'. *Arch. Math.* **23** (1972) 435–441.

[Str76] Strecker, G. E. 'Perfect sources'. In E. Binz and H. Herrlich (eds.), 'Categorical Topology, Proceedings 1975', *Lecture Notes in Math.*, Volume 540. Springer-Verlag, Heidelberg (1976), 605–624.

[SW57] Spanier, E. H. and Whitehead, J. H. C. 'Carriers and S-theory'. In 'Algebraic Geometry and Topology, a Symposium in Honour of S. Lefschetz', Princeton Univ. Press, Princeton (1957), 330–360.

[SW79] Scott, G. P. and Wall, C. T. C. 'Topological methods in group theory'. In C. T. C. Wall (ed.), 'Homological group theory (Proc. Sympos., Durham, 1977)', *London Math. Soc. Lecture Note Ser.*, Volume 36. Cambridge Univ. Press, Cambridge (1979), 137–203.

[SW83] Spencer, C. B. and Wong, Y. L. 'Pullback and pushout squares in a special double category with connection'. *Cah. Yop. Géom. Diff.* **24** (1983) 161–192.

[Swa64] Swan, R. G. *The theory of sheaves*. University of Chicago Press, Chicago (1964). Notes by R. Brown of lectures given at Oxford in 1958 by R. G. Swan.

[Tay54] Taylor, R. L. 'Covering groups of non-connected topological groups'. *Trans. Amer. Math. Soc.* (1954) 753–768.

[Tay82] Taylor, J. *Group actions on ω-groupoids and crossed complexes, and the homotopy groups of orbit spaces*. Ph.D. thesis, University of Durham (1982).

[Tay88] Taylor, J. 'Quotients of groupoids by the action of a group'. *Math. Proc. Camb. Phil. Soc.* **103** (1988) 239–249.

[Tay99] Taylor, P. *Practical foundations of mathematics, Cambridge Studies in Advanced Mathematics*, Volume 59. Cambridge University Press, Cambridge (1999).

[Tod62] Toda, H. *Composition methods in homotopy groups of spheres, Annals of Mathematics Studies*, Volume 49. Princeton Univ. Press, Princeton (1962).

[Vai60] Vaidyanathaswamy, R. *Set topology*. Chelsea Publishing Co., New York (1960).

[Č32] Čech, E. 'Höherdimensionale Homotopiegruppen'. In 'Verhandlungen des Internationalen Mathematiker-Kongresses Zurich 1932', Volume 2. International Congress of Mathematicians (4th: 1932: Zurich, Switzerland), Walter Saxer, Zurich (1932), 203. Reprint Kraus, Nendeln, Liechtenstein (1967).

[vE86] van Est, W. T. 'Revêtements ramifiés'. *Bull. Soc. Math. Belgique* **38** (1986) 151–169.

[Vog71] Vogt, R. M. 'Convenient categories of topological spaces for homotopy theory'. *Arch. Math.* **22** (1971) 545–555.

[Vog72] Vogt, R. M. 'A note on homotopy equivalences'. *Proc. Amer. Math. Soc.* **32** (1972) 627–629.

[Vog73] Vogt, R. M. 'Homotopy limits and colimits'. *Math. Z.* **134** (1973) 11–52.

[Wal65] Wall, C. T. C. 'Topology of smooth manifolds'. *J. London Math. Soc.* **40** (1965) 1–20.

[Wal74] Walker, R. C. *The Stone-Čech compactification, Erg. d. Mat. u.i. Grendz.*, Volume 82. Springer-Verlag, Berlin-Heidelberg-New York (1974).

[Wei61] Weinzweig, A. E. 'The fundamental group of a union of spaces'. *Pacific J. Math.* **11** (1961) 763–776.

[Whi49] Whitehead, J. H. C. 'Combinatorial homotopy I'. *Bull. Amer. Math. Soc.* **55** (1949) 213–245.

[Whi78] Whitehead, G. W. *Elements of homotopy theory, Graduate Texts in Mathematics*, Volume 61. Springer-Verlag, Berlin-Heidelberg-New York (1978).

[Whi84] White, A. T. *Graphs, groups and surfaces, North-Holland Mathematics Series,* Volume 8. North-Holland, Amsterdam-New York (1984).

[Why42] Whyburn, G. T. *Analytic topology, AMS Colloquium Publications,* Volume 28. American Mathematical Society, New York (1942).

[Why66] Whyburn, G. T. 'Compactification of mappings'. *Math. Ann.* **166** (1966) 168–174.

[Wig67] Wigner, E. P. 'The unreasonable effectiveness of mathematics in the natural sciences'. In 'Symmetries and Reflections: Scientific Essays of Eugene P. Wigner', Bloomington Indiana Press (1967), 222–237.

[Wil49] Wilder, R. L. *Topology of manifolds, AMS Colloquium Publications,* Volume 32. American Mathematical Society, New York (1949).

[Wus84] Wussing, H. *The genesis of the abstract group concept.* MIT Press, Cambridge, Mass. (1984). Translated by A. Shenitzer.

[ZS60] Zariski, O. and Samuel, P. *Commutative algebra.* van Nostrand, Princeton, NJ (1958–1960).

Glossary of symbols

Standard spaces

$\mathbb{C}$	complex numbers	15
$\mathbb{I}$	unit interval	11
$\mathbb{H}$	quaternions	141
$\mathbb{L}$	$\{0\} \cup \{n^{-1} : n = 1, 2, \ldots\}$	62
$\mathbb{N}$	natural numbers (including 0)	468
$\mathbb{Q}$	rational numbers	469
$\mathbb{R}$	real numbers	469
$\mathbb{Z}$	integers	467
$\mathbb{K}$	one of $\mathbb{C}$, $\mathbb{R}$, $\mathbb{H}$	45
$\dot{\mathbb{I}}$	$\{0, 1\}$	160
$\mathbb{K}^n$	standard n-dimensional space over $\mathbb{K}$	46
$\mathbb{B}^n$	unit Euclidean ball	48
$\mathbb{E}^n$	unit Euclidean disc	48
$\mathbb{S}^{n-1}$	unit Euclidean sphere	48
$\mathbb{J}^n$	n-cube	48
T^2	torus	98
$M_n(\mathbb{K})$	$n \times n$ matrices	154
Δ^n	standard n-simplex	169
$\dot{\Delta}^n$	simplicial boundary of Δ^n	160
$P^n(\mathbb{K})$	projective space	147
$O(n)$	orthogonal group	155
$SO(n)$	special orthogonal group	155
$U(n)$	unitary group	155
$SU(n)$	special unitary group	155
$Sp(n)$	symplectic group	155

Subsets

Topological operations on subsets

Operations on spaces and sets

$X \divideontimes Y$	smash product	177
$X \vee Y$	wedge	123, 177
$X_1 \, {}_{i_1}\!\sqcup_{i_2} X_2$	pushout of i_1, i_2	120, 236
$B_f \sqcup X$	adjunction space	119
X/A	X with A shrunk to a point	107
$M(f)$	mapping cylinder	129
$C(f)$	mapping cone	129
$M(f) \cup X$	mapping cylinder union X	291
$X \cup e^n$	X union a cell	138
X^n	n-fold Cartesian product	
K^n	n-skeleton of a cell complex	132
$C(x)$	component of the point x	72
$\mathcal{B}_r$	cover by balls of radius r	91
X^Y	function space	181
$X \times_W Y$	weak product	112
X_Σ	X with a weak topology	105
$\widetilde{X}$	covering of X	359
$G(V)$	isometry group	154
$P(V)$	projective space	147
$\hat{f}$	map to a function space	181
kX	k-space of a space	184
$K(Y, Z)$	space of k-maps	185
$W(t, U)$	sub-basic set in test-open topology	185
$W(C, U)$	sub-basic set in compact-open topology	185
ε	evaluation map	187
e	exponential map	189
$\mathbf{K}(X, Y)$		190
$X \times_k Y$	k-product	190
Y^+		195
$Y^{\sim}$		195
$Y^{\wedge}$		195
$\mathcal{K}_{\mathcal{PC}}(X, Y)$	partial k-maps with closed domain	195
$\mathcal{PO}(X, Z)$	partial k-maps with open domain	196
$\mathcal{O}(Y)$	set of open sets of Y	196

Functions

$f : A \to B$	function from A	443
$f : A \rightarrowtail B$	partial function	164, 447
$f \mid A', B'$	restriction of f	445
$f \mid A'$	restriction of f	445

Paths, homotopies

Categories, groupoids

$\mathsf{Ob}(\mathsf{C})$	Objects of a category	202
Set	sets	204
Top	topological spaces	204
Top^A	the category of spaces under A	258
Grp	groups	204
Grpd	groupoids	220
Top.	pointed spaces	221
HTop	homotopy category	233
$\mathsf{C} \times \mathsf{D}$	product of categories	222
I	tree groupoid on two objects	217
$\mathsf{I}^{\to}$	a subcategory of I	233
$\mathsf{Fun}(\mathsf{C}, \mathsf{D})$	functor category	233
$G(x)$	object group at x	216
C_x	component of x	216
$\Gamma : \mathsf{C} \to \mathsf{D}$	functor	219
a_x	conjugacy function	217
C^x, C_x	functors to Set	224
$\widehat{C}$	functor to Set	234
$U_\sigma(G)$	universal groupoid	314
$\overline{\sigma}$	giving a final structure; 0-identification morphism	315
$U(G)$	universal group	318
FA	free group on A	318
$G * H$	free product of G and H	319
$\mathrm{Im}\, f$	Image of f	324
$\mathrm{Ob}(\Gamma)$	objects of a graph Γ	323
$F\Gamma$	free groupoid on Γ	326
$\mathrm{Ker}\, f$	Kernel of f	329
G/N	quotient groupoid	329
$N(R)$	groupoid of consequences	331
$\mathrm{St}_G\, x$	star of x in G	365
$\widetilde{G}$	covering groupoid	365
$\mathrm{Tr}(G, H)$	translation groupoid	375
χ_f	characteristic groupoid of f	379
G_S	group of stability of s	374
$S \rtimes G$	semidirect product groupoid	374
$\square_N H$	$(H/N) \rtimes (H \times H)$	377
$C_G(S)$	centraliser of s	378
$[G]$	set of conjugacy classes	378
$\mathsf{CovTop}(X)$	category of covering maps of X	387

$\mathsf{CovGrp}(G)$	category of covering morphisms of G	387
$\pi_!$	functor induced by π	388
$M_A(F, C)$		389
$N_A(F)$	normaliser of F in A	390
$L(f, p)$	set of liftings of f	390
$\mathrm{Aut}(p)$	group of covering isomorphisms of p	392
$A\,{}_f{\times}_p\,Y$	pullback space	394
$A \times_X Y$	pullback space	394
$\underline{p}, \underline{f}$	maps from pullback	394
$L \times_G H$	pullback groupoid	395
S_3	symmetric group on $\{1, 2, 3\}$	400
$g_\#$	action of g	411
X/G	orbit space	411
$\Gamma//G$	orbit groupoid	415
Q^2X	symmetric space	431
H^{ab}	abelianisation of H	430
Q^nX	symmetric product	435

Miscellaneous

$\mathrm{cls}\, x$	equivalence class of x	467
$x \in\in E$	$x \in \mathrm{cls}\, x \in E$	467
$\mathrm{Re}(q)$	Real part of a quaternion	141
$\mathrm{Ve}(q)$	Vector part of a quaternion	141
$\bar{q}$	conjugate of q	143
$\lvert x \rvert$	Euclidean norm of x	46
$\lVert x \rVert$	norm of x	46
$d(\mathbb{K})$	dimension of $\mathbb{K}$ over $\mathbb{R}$	146
$(x \mid y)$	inner product	153
$U^\perp$	orthogonal space of U	154
$U^\dagger$	subset of $P(V)$	148
$\mathrm{dist}(x, A)$	distance of x from A	57
X	cardinality of X	448
$\mathbb{Z}_n$	integers mod n	263

Index

This book is typeset by LaTeX 2E using the math fonts given by the package *eulervm*. These fonts do not go well with the standard Computer Modern text fonts. Of the ones which do match well, we have chosen Charter, from Micropress. We hope you like the result.
The following is extracted from the documentation on Euler-VM:

> With Donald Knuth's assistance and encouragement, Hermann Zapf, one of the premier font designers of this century, was commissioned to create designs for Fraktur and script, and for a somewhat experimental, upright cursive alphabet that would represent a mathematician's handwriting on a blackboard and that could be used in place of italic. The designs that resulted were named Euler, in honor of Leonhard Euler, a prominent mathematician of the eighteenth century. Zapf's designs were rendered in METAFONT code by graduate students at Stanford, working under Knuth's direction.
>
> Euler Virtual Math (Euler-VM) is a set of virtual fonts based primarily on the Euler fonts.

Diagrams were created using xypic. Other figures were created in CoRelDraw™, and eps files were lettered using overpic.

Ronald Brown and Gareth Evans

CPSIA information can be obtained
at www.ICGtesting.com
Printed in the USA
LVHW022238140721
692685LV00011B/872